Fungal Biotechnology and Application 3.0

Fungal Biotechnology and Application 3.0

Guest Editor

Baojun Xu

MDPI

Basel • Beijing • Wuhan • Barcelona • Belgrade • Novi Sad • Cluj • Manchester

Guest Editor
Baojun Xu
Department of Life Sciences
Beijing Normal-Hong Kong
Baptist University
Zhuhai
China

Editorial Office
MDPI AG
Grosspeteranlage 5
4052 Basel, Switzerland

This is a reprint of the Special Issue, published open access by the journal *Journal of Fungi* (ISSN 2309-608X), freely accessible at: https://www.mdpi.com/journal/jof/special_issues/ZDXV611U2N.

For citation purposes, cite each article independently as indicated on the article page online and as indicated below:

Lastname, A.A.; Lastname, B.B. Article Title. *Journal Name* **Year**, *Volume Number*, Page Range.

ISBN 978-3-7258-8279-3 (Hbk)
ISBN 978-3-7258-8280-9 (PDF)
https://doi.org/10.3390/books978-3-7258-8280-9

Contents

About the Editor

Baojun Xu

Baojun Xu is a Chair Professor at Beijing Normal-Hong Kong Baptist University (BNBU, a full English teaching university in China), a Fellow of the Royal Society of Chemistry, a Chartered Chemist (CChem), a Zhuhai Scholar Distinguished Professor, Department Head of the Department of Life Sciences, Program Director of the Food Science and Technology Program, and author of over 460 peer-reviewed papers. Dr. Xu received a Ph.D. in Food Science from Chungnam National University, South Korea. He conducted postdoctoral research work at North Dakota State University (NDSU), Purdue University, and the Gerald P. Murphy Cancer Foundation in the USA from 2005 to 2009. He carried out short-term visiting research at NDSU in 2012 and the University of Georgia in 2014, followed by visiting research during his sabbatical leave (7 months) at Pennsylvania State University in the USA in 2016. Dr. Xu is serving as Associate Editor-in-Chief of *Food Science and Human Wellness*, Associate Editor of *Food Frontiers*, and an Editorial Board Member of around 10 international journals. He received the inaugural President's Award for Outstanding Research at the UIC in 2016 and the President's Award for Outstanding Service at the UIC in 2020. Dr. Xu has been listed as one of the world's top 2% of scientists by Stanford University for the past six consecutive years and has been listed as the Best Scientist in the world in the field of Biology and Biochemistry by Research.com in 2023, 2024, and 2025. Prof. Xu was named a Highly Ranked Scholar (top 0.05%) by ScholarGPS^{TM}, ranking at #8 in Food Science and Technology in the world and #14 in Agricultural and Natural Sciences.

Preface

Fungi represent one of the most diverse and functionally versatile groups of organisms on Earth, with profound impacts on human health, food security, industrial manufacturing, environmental remediation, and ecological sustainability. In recent decades, fungal biotechnology has evolved from traditional fermentation practices to cutting-edge interdisciplinary research spanning synthetic biology, multi-omics, material science, and precision medicine. Novel fungal strains, secondary metabolites, enzymatic systems, and mycelium-based materials continue to unlock new applications in cosmetics, food, pharmaceuticals, agriculture, and environmental engineering.

Baojun Xu
Guest Editor

Journal of
Fungi

MDPI

ditorial

ungal Biotechnology and Application 3.0

aojun Xu

Food Science and Technology Program, Department of Life Sciences, Beijing Normal-Hong Kong Baptist University, Zhuhai 519087, China; baojunxu@bnbu.edu.cn; Tel.: +86-756-3620636

1. Preface

Fungi represent one of the most diverse and functionally versatile groups of organisms on Earth, with profound impacts on human health, food security, industrial manufacturing, environmental remediation, and ecological sustainability. In recent decades, fungal biotechnology has evolved from traditional fermentation practices to cutting-edge interdisciplinary research spanning synthetic biology, multi-omics, material science, and precision medicine. Novel fungal strains, secondary metabolites, enzymatic systems, and mycelium-based materials continue to unlock new applications in cosmetics, food, pharmaceuticals, agriculture, and environmental engineering.

This Special Issue "Fungal Biotechnology and Application 3.0" in the Section of "Fungi in Agriculture and Biotechnology" published in *Journal of Fungi* continues the successful series of Fungal Biotechnology and Application 1.0 and 2.0, aiming to gather frontier advances in fungal biology, biotechnology, and translational applications. It welcomes original research and comprehensive reviews covering beneficial and pathogenic fungi, with a focus on industrial utilization, secondary metabolism, novel strain prospecting, pathogenic mechanisms, fungal ecology, and the cultivation of edible and medicinal mushrooms. After rigorous peer review, this Special Issue has published 13 high-quality papers, including original research and review articles, reflecting the latest progress and trends in fungal biotechnology.

2. Overview of Contributions

The 13 publications in this Special Issue cover a broad spectrum of topics, including fungal biocontrol and plant growth promotion, the discovery of bioactive secondary metabolites, the development and functional regulation of edible mushrooms, biodegradation of environmental pollutants, mycelium-based composite materials, lignin degradation and lignocellulose utilization, multi-omics applications, and the health-promoting functions of edible mushrooms.

2.1. Fungal Biocontrol and Agricultural Applications

Several studies highlight the value of fungi as sustainable biocontrol agents. Li et al. (2026) reported that *Trichoderma nordicum* V1 exhibits strong antagonism against multiple oomycete pathogens and effectively controls cucumber damping-off and pepper blight while promoting plant growth, supporting its potential as a biopesticide in vegetable production [Contribution 1]. Similarly, Zhou et al. (2025) demonstrated that *Trichoderma brevicompactum* 6311 strongly inhibits *Phytophthora capsici* and enhances pepper growth via induced resistance and growth-promoting metabolites, providing a green strategy for managing *Phytophthora blight* [Contribution 2].

2.2. Secondary Metabolites and Drug Discovery

Yuan et al. (2026) established an HSQC–DeepSAT-guided strategy to rapidly discover α-pyrone analogs from endophytic *Diaporthe kyushuensis*, identifying new compounds with selective antifungal activity [Contribution 3]. This work underscores the efficiency of spectrum-based strategies in mining cryptic natural products from fungi. Zhao et al. (2025) reviewed global transcriptional regulator engineering in *Aspergillus*, a powerful approach to activate silent biosynthetic gene clusters and expand the discovery of fungal secondary metabolites [Contribution 4].

2.3. Developmental Biology and Multi-Omics of Edible Fungi

Jiang et al. (2025) performed transcriptomic analysis across three developmental stages of *Sarcomyxa edulis*, revealing key genes and pathways regulating fruiting-body formation [Contribution 5]. Im et al. (2024) investigated the effects of light on *Flammulina velutipes*, showing that blue and green light strongly modulated pigmentation and gene expression related to cytochrome P450 and monooxygenases [Contribution 6]. Xie et al. (2025) systematically reviewed genomic, transcriptomic, and proteomic applications in edible fungi, providing a framework for molecular breeding and quality improvement [Contribution 7]. Kumar and Xu (2025) reviewed the chemical profiles and health benefits of *Dictyophora indusiata*, highlighting its polysaccharides, terpenoids, and other bioactives with antioxidant, immunomodulatory, and neuroprotective functions [Contribution 8].

2.4. Environmental Biotechnology and Bioremediation

Nóbrega et al. (2025) demonstrated that *Penicillium guaibinense* CBMAI 2758 efficiently biodegradeDs chloroquine, producing less toxic metabolites and offering a green approach for pharmaceutical pollutant remediation [Contribution 9]. Nilza et al. (2024) reported that *Aspergillus ochraceus* DY1 efficiently degraded alkali lignin and lignocellulosic biomass, supporting applications in biofuels, pulp waste treatment, and biorefinery [Contribution 10].

2.5. Mycelium-Based Green Materials

Jiménez-Obando et al. (2025) developed insulating mycelium composites using agro-industrial waste and *Ganoderma lucidum*, with low thermal conductivity comparable to commercial insulation [Contribution 11]. Jinanukul et al. (2024) optimized substrate ratios of corn husk and sawdust for *Lentinus sajor-caju* mycelium boards, tailoring mechanical, physical, and sound-absorbing properties for eco-friendly construction and packaging [Contribution 12]. Onomu et al. (2024) reviewed fungal applications in aquaculture, including feed improvement, immunity enhancement, and bioremediation [Contribution 13].

3. Outlook

Collectively, the 13 papers in this Special Issue illustrate the vitality and interdisciplinary nature of modern fungal biotechnology. Fungi not only serve as cell factories for food, medicines, and materials, but also provide eco-friendly solutions for disease control, pollution remediation, and carbon neutrality. Future directions include: (1) high-throughput discovery and functional characterization of novel fungal strains and metabolites; (2) the precision engineering of fungal chassis for high-value compound production; (3) the mechanistic elucidation of fungal–host and fungal–environment interactions; (4) the scale-up and commercialization of mycelium materials and biocontrol products; and (5) the integration of multi-omics and artificial intelligence to accelerate fungal biotech innovation.

Funding: This editorial received no external funding.

https://doi.org/10.3390/jof1206040

Acknowledgments: I sincerely thank the Editor-in-Chief and editorial office of *Journal of Fungi* for supporting this Special Issue. I am grateful to all authors for their valuable contributions and to the reviewers for their rigorous evaluation. I hope this Special Issue inspires further research and collaboration in fungal biotechnology and promotes its translation into industrial and societal benefits.

Conflicts of Interest: The author declares no conflicts of interest.

List of Contributions:

1. Li, S.; Wen, X.; Chen, S.; Zhao, Y.; Chen, J.; Li, W.; Chen, Y.; Ding, M.; Jiang, S.; Anjago, W.M.; Zhou, D.; Cai, F.M.; Druzhinina, I.S.; Jiu, M.; Wei, L.; Daly, P. Dual biocontrol and plant growth-promoting effects of *Trichoderma nordicum* V1 against Oomycete plant pathogens. *J. Fungi* **2026**, *12*, 292. https://doi.org/10.3390/jof12040292.
2. Zhou, J.; Liang, J.; Zhang, X.; Wang, F.; Qu, Z.; Gao, T.; Yao, Y.; Luo, Y. *Trichoderma brevicompactum* 6311: Prevention and control of *Phytophthora capsici* and its growth-promoting effect. *J. Fungi* **2025**, *11*, 105. https://doi.org/10.3390/jof11020105.
3. Yuan, S.; Zheng, J.; Wu, Y.; Wang, X.; Wang, H.; Yan, M.; Yin, T. Rapid discovery of antifungal α-pyrone analogs from *Diaporthe kyushuensis* ZMU-48-1 via an HSQC-based DeepSAT strategy. *J. Fungi* **2026**, *12*, 161. https://doi.org/10.3390/jof12030161.
4. Zhao, Y.; Gong, Q.; Zhang, H. Engineering of global transcriptional regulators (GTRs) in *Aspergillus* for natural product discovery. *J. Fungi* **2025**, *11*, 449. https://doi.org/10.3390/jof11060449.
5. Jiang, W.; Huang, X.; Wang, P.; Ahmad, B.; Yang, T.; Wang, Z.; Ren, T.; Xu, J. Transcriptome analysis and its application in screening genes related to the growth and development of *Sarcomyxa edulis*. *J. Fungi* **2025**, *11*, 750. https://doi.org/10.3390/jof11100750.
6. Im, J.-H.; Park, C.-H.; Shin, J.-H.; Oh, Y.-L.; Oh, M.; Paek, N.-C.; Park, Y.-J. Effects of light on the fruiting body color and differentially expressed genes in *Flammulina velutipes*. *J. Fungi* **2024**, *10*, 372. https://doi.org/10.3390/jof10060372.
7. Xie, M.; Wang, J.; Wang, F.; Wang, J.; Yan, Y.; Feng, K.; Chen, B. A review of genomic, transcriptomic, and proteomic applications in edible fungi biology: Current status and future directions. *J. Fungi* **2025**, *11*, 422. https://doi.org/10.3390/jof11060422.
8. Kumar, Y.; Xu, B. New insights into chemical profiles and health-promoting effects of edible mushroom *Dictyophora indusiata* (Vent ex. Pers.) Fischer: A review. *J. Fungi* **2025**, *11*, 75. https://doi.org/10.3390/jof11010075.
9. de Almeida Nóbrega, P.; Lopes, S.Q.; Sá, L.S.; da Silva Ramos, R.; Holanda, F.H.e.; de Araújo, I.F.; Porto, A.L.M.; Birolli, W.G.; Ferreira, I.M. Biodegradation of chloroquine by a fungus from Amazonian soil, *Penicillium guaibinense* CBMAI 2758. *J. Fungi* **2025**, *11*, 579. https://doi.org/10.3390/jof11080579.
10. Nilza, N.; Prasad, R.; Varma, A.; Salam, M.D. Deconstruction of alkali lignin and lignocellulosic substrates by *Aspergillus ochraceus* DY1 isolated from rotten wood. *J. Fungi* **2024**, *10*, 810. https://doi.org/10.3390/jof10120810.
11. Jiménez-Obando, G.; Arcila, J.S.; Tolosa-Correa, R.A.; Valencia-Cardona, Y.L.; Montoya, S. Development and characterization of mycelium-based composite using agro-industrial waste and *Ganoderma lucidum* as insulating material. *J. Fungi* **2025**, *11*, 460. https://doi.org/10.3390/jof11060460.
12. Jinanukul, P.; Kumla, J.; Aiduang, W.; Thamjaree, W.; Oranratmanee, R.; Shummadtayar, U.; Tongtuam, Y.; Lumyong, S.; Suwannarach, N.; Waroonkun, T. Comparative evaluation of mechanical and physical properties of mycelium composite boards made from *Lentinus sajor-caju* with various ratios of corn husk and sawdust. *J. Fungi* **2024**, *10*, 634. https://doi.org/10.3390/jof10090634.
13. Onomu, A.J.; Okuthe, G.E. The application of fungi and their secondary metabolites in aquaculture. *J. Fungi* **2024**, *10*, 711. https://doi.org/10.3390/jof10100711.

Journal of *Fungi*

MDPI

Article

Dual Biocontrol and Plant Growth-Promoting Effects of *Trichoderma nordicum* V1 Against Oomycete Plant Pathogens

Songrong Li [1,2,†], Xian Wen [1,2,†], Siqiao Chen [2,3], Yishen Zhao [1,2], Jinhao Chen [1,2], Wanrong Li [2,4], Yajuan Chen [2,4], Mingyue Ding [3], Siqi Jiang [3], Wilfred Mabeche Anjago [2], Dongmei Zhou [2], Feng M. Cai [5], Irina S. Druzhinina [6], Min Jiu [1,*], Lihui Wei [2] and Paul Daly [2,*]

1 College of Food and Bioengineering, Henan University of Science and Technology, Luoyang 471023, China; 230320070635@stu.haust.edu.cn (S.L.); 210321090590@stu.haust.edu.cn (X.W.); 210321090599@stu.haust.edu.cn (Y.Z.); 230320070648@stu.haust.edu.cn (J.C.)

2 Institute of Plant Protection, Jiangsu Academy of Agricultural Sciences, Nanjing 210014, China; 33315220@njau.edu.cn (S.C.); ts23040231p31@cumt.edu.cn (W.L.); ts22040220p31@cumt.edu.cn (Y.C.); 20230010@jaas.ac.cn (W.M.A.); dongmeizhou@jaas.ac.cn (D.Z.); weilihui@jaas.ac.cn (L.W.)

3 Key Lab for Organic-Based Fertilizers of China and Jiangsu Provincial Key Lab for Solid Organic Waste Utilization, Nanjing Agricultural University, Nanjing 210095, China; 2020203025@stu.njau.edu.cn (M.D.); 2019203045@njau.edu.cn (S.J.)

4 Key Laboratory of Coal Processing and Efficient Utilization, Ministry of Education, China University of Mining and Technology, Xuzhou 221116, China

5 School of Ecology, Sun Yat-sen University, Shenzhen 518000, China; caif8@mail.sysu.edu.cn

6 Department of Accelerated Taxonomy, The Royal Botanic Gardens Kew, London TW9 3AE, UK; i.druzhinina@kew.org

* Correspondence: jiumin0912@haust.edu.cn (M.J.); paul.daly@jaas.ac.cn (P.D.)

† These authors contributed equally to this work.

Abstract

The potential of *Trichoderma nordicum* (*Hypocreales*, *Ascomycota*), a recently described species, for antagonism and use in the biocontrol of oomycete-caused plant diseases is unknown. *Trichoderma* is a well-known genus for containing microbial antagonists and biocontrol agents. The *T. nordicum* in this study was isolated from decomposing wood, and *rpb2* and *tef1* barcode sequencing demonstrated that the isolates were a match to the reference *T. nordicum* and *T. nigricans* strains. Since *T. nordicum* was described before *T. nigricans*, the isolates were assigned to *T. nordicum*, although taxonomic uncertainty between these species requires future clarification. In dual-culture confrontation assays, *T. nordicum* overgrew five economically important oomycete plant pathogens (*Phytophthora capsici*, *P. sojae*, *Pythium aphanidermatum*, *P. myriotylum*, and *Globisporangium ultimum*). The inability to recover viable *P. aphanidermatum* and *P. capsici* from the parts of the plate overgrown by *T. nordicum*, coupled with protease and endo-cellulase activities, correlates with *T. nordicum* having antagonistic abilities. Inoculation with *T. nordicum* preventively reduced the levels of cucumber seedling damping-off caused by *P. aphanidermatum* by up to 70%. The *T. nordicum* biocontrol effects against pepper blight caused by *P. capsici* were greater than 80%, compared to an autoclaved *T. nordicum* spore control. *T. nordicum* could also significantly promote the growth of pepper, with plant weight increased by up to 40%, compared to an autoclaved-spore control. In contrast, *T. nordicum* could not be used to control *Pythium* soft rot of ginger caused by *P. myriotylum*, even though *P. myriotylum* was overgrown by *T. nordicum*, suggesting host- or pathosystem-specific factors influence biocontrol efficacy. In summary, *T. nordicum* is a promising biocontrol agent for use in the control of pepper blight caused by *P. capsici*, and also has potential for use in the control of other oomycete-caused plant diseases in vegetable production systems.

Keywords: pepper blight; *Trichoderma nordicum*; antagonism; biocontrol

1. Introduction

Developing effective biological control is important as there are objectives to reduce or phase out certain conventional pesticides, especially in regions such as Europe, due to the potential unintended environmental damage of pesticides [1]. The *Trichoderma* genus includes important microbial antagonists, plant-growth-promoting, and plant immunity-inducing strains that have applications in the biological control of plant diseases [2]. The ancestral lifestyle of the *Trichoderma* genus is considered to be mycoparasitism [3], and *Trichoderma* genomes encode inventories of antagonism-related genes encoding such enzymes as chitinases, proteases, as well as secondary metabolite gene clusters [4]. A key part of the successful use of *Trichoderma* species in the biocontrol of crop diseases is that their use can combine microbial antagonism, parasitism, plant growth promotion, and plant immunity-inducing effects [5]. It is challenging to develop effective biocontrol of oomycete plant pathogens, and further screening and analysis of appropriate potential biocontrol agents are required [6].

Oomycete plant pathogens cause devastating diseases across diverse cropping systems, with species in the genera *Phytophthora*, *Pythium*, and *Globisporangium* responsible for billions of dollars in annual crop losses worldwide [7]. Oomycete plant pathogens are notoriously difficult to control due to their production of resilient oospores, rapid asexual reproduction via sporangia and zoospores, and broad host ranges [6]. *Phytophthora capsici* causes blight diseases on solanaceous and cucurbit crops [8], and *Pythium* and *Globisporangium* species cause seedling damping-off and root rots across numerous crop species [9]. The effectiveness of biocontrol agents against oomycetes can vary substantially depending on the specific pathogen-host-biocontrol agent combination [6], necessitating a comprehensive evaluation of biocontrol candidates against multiple pathosystems. Interestingly, oomycete cell walls contain cellulose rather than chitin [10], which requires biocontrol agents to produce cellulases in addition to, or instead of, chitinases for effective parasitism or predation. The *Trichoderma* genus is well-known for producing cellulases [11], especially *T. reesei* [12], making *Trichoderma* particularly suitable for parasitizing oomycetes [13].

Among crops infected by oomycete plant pathogens, chili pepper (*Capsicum annuum*) represents a high-value target for biocontrol deployment. Chili pepper fruits have uses as vegetables, ingredients, coloring agents, and in traditional medicine [14]. Global fresh pepper production is approximately 40 M metric tons [15] and has an economic value of approximately USD 4 billion. Although pepper faces multiple disease challenges—including anthracnose caused by *Colletotrichum* spp. [16], bacterial spot of pepper caused by *Xanthomonas* spp. [17], southern blight of pepper caused by *Sclerotium rolfsii* [18], and tomato spotted wilt virus [19]—Phytophthora blight caused by *P. capsici* is a particularly important threat [8]. *P. capsici* is a hemibiotrophic, broad-host-range oomycete plant pathogen with a global distribution and is well-known for its ability to evade disease management strategies [20,21]. Apart from *P. capsici*, other oomycete species cause economically significant diseases across diverse vegetable crops. *P. aphanidermatum* is a major cause of cucumber seedling damping-off and root rot, particularly in greenhouse production systems [22], while *P. myriotylum* causes devastating soft rot of ginger rhizomes, leading to substantial losses [23].

P. capsici-caused pepper blight is difficult to control, partly because of the broad-host range of *P. capsici*, the high genetic variation in *P. capsici* populations, and the persistence in soils [20]. Disease control of *P. capsici*-caused diseases includes cultural practices, water

https://doi.org/10.3390/jof12040292

management, host resistance, conventional fungicides, and biocontrol methods [20]. There are several examples of *Trichoderma* species being used in the biocontrol of pepper blight caused by *P. capsici*. *T. brevicompactum* 6311 (sect. *Pachybasium*) inhibited *P. capsici* mycelial growth, inhibited fruit infection, and its use led to control of *P. capsici*-caused disease in a pot trial [24]. Both *T. aggressivum* f. sp. *europaeum* (*Harzianum/Virens* clade) and *T. longibrachiatum* (sect. *Longibrachiatum*) reduced the mycelial growth of *P. capsici* and their use as part of a biocontrol treatment reduced the disease levels on pepper caused by *P. capsici* [25]. *T. asperellum* strain T34 (sect. *Trichoderma*) was used for control of *P. capsici*-caused pepper disease, and this strain is the active agent in the commercial T34 Biocontrol® product [26]—an excellent example of the commercial viability of *Trichoderma*-containing biocontrol methods. While some *Trichoderma* strains demonstrate broad-spectrum activity against multiple oomycetes, others show pathogen- or crop-specific efficacy, highlighting the importance of evaluating biocontrol candidates across multiple pathosystems rather than relying solely on *in vitro* antagonism assays or single-disease trials.

Recently, the *T. nordicum* (sect. *Trichoderma*) species was identified from northern China, where it had distinct barcode marker sequences compared to the related *T. atroviride* species [27]. The novelty of our study and a key knowledge gap is how the biocontrol potential of *T. nordicum* remains unexplored. We identified *T. nordicum* isolates from our own screening, and our two key aims were to determine, firstly, whether the *T. nordicum* isolates had antagonistic abilities comparable to those of other *Trichoderma* species, and, secondly, whether the *T. nordicum* isolates could be used as part of the biocontrol of multiple oomycete-caused plant diseases.

2. Materials and Methods

2.1. Isolation and Identification of Trichoderma Strains

Samples with green-colored spores were collected from the stump of a *Magnolia grandiflora* tree in Xianlin district, Nanjing, China, in May 2020, and were single-spore purified and stored as spore glycerol stocks at −80 °C.

The molecular identification of the *Trichoderma* isolates was carried out following the methods described by Cai and Druzhinina [28]. The internal transcribed spacer (ITS) region was amplified with the ITS5 (5′ GGAAGTAAAAGTCGTAACAAGG 3′) and ITS4 (5′ TCCTCCGCTTATTGATATGC 3′) primer pair [29]. The partial sequences of *tef1* were amplified with Tef1-728-F (5′ CATCGAGAAGTTCGAGAAGG 3′) and Tef1-R (5′ GCCATCCTTGGGAGATACCAGC 3′), and *rpb2* with rpb2-F (5′ CCGGCTGAGACCCCTGAAG 3′) and rpb2-R (5′ CCCATGGCTTGTTTACCCAT 3′). Note that the above *rpb2* primers were designed using the *T. atroviride rpb2* sequence because the partially degenerate primers fRpb2-5F and fRpb2-7cR [30] could not amplify successfully from the gDNA samples. The PCR was performed using standard methods, and the products were sequenced using Sanger sequencing. The ITS region, *rpb2*, and *tef1* sequences were trimmed to improve identification accuracy, according to Cai and Druzhinina [28], before using BLAST (version 2.17.0) or pairwise sequence alignment.

For phylogenetic analysis, sequences were aligned using MUSCLE [31] in MEGA12 [32], and the maximum likelihood method was used to construct the phylogenetic trees with the Tamura–Nei substitution model with a gamma distribution and invariable sites. All positions with less than 95% site coverage were excluded, and 500 bootstrap replicates were performed.

The three barcode sequences from the *T. nordicum* V1 isolate were deposited in GenBank with the accession numbers PQ489411.1 (ITS region), PQ510043.1 (*rpb2*), and PQ510044.1 (*tef1*). The *T. nordicum* isolates were deposited in the China General Microbio-

https://doi.org/10.3390/jof1204029

logical Culture Collection (CGMCC) with deposit numbers CGMCC No. 41478 (*T. nordicum* V1) and CGMCC No. 41477 (*T. nordicum* V4).

2.2. *In Vitro Assays for Antagonism Towards Oomycetes*

For plate confrontation assays, as well as for the *T. nordicum* isolates, three other *Trichoderma* reference species were used for comparison: *T. asperellum* CBS433.97, *T. atroviride* IMI 206040, and *T. virens* Gv29-8. These strains were confronted with the oomycete plant pathogens *P. capsici* LT263 [33], *P. sojae* P6497 [34], *P. myriotylum* SWQ7 [35], *P. aphanidermatum* HBT1 [13], and *Globisporangium ultimum* G001 [13]. Stocks of oomycetes were maintained on 10% cV8 juice agar [36] slants at 12 °C.

For the confrontation assays, both *Trichoderma* and the oomycete species were pre-cultured on cV8 agar at 25 °C in the dark. Each 5 mm agar plug covered with hyphae from the hyphal front was transferred onto the opposite side of the 9 cm Petri dish containing 12.5 mL cV8 medium to initiate the confrontation assays (at least three replicate plates were used for each confrontation). Single-species cultures were also inoculated to determine the growth rate under the same conditions as the confrontation assay. Whether both species in the confrontation are inoculated simultaneously or at different times depends on their growth rates, with the aim of inoculating so that the colonies of the two species confront each other approximately in the middle of the plate. These cultures were incubated at 25 °C for up to 15 d under diffuse light (12 h/d). The scoring system was based on the scale of Bell [37]. On every odd-numbered day after hyphal contact in the dual culture, an antagonistic score from 1 to 5 was given based on the coverage of the *Trichoderma* colony where 1 = complete overgrowth by *Trichoderma*, 2 = partial overgrowth of the oomycete and the 2/3 coverage of Petri dish by *Trichoderma*, 3 = a type of deadlock where neither of the species overgrows each other, 4 = a partial overgrowth of *Trichoderma* and the 2/3 coverage of Petri dish by the oomycete species, and 5 = a complete overgrowth of *Trichoderma* by the oomycete species. The proportion of overgrowth was determined by the proportion of the *Trichoderma* colony that appeared to be on top of the oomycete colony, as viewed from the side of the Petri dish under back illumination at this time point. The raw scores from the confrontation are included in Table S1.

To test if the hyphae of the host/prey were still viable, mycelial plugs from various parts of the *P. aphanidermatum* and *P. capsici* colonies that were overgrown by the *T. nordicum* V1 isolate were transferred to cV8 agar and incubated at 33 °C. A temperature of 33 °C was found suitable for *P. aphanidermatum* and *P. capsici* to grow (if viable), but *T. nordicum* growth was inhibited.

2.3. *Microscopy on In Vitro Confrontation Assays and Trypan Blue Staining*

A mycelial plug of *P. capsici* LT263 or *P. myriotylum* SWQ7 was inoculated on one side of 3.5 cm plates (Nest® (Wuxi NEST Biotechnology, Wuxi, China) Cat# 706001) containing cV8 agar, and *T. nordicum* V1 or V4 was inoculated on the opposite side at distances or times so that the hyphae of both species would meet in the middle of the plate. The plates were incubated at 25 °C, and after the hyphae contacted, the plates were stained with 1 mL of 0.01% trypan blue for 3 min, then rinsed with water until the background color was mostly removed. Hyphal morphology and color were imaged using a Nexcope NE900 microscope (Ningbo Yongxin Optics, Ningbo, China). The clear presence of septa in the hyphae of *T. nordicum* was used to distinguish it from the aseptate oomycete hyphae.

2.4. *Measurement of Cellulase and Protease Activities*

To measure endo-cellulase activity, mycelial plugs from the *T. nordicum* V1 isolate and *T. atroviride* IMI206040 were inoculated onto plates containing either 0.4% *w/v* inactivated *P. myriotylum* mycelial powder, 1% *w/v* cellulose (from cotton linters, Sigma, St. Louis,

https://doi.org/10.3390/jof12040292

MO, USA, Cat# 435236), 25 mM fructose, or without an added carbon source in a minimal medium with 0.8% *w/v* agar described previously [13], and incubated at 25 °C. These plates contained 0.1% AZCL HE-cellulose (Megazyme, Bray, Ireland) to measure the endo-cellulase activity, where the size of the blue-dyed zone from the degradation of the AZCL-HE-cellulose was used to semi-quantify endo-cellulase activity. The *P. myriotylum* SWQ7 mycelial powder was prepared according to a previous study, where compositional analysis of the powder was also performed [13,38]. To measure protease activity, 1% *w/v* skim milk in 0.8% *w/v* water agar plates were used, and a polycarbonate membrane with a 0.1 μm pore size (GVS filter technology, Bologna, Italy, Cat# 1215304) was used to separate the colony from the skim milk agar to facilitate measurement of the clearing zone beneath. The skim milk agar plates were inoculated and incubated in the same way as the AZCL-HE-cellulose containing plates above.

2.5. Fungal Inoculum Preparation Methods

The solid-state fermentation medium consisted of 17.1 g wheat bran, 11.4 g rice hulls, 0.5 g ammonium sulfate, 14 g sucrose, 0.5 g potassium dihydrogen phosphate (KH_2PO_4), 0.57 g soybean meal powder, and 43 mL distilled water. The medium was sterilized by autoclaving and inoculated with 5.2×10^5 *Trichoderma* spores per gram of medium, using spores harvested from a 6-day-old PDA plate culture, and incubated at 25 °C under light. The culture was stirred at 60 h after inoculation, and cultured statically until 8 d, when spores were harvested with sterile water.

The liquid-fermentation medium consisted of maize powder 4.5% *w/v*, glucose 2% *w/v*, yeast extract 0.5% *w/v*, at an initial pH 5, with or without Shengli lignite 2.5% *w/v* (~2 mm diameter). The source and preparation of Shengli lignite were described previously [39]. The liquid-fermentation medium (75 mL in 250 mL flasks) was inoculated with 3 mL (total of 8×10^8 spores) of *T. nordicum* V1 isolate spore suspension, and shaken at 180 RPM, at 28 °C in darkness for 7 d. During the culture period, samples were initially collected every 12 h to monitor spore germination, and then, after germination and the onset of sporulation, samples were collected every 24 h to determine spore concentration using a hemocytometer. To test spore viability, the conidial suspensions collected at days 6 and 7 after inoculation were filtered through a 30 μm pore-size nylon membrane to remove hyphal fragments and plated on PDA.

2.6. Cucumber Seedling Damping-Off Biocontrol Assays

Cucumber (*Cucumis sativus*) "Lufeng" variety seeds were surface-sterilized by immersing in 75% ethanol for 2 min, then washed three times with sterile water, followed by 1% NaClO, and shaken at 200 RPM and 25 °C for 30 min, followed by extensive washing to remove the bleach. The air-dried cucumber seeds were sown on 1% *w/v* water agar, with eight seeds per plate. Seedlings were grown in a growth cabinet at 25 °C (12 h light and 12 h dark cycle). When the cucumber seedling roots reached approximately 1 cm in length, six seedlings were retained per plate. *T. nordicum* was inoculated onto the seedling in 85 μL of 10^7 spores/mL spore suspension, and *P. aphanidermatum* HBT1 was inoculated using a mycelial plug from a cV8 culture at 1 cm below the seedling root. On the same day, seedlings were inoculated with *T. nordicum* for the treatments where *T. nordicum* was inoculated first, and seedlings were inoculated with *P. aphanidermatum* in the treatment where *T. nordicum* was inoculated second. Two days later, *P. aphanidermatum* was inoculated into treatments previously inoculated by *T. nordicum*, and *T. nordicum* was inoculated into treatments previously inoculated by *P. aphanidermatum*. Single-culture inoculated controls of *P. aphanidermatum* and *T. nordicum*, and non-inoculated controls were also included. Note that a mock inoculation of autoclaved *T. nordicum* spores was also used. Each treatment included 18 seedlings. At day 9 after the first inoculations, the disease severity of cucumber seedlings was recorded using the scale 0 = Seedlings are healthy; 1 = Seedling roots show slight brown lesions; 2 = Seedling roots show large brown lesions; and 3 = Seedlings are dead. The disease index and control efficacy were calculated from these severity scores.

2.7. Pepper Phytophthora Blight Disease Biocontrol Assays

Pepper (*Capsicum annuum*) cv. Sujiao No. 5 seeds were germinated in a soil:vermiculite (5:2) mixture in a growth chamber at 25 °C with 12 h of light (~5000 Lux) per day. Pepper seedlings at the 4–6 leaf stage were transplanted to 240 mL plastic pots containing 180 mL of a soil:vermiculite (5:2) mixture, and placed on an inverted Petri dish lid to prevent the spread of inoculum in the tray. The pot was covered with a larger plastic pot to maintain high humidity and prevent the spread of inoculum and cross-contamination. Pots were randomly arranged across three shelves in the growth cabinet. To minimize positional bias, pot positions were rotated each time they were removed for watering, inoculation, or assessment. Approximately 3 and 6 days after transplanting, the seedlings were fertilized with an NPK (20-20-20) fertilizer containing trace elements. On day 7 after transplanting (approximately four-leaf stage), a frozen 4 mL *T. nordicum* spore suspension, to give a final spore concentration of 1×10^6/mL in the soil:vermiculite mixture, was applied around the seedling stem base and allowed to thaw and slowly drip down into the soil:vermiculite mixture. The spores from solid-state fermentation were frozen horizontally overnight at −20 °C (the purpose of freezing is to prevent too many spores from collecting at the base of the pot by slowing their passage through the soil:vermiculite mixture). A control group received 4 mL of autoclaved V1 or V4 spore suspension. One week later (approximately six-leaf and one heart stage), seedlings were inoculated ~2 cm below the stem base using a syringe with 5 mL of a *P. capsici* LT263 blended mycelium suspension or an autoclaved suspension. The *P. capsici* LT263 inoculum was prepared from a 7-day 10% cV8 shake flask culture inoculated with mycelium-covered agar plugs from the colony edge of the cV8 plate culture (~ 20 × 3 mm plugs per 300 mL cV8). The shake flask culture was incubated at 25 °C in the dark at 100 RPM for 7 d, and the pellet-form culture was blended to form a cloudy suspension. Each treatment consisted of 10 pots per replicate. The disease severity was recorded from 7 DAI to 21 DAI. Disease severity was assessed by two independent evaluators, and while complete blinding was not feasible due to visible treatment differences, the criteria were strictly followed to maintain objectivity. The disease severity rating scale was as follows: 0 = no symptoms; 1 = slight blackening of root and stem base in seedlings, leaves not wilted or showing recoverable wilting; 2 = blackening of root and stem base extending 1–2 cm, irreversible leaf wilting, occasional lower leaf abscission; 3 = blackening of root and stem base exceeding 2 cm, obvious leaf wilting or significant leaf abscission; 4 = root and stem base blackened and constricted, complete defoliation except for growing point, or plant wilting; and 5 = plant death. Disease index = Σ(number of plants at each disease grade × representative value of each grade)/(total number of plants surveyed × maximum disease grade value) × 100%. For growth promotion effects, pepper plant height, root length, plant fresh weight, and dry weight were measured 28 d after inoculation with *T. nordicum*.

For the re-isolation of *P. capsici* and *T. nordicum*, pepper root samples were collected from pots, gently washed with sterile tap water to remove soil and vermiculite, and surface-sterilized with 1% NaClO for 3 min in a small dish, and washed several times to remove the bleach. The surface-sterilized root samples were cut into ~0.5 cm pieces and placed on 10% cV8 agar (containing 100 µg/mL ampicillin) and incubated at 25 °C for 2 d. Agar plugs were subcultured from the edges of colonies that grew out from the root pieces onto another cV8 plate, and mycelia were sampled for gDNA extraction. The identity of

https://doi.org/10.3390/jof12040292

the colonies was determined by PCR amplification of the ITS region using ITS5 and ITS4 primers, sequencing, and pairwise alignment, as described in Section 2.1.

2.8. *Pythium* Soft Rot of Ginger Disease Biocontrol Assays

The method for the ginger pot trial was based on that described previously [13]. Tissue-cultured-derived ginger seedlings ("Laiwu big ginger" variety) were transplanted into sterilized vermiculite in a growth chamber at 25 °C with 12 h of light (~5000 Lux) per day. At day 7 after transplantation, 4 mL of *T. nordicum* V1 frozen spore suspension (prepared as described in the previous section for the pepper pot trial) at final concentrations in the vermiculite of 1×10^5 spores/mL and 1×10^6 spores/mL were inoculated into the vermiculite around each ginger seedling. The control group received 4 mL of heat-inactivated (autoclaved) V1 spore suspension. To prepare the *P. myriotylum* inoculum, wheat seeds were soaked in distilled water for 24 h, then autoclaved twice and spread evenly on plates containing *P. myriotylum* SWQ7, and cultured for a further ~5 d. Four mycelium-covered wheat seeds were inoculated into the vermiculite around the roots of each ginger plant. Photos were taken, and disease severity was recorded 15 DAI of *P. myriotylum* SWQ7. Each treatment included 12 ginger seedlings, and the experiment was repeated two times. Disease severity rating scale: 0 = healthy ginger seedling, 1 = yellowing of lower leaves, 2 = all leaves yellowed but plant not dead, and 3 = plant dead. The disease index was calculated as: Disease index = Σ(number of plants at each disease grade × representative value of each grade)/(total number of plants surveyed × maximum disease grade value) × 100%. For growth promotion effects, ginger plant fresh weight and dry weight were measured 21 d after inoculation with *T. nordicum*. For the re-isolation of *P. myriotylum* and *T. nordicum*, ginger root samples were collected from pots following the same procedure as described for the pepper pot trial in Section 2.7.

2.9. Statistical Analysis

Statistical analysis was carried out using GraphPad Prism 9.5.0 (GraphPad Software, San Diego, CA, USA). The Kruskal–Wallis non-parametric test and Dunn's post hoc test (this post hoc test was chosen because it can compare treatment ranks to the rank values of a single control after a non-parametric test) were used for disease indices (disease indices are ordinal and non-normally distributed and thus require a non-parametric test). ANOVA with Dunnett's post hoc test (this post hoc test was chosen because it can compare treatment means to a single control after a parametric test) was used for growth-related measurements. Student's *t*-tests with an FDR correction were used for spore yields and viability measurements. For the calculation of AUDPC, the area-under-the-curve function in GraphPad Prism was used, which uses the trapezoidal rule.

3. Results

3.1. Identification of *Trichoderma* Isolates as Belonging to the *T. nordicum*/*T. nigricans* Species Complex

Isolates were collected from the stump of a *Magnolia grandiflora* tree in Nanjing, China, and two of the isolates were identified as *T. nordicum*/*T. nigricans* species. The ITS region sequences of the V1 and V4 isolates were identical to each other (GenBank accession: PQ489411.1) and were a match of 100% identity to species from the *Trichoderma* genus in GenBank, and therefore met the initial criteria to continue with more accurate identification of the strain using *rpb2* and *tef1* sequences. The trimmed V1 and V4 *rpb2* sequences were identical to each other (GenBank accession: PQ510043.1) and were identical to the *rpb2* sequence (MH287502.1) of the *T. nordicum* type strain ACCC 39713 (WT13001) [27] and the *rpb2* (OP357959) of the *T. nigricans* type strain T32781 = CGMCC40314 [40]. The trimmed V1 and V4 *tef1* sequences were also identical to each other (GenBank accession: PQ510044.1) and had a similarity of 99% with the *tef1* sequence (MH287501.1) of the same *T. nordicum* type strain and 99% with the *tef1* sequence (OP357974) of the same *T. nigricans* type strain (Table 1). Importantly, the V1 and V4 sequences had lower similarity with the *T. atroviride* reference strain CBS142.95 sequences, supporting that V1 and V4 were *T. nordicum* or *T. nigricans* isolates and not *T. atroviride*. Since *T. nordicum* was described earlier [27] than *T. nigricans* [40], the *T. nordicum* species was assigned to the V1 and V4 isolates. See the Discussion section regarding a discussion of the taxonomic assignment of the V1 and V4 isolates, and how there is uncertainty in the assignment of *T. nordicum* compared to *T. nigricans*.

Table 1. Summary of the nucleotide sequence similarity of the *T. nordicum* V1 strain barcode sequences with those of the type strains for *T. nordicum*, *T. nigricans*, and *T. atroviride*.

Barcode	V1 Sequence ID (bp)	Trichoderma Strain	% Similarity (Sites)
		T. nordicum ACCC 39713 (WT13001) (bp)	
ITS	PQ489411.1 (571 bp)	NR_184883.1 (607 bp)	100% (571/571)
tef1	PQ510044.1 (551 bp)	MH287501.1 (1217 bp)	99% (526/530)
rpb2	PQ510043.1 (805 bp)	MH287502.1 (1175 bp)	100% (805/805)
		T. nigricans (T32781 = CGMCC40314) (bp)	
ITS	PQ489411.1 (571 bp)	OP341437.1 (533 bp)	100% (524/524)
tef1	PQ510044.1 (551 bp)	OP357974 (849 bp)	99% (482/484)
rpb2	PQ510043.1 (805 bp)	OP357959 (813 bp)	100% (614/614)
		T. atroviride CBS142.95 (bp)	
ITS	PQ489411.1 (571 bp)	MH862505.1 (633 bp)	99% (569/571)
tef1	PQ510044.1 (551 bp)	AF456891.1 (548 bp)	91% (449/491)
rpb2	PQ510043.1 (805 bp)	EU341801.1 (1009 bp)	97% (599/615)

The phylogenetic trees constructed based on the *tef1* and *rpb2* genes are shown in Figure 1, and there was high bootstrap support for the clustering of the *T. nordicum* V1 and V4 *tef1* and *rpb2* sequences with the sequences from the *T. nordicum* type strain ACCC 39713 (WT13001) and *T. nigricans* type strain T32781 = CGMCC40314, and clustering separately from the *T. atroviride* reference strain CBS142.95.

3.2. T. nordicum Isolates Overgrow a Range of Oomycete Plant Pathogens and Have Antagonistic Abilities

The *T. nordicum* and *T. nigricans* species were only recently described, and since there were no reports of antagonism by these species, the antagonism of the V1 and V4 isolates towards oomycete plant pathogens and their ability to be used in the control of oomycete-caused diseases were investigated. The ability of the *T. nordicum* isolates to overgrow two *Phytophthora* (*P. capsici* and *P. sojae*) and two *Pythium* (*P. myriotylum* and *P. aphanidermatum*) plant pathogens, and *G. ultimum* was tested. Both *T. nordicum* isolates overgrew all five of the tested plant pathogens (Figure 2A). The overgrowth of *G. ultimum* by both *T. nordicum* isolates was generally less than the overgrowth of the other plant pathogens, while the overgrowth of *P. capsici* by both *T. nordicum* isolates tended to be the most. The relatively faster growth rate of *G. ultimum* may partly explain the weaker overgrowth by the *T. nordicum* isolates (Figure 2B). Generally, the *T. nordicum* V1 isolate tended to overgrow the plant pathogens more than the V4 isolate, and in subsequent assays in which only a single isolate was used, preference was given to the V1 isolate. The *T. nordicum* isolates had comparable levels of overgrowth to those of reference strains of *T. atroviride*, *T. virens*, and *T. asperellum* (Figure 2). The raw data of the lengths of colonies, alongside the dates when those lengths were recorded (which shows the time taken to overgrow the host), and corresponding scores from Bell's scale for individual replicate confrontations, are shown in Table S1.

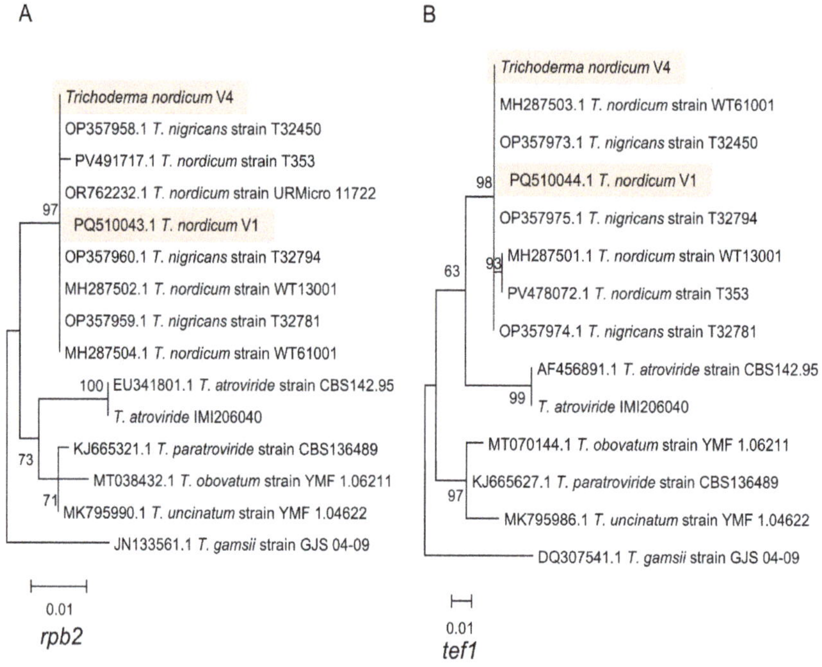

Figure 1. Maximum likelihood phylogenetic trees for (**A**) *rpb2* and (**B**) *tef1* showing the clustering of the *T. nordicum* V1 and V4 isolates (highlighted with the colored boxes) with the type strain *T. nordicum* WT13001 and the type strain *T. nigricans* T32781. Bootstrap values > 60% are shown on trees. There were 566 positions in the alignment for the *rpb2* tree, and there were 262 positions in the alignment for the *tef1* tree. Note that the *T. atroviride* IMI206040 sequences were obtained from the genome sequence of the strain.

The overgrowth by the *T. nordicum* isolates of the *Pythium*, *Phytophthora*, and *Globisporangium* plant pathogens can be due to various factors, such as competition for nutrients, predation, or parasitism. To test whether the plant pathogen was still viable when overgrown by the *T. nordicum* isolates, the ability of *P. aphanidermatum* and *P. capsici* to grow at higher temperatures than the *T. nordicum* isolates was taken advantage of. The lack of growth at 33 °C of mycelium-covered plugs taken from parts of the plate where *T. nordicum* had overgrown *P. aphanidermatum* indicated that *P. aphanidermatum* was dead in the parts of the plate where it was overgrown by *T. nordicum* (Figure 3A). Control viable mycelium plugs of *P. aphanidermatum* grew readily at 33 °C. There was a similar trend using mycelium-covered plugs from parts of the plate where *T. nordicum* had overgrown *P. capsici*, whereby no *P. capsici* growth was observed, although the growth was more difficult to image because of the slower growth rate of *P. capsici* at 33 °C (Figure S1). The lack of viability of the mycelia of *P. aphanidermatum* and *P. capsici* from the confrontation supported an antagonistic mechanism from *T. nordicum*. In the microscopy analysis of hyphae in the confrontation zone between *T. nordicum* and *P. capsici* or *P. myriotylum*, the clear presence of septa in the hyphae of *T. nordicum* was used to distinguish it from the aseptate oomycete hyphae (Figure 3B). For both *P. capsici* and *P. myriotylum* hyphae stained with the viability stain trypan blue, there were examples of blue-stained hyphae indicating a loss of viability (Figure 3B). There were no clear examples of coiling of *T. nordicum* hyphae around the hyphae of *P. myriotylum* or *P. capsici*, although it was difficult to observe the confrontation other than at initial contact points, and reliably distinguish *T. nordicum* from oomycete hyphae. Hyphal coiling is evidence supporting a parasitic mechanism, and the lack of observation of hyphal coils, as well as the lack of other direct evidence of parasitism, cautions against interpreting the loss of plant pathogen viability after *T. nordicum* overgrows as parasitism, when other mechanisms, such as antibiosis, could be responsible.

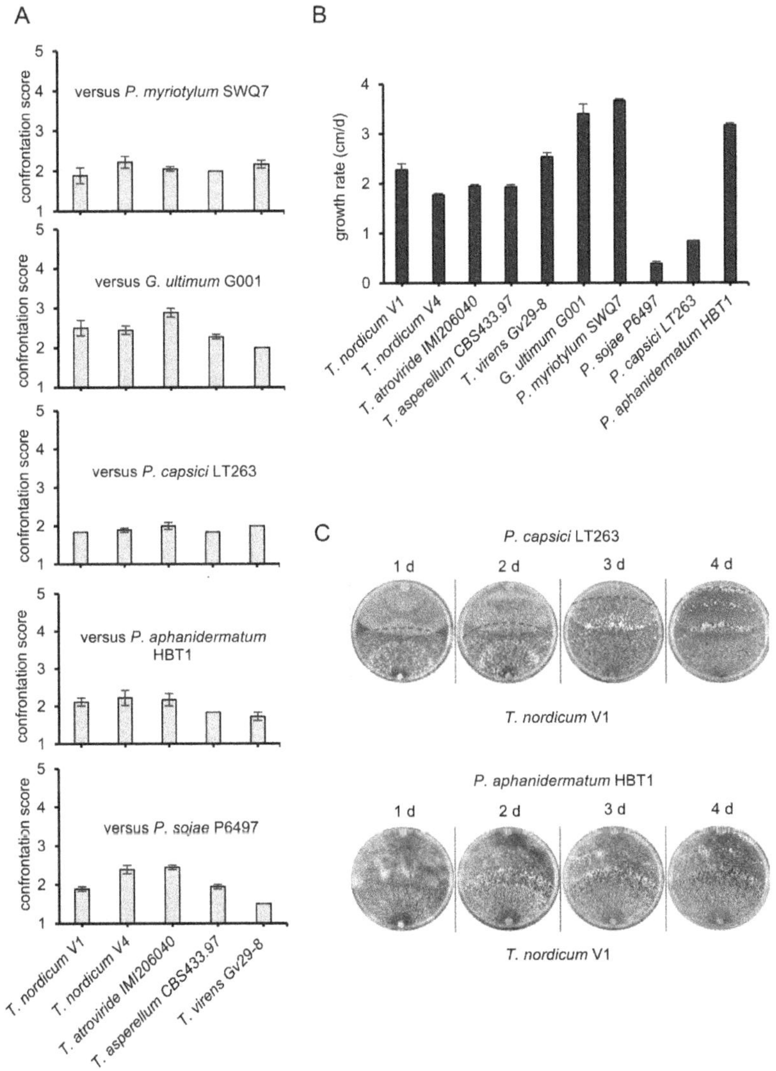

Figure 2. (**A**) Confrontation score from Bell's scale (1 = the strongest level of *Trichoderma* antagonism) from the in vitro confrontation assay of *Trichoderma* strains with five oomycete plant pathogens. The score is the average of the scores given to three time points after colonial contact. (**B**) Growth rates of single cultures of *Trichoderma* and oomycete strains. (**C**) Representative images of the confrontation from days 1 to 4 after contact between *T. nordicum* and *P. capsici*, and *T. nordicum* and *P. aphanidermatum*. The dotted line marks the extent of the *T. nordicum* overgrowth. The error bars represent standard errors (*n* (plate) = 3). The raw scores from the confrontation are included in Table S1.

https://doi.org/10.3390/jof12040292

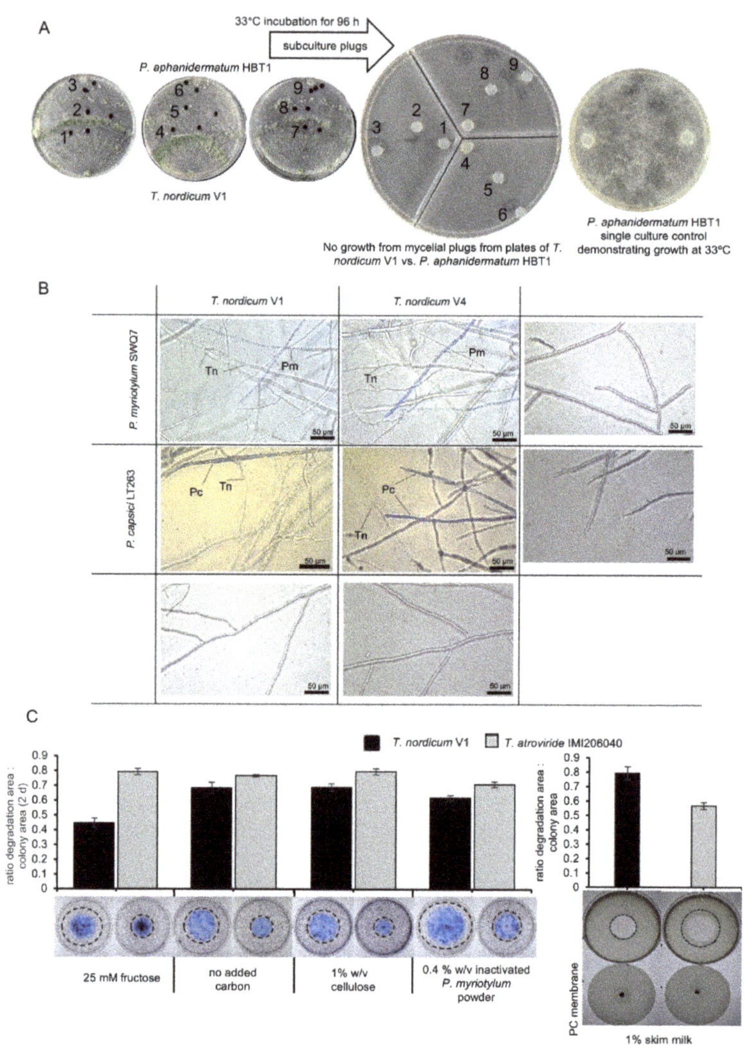

Figure 3. Data supporting the antagonistic ability of *T. nordicum*. (**A**) Test of viability of *P. aphanidermatum* HBT1 mycelia from parts overgrown by *T. nordicum* V1 isolate (numbers 1–9 correspond to the plugs transferred from the confrontation plate to the plate to test for plant pathogen viability); (**B**) brightfield microscopy of trypan blue-stained hyphae of confrontation between *T. nordicum* isolates and either *P. myriotylum* SWQ7 or *P. capsici* LT263; and (**C**) endo-cellulase and caseinase activity of *T. nordicum* V1 isolate and *T. atroviride* IMI206040. The dotted lines mark the edge of the colony. The error bars represent standard errors (n (plate) = 4).

Enzymatic activities that can contribute to *Trichoderma* antagonism of oomycetes include proteases and cellulases, as reviewed previously [41]. Protease activity, measured by the hydrolysis of skim milk agar, was detected for *T. nordicum* and had similar levels of activity to the *T. atroviride* IMI206040 strain. Endo-cellulase activity, measured using dyed AZCL-cellulose, was detected for *T. nordicum* in medium supplemented with cellulose and also in medium supplemented with inactivated oomycete mycelial powder, and the cellulase activity levels were similar in the *T. atroviride* IMI206040 strain (Figure 3C). The endo-cellulase activity was also detected from *T. nordicum* and *T. atroviride* in medium

https://doi.org/10.3390/jof1204029

supplemented with fructose and in medium without an added carbon source, suggesting that both strains have a constitutive level of endo-cellulase activity and may also use the agar as a carbon source. Considering this, the endo-cellulase activity detected from *T. nordicum* cultured on the medium with inactivated mycelial powder or cellulose may not necessarily be due to induction by these carbon sources.

3.3. Preventative but Not Curative Inoculation of T. nordicum Reduced Cucumber Seedling Damping-off Compared to an Autoclaved-Spore Control

Compared to the control group inoculated with autoclaved *T. nordicum* spores followed by *P. aphanidermatum* HBT1, when V1 isolate was inoculated first (preventative inoculation), followed by *P. aphanidermatum*, the V1 isolate significantly reduced the seedling damping-off on cucumber seedlings, with significant biocontrol efficacy of 70% ($p < 0.05$, Dunn's test following Kruskal–Wallis test) (Figure 4). There was also a substantial biocontrol effect of the V4 isolate when inoculated first, but it was not statistically significant using a non-parametric statistical method in which the plate, not individual seedlings, was considered the unit of replication ($n = 3$), which likely reduced statistical power. However, when *P. aphanidermatum* was inoculated first, followed by V1 or V4 isolates (curative inoculation), the biocontrol effects were much weaker, with neither biocontrol efficacy statistically significant. These results indicate that when using *T. nordicum* to control cucumber seedling damping-off caused by *P. aphanidermatum*, it is necessary to apply it as a preventative treatment so that *T. nordicum* can establish a competitive advantage for significant biocontrol efficacy.

3.4. Greater than 80% Biocontrol of Pepper Blight Disease Caused by P. capsici Using T. nordicum Compared to an Autoclaved-Spore Control

With the ability of *T. nordicum* to antagonize *P. capsici* established in the confrontation assays (Figure 2), the use of *T. nordicum* for the biocontrol of Phytophthora pepper blight caused by *P. capsici* was tested. *T. nordicum* spores were inoculated onto a subset of the pepper seedlings, followed by inoculation with *P. capsici* 7 d later. In the pepper plants inoculated with *P. capsici* and inactivated (autoclaved) spores of *T. nordicum*, there were clear signs of symptoms associated with Phytophthora-caused pepper blight, such as wilting leaves, yellowing leaves, wilting stems, and a blackening appearance of roots and blackening at the base of the stem, and the disease index was 0.62 (Figure 5). In contrast, in the pepper plants that were inoculated with *T. nordicum*, there were a few signs of pepper blight disease symptoms from the pepper plants inoculated with V4 isolate (biocontrol rate of 84%), and no symptoms were observed on pepper plants inoculated with V1 isolate (biocontrol rate of 100%). At the end of the trial, *P. capsici* was re-isolated from infected plants, and *T. nordicum* was re-isolated from healthy plants that had been inoculated with both *P. capsici* and *T. nordicum* (Figure S3).

In contrast to the use of *T. nordicum* for biocontrol of pepper blight caused by *P. capsici*, there was no clear control of Pythium soft rot of ginger caused by *P. myriotylum* compared to an autoclaved *T. nordicum* spore control (Figure S5). The above-ground disease symptoms included yellowing of ginger leaves, and there were many dead and dying ginger plants at 16 DAI of *P. myriotylum* across all treatments (Figure S5). There were no significant reductions ($p > 0.05$, Dunn's test following Kruskal–Wallis test) in disease severity from the *T. nordicum* spore treatments on Pythium soft rot disease of ginger here. At the end of the trial, *P. myriotylum* was easily re-isolated from infected plants, and while *T. nordicum* was re-isolated from infected plants inoculated with *T. nordicum* before inoculation with *P. myriotylum*, it was easier to re-isolate *P. myriotylum* from these infected plants (Figure S6).

https://doi.org/10.3390/jof12040292

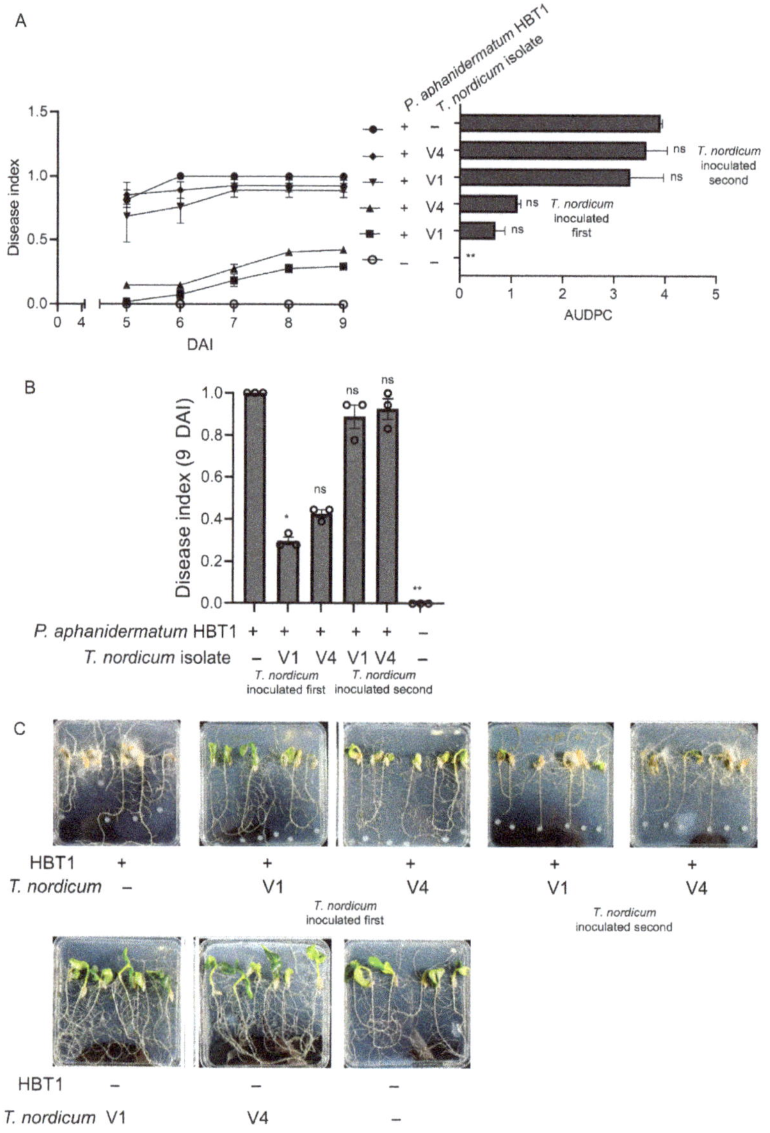

Figure 4. (**A**) Time course of disease index and AUDPC, and (**B**) disease index from 9 DAI from control of cucumber seedling damping-off caused by *P. aphanidermatum* HBT1 using spore inoculum of *T. nordicum* V1 and V4 isolates. The error bars represent standard errors (*n* (plates) = 3 (each plate contained six seedlings)). ** = $p < 0.01$, * = $p < 0.05$, ns = non-significant from Dunn's post hoc test after Kruskal–Wallis test in comparison to the disease index or AUDPC of seedlings inoculated with *P. aphanidermatum* HBT1 and autoclaved *T. nordicum* spores. (**C**) Representative images from different treatments. Note that the images include two treatments not included in the graphs because only a single plate was tested to demonstrate that *T. nordicum* alone does not cause disease symptoms.

https://doi.org/10.3390/jof12040292

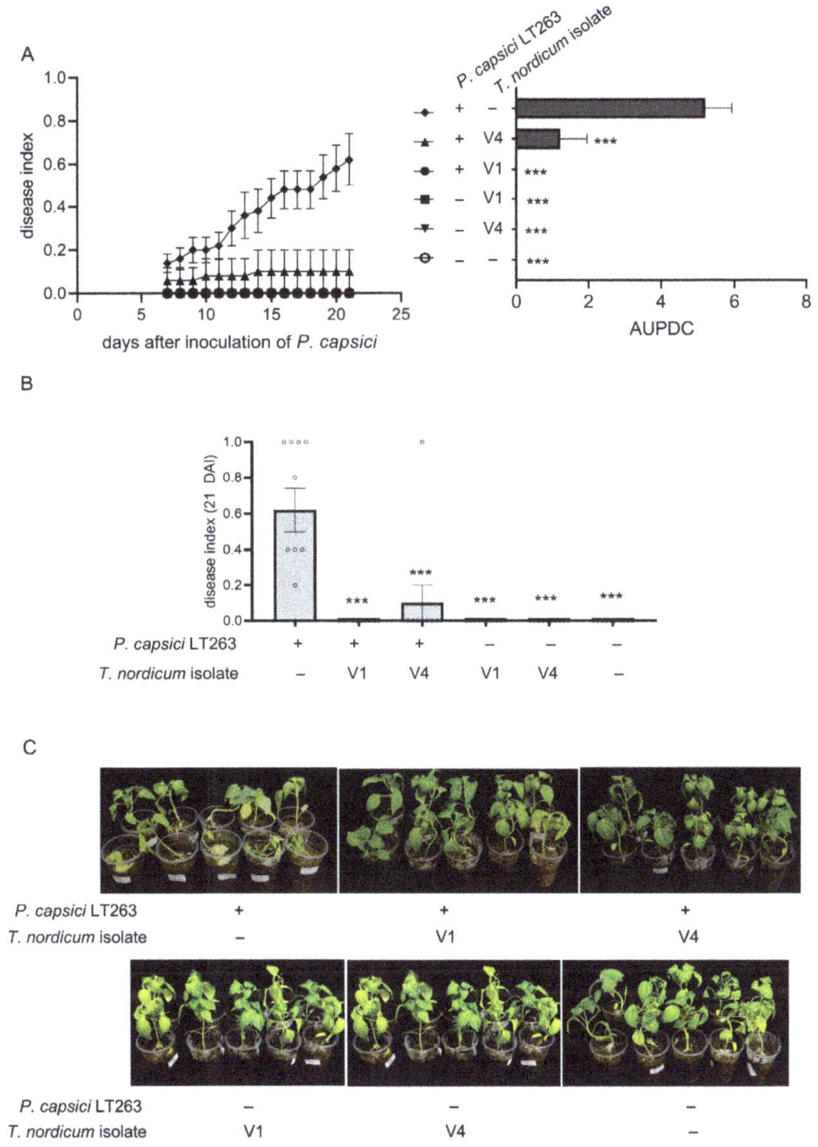

Figure 5. (**A**) Time course of disease index and AUDPC and (**B**) disease index from 21 DAI from control of *T. nordicum* on pepper blight caused by *P. capsici*. The error bars represent standard errors (n (plants) = 10). *** = $p < 0.001$ from Dunn's post hoc test after Kruskal–Wallis test in comparison to the disease index or AUDPC of seedlings inoculated with *P. capsici* LT263 and autoclaved *T. nordicum* spores. (**C**) Images of pepper plants 21 d after inoculation of *P. capsici*. The pot trial was repeated three times with similar results (see Figure S2 for the results of the other two repeats).

3.5. T. nordicum Promoted the Growth of Pepper by More than 40% Compared to an Autoclaved-Spore Control

Trichoderma is known to function via microbial antagonism-mediated mechanisms, as well as via plant-mediated mechanisms. The ability of *T. nordicum* to promote the growth of pepper plants was investigated.

https://doi.org/10.3390/jof12040292

The pot trial with pepper showed that inoculation with *T. nordicum* V1 and V4 isolates had significant growth-promoting effects on pepper compared to an autoclaved-spore control. At 4 weeks after V1 inoculation, significant increases ($p < 0.05$, Dunnett's test following ANOVA) in plant height (37%), root length (65%), fresh weight (87%), and dry weight (53%) were measured compared to the control of autoclaved *T. nordicum* spores (Figure 6). The *T. nordicum* V4 isolate also promoted the growth of pepper plants, but to a lesser level than the V1 isolate. At 4 weeks after V4 inoculation, there were significant increases ($p < 0.05$, Dunnett's test following ANOVA) in root length (26%) and fresh weight (44%) compared to the control, whereas there was no significant difference in plant height or dry weight ($p > 0.05$) (Figure 6). In two independent repeat experiments, there were similar trends, although there were fewer statistically significant differences, e.g., the increase in dry weight of pepper plants inoculated with *T. nordicum* V4 was significant in one of the repeats but marginally non-significant in the other repeat ($p = 0.057$, Dunnett's test following ANOVA) (Figure S4). Of note, re-isolation of *T. nordicum* was not attempted from pepper plants from the growth promotion experiment, but it was done from roots inoculated with *T. nordicum* followed by the plant pathogen *P. capsici*, and here, *T. nordicum* was readily recovered (Figure S3), supporting that *T. nordicum* can colonize pepper roots, which is likely a factor contributing to the growth promotion effect.

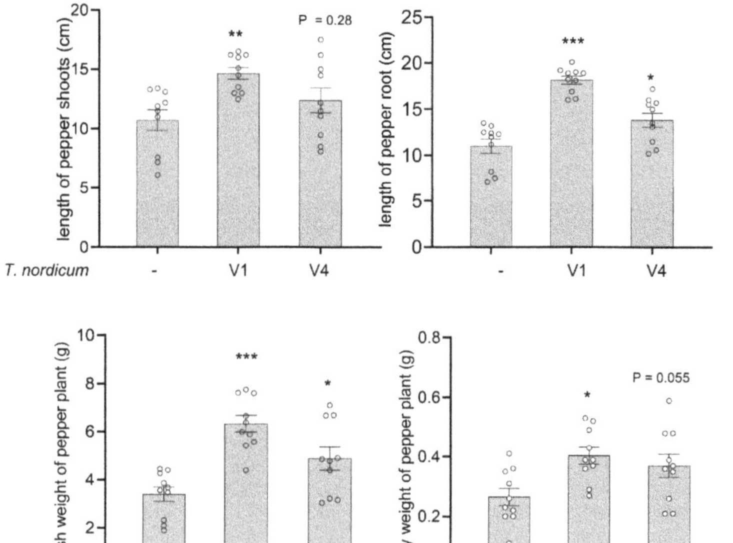

Figure 6. Growth promotion effect of *T. nordicum* isolates on pepper 28 d after inoculation with *T. nordicum*. The error bars represent standard errors (n (plants) = 10). *** = $p < 0.001$, ** = $p < 0.01$, * = $p < 0.05$ from Dunnett's post hoc test (after ANOVA) in comparison to the seedlings inoculated with autoclaved *T. nordicum* spores (-). The pot trial was repeated three times with similar results (see Figure S4 for the results of the other two repeats).

In contrast, there was no clear growth promotion effect on ginger due to inoculation with *T. nordicum* V1, whereby there were no significant differences ($p > 0.05$, Dunnett's test following ANOVA) in fresh weight or dry weight of the ginger plants inoculated to give a final spore concentration in vermiculite of either 1×10^5/mL or 1×10^6/mL (Figure S5).

https://doi.org/10.3390/jof1204029

3.6. Culturing T. nordicum with Lignite Increases Spore Yield

In the previous pot-trial experiments, the *T. nordicum* spore inoculum came from a solid-state type fermentation. Another liquid fermentation method was also tested, partly to assess the effect of lignite addition to the liquid fermentation medium. Lignite can function as a suitable carrier for fungal spores in biocontrol applications. After inoculating the media with V1 spores, samples were taken every 24 h to initially observe spore germination and later count sporulation levels. Spores began to germinate after 24 h, and almost all spores had germinated by 72 h, with some hyphae differentiating into conidiophores and beginning sporulation. On days 4, 5, and 7, the spore concentrations in the medium supplemented with lignite were ~5-fold, ~3-fold, and ~1.5-fold significantly ($q < 0.01$, multiple *t*-tests with an FDR threshold of 1%) higher, respectively, than the medium without lignite (Figure 7A). These spore concentrations indicated that the addition of lignite significantly increased *T. nordicum* V1 sporulation levels. The germination rate of the spores from these cultures was determined to confirm that there were no adverse effects on the spore germination rate that would obviate the benefit of higher spore concentrations. From day 6, the germination rate of spores from the medium supplemented with lignite was 78.3 ± 9.8%, which was significantly higher ($q < 0.01$, multiple *t*-tests with an FDR threshold of 1%) than the germination rate of spores from the medium without lignite at 35.2 ± 3.9%. This increased germination rate was also measured at day 7, when it was 31.9 ± 8.4%, which was marginally non-significant ($q > 0.01$) than the control group at 11.3 ± 3.3% (Figure 7B). In both days where germination rate was measured, it was clearly not significantly lower from the medium supplemented with lignite, indicating that the increased spore concentrations in the lignite-supplemented medium will potentially be beneficial for biocontrol application of this spore inoculum to plants.

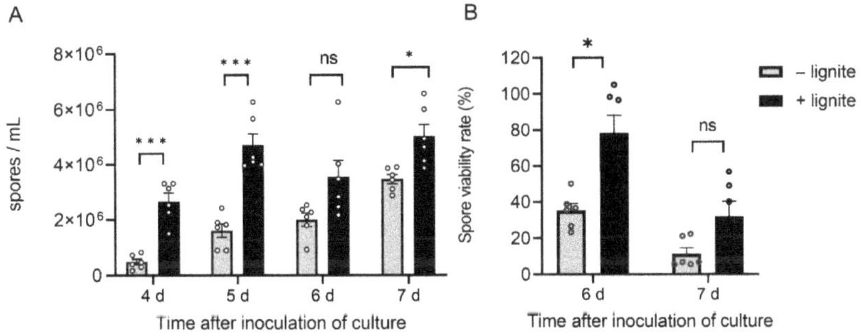

Figure 7. Increase in *T. nordicum* V1 spore yield in cultures supplemented with lignite. (**A**) Spore counts and (**B**) spore viability of spores cultured with and without lignite in shake flasks in liquid fermentation medium. The error bars represent standard error (*n* (flasks) = 6). *** = $q < 0.0001$, * = $q < 0.01$, ns = non-significant from multiple *t*-tests with an FDR threshold of 1%. The experiment was repeated twice with similar trends.

4. Discussion

Here, we show that isolates of *T. nordicum* can antagonize oomycete plant pathogens and control Phytophthora blight of pepper caused by *P. capsici*. The efficient disease control of pepper blight is likely due to antagonism of *P. capsici*, as well as stimulation of pepper growth. Overgrowth of the plant pathogen in a plate confrontation assay did not necessarily correlate with control of disease, because while *T. nordicum* could control pepper blight caused by *P. capsici*, and cucumber seedling damping-off caused by *P. aphanidermatum*, it could not control Pythium soft rot disease caused by *P. myriotylum* (even though *P. myriotylum* was overgrown by *T. nordicum*).

https://doi.org/10.3390/jof12040292

The Phytophthora blight of pepper disease biocontrol was greater than 80% and comparable to or better than the highest disease control levels reported in other studies using *Trichoderma*. The use of *T. harzianum* 2413 (*Harzianum/Virens* clade) led to a reduction in pepper root rot caused by *P. capsici* by approximately 60–80%, depending on the inoculum size of *P. capsici* that was inoculated at the same time as *T. harzianum* [42]. *T. asperellum* strain T34 (sect. *Trichoderma*) led to a reduction in *P. capsici*-caused pepper disease by up to 71% in growth chamber trials where *T. asperellum* T34 was applied before *P. capsici* inoculation and again at the appearance of disease symptoms [26]. Both *T. aggressivum* f. sp. *europaeum* (*Harzianum/Virens* clade) and *T. longibrachiatum* (sect. *Longibrachiatum*) (applied before *P. capsici* inoculation) led to a reduction in the disease levels on pepper caused by *P. capsici* in a greenhouse, whereby 60% of plants were without symptoms in the *T. aggressivum* f. sp. *europaeum* treatment and 80% in the *T. longibrachiatum* treatment [25]. *T. brevicompactum* 6311 spore treatment applied before *P. capsici* inoculation was used to partially biocontrol *P. capsici*-caused disease in a pot trial, whereby the disease level was halved [24]. Several factors may explain differences in control rates compared with other studies involving *Trichoderma*, and comparisons are complicated by different growth conditions, disease scoring systems, and *P. capsici* strains and pepper varieties. For *Trichoderma*, most studies described here used a preventive method in which the biocontrol agent is inoculated before the plant pathogen. In the future, other strains/isolates of *P. capsici*, alongside other pepper varieties, could be used to test whether similar control effects from *T. nordicum* are achieved.

The results with the *T. nordicum* strains suggest several mechanisms that contribute to its use in a biocontrol treatment of pepper blight. Firstly, *T. nordicum* overgrows *P. capsici* in plate confrontation assays, and *P. capsici* was no longer viable in parts of the plate where it was overgrown by *T. nordicum*, demonstrating an antagonistic effect. Of note, the plate confrontation assays do not distinguish mycelial contact-dependent effects from those independent of contact, such as volatiles or compounds secreted into the agar medium. In the future, other plate assays could tease apart these effects, such as by using bi-partite plates to test the effects of volatiles and culturing the plant pathogens on cell-free compounds secreted by *T. nordicum*. This antagonistic effect may also be occurring in the pepper root rhizosphere, whereby *T. nordicum* antagonizes the *P. capsici* that is inoculated onto the pepper after the inoculation of *T. nordicum*. The protease and cellulase activities detected from *T. nordicum* are likely contributing to this antagonism in the plate confrontation assays. The *T. nordicum* endo-cellulase activity appears largely constitutive, as there were similar levels of endo-cellulase activity on all the tested medium supplementations, and therefore, if constitutive, there is a greater likelihood of endo-cellulase activity being present in the plate confrontation and pepper rhizosphere than an activity that would require the presence of an inducer. Previously, it was shown that cellulases were contributing to the antagonism from *T. reesei* towards *Pythium* plant pathogens [13], and *T. harzianum* strains with higher protease activity were shown to have higher antagonism levels in plate confrontation assays [43]. Secondly, there was a clear effect on pepper growth promotion after inoculation with *T. nordicum* compared to an autoclaved-spore control. This increased growth may lead to more robust plants, which are less susceptible to *P. capsici* and, in that way, also contribute to the pepper disease control effect of *T. nordicum*. The growth-promoting effects of *Trichoderma* on pepper have been reported previously, such as by *T. harzianum* and *T. asperellum* [44]. Thirdly, the control effect has a level of pathosystem-dependent performance involved in it because *T. nordicum* did not control soft rot disease of ginger caused by *P. myriotylum*, even though *T. nordicum* also overgrew *P. myriotylum* in the plate confrontation assays. Previously, another *Trichoderma* species, *T. asperellum* CBS433.97 (which also overgrew *P. myriotylum* in the plate confrontation assays), was used to partially biocontrol soft rot disease of

https://doi.org/10.3390/jof12040292

ginger caused by *P. myriotylum*, where there was a significant control rate of ~50% [13]. Differences in *T. nordicum* biocontrol efficacy across pathosystems do not necessarily reflect plant-pathogen-specific antagonism but may result from pathosystem-dependent factors, including plant-microbe interactions and other environmental conditions that vary among the ginger, pepper, and cucumber systems tested. For example, the ginger rhizosphere may be less favorable to *T. nordicum* colonization than the pepper rhizosphere, or *P. myriotylum* genes involved in antagonism during plate confrontation may be differentially expressed in the ginger rhizosphere. Also, biochemical conditions in the ginger rhizosphere, such as pH, may be less favorable for the activity of enzymes or other metabolites that *T. nordicum* may use to antagonize *P. myriotylum* in plate culture. Of note with regard to the use of autoclaved spore inoculum as the control, it is possible that the inactivated spore suspension also has some growth-promoting effects (e.g., some polymers from spores, such as chitin, may have a type of MAMP or immune inducing function or the nutrients from the medium may alter microbiota), and in future work, it would be beneficial to compare with other controls such as water or the medium used for culturing *T. nordicum*.

The isolation source for the *T. nordicum* isolates was wood from a tree stump, which contrasts with more commonly used isolation sources for biocontrol, such as the rhizosphere of healthy plants in an area with many diseased plants [45]. On dead wood, *T. nordicum* may still function as a microbial antagonist through parasitism of other wood-associated microorganisms rather than being a primary wood decomposer. This is supported by how *Trichoderma* species can be used in the control of wood-decaying fungi in urban trees, whereby *Trichoderma* protects rather than decomposes the wood [46]. Although *T. nordicum* may not be a primary wood decomposer, its isolation from a woody substrate may mean it is tolerant to inhibitory compounds that can be found in wood, and this might contribute to its ability to be cultured with lignite, which contains phenolic and other aromatic compounds that may overlap with inhibitory compounds found in wood lignin [47]. Also, the higher *T. nordicum* spore yield in shake flask cultures with lignite may be related to the lignite pieces providing more solid surfaces for sporulation and to increased oxygen transfer due to the enhanced mixing and potential bubble formation caused by the movement of the lignite pieces. Lignite is a humic-rich substance, and this can promote plant growth, which is an additional benefit, alongside increasing *T. nordicum* spore yield, of using lignite. The potential of *T. nordicum* V1 to degrade lignite in an agricultural environment is worthwhile testing, as it may function as a carbon source as well as a carrier, and beneficially contribute to the release of humic acids from the lignite, but perhaps, adversely, it could reduce bioformulation shelf-life if the lignite can be degraded during storage. For use as a bioformulation, it may require separating the mycelial fragments from the spores and lignite pieces. Testing with smaller lignite pieces than the ~2 mm size used may be beneficial for downstream separation from the larger mycelial fragments. From a regulatory perspective, the use of lignite may require separate approval applications beyond those required for *T. nordicum* itself, especially as lignite may contain levels of metals or aromatic compounds that are environmentally toxic.

Initially, analysis of ITS region sequences indicated that the V1 and V4 isolates could be identified as *T. atroviride* or *T.* cf. *atroviride*, but further analysis indicated that the best species identification was to two species (*T. nordicum* and *T. nigricans*). Since *T. nordicum* was described earlier by [27] than *T. nigricans* (type strain: T32781 = CGMCC40314) by [40], the *T. nordicum* description, rather than *T. nigricans*, was used for the V1 and V4 isolates. In the future, further taxonomic analysis (e.g., whole genome sequencing and morphological analysis) is required to delineate the species boundaries of the *T. nordicum*/*T. nigricans* species complex, and for now, we consider V1 and V4 isolates part of the *T. nordicum*/*T. nigricans* species complex and adopt the name *T. nordicum* for the V1 and V4 isolates.

https://doi.org/10.3390/jof12040292

Notably, a genome sequence is available for *T. nigricans* (Zhang, Qiu et al. 2025 [48]), but none is available yet for *T. nordicum*. While there are no reports in the literature of *T. nordicum* antagonizing plant pathogens, there are sequence data in GenBank from at least two *T. nordicum* strains that may be involved in this. *T. nordicum* strain URMicro 11722 (GenBank: OR762232.1) is from an entry titled about novel species of *Trichoderma* antagonistic to sclerotia-producing plant pathogens, and *T. nordicum* isolate T353 (GenBank: PV491717.1) from an entry titled identification of *Trichoderma* spp. from ancient *Platycladus orientalis* (a coniferous cypress tree) and their biocontrol for Alternaria leaf blight. There are no reports in the literature of *T. nigricans* antagonizing plant pathogens or being used as a biocontrol treatment of crop diseases. *T. nigricans* T32781 is tolerant to the metal cadmium, which can lead to reduced levels of cadmium availability and contamination in soils colonized by *T. nigricans* T32781 [48].

Future work with *T. nordicum* could involve strain engineering, chlamydospore production, testing spore carriers, and analyzing the control potential for other oomycete-caused diseases. Genome sequencing and gene expression analysis could be used in the future to identify key genetic factors for the biocontrol of *T. nordicum* strains and to engineer *T. nordicum* for improved biocontrol. Conidia from the *T. nordicum* strains were used here, but in other studies, such as with *T. longibrachiatum*, it was found that chlamydospores had greater survival ability and biocontrol potential than conidia [49], and the potential for future chlamydospore production has implications for field applications and product shelf life. The ability of the *T. nordicum* strains to be cultured with lignite also leads to the potential to use lignite as a carrier for *T. nordicum* spores, as lignite can be an effective carrier for bioformulations [50]. An important point to note is the potential biosecurity concerns with the use of *T. nordicum*-based biocontrol solutions. Another species, *T. afroharzianum*, has been reported to cause ear rot in maize [51] and is listed on the European and Mediterranean Plant Protection Organization alert list [52]. There is a need for careful evaluation of the potential unintended effects of *T. nordicum* before use in biocontrol solutions.

5. Conclusions

T. nordicum V1 can antagonize several oomycete plant pathogens and promote pepper growth, and can be used for biocontrol of seedling damping-off caused by *P. aphanidermatum*, and Phytophthora blight of pepper caused by *P. capsici*. The mechanisms underlying the antagonistic interactions between *T. nordicum* V1 and oomycetes, such as parasitic or predatory mechanisms, have not yet been demonstrated and require further investigation. Overall, *T. nordicum* V1 is a promising candidate to develop into a biocontrol agent for the biocontrol of oomycete-caused plant diseases. In the future, further taxonomic analysis is required to delineate the species boundaries of the *T. nordicum*/*T. nigricans* species complex.

Supplementary Materials: The following supporting information can be downloaded at https://www.mdpi.com/article/10.3390/jof12040292/s1: Figure S1. (A) Test of viability of *P. capsici* LT263 mycelia from parts overgrown by *T. nordicum* V1 isolate. Note that the growth of *P. capsici* plugs at 33 °C is slow, with growth mainly around the plug in the photo. (B) Summary of the numbers of plugs with growth after transfer from the indicated confrontation or single culture control plates. The viability tests were repeated two times with similar results; Figure S2. Time course of disease index and AUDPC for two repeat experiments (A,B) from control of *T. nordicum* on pepper blight caused by *P. capsici*. The error bars represent standard errors (n = 10 plants). *** = $p < 0.001$ from Dunn's post hoc test after Kruskal–Wallis test in comparison to the disease index or AUDPC of seedlings inoculated with *P. capsici* LT263 and autoclaved *T. nordicum* spores; Figure S3. Re-isolation of *P. capsici* and *T. nordicum* from the pepper pot trial. Agarose gel of PCR reactions from gDNA from mycelia grown out from pepper root samples from indicated treatments from the pot trial, with information below the gel on the match of the sequencing of the ITS PCR products. (A) is from the first repeat and (B) is from the second repeat of the pepper pot-trial; Figure S4. Growth promotion effect of *T. nordicum* isolates on pepper from two repeat experiments (A,B). The error bars represent standard errors ($n = 10$ plants). * = $p < 0.05$, ns = non-significant from Dunnett's post hoc test (after ANOVA) in comparison to the seedlings inoculated with autoclaved *T. nordicum* spores (-); Figure S5. Disease index of *P. myriotylum*-caused Pythium soft rot of ginger and biocontrol attempts using *T. nordicum* V1 isolate, whereby disease scores were recorded from 8 to 16 DAI with *P. myriotylum* SWQ7 where (A) shows the time course of disease index for all days recorded and AUDPC and (B) shows scores at 16 DAI. The pot trial was repeated twice with similar results. The error bars represent standard errors ($n = 12$). *** = $p < 0.001$, ns = non-significant from Dunn's post hoc test after Kruskal–Wallis test in comparison to the disease index or AUDPC of seedlings inoculated with *P. myriotylum* SWQ7 and autoclaved *T. nordicum* spores. (C) Images of the ginger seedlings from the disease control pot trial. (D) Lack of growth promotion effect of *T. nordicum* V1 on ginger seedlings. The error bars represent standard error ($n = 12$). ns = non-significant from Dunnett's post hoc test (after ANOVA) in comparison to the seedlings inoculated with autoclaved *T. nordicum* spores (-); Figure S6. Re-isolation of *P. myriotylum* and *T. nordicum* from the ginger pot trial. Agarose gel of PCR reactions from gDNA from mycelia grown out from ginger root samples from indicated treatments from the pot trial, with information below the gel on the match of the sequencing of the ITS PCR products; Table S1. Individual scores from the plate confrontation assay to screen the ability of *T. nordicum* V1 and V4 isolates to overgrow a range of oomycete plant pathogens. The following Excel nested IF function was used to categorize the *Trichoderma* colony coverage for Bell's scale where LoT = the length of Trichoderma colony coverage: "=IF(LoT > 8.1,1,IF(LoT > 5.9,2,IF(LoT > 4.4,2.5,IF(LoT > 2.9,3,IF(LoT > 0.5,4,5)))))".

Author Contributions: S.L.: Investigation; writing—original draft. X.W.: Investigation; writing—original draft. S.C.: Investigation; writing—original draft. Y.Z.: Investigation. J.C.: Investigation. W.L.: Investigation. Y.C.: Investigation. M.D.: Investigation. S.J.: Investigation. W.M.A.: Investigation; funding acquisition. D.Z.: Conceptualization; supervision; writing—review and editing. F.M.C.: Writing—review and editing; supervision. I.S.D.: Writing—review and editing; supervision. L.W.: Conceptualization; writing—review and editing; funding acquisition. M.J.: Project administration; supervision; writing—review and editing. P.D.: Conceptualization; supervision; project administration; writing—review and editing. All authors have read and agreed to the published version of the manuscript.

Funding: This research was funded by the China Agriculture Research System (CARS-24-C-01) of MOF and MARA, and the National Natural Science Foundation of China (W2433090).

Institutional Review Board Statement: Not applicable.

Informed Consent Statement: Not applicable.

Data Availability Statement: The data supporting the conclusions of the study are included in the main text or Supplementary Materials.

Acknowledgments: We acknowledge members of the Institute of Plant Protection, JAAS, for technical support.

Conflicts of Interest: The authors declare a potential conflict of interest. Lihui Wei, Siqiao Chen, Paul Daly, Dongmei Zhou, Irina S. Druzhinina, Yishen Zhao, and Xian Wen are co-inventors of a patent for the application of the *T. nordicum* V1 isolate to biologically control crop diseases. China patent application number: CN202411738888.9.

Abbreviations

The following abbreviations are used in this manuscript:

AUDPC	Area under the disease progress curve
AZCL	Azurine cross-linked
cV8	Clarified V8 juice
DAI	Days after inoculation
ITS	Internal transcribed spacer region
PDA	Potato dextrose agar

References

1. Scheringer, M.; Schulz, R. The state of the world's chemical pollution. *Annu. Rev. Environ. Resour.* **2025**, *50*, 381–408. [CrossRef]
2. Woo, S.L.; Hermosa, R.; Lorito, M.; Monte, E. *Trichoderma*: A multipurpose, plant-beneficial microorganism for eco-sustainable agriculture. *Nat. Rev. Microbiol.* **2023**, *21*, 312–326. [CrossRef] [PubMed]
3. Kubicek, C.; Herrera-Estrella, A.; Seidl-Seiboth, V.; Martinez, D.; Druzhinina, I.; Thon, M.; Zeilinger, S.; Casas-Flores, S.; Horwitz, B.; Mukherjee, P.; et al. Comparative genome sequence analysis underscores mycoparasitism as the ancestral life style of Trichoderma. *Genome Biol.* **2011**, *12*, R40. [CrossRef]
4. Kubicek, C.P.; Steindorff, A.S.; Chenthamara, K.; Manganiello, G.; Henrissat, B.; Zhang, J.; Cai, F.; Kopchinskiy, A.G.; Kubicek, E.M.; Kuo, A.; et al. Evolution and comparative genomics of the most common *Trichoderma* species. *BMC Genom.* **2019**, *20*, 485. [CrossRef]
5. Benitez, T.; Rincon, A.M.; Limon, M.C.; Codon, A.C. Biocontrol mechanisms of Trichoderma strains. *Int. Microbiol.* **2004**, *7*, 249–260. [PubMed]
6. Taboadela-Hernanz, J.; Hieno, A.; Shimizu, M. Toward effective biocontrol of oomycete plant pathogens: Traits and modes of action of biocontrol agents, and their screening approaches. *Rev. Agric. Sci.* **2025**, *13*, 32–51. [CrossRef]
7. Kamoun, S.; Furzer, O.; Jones, J.D.G.; Judelson, H.S.; Ali, G.S.; Dalio, R.J.D.; Roy, S.G.; Schena, L.; Zambounis, A.; Panabières, F. et al. The Top 10 oomycete pathogens in molecular plant pathology. *Mol. Plant Pathol.* **2015**, *16*, 413–434. [CrossRef] [PubMed]
8. Lamour, K.H.; Stam, R.; Jupe, J.; Huitema, E. The oomycete broad-host-range pathogen *Phytophthora capsici*. *Mol. Plant Pathol.* **2012**, *13*, 329–337. [CrossRef] [PubMed]
9. Ho, H. The taxonomy and biology of *Phytophthora* and *Pythium*. *J. Bacteriol. Mycol.* **2018**, *6*, 40–45. [CrossRef]
10. Mélida, H.; Sandoval-Sierra, J.V.; Diéguez-Uribeondo, J.; Bulone, V. Analyses of extracellular carbohydrates in oomycetes unveil the existence of three different cell wall types. *Eukaryot. Cell* **2013**, *12*, 194–203. [CrossRef]
11. Druzhinina, I.S.; Chenthamara, K.; Zhang, J.; Atanasova, L.; Yang, D.; Miao, Y.; Rahimi, M.J.; Grujic, M.; Cai, F.; Pourmehdi, S. et al. Massive lateral transfer of genes encoding plant cell wall-degrading enzymes to the mycoparasitic fungus *Trichoderma* from its plant-associated hosts. *PLoS Genet.* **2018**, *14*, e1007322. [CrossRef]
12. Bischof, R.H.; Ramoni, J.; Seiboth, B. Cellulases and beyond: The first 70 years of the enzyme producer *Trichoderma reesei*. *Microb Cell Fact.* **2016**, *15*, 106. [CrossRef]
13. Chen, S.; Daly, P.; Anjago, W.M.; Wang, R.; Zhou, D.; Feng, C.; Wen, X.; Zhao, Y.; Deng, S.; Voglmeir, J.; et al. Genus-wide analysis of *Trichoderma* antagonism towards *Pythium* and *Globisporangium* plant pathogens and the contribution of cellulases to the antagonism. *Appl. Environ. Microbiol.* **2024**, *90*, e0068124. [CrossRef] [PubMed]
14. Hernández-Pérez, T.; Gómez-García, M.D.R.; Valverde, M.E.; Paredes-López, O. *Capsicum annuum* (hot pepper): An ancient Latin-American crop with outstanding bioactive compounds and nutraceutical potential. A review. *Compr. Rev. Food Sci. Food Saf* **2020**, *19*, 2972–2993. [CrossRef] [PubMed]
15. FAOSTAT. Food and Agricultural Organisation of the United Nations Statistical Database. 2023. Available online http://faostat.fao.org/ (accessed on 1 December 2025).
16. Ro, N.; Haile, M.; Hur, O.; Ko, H.-C.; Yi, J.-Y.; Woo, H.-J.; Choi, Y.-M.; Rhee, J.; Lee, Y.-J.; Kim, D.-A.; et al. Genome-wide association study of resistance to anthracnose in pepper (*Capsicum chinense*) germplasm. *BMC Plant Biol.* **2023**, *23*, 389. [CrossRef]
17. Larrahondo-Rodríguez, E.; Liao, Y.-Y.; Huerta, A.I. Diagnostic guide for bacterial spot of tomato and pepper. *Plant Health Progress* **2022**, *23*, 355–361. [CrossRef]
18. Remesal, E.; Lucena, C.; Azpilicueta, A.; Landa, B.B.; Navas-Cortés, J.A. First report of southern blight of pepper caused by *Sclerotium rolfsii* in southern Spain. *Plant Dis.* **2010**, *94*, 280. [CrossRef]
19. Mensah, E.O.; Oh, H.; Song, J.; Baek, J. Exploring imaging techniques for detecting Tomato spotted wilt virus (TSWV) infection in pepper (*Capsicum* spp.) germplasms. *Plants* **2024**, *13*, 3447. [CrossRef]

https://doi.org/10.3390/jof12040292

Quesada-Ocampo, L.M.; Parada-Rojas, C.H.; Hansen, Z.; Vogel, G.; Smart, C.; Hausbeck, M.K.; Carmo, R.M.; Huitema, E.; Naegele, R.P.; Kousik, C.S.; et al. *Phytophthora capsici*: Recent progress on fundamental biology and disease management 100 years after its description. *Annu. Rev. Phytopathol.* **2023**, *61*, 185–208. [CrossRef] [PubMed]

Parada-Rojas, C.H.; Granke, L.L.; Naegele, R.P.; Hansen, Z.; Hausbeck, M.K.; Kousik, C.S.; McGrath, M.T.; Smart, C.D.; Quesada-Ocampo, L.M. A diagnostic guide for *Phytophthora capsici* infecting vegetable crops. *Plant Health Prog.* **2021**, *22*, 404–414. [CrossRef]

Al-Sadi, A.M.; Al-Ghaithi, A.G.; Al-Balushi, Z.M.; Al-Jabri, A.H. Analysis of diversity in *Pythium aphanidermatum* populations from a single greenhouse reveals phenotypic and genotypic changes over 2006 to 2011. *Plant Dis.* **2012**, *96*, 852–858. [CrossRef]

Le, D.P.; Smith, M.; Hudler, G.W.; Aitken, E. Pythium soft rot of ginger: Detection and identification of the causal pathogens, and their control. *Crop Prot.* **2014**, *65*, 153–167. [CrossRef]

Zhou, J.; Liang, J.; Zhang, X.; Wang, F.; Qu, Z.; Gao, T.; Yao, Y.; Luo, Y. *Trichoderma brevicompactum* 6311: Prevention and control of *Phytophthora capsici* and its growth-promoting effect. *J. Fungi* **2025**, *11*, 105. [CrossRef] [PubMed]

Santos, M.; Diánez, F.; Sánchez-Montesinos, B.; Huertas, V.; Moreno-Gavira, A.; Esteban García, B.; Garrido-Cárdenas, J.A.; Gea, F.J. Biocontrol of diseases caused by *Phytophthora capsici* and *P. parasitica* in pepper plants. *J. Fungi* **2023**, *9*, 360. [CrossRef] [PubMed]

Segarra, G.; Aviles, M.; Casanova, E.; Borrero, C.; Trillas, I. Effectiveness of biological control of *Phytophthora capsici* in pepper by *Trichoderma asperellum* strain T34. *Phytopathol. Mediterr.* **2013**, *52*, 77–83.

Zhang, G.Z.; Yang, H.T.; Zhang, X.J.; Zhou, F.Y.; Wu, X.Q.; Xie, X.Y.; Zhao, X.Y.; Zhou, H.Z. Five new species of *Trichoderma* from moist soils in China. *MycoKeys* **2022**, *87*, 133–157. [CrossRef]

Cai, F.; Druzhinina, I.S. In honor of John Bissett: Authoritative guidelines on molecular identification of *Trichoderma*. *Fungal Divers.* **2021**, *107*, 1–69. [CrossRef]

White, T.J.; Bruns, T.; Lee, S.; Taylor, J. Amplification and direct sequencing of fungal ribosomal RNA genes for phylogenetics. In *PCR Protocols*; Innis, M.A., Ed.; Academic Press: Cambridge, MA, USA, 1990; pp. 315–322.

Liu, Y.J.; Whelen, S.; Hall, B.D. Phylogenetic relationships among ascomycetes: Evidence from an RNA polymerse II subunit. *Mol. Biol. Evol.* **1999**, *16*, 1799–1808. [CrossRef]

Edgar, R.C. MUSCLE: Multiple sequence alignment with high accuracy and high throughput. *Nucleic Acids Res.* **2004**, *32*, 1792–1797. [CrossRef]

Kumar, S.; Stecher, G.; Suleski, M.; Sanderford, M.; Sharma, S.; Tamura, K. MEGA12: Molecular evolutionary genetic analysis version 12 for adaptive and green computing. *Mol. Biol. Evol.* **2024**, *41*, msae263. [CrossRef]

Cui, C.; Herlihy, J.H.; Bombarely, A.; McDowell, J.M.; Haak, D.C. Draft assembly of *Phytophthora capsici* from long-read sequencing uncovers complexity. *Mol. Plant-Microbe Interact.* **2019**, *32*, 1559–1563. [CrossRef]

Forster, H.; Tyler, B.; Coffey, M. *Phytophthora sojae* races have arisen by clonal evolution and by rare outcrosses. *Mol. Plant-Microbe Interact.* **1994**, *7*, 780–791. [CrossRef]

Daly, P.; Chen, Y.; Zhang, Q.; Zhu, H.; Li, J.; Zhang, J.; Deng, S.; Wang, L.-J.; Zhou, D.; Tang, Z.; et al. *Pythium myriotylum* is recovered most frequently from Pythium soft-rot infected ginger rhizomes in China. *Plant Dis.* **2022**, *106*, 510–517. [CrossRef]

Taylor, R.J.; Salas, B.; Secor, G.A.; Rivera, V.; Gudmestad, N.C. Sensitivity of North American isolates of *Phytophthora erythroseptica* and *Pythium ultimum* to mefenoxam (metalaxyl). *Plant Dis.* **2002**, *86*, 797–802. [CrossRef]

Bell, D.K.; Wells, H.D.; Markham, C.R. In vitro antagonism of *Trichoderma* species against six fungal plant pathogens. *Phytopathology* **1982**, *72*, 379–382. [CrossRef]

Anjago, W.M.; Chen, S.; Zhao, Y.; Lu, J.; Chen, Y.; Zhou, D.; Druzhinina, I.S.; Wei, L.; Daly, P. Regulation of *Trichoderma harzianum* gene expression and protein production in submerged cultures with inactivated oomycete mycelium. *Microbiol. Res.* **2026**, *303*, 128378. [CrossRef] [PubMed]

Yao, J.; Chen, Y.; Zhuo, D.; Chen, S.; Xu, B.; Yan, C.; Li, W.; Feng, H.; Deng, S.; Cai, F.M.; et al. Genome and transcriptome analysis of the lignite-degrading *Trichoderma* cf. *simile* WF8 strain highlights potential degradation mechanisms. *Int. Biodeterior. Biodegrad.* **2025**, *198*, 105997. [CrossRef]

Zhao, R.; Mao, L.-J.; Zhang, C.-L. Three new species of *Trichoderma* (Hypocreales, Hypocreaceae) from soils in China. *MycoKeys* **2023**, *97*, 21–40. [CrossRef]

Chen, S.; Daly, P.; Zhou, D.; Li, J.; Wang, X.; Deng, S.; Feng, H.; Wang, C.; Sheikh, T.M.M.; Chen, Y.; et al. The use of mutant and engineered microbial agents for biological control of plant diseases caused by *Pythium*: Achievements versus challenges. *Fungal Biol. Rev.* **2022**, *40*, 79–90. [CrossRef]

Sid Ahmed, A.; Pérez-Sánchez, C.; Egea, C.; Candela, M.E. Evaluation of *Trichoderma harzianum* for controlling root rot caused by *Phytophthora capsici* in pepper plants. *Plant Pathol.* **1999**, *48*, 58–65. [CrossRef]

Szekeres, A.; Kredics, L.; Antal, Z.; Kevei, F.; Manczinger, L. Isolation and characterization of protease overproducing mutants of *Trichoderma harzianum*. *FEMS Microbiol. Lett.* **2004**, *233*, 215–222. [CrossRef]

44. Yadav, M.; Divyanshu, K.; Dubey, M.K.; Rai, A.; Kumar, S.; Tripathi, Y.N.; Shukla, V.; Upadhyay, R.S. Plant growth promotion and differential expression of defense genes in chilli pepper against *Colletotrichum truncatum* induced by *Trichoderma asperellum* and *T. harzianum*. *BMC Microbiol.* **2023**, *23*, 54. [CrossRef]
45. Ciancio, A.; Pieterse, C.M.J.; Mercado-Blanco, J. Editorial: Harnessing useful rhizosphere microorganisms for pathogen and pest biocontrol—Second edition. *Front. Microbiol.* **2019**, *10*, 1935. [CrossRef]
46. Schubert, M.; Fink, S.; Schwarze, F.W.M.R. Evaluation of *Trichoderma* spp. as a biocontrol agent against wood decay fungi in urban trees. *Biol. Control.* **2008**, *45*, 111–123. [CrossRef]
47. Wang, Y.; Hou, Y.; Li, H.; Wu, W.; Ren, S.; Li, J. New insights into the coalification of lignite from lignin: A comparison of their chemical structures. *Fuel* **2022**, *326*, 125017. [CrossRef]
48. Zhang, S.; Qiu, C.W.; Shi, S.H.; Sreesaeng, J.; Gao, Z.F.; Zhang, C.; Mao, W.; Ali, M.A.; Chen, Z.H.; Wu, F. *Trichoderma* genome and multiomics insight into promoting yield and reducing grain cadmium in barley and wheat. *J. Integr. Plant Biol.* **2025**, *68*, 39–56. [CrossRef]
49. Cao, Q.; Liang, Y.; Tian, Y.; Lian, H.; Jiang, X.; Li, M. Survival dynamics of *Trichoderma longibrachiatum* Tr58 in conidia- and chlamydospore-amended soils with different moisture levels. *Agriculture* **2023**, *13*, 238. [CrossRef]
50. Khan, A.; Singh, A.V.; Gautam, S.S.; Agarwal, A.; Punetha, A.; Upadhayay, V.K.; Kukreti, B.; Bundela, V.; Jugran, A.K.; Goel, R. Microbial bioformulation: A microbial assisted biostimulating fertilization technique for sustainable agriculture. *Front. Plant Sci.* **2023**, *14*, 1270039. [CrossRef] [PubMed]
51. Pfordt, A.; Schiwek, S.; Karlovsky, P.; von Tiedemann, A. *Trichoderma afroharzianum* ear rot—A new disease on maize in europe. *Front. Agron.* **2020**, *2*, 547758. [CrossRef]
52. EPPO. *EPPO Alert List—Trichoderma afroharzianum Causing Ear Rot on Maize*; EPPO Global Database: Paris, France, 2022.

https://doi.org/10.3390/jof1204029

Journal of
Fungi

MDPI

Article

Rapid Discovery of Antifungal α-Pyrone Analogs from *Diaporthe kyushuensis* ZMU-48-1 via an HSQC-Based DeepSAT Strategy

Siwen Yuan *,†, Jiaqi Zheng †, Yueling Wu, Xijing Wang, Haiwen Wang, Min Yan and Tianpeng Yin *

School of Bioengineering, Zunyi Medical University, Zhuhai 519041, China; zhengjiaqi@zmuzh.edu.cn (J.Z.); wuyuelin@zmuzh.edu.cn (Y.W.); wangxijing@zmuzh.edu.cn (X.W.); wanghaiwen@zmuzh.edu.cn (H.W.); yanmin@zmuzh.edu.cn (M.Y.)
* Correspondence: yuansiwen@zmuzh.edu.cn (S.Y.); ytp@zmu.edu.cn (T.Y.)
† These authors contributed equally to this work.

Abstract

Endophytic fungi represent an extensive source of chemically diverse and bioactive polyketides. Herein, an HSQC-based DeepSAT-guided strategy was employed for the scaffold-prioritized isolation of metabolites from *Diaporthe kyushuensis* ZMU-48-1, an endophyte isolated from *Acacia confusa*. DeepSAT analysis of the semi-purified fractions identified an α-pyrone chemotype, which facilitated the targeted isolation of thirteen pyranone-type polyketides, including eight previously undescribed analogues, diaporthopyranone A–H (**1**–**8**). The structures of the new compounds were rigorously elucidated via comprehensive spectroscopic analysis (1D and 2D NMR and HRESIMS), while their absolute configurations were established by comparing experimental and TDDFT-calculated ECD data. All isolates (**1**–**13**) were screened for antifungal activity against nine plant pathogens. While most metabolites were inactive at 200 μg/mL, compound 4 inhibited *Alternaria alternata* and *Valsa mali* (MICs = 100 and 80 μg/mL, respectively), and compounds **6** and **8** exhibited selective inhibition against *Colletotrichum musae* (MICs = 60 and 80 μg/mL). These results expand the pyranone repertoire of the genus *Diaporthe* and underscore the utility of the HSQC–DeepSAT platform for the streamlined discovery of specific natural product scaffolds.

Keywords: DeepSAT; α-pyrone; antifungal; *Diaporthe kyushuensis*

1. Introduction

Fungal diseases of crops pose a persistent threat to global food security, resulting in substantial yield losses and compromising the quality of agricultural products. The emergence of resistant plant pathogenic strains, driven by the prolonged application of a limited set of commercial fungicides, necessitates the continuous discovery of novel antifungal lead structures with diverse modes of action [1]. Endophytic fungi have increasingly been recognized as a prolific and under-explored repository of structurally unique polyketides, many of which exhibit significant agrochemical potential [2,3]. Among these, species of the genus *Diaporthe* are notable for their ability to produce a diverse array of secondary metabolites, particularly polyketide-derived lactones and oxygenated heterocycles with potent antimicrobial and phytotoxic properties [4,5].

α-Pyrones (2*H*-pyran-2-one derivatives) represent a prominent class of simple yet versatile polyketide scaffolds characterized by a six-membered unsaturated lactone ring. These metabolites are widely distributed across the fungal and plant kingdoms, often serving as key biosynthetic intermediates or end-products with broad-spectrum biological activities, including antifungal, antibacterial, and cytotoxic effects [6,7]. Despite their structural simplicity, the -pyrone framework provides a scaffold for extensive functionalization through hydroxylation, methylation, and side-chain elongation, leading to a rich chemical space. Although several pyranones have been isolated from the genus *Diaporthe* [8–10], their structural diversity within this genus remains incompletely mapped, and their specific roles as antifungal agents against specialized plant pathogens are only partially characterized.

The traditional "isolate-then-characterize" workflow for discovering such metabolites is often hindered by the presence of complex metabolic profiles and the repetitive isolation of known compounds. To address these challenges, Nuclear Magnetic Resonance (NMR)-based metabolite profiling has emerged as a robust tool for the rapid dereplication and prioritization of extracts. Specifically, Heteronuclear Single Quantum Coherence (HSQC)-guided analysis allows for the organization of semi-purified fractions based on chemical class and scaffold characteristics. The development of DeepSAT—a deep learning-based system trained to predict chemical fingerprints and scaffold classes directly from HSQC spectra—has revolutionized the annotation of dominant chemotypes in complex mixtures [11]. By enabling scaffold-level prioritization prior to exhaustive purification, DeepSAT streamlines the discovery process, allowing researchers to focus on specific, high-interest chemical families such as α-pyrones within diverse fungal extracts.

In the present study, we describe a structure-directed investigation of the endophytic fungus *Diaporthe kyushuensis* ZMU-48-1, isolated from *Acacia confusa* [12]. Utilizing an HSQC-based DeepSAT strategy, we successfully prioritized fractions exhibiting an α-pyrone dominated chemotype, which directed the subsequent targeted isolation. This approach afforded thirteen pyranone derivatives, including eight previously undescribed analogues, diaporthopyranone A–H (**1–8**). (Figure 1). Herein, we report the DeepSAT-assisted isolation, the rigorous structural elucidation of the new metabolites via extensive spectroscopic analysis and Time-Dependent Density Functional Theory-Electronic Circular Dichroism (TDDFT-ECD) calculations, and the evaluation of their antifungal activities against a panel of nine economically significant plant pathogens (*Fusarium graminearum, Botryosphaeria dothidea, Fusarium oxysporum, Colletotrichum gloeosporioides, Valsa mali, Colletotrichum musae, Alternaria alternata, Colletotrichum agenarium,* and *Bipolaris sorokiniana*).

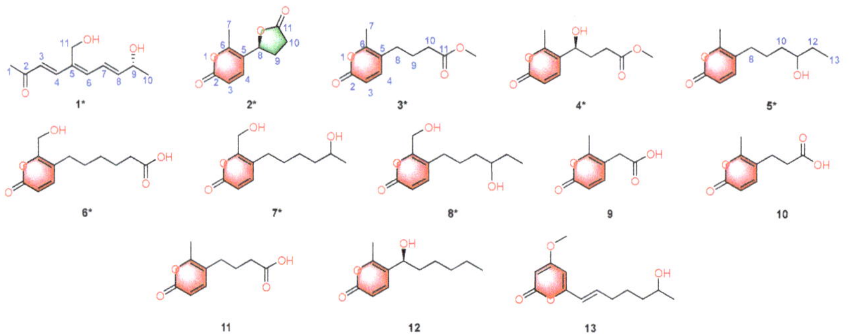

Figure 1. Structures of compounds **1–13** (* denotes new compounds).

2. Materials and Methods

2.1. General Experimental Procedure

1D and 2D NMR spectra were recorded on a Bruker AVANCE III 600 spectrometer (Bruker, Karlsruhe, Germany) equipped with a 5 mm cryoprobe (CPP BBO600S3 BB-H&F-D-05 Z XT). The deuterated solvents $CDCl_3$ and CD_3OD were purchased from Cambridge

https://doi.org/10.3390/jof1203016

Isotope Laboratories (Tewksbury, MA, USA). High-resolution electrospray ionization mass spectra (HR-ESI-MS) were acquired on an Orbitrap Fusion Lumos mass spectrometer (Thermo Fisher Scientific, Waltham, MA, USA). Infrared (IR) spectra were recorded on an Agilent Cary 630 FT-IR spectrometer (Agilent, Santa Clara, CA, USA). Optical rotations were measured on an MCP 200 polarimeter (Anton Paar, Ryde, Austria).

Preparative high-performance liquid chromatography (HPLC) was carried out on a Waters system (Waters, Milford, MA, USA) comprising a 1525 binary gradient pump and a 2998 photodiode-array detector. Separations were performed on Xtimate C18 (10 × 250 mm, 5 μm; Welch Materials, Shanghai, China), Ultimate XB-Phenyl (10 × 250 mm, 5 μm; Welch Materials, Stamford, CT, USA), and ACE C18 (4.6 × 250 mm, 5 μm; Advanced Chromatography Technology, Aberdeen, UK) columns. HPLC-grade acetonitrile and methanol were obtained from Tianjin Concord Technology Co., Ltd. (Tianjin, China), and chromatograms were monitored at 220, 254, 275, and 310 nm.

Column chromatography (CC) was performed on silica gel (300–400 mesh; Qingdao Haiyang Chemical Factory, Qingdao, China). Thin-layer chromatography (TLC) was conducted on GF_{254} plates from the same manufacturer. The spots were visualized under UV light and/or by spraying with vanillin–sulfuric acid reagent followed by heating. Analytical-grade methanol (MeOH), dichloromethane (DCM), petroleum ether (PE), ethyl acetate (EtOAc), and ethanol (EtOH) were purchased from Tianjin Damao Chemical Reagent Co., Ltd. (Tianjin, China).

2.2. Fungal Material and Identification

The fungal strain *Diaporthe kyushuensis* ZMU-48-1 was obtained from decayed leaves of *Acacia confusa* Merr. collected from Huanglübei Mountain, Jinwan District, Zhuhai, Guangdong Province, China, in October 2023. Briefly, the collected leaves were surface-sterilized with 2% (*v*/*v*) sodium hypochlorite for 2 min and 75% (*v*/*v*) ethanol for 1 min, followed by multiple washes with sterile water. The sterilized leaves were then cultured on Potato Dextrose Agar (PDA) medium at 30 °C. Pure cultures were obtained via the hyphal tip method, where actively growing hyphae at the colony edge were repeatedly transferred to fresh medium until a single, morphologically uniform isolate was secured [13]. The strain was deposited in the Microbial Resource Bank of the Zhuhai Campus, Zunyi Medical University (Zhuhai, China). Molecular identification was conducted via internal transcribed spacer (ITS) rDNA sequencing (Beijing Qingke Biotechnology Co., Ltd., Beijing, China) using the universal primers ITS1 (5′-TCCGTAGGTGAACCTGCGG-3′) and ITS4 (5′-TCCTCCGCTTATTGATATGC-3′). The ITS rDNA sequence of strain ZMU-48-1 has been deposited in the GenBank database under the accession number PV259067. Whole-genome sequencing of strain ZMU-48-1 was performed by Sangon Biotech Co., Ltd. (Shanghai, China).

Mycelial plugs were cultured in 50 mL PDB at 28 °C and 180 rpm for 6 days. Approximately 20 mg of dried mycelium was ground in liquid nitrogen and incubated with 400 μL Buffer Digestion at 65 °C for 1 h. Following RNase A treatment and protein precipitation with Buffer PF at −20 °C, the DNA was purified via centrifugation and isopropanol precipitation. PCR amplification was conducted using the following thermal profile: initial denaturation at 98 °C for 1 min; followed by 12 cycles of 98 °C for 10 s, 60 °C for 30 s, and 72 °C for 30 s; and a final extension at 72 °C for 5 min [12].

2.3. Fermentation

Seed cultures of *D. kyushuensis* ZMU-48-1 were obtained by transferring mycelial plugs from 2–3-day-old PDA plates into 250 mL Erlenmeyer flasks containing 100 mL PDB and incubating at 28 °C and 180 rpm for 3 days. For submerged production cultures, 10 mL aliquots of the seed culture were inoculated into 500 mL Erlenmeyer flasks containing 200 mL PDB supplemented with either 3% (*w*/*v*) NaCl or 3% (*w*/*v*) NaBr. The flasks were shaken at 28 °C and 180 rpm for 14 days, affording a total culture volume of 4.0 L for each condition. For rice solid-state fermentation, 100 g of rice and 100 mL of 3% (*w*/*v*) sea-salt solution were placed in 600 mL tissue culture bottles (20 kg of rice in total) and sterilized at 121 °C for 30 min. After cooling, each bottle was directly inoculated with several mycelial plugs from 2–3-day-old PDA cultures of *D. kyushuensis* ZMU-48-1 and incubated statically at 28 °C for 43 days. The resulting liquid and solid cultures were harvested separately for extraction and isolation.

2.4. Extraction and Isolation

The fermented rice medium was extracted three times with EtOH (50 L in total); the combined EtOH extracts were concentrated and partitioned with EtOAc (45 L) to afford the crude rice extract (66.7 g). The crude rice extract was fractionated by silica gel CC using a PE–EtOAc gradient (100:0 to 0:100, *v*/*v*) to give seven fractions (Fr. B1–B7). Fr. B4 and Fr. B5 were combined and further separated on silica gel with a CH_2Cl_2–MeOH gradient (20:1 to 2:1, *v*/*v*) to yield subfractions, among which Fr. D4 and Fr. D5 were prioritized for purification.

Fr. D4 was purified by preparative HPLC using MeOH–H_2O and CH_3CN–H_2O systems to afford compounds **10** (7 mg, 25% MeOH, t_R = 7.54 min), **2** (36 mg, 18% MeOH, t_R = 11.40 min), **11** (10 mg, 25% MeOH, t_R = 11.74 min), **13** (10 mg, 27% CH_3CN, t_R = 17.47 min), **3** (12 mg, 30% CH_3CN, t_R = 14.90 min), and **12** (9 mg, 40% CH_3CN, t_R = 9.39 min). Fr. D5 was purified on the same system with CH_3CN–H_2O to yield compounds **1** (6 mg, 27% CH_3CN, t_R = 8.77 min), **4** (17 mg, 28% CH_3CN, t_R = 7.56 min), **5** (8 mg, 23% CH_3CN, t_R = 9.82 min), **7** (15 mg, 17% CH_3CN, t_R = 14.64 min) and **9** (12 mg, 22% CH_3CN, t_R = 8.13 min); further elution with 20% CH_3CN furnished compounds **6** (7 mg, t_R = 9.15 min) and **8** (5 mg, t_R = 12.33 min). The overall separation scheme and the chromatographic peaks corresponding to compounds **1–13** in the crude rice extract is summarized in Figure 2.

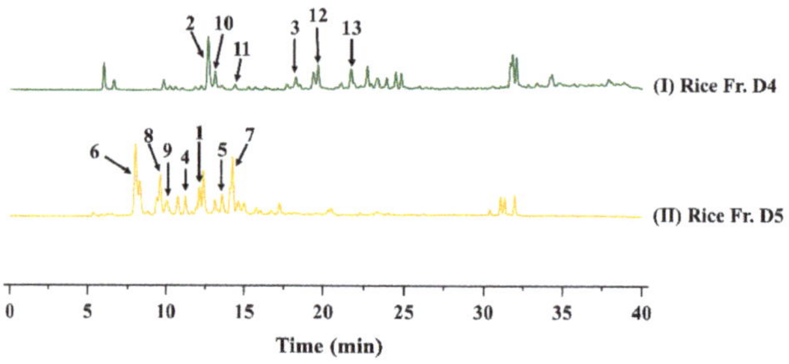

Figure 2. The HPLC analysis of different secondary metabolites in different medium.

2.4.1. Diaporthopyranone A (**1**)

Brownish yellow oil; [α]25 D −1.13 (c 0.1, MeOH); IR (KBr, ν_{max} cm^{-1}): 3336, 2929, 2856, 2837, 1716, 1683, 1670, 1506, 1456, 1018, 669, 592, 547. For the ^{1}H and ^{13}C NMR spectroscopic data, see Table 1. HR-ESI-MS *m*/*z* 219.0992 [M + Na]$^+$ (calcd for $C_{11}H_{16}NaO_3$, 219.0997).

2.4.2. Diaporthopyranone B (**2**)

Brown oil; IR (KBr, ν_{max} cm^{-1}): 3383, 2929, 2856, 2505, 2077, 1714, 1647, 1456, 1338, 1188, 974, 659, 592. For the ^{1}H and ^{13}C NMR spectroscopic data, see Table 1. HR-ESI-MS *m*/*z* 195.0647 [M + H]$^+$ (calcd for $C_{10}H_{11}O_4$, 195.0657).

https://doi.org/10.3390/jof1203016

2.4.3. Diaporthopyranone C (**3**)

Brown solid; IR (KBr, ν_{max} cm^{-1}): 3361, 2929, 2856, 1732, 1716, 1635, 1506, 1456, 1338, 1085, 1029, 867, 827. For the ^1H and ^{13}C NMR spectroscopic data, see Table 1. HR-ESI-MS *m/z* 211.0961 $[M + H]^+$ (calcd for $C_{11}H_{15}O_4$, 211.0970).

2.4.4. Diaporthopyranone D (**4**)

Brown solid; IR (KBr, ν_{max} cm^{-1}): 3340, 2943, 2833, 1716, 1456, 1417, 1118, 1020, 667, 599. For the ^1H and ^{13}C NMR spectroscopic data, see Table 1. HR-ESI-MS *m/z* 241.1069 $[M + H]^+$ (calcd for $C_{12}H_{17}O_5$, 241.1076).

2.4.5. Diaporthopyranone E (**5**)

Brownish yellow oil; [α]25 D −7.50 (c 0.1, MeOH); IR (KBr, ν_{max} cm^{-1}): 3365, 2929, 2899, 2864, 2505, 2320, 2077, 1734, 1683, 1541, 1456, 1116, 1082, 972, 831, 667, 603. For the ^1H and ^{13}C NMR spectroscopic data, see Table 1. HR-ESI-MS *m/z* 227.0911 $[M + H]^+$ (calcd for $C_{11}H_{15}O_5$, 227.0919).

2.4.6. Diaporthopyranone F (**6**)

Brownish yellow oil; [α]25 D +3.85 (c 0.168, MeOH); IR (KBr, ν_{max} cm^{-1}): 3325, 2929, 2854, 2835, 1716, 1647, 1456, 1417, 1015, 659, 599, 555. For the ^1H and ^{13}C NMR spectroscopic data, see Table 1. HR-ESI-MS *m/z* 211.1326 $[M + H]^+$ (calcd for $C_{12}H_{19}O_3$, 211.1334).

2.4.7. Diaporthopyranone G (**7**)

Amorphous powder; [α]25 D −3.08 (c 0.184, MeOH); IR (KBr, ν_{max} cm^{-1}): 3373, 2935, 2866, 1770, 1714, 1699, 1635, 1558, 1456, 1417, 1338, 1118, 1029, 929, 887, 833, 667, 592, 547. For the ^1H and ^{13}C NMR spectroscopic data, see Table 1. HR-ESI-MS *m/z* 227.1276 $[M + H]^+$ (calcd for $C_{12}H_{19}O_4$, 227.1283).

2.4.8. Diaporthopyranone H (**8**)

Amorphous powder; [α]25 D −10.57 (c 0.1, MeOH); IR (KBr, ν_{max} cm^{-1}): 3363, 2941, 2833, 1714, 1635, 1456, 1417, 1116, 1022, 833, 667, 599, 547. For the ^1H and ^{13}C NMR spectroscopic data, see Table 1. HR-ESI-MS *m/z* 227.1277 $[M + H]^+$ (calcd for $C_{12}H_{19}O_4$, 227.1283).

2.5. ECD Calculations

Conformational searches were performed using the MMFF force field in Spartan'14 (Version 1.1.4). The resulting low-energy conformers within a 10 kcal mol^{-1} window were further optimized via DFT at the B3LYP/6-31G(d) level in Gaussian 09 (Revision B.01). Vibrational frequency calculations at the same level confirmed the absence of imaginary frequencies and provided Gibbs free energies (ΔG) at 298.15 K. ECD spectra were subsequently calculated for the stable conformers at the B3LYP/6-311G(d,p) level, employing the polarizable continuum model (PCM) for solvent effects. The Boltzmann-weighted ECD curves were generated using SpecDisth (Version 1.71), and Origin (Version 2018), and compared with experimental data. Computational resources were provided by the National Supercomputer Center in Guangzhou (Tianhe-2).

2.6. Antifungal Activity

Minimum inhibitory concentrations (MICs) were determined in liquid culture via a serial dilution method as described previously with minor modifications [14]. Briefly, six microtubes were prepared, each containing 0.5 mL of broth medium. Test compounds were dissolved in a 5% (*v/v*) aqueous DMSO solution. An initial screening was performed at 200 μg/mL for all compounds, and active ones were further evaluated using a targeted gradient (100, 80, 60, 40, and 20 μg/mL) to determine the precise MIC. A negative control containing 5% (*v*/*v*) DMSO was included in parallel and showed no inhibitory effect. Subsequently, 0.5 mL of a standardized spore suspension was added to each tube, followed by incubation at 28 °C for 24 h under identical conditions. Fungal growth was inspected visually, and the MIC was defined as the lowest concentration at which no visible growth was observed. Carbendazim was used as a positive control to validate the assay. All antifungal assays were performed in three independent biological replicates to ensure the reproducibility of the MIC values.

3. Results

3.1. HSQC-Based DeepSAT Analysis

After silica gel fractionation of the crude rice EtOAc extract (Fr. B1–B7), the combined Fr. B4 and Fr. B5 were further separated into subfractions Fr. D1–D7. To rapidly assess the dominant chemotypes, the ^{1}H–^{13}C HSQC spectra of all seven subfractions (Fr. D1–D7) were acquired and analyzed using the DeepSAT platform (https://www.deepsat.com/, accessed on 12 July 2025) [11]. DeepSAT independently suggested that only Fr. D4 and Fr. D5 were overwhelmingly enriched in cyclic polyketide features and returned top-ranked scaffold predictions centered on an α-pyranone (2*H*-pyran-2-one) type core, with a small set of closely related analogs differing mainly in side-chain substitution (Figure 3).

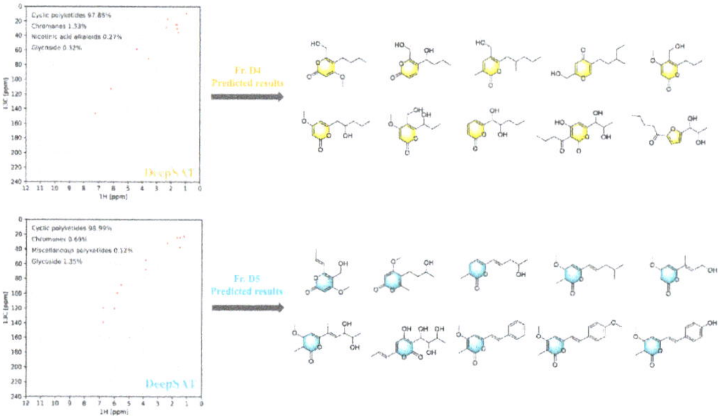

Figure 3. DeepSAT predicted the pyranone scaffold family from fraction Fr. D4 and Fr. D5.

Guided by these HSQC–DeepSAT predictions, Fr. D4 and Fr. D5 were prioritized for targeted, repeated purification. This scaffold-focused workflow furnished a coherent series of structurally related α-pyranone derivatives, consistent with the predicted chemotype distribution. Collectively, the HSQC-based DeepSAT analysis provided a rapid, fraction-level rationale for prioritization and enabled efficient isolation of an α-pyranone analogue set from the rice culture extract.

3.2. Structural Characterization

Diaporthopyranone A (**1**) was obtained as a brownish yellow oil. The HR-ESI-MS spectrum of **1** revealed a pseudomolecular ion peak at *m*/*z* 219.0992 [M + Na]$^{+}$, which is in accordance with the molecular formula $C_{11}H_{16}NaO_3$ and indicates four degrees of unsaturation. Its IR spectrum exhibited absorptions for hydroxyl groups (3336 cm^{-1}) and conjugated carbonyl groups (1670 cm^{-1}). The ^{1}H NMR spectrum revealed five olefinic protons at δ_H 6.41 (1H, d, *J* = 16.1 Hz, H-3), δ_H 7.27 (1H, d, *J* = 16.1 Hz, H-4), δ_H 6.64 (1H, d, *J* = 11.5 Hz, H-6), δ_H 6.78 (1H, dd, *J* = 15.0, 11.5 Hz, H-7), and δ_H 6.12 (1H, dd, *J* = 15.0, 5.8 Hz, H-8), one oxygenated methine δ_H 4.38 (1H, m, H-9), one oxygenated methylene δ_H 4.41 (2H, s, H-11), and two methyls δ_H 2.30 (3H, s, H-1) and δ_H 1.27 (3H, d, *J* = 6.5 Hz, H-10).

https://doi.org/10.3390/jof12030161

The ^{13}C and HSQC spectra revealed eleven signals, including two methyl signals at δ_C 23.3 (C-1) and δ_C 27.3 (C-10), one oxygenated methylene signal at δ_C 56.7 (C-11), five olefinic signals (δ_C 127.5, δ_C 147.9, δ_C 142.4, δ_C 125.2 and δ_C 146.4), one oxygenated methine signal at δ_C 68.7 (C-9), and two quaternary carbons (one olefinic at δ_C 137.4 and one carbonyl at δ_C 201.5) (Table 1). The presence of three double bonds and one carbonyl group results in a total of four degrees of unsaturation, indicating that compound **1** is a chain compound. A five-carbon chain consisting of one methyl, one oxygenated methine, and three olefinic methines could be identified by consecutive ^1H-^1H COSY correlations through H-10 to H-6, along with HMBC correlations from H-10 to C-8 and H-8 to C-6. Moreover, the HMBC correlations from H-6 to C-5 and C-4, and from H-3 to C-5 and C-1, and from H-1 to C-2 and C-3 established the long ten-carbon chain. Finally, the oxygenated methylene was connected to C-5 according to the HMBC correlations from H-11 to C-5, C-4, and C-6 (Figure 4). Thus, the planar structure of **1** was established. The *trans* configurations of $\Delta^{3,4}$ and $\Delta^{7,8}$ were confirmed by their larger coupling constants. And the *cis* configuration of $\Delta^{5,6}$ was established on the basis of NEOSY correlations between H-4 and H-6, H-7 and H-11. In addition, the configuration of C-9 was identified to be *R* according to the ECD calculations and comparisons of the experimental data (Figure 5). Therefore, the structure of **1** was established.

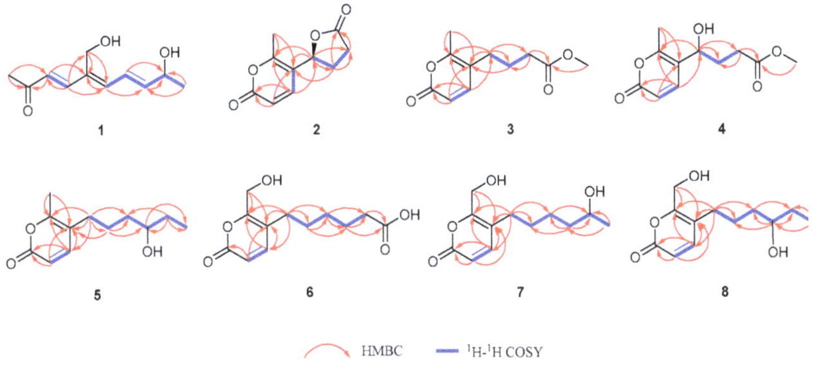

Figure 4. Key ^1H-^1H COSY and HMBC correlations of **1–8**.

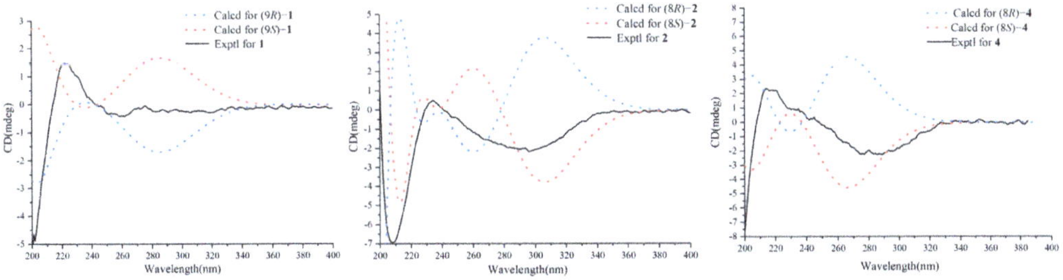

Figure 5. Experimental and predicted ECD spectra of compounds **1**, **2** and **4**.

Diaporthopyranone B (**2**) was isolated as a brown oil, whose molecular formula was $C_{10}H_{10}O_4$ with six degrees of unsaturation on the basis of the pseudomolecular ion peak at m/z 195.0647 $[M + H]^+$ in the HR-ESI-MS spectrum. The IR spectrum exhibited absorption bands attributable to hydroxyl groups (3383 cm^{-1}) and conjugated lactone carbonyl groups (1714 cm^{-1}). ^1H NMR revealed two neighboring coupling olefinic protons (δ_H 6.28, 7.61, d, J = 9.7 Hz, each 1H), an oxygenated methine (δ_H 5.56 dd, J = 9.7, 6.5 Hz, 1H), and a methyl (δ_H 2.35 s, 3H). Its ^{13}C and DEPT spectra displayed ten carbon signals, including one methyl, two methylene, three methine (two olefinic, one oxygenated), and four quaternary carbon signals (two olefinic, two carbonyl) (Table 1). The four olefinic carbons, one carbonyl carbon, and one methyl form a methyl-substituted α-pyrone, which has been reported to be the main metabolite of *Diaporthe* fungi. The existence of a methyl-substituted α-pyrone could be further supported by the ^{1}H-^{1}H COSY correlation between H-3 and H-4 and the HMBC correlations from the methyl group to C-5 and C-6, from H-3 to C-2 and C-5, and from H-4 to C-5 and C-6 (Figure 3). In addition, an independent spin system consisting of one oxygenated methine and two methylenes could be identified in the ^{1}H-^{1}H COSY spectrum, along with a carbonyl carbon at δ_C 179.2 s (C-11), which formed a γ-butyrolactone. This butyrolactone ring was connected to C-5 in the pyrone ring on the basis of the HMBC correlations from H-4 to C-8 and from H-8 to C-4 and C-6 (Figure 4). Thus, the planar structure of **2** was established. The absolute configuration of 2 was assigned as 8*S* based on the good agreement between its experimental ECD spectrum and the calculated spectrum of the (8*S*)-2 isomer (Figure 5).

Diaporthopyranone C (**3**) was obtained as a brown solid. HR-ESI-MS revealed a pseudomolecular ion at m/z 211.0961 $[M + H]^+$, which is consistent with the molecular formula $C_{11}H_{14}O_4$ and indicates five degrees of unsaturation. Its IR spectrum showed characteristic absorptions for hydroxyl groups (3361 cm^{-1}) and conjugated carbonyl groups (1716 cm^{-1}). Its ^{1}H and ^{13}C NMR spectra also showed characteristic signals for a methyl-substituted α-pyrone (δ_H 6.17, 7.43, d, J = 9.4 Hz, each 1H, δ_H 2.26 s, 3H; δ_C 165.1 s, 113.8 d, 149.5 d, 116.7 s, 160.5 s, and 17.2 q) (Table 1) [15]. In addition, a spin system consisting of three methylenes was present in the ^{1}H-^{1}H COSY spectrum. The termini of this chain were linked with a formic ether according to the HMBC correlation from H-9 and H-10 to C-11, which formed a methyl butyrate. This side chain was connected to C-5 in the pyrone ring on the basis of the HMBC correlations from H-4 to C-8 and from H-8 to C-6 and C-4 (Figure 4). Therefore, the structure of **3** was established.

Diaporthopyranone D (**4**) was obtained as a brownish yellow oil, and its molecular formula was determined to be $C_{11}H_{14}O_5$ via HR-ESI-MS analysis, which revealed a pseudomolecular ion peak at m/z 227.0911 $[M + H]^+$. Its IR spectrum exhibited absorptions for hydroxyl groups (3365 cm^{-1}) and conjugated carbonyl groups (1716 cm^{-1}). The diagnostic signals, including a pair of coupled olefinic protons (δ_H 6.23, 7.63, d, J = 9.6 Hz, each 1H) and a methyl singlet (δ_H 2.29, s, 3H) in the ^{1}H NMR spectrum, four olefinic (δ_C 114.1 d, 145.8 d, 120.3 s, and 160.1 s) and one carbonyl (δ_C 164.8 s) carbon, and one methyl (δ_C 17.0 q) in the ^{13}C and DEPT spectra, revealed that **4** possesses a methyl-substituted α-pyrone nucleus (Table 1). In addition, the ^{1}H-^{1}H COSY spectrum revealed the existence of a spin system consisting of an oxygenated methine (δ_H 4.60, m, 1H) and two methylenes. The termini of this chain were linked with a formic ether according to the HMBC correlation from H-9 and H-10 to C-11, whereas the other end of this chain was connected to C-5 on the basis of the HMBC correlations from H-4 to C-8 and from H-8 to C-5 and C-6 (Figure 4). Thus, the planar structure of **4** was established. Comparison of the experimental ECD spectrum of **4** with the calculated ECD spectra indicated that it corresponds to the (8*S*)-**4** isomer, thereby establishing the absolute configuration at C-8 as *S* (Figure 5).

Diaporthopyranone E (**5**) was obtained as a brownish yellow oil. HR-ESI-MS revealed a pseudomolecular ion peak at m/z 211.1326 $[M + H]^+$, which is consistent with the molecular formula $C_{12}H_{18}O_3$. Its IR spectrum showed absorptions for a hydroxyl group (3325 cm^{-1}), conjugated carbonyl groups (1647 cm^{-1}), and a C–O stretching vibration at 1016 cm^{-1}. The ^{1}H, ^{13}C and DEPT spectra also revealed diagnostic signals for a methyl-substituted α-pyrone nucleus (Table 1) [16]. In addition, a six-carbon chain consisting of four methylenes, one methyl, and one oxygenated methine could be identified by consecutive ^{1}H-^{1}H COSY correlations through H-8 to H-13, along with HMBC correlations from H-8 to C-10, from the oxygenated methine (δ_H 3.54, dt, J = 7.7, 3.7 Hz, 1H) to C-13 and C-8. This chain was linked to C-5 in the α-pyrone ring on the basis of the HMBC correlations from H-4 to C-8 and from H-8 to C-4, C-5 and C-6 (Figure 4). Thus, the planar structure of **5** was established. However, the limited availability of the isolate precluded the application of Mosher's method, leaving its absolute configuration unassigned at this stage.

Diaporthopyranone F (**6**) was isolated as a brown solid. Its molecular formula was established as $C_{12}H_{16}O_5$ with five degrees of unsaturation according to the pseudomolecular ion peak at m/z 241.1069 $[M + H]^+$ in the HR-ESI-MS spectrum. The IR spectrum showed absorptions for a hydroxyl group (3340 cm^{-1}), conjugated carbonyl groups (1683 cm^{-1}), and a C–O stretching vibration at 1020 cm^{-1}. The ^1H NMR spectrum revealed two coupled olefinic protons (δ_H 6.27, 7.48, d, J = 9.4 Hz, each 1H) and an oxygenated methylene (δ_H 4.39, s, 2H). The ^{13}C and DEPT spectra revealed four olefinic carbons (δ_C 116.0 d, 149.5 d, 118.8 s, and 160.1 s), a carbonyl carbon (δ_C 164.6 s), and an oxygenated methylene (δ_C 59.2 t) (Table 1). These data suggested that **6** possesses a hydroxymethyl substituent in the α-pyrone ring [17]. In addition, n-hexanoic acid was identified on the basis of consecutive ^1H-^1H COSY correlations through H-8 to H-12, along with HMBC correlations from H-8 to C-10, from H-9 to C-11, and from H-11 and H-12 to C-13 [18]. This side chain was connected to C-5 according to the HMBC correlations from H-4 to C-8 and from H-8 to C-6, C-5 and C-4 (Figure 4). Thus, the structure of 6 was established.

Diaporthopyranone G (**7**) was obtained as an amorphous powder. HR-ESI-MS revealed a pseudomolecular ion peak at m/z 227.1276 $[M + H]^+$, which is consistent with the molecular formula $C_{12}H_{18}O_4$. Its IR spectrum showed absorption bands for a hydroxyl group (3373 cm^{-1}) and conjugated carbonyl groups (1699 cm^{-1}). Its NMR spectra showed characteristic signals for a hydroxymethyl-substituted α-pyrone ring (δ_H 6.28, 7.48, d, J = 9.4 Hz, each 1H, δ_H 4.39 s, 2H; δ_C 164.6 s, 116.0 d, 149.5 d, 118.8 s, 160.1 s, and 59.2 t) (Table 1). Similarly, a six-carbon chain consisting of four methylenes, one methyl, and one oxygenated methine was identified by consecutive ^1H-^1H COSY correlations through H-8 to H-13, along with HMBC correlations from H-8 to C-10, from H-9 to C-11, and from H-13 to the oxygenated methine and C-11. Finally, the location of this side chain was determined by the HMBC correlations from H-4 to C-8 and from H-8 to C-6, C-5 and C-4 (Figure 4). With the planar scaffold of **7** firmly established, its absolute stereochemistry remained unassigned, as the scarcity of the isolate precluded the requisite chemical derivatization.

Diaporthopyranone H (**8**) was obtained as an amorphous powder. HR-ESI-MS revealed a pseudomolecular ion peak at m/z 227.1277 $[M + H]^+$, which is consistent with the molecular formula $C_{12}H_{18}O_4$ and indicates four degrees of unsaturation. Its IR spectrum exhibited absorption bands for a hydroxyl group (3363 cm^{-1}), conjugated carbonyl groups (1635 cm^{-1}), and a C–O stretching vibration at 1022 cm^{-1}. The ^1H, ^{13}C, and DEPT NMR spectra revealed signals for a hydroxymethyl-substituted α-pyrone ring (δ_H 6.27, 7.48, d, J = 9.4 Hz, each 1H; δ_H 4.39 s, 2H; δ_C 164.6 s, 116.0 d, 149.5 d, 118.8 s, 160.1 s and 59.2 t) (Table 1). Similarly, a six-carbon chain was identified by consecutive ^1H-^1H COSY correlations through H-8 to H-13, along with HMBC correlations from H-8 to C-10, from H-9 to C-11, and from the oxygenated methine H-11 (δ_H 3.44, m, 1H) to C-9, C-10, and C-13. This side chain was located at C-5 on the basis of the HMBC correlations from H-4 to C-8 and from H-8 to C-6, C-5 and C-4 (Figure 4). The planar structures of compounds **5**, **7**, and **8** were thus defined. However, their absolute configurations remain unassigned at this stage. Because their stereocenters are located on highly flexible, saturated aliphatic chains distant from the primary α-pyrone chromophore, ECD analysis would yield highly conformationally averaged and unreliable data. Furthermore, the limited supply of these minor isolates precluded the necessary chemical derivatization (e.g., Mosher's method) required for assigning such flexible distal alcohols.

https://doi.org/10.3390/jof12030161

Table 1. ^{1}H (600 MHz) and ^{13}C (150 MHz) NMR spectroscopic data of compounds **1–8**.

Nos.	1 [b]		2 [b]		3 [b]		4 [b]		5 [b]		6 [a]		7 [b]		8 [b]	
	δ_H Type (J)	δ_C Type	δ_H Type (J)	δ_C Type	δ_H Type (J)	δ_C Type	δ_H Type (J)	δ_C Type	δ_H Type (J)	δ_C Type	δ_H Type (J)	δ_C Type	δ_H Type (J)	δ_C Type	δ_H Type (J)	δ_C Type
1	2.30 s	23.3 q	/	/	/	/	/	/	/	/	/	/	/	/	/	/
2		201.5 s		164.0 s		165.1 s		164.8 s		163.0 s		164.6 s		164.6 s		164.6 s
3	6.41 d (16.1)	127.5 d	6.28 d (9.7)	114.6 d	6.17 d (9.4)	113.8 d	6.23 d (9.6)	114.1 d	6.14 d (9.4)	113.5 d	6.27 d (9.4)	116.0 d	6.28 d (9.4)	116.0 d	6.27 d (9.4)	116.0 d
4	7.27 d (16.1)	147.9 d	7.61 d (9.7)	144.5 d	7.43 d (9.4)	149.5 d	7.63 d (9.6)	145.8 d	7.17 d (9.4)	147.1 d	7.48 d (9.4)	149.5 d	7.48 d (9.4)	149.5 d	7.48 d (9.4)	149.5 d
5	/	137.4 s		115.7 s		116.7 s		120.3 s		115.3 s		118.8 s		118.8 s		118.8 s
6	6.64 d (11.5)	142.4 d	/	162.3 s	/	160.5 s	/	160.1 s	/	158.6 s	/	160.1 s	/	160.1 s	/	160.1 s
7	6.78 dd (15.0, 11.5)	125.2 d	2.35 s	17.2 q	2.26 s	17.2 q	2.29 s	17.0 q	2.23 s	17.4 q	4.39 s	59.2 t	4.39 s	59.2 t	4.39 s	59.2 t
8	6.12 dd (15.0, 5.8)	146.4 d	5.56 dd (9.7, 6.5)	78.0 d	2.40 m	29.4 t	4.60 m	67.7 d	2.31 q (7.4)	29.6 t	2.45 m	29.5 t	2.45 t (7.7)	29.6 t	2.45 m	29.6 t
9	4.38 m	68.7 d	2.52 m	29.7, t	1.80 m	25.9 t	1.95 m	33.0 t	1.51 m	26.0 t	1.55 m	31.1 t	1.54 m	31.4 t	1.65 m	27.6 t
10	1.27 d (6.5)	27.3 q	2.73 m	30.0, t	2.37 m	33.7 t	2.42 t (7.1)	30.8 t	/	36.2 t	1.40 m	29.8 t	1.45 m	26.3 t	1.41 m	37.1 t
11	4.41 s	56.7 t	/	179.2 s		175.3 s		175.4 s	3.54 dt (7.7, 3.7)	73.1 d	1.63 m	26.5 t	1.31 m	39.9 t	3.44 m	73.6 t
12									1.41 m	30.5 t	2.24 m	36.7 d	3.71 dd (6.4, 5.0)	68.4 d	/	31.1 d
13									0.94 t (7.5)	10.0 q	/	180.0 s	1.15 d (6.2)	23.5 q	/	10.4 q
OCH_3-11						52.1 q		52.1 q								

Note: [a] Measured in $CDCl_3$; [b] Measured in CD_3OD.

https://doi.org/10.3390/jof12030161

The planar structures of compounds **5**, **7**, and **8** were thus defined. However, their absolute configurations remain unassigned. Extensive computational efforts, including TDDFT-ECD, DP4+ NMR analysis, and optical rotation calculations, proved inadequate for reliable assignment due to severe conformational averaging arising from their highly flexible, saturated aliphatic side chains. Definitive determination via chemical derivatization, such as the Mosher's method, was also prohibited by the limited supply of these minor isolates.

Although the isolated compounds share a well-known α-pyrone core scaffold characteristic of the genus *Diaporthe*, compounds **1–8** represent new chemical entities that significantly expand the structural diversity of this class. The precise structural novelties of these analogues lie primarily in their diverse side-chain functionalizations relative to previously reported metabolites. For instance, compound **2** is distinguished by an advanced fused γ-butyrolactone ring, whereas compound **4** features a specific C-9 hydroxylation that differentiates it from its known deoxy analogue **3**. Furthermore, compounds **6–8** exhibit distinct alkyl chain elongations coupled with unique terminal or sub-terminal oxidation states. These atom-level modifications collectively enrich the known chemodiversity of *Diaporthe*-derived pyranones.

On the basis of spectroscopic analyses and comparisons with reported data, the remaining metabolites were identified as five known pyranone derivatives, namely, diaporpyrone D (**9**) [8], diaporpyrone F (**10**) [8], diaporpyrone A (**11**) [9] and 5-[(1*S*)-1-hydroxyhexyl]-6-methyl-2*H*-pyran-2-one (**12**) [10], phomopyrone D (**13**) [19].

3.3. Putative Biosynthetic Pathways

The putative biosynthetic assembly of these pyrone polyketides is hypothesized to begin with the iterative condensation of acetyl-CoA and malonyl-CoA units, initiated by a putative non-reducing polyketide synthase PKS-1 to generate the foundational 6-methyl-α-pyrone core. This scaffold is proposed to serve as the priming substrate for a secondary polyketide synthase, PKS-2, which likely orchestrates further chain elongation and programmed ketoreduction to yield diversified intermediates such as the carboxylic acids **9** and **10** and the reduced analogs **11** and **12**. The structural complexity of the library is subsequently elaborated through a proposed suite of post-PKS tailoring enzymes: an O-methyltransferase (MeT) may catalyze the esterification of terminal carboxylic acids to produce methyl esters **3** and **4**, while the formation of the bicyclic lactone **2** likely proceeds via the enzymatic or spontaneous cyclization of a hydroxylated side-chain precursor. Furthermore, late-stage oxidative processing, presumably mediated by cytochrome P450 monooxygenases, is hypothesized to introduce significant functional diversity through ω- or (ω−1)-oxidation of the aliphatic side chains and the hydroxylation of the C6-methyl group to a hydroxymethyl moiety [20]. These modifications would lead to the production of terminal dicarboxylates like **6** and polyhydroxylated derivatives such as **7** and **8**, illustrating a highly branched and modular biosynthetic logic characteristic of fungal specialized metabolism (Figure 6).

3.4. Antifungal Activities

All the isolated compounds (**1–13**) were evaluated for their antifungal activity against nine plant pathogenic fungi, namely, *Fusarium graminearum*, *Botryosphaeria dothidea*, *Fusarium oxysporum* Schltdl., *Colletotrichum gloeosporioides*, *Valsa mali*, *Colletotrichum musae*, *Alternaria alternata*, *Colletotrichum agenarium*, and *Bipolaris sorokiniana*. As shown in Figure 7, compound **4** showed moderate inhibitory activity against *Alternaria alternata* and *Valsa mali*, with MIC values of 100 and 80 μg/mL, respectively. In comparison, the positive control carbendazim had an MIC of 6.25 μg/mL against *C. musae* and 1.0625 μg/mL against

C. gloeosporioides. In addition, compounds **6** and **8** inhibited *Colletotrichum musae*, with MIC values of 60 and 80 μg/mL, respectively, whereas carbendazim exhibited an MIC of 12.5 μg/mL against this pathogen. The remaining compounds displayed only weak or negligible activity under the same conditions. Overall, these data indicate that the pyranone derivatives, particularly compound **4** against *V. mali* and *A. alternata*, and compounds **6** and **8** against *C. musae*, possess measurable antifungal potential (representative images of the growth inhibition are provided in Supplementary Figure S1), although their potency is still inferior to that of the commercial fungicide carbendazim.

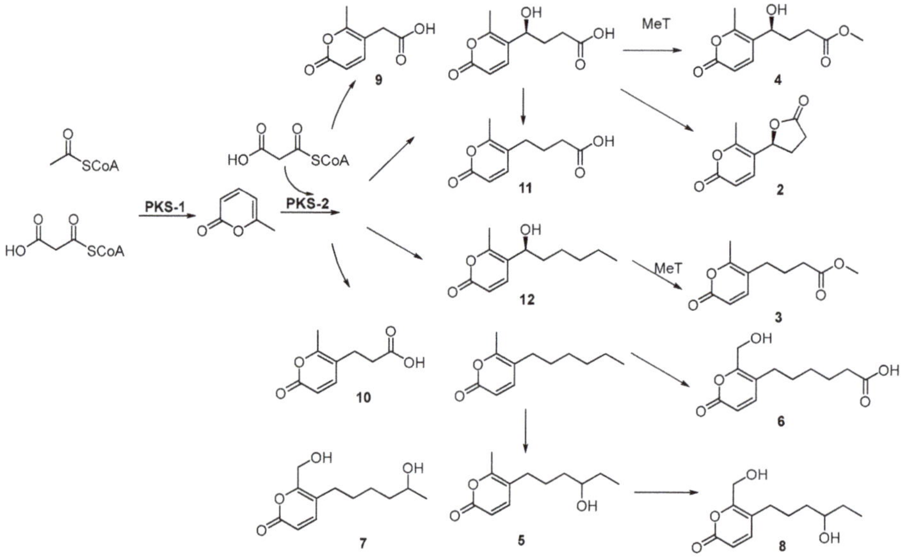

Figure 6. Proposed biosynthetic pathway of compounds **2–12**.

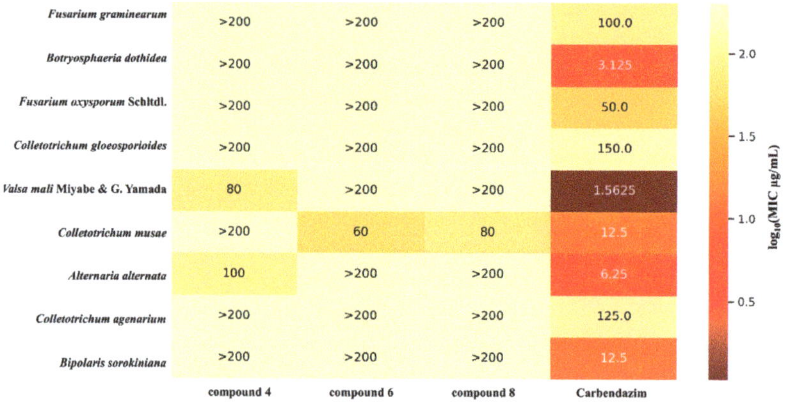

Figure 7. Antifungal activities (MICs, μg/mL) of compounds **4**, **6** and **8** against plant pathogenic fungi.

4. Discussion

The identification of bioactive secondary metabolites from endophytic fungi is frequently impeded by the rediscovery of known compounds and the masking of minor constituents by major ones. To overcome these bottlenecks, advanced NMR-based dereplication tools have been developed. For instance, computational tools like MixONat have proven highly effective in dereplicating complex mixtures by matching experimental ^{13}C and DEPT NMR signals against predefined structural databases [21]. However, their inherent reliance on existing databases can potentially limit the identification of truly novel scaffolds. In contrast, the HSQC-based DeepSAT strategy leverages convolutional neural networks (CNNs) to directly predict chemical fingerprints and scaffold classes from 2D HSQC spectra, circumventing the strict requirement for exact database matches [11]. Recently, this cutting-edge artificial intelligence platform has been successfully employed by researchers to rapidly annotate structural families within unpurified marine cyanobacterial extracts [22] and to guide the discovery of novel polyketides from mangrove-derived fungi [23], demonstrating its robust capability in early-stage chemotype prioritization. Building upon these emerging applications, in this study, we effectively adapted the DeepSAT workflow to prioritize the isolation of an α-pyrone chemotype from the terrestrial endophytic fungus *Diaporthe kyushuensis* ZMU-48-1. This targeted approach successfully facilitated the recovery of thirteen pyranone analogs, including eight novel compounds (**1–8**), thereby validating the utility of deep learning-assisted profiling in navigating endophytic metabolomes to access specific chemical spaces.

Interpreted within the context of antifungal discovery, the biological evaluation indicated that while the overall potency of the isolated pyranones was moderate compared to the commercial fungicide carbendazim, specific analogues exhibited distinct species-selectivity. A preliminary analysis of the structure-activity relationships (SAR) suggests that oxidative modifications on the pyrone scaffold may influence this selectivity. Notably, compounds **6** and **8**, which feature a C-6 hydroxymethyl group, showed selective inhibition against *C. musae*, whereas analogues with a C-6 methyl group (such as **1–5**) were inactive against this specific pathogen. Furthermore, compound **4** differs from the inactive compound **3** primarily by the presence of a hydroxyl group at the C-9 position, which appears to be crucial for its activity against *V. mali* and *A. alternata*. While the sparse data and weak-to-moderate MIC values (60–100 μg/mL) limit broad mechanistic interpretations, such oxidative functionalization is a well-documented strategy in fungal secondary metabolism to alter target specificity [24]. From a broader ecological perspective, the production of these specific metabolites, albeit with low in vitro potency, parallels the chemical defense mechanisms of other bioactive endophytic fungi (e.g., *Trichoderma* and *Pestalotiopsis* spp.), which deploy diverse functionalized polyketides against competing phytopathogens [25,26]. Ultimately, rather than representing immediate agricultural leads, these specific α-pyrone scaffolds serve as a preliminary proof-of-concept for our targeted discovery methodology. They highlight the necessity of future structural optimization (e.g., at C-6 and C-9) to enhance antifungal efficacy.

5. Conclusions

In summary, the implementation of an HSQC-based DeepSAT workflow successfully streamlined the metabolic profiling and prioritization of the endophytic fungus *D. kyushuensis* ZMU-48-1. By rapidly identifying an α-pyrone-dominated chemotype within complex fractions, this structure-guided approach facilitated the targeted isolation of a focused library of thirteen pyranone congeners, including eight novel metabolites, diaporthopyranone A–H (**1–8**). Biological evaluation revealed that these analogues exhibit only preliminary, low-potency antifungal activity (MICs 60–100 μg/mL) compared to the highly potent commercial fungicide carbendazim. Therefore, rather than establishing these compounds as promising antifungal leads, these findings more broadly validate the efficacy of HSQC-DeepSAT analysis as an efficient, structure-informed strategy to rapidly prioritize

https://doi.org/10.3390/jof12030161

specific chemical spaces and accelerate the targeted discovery of natural product classes from complex fungal metabolomes.

6. Patents

The work reported in this manuscript has resulted in two patent applications filed with the China National Intellectual Property Administration (CNIPA). These include "A pyranone natural product with antifungal activity, its preparation method, and application" (Application No. 202511541679.X) and "A preparation method of pyranone natural products and their use in agricultural antifungal agents" (Application No. 202511541673.2).

Supplementary Materials: The following supporting information can be downloaded at: https://www.mdpi.com/article/10.3390/jof12030161/s1, Figure S1: Visual assessment of the antifungal activities of compounds **4**, **6**, and **8** against *Botryosphaeria dothidea* (A–C), *Colletotrichum gloeosporioides* (D–E), and *Colletotrichum musae* (F–G); Figure S2: HRESIMS spectrum of compound **1**; Figure S3: IR spectrum of compound **1**; Figure S4: ECD spectra of compound **1** in MeOH; Figure S5: ^{1}H-NMR (600 MHz, CD_3OD) spectrum of compound **1**; Figure S6: ^{13}C-NMR (150 MHz, CD_3OD) spectrum of compound **1**; Figure S7: HSQC spectrum of compound **1**; Figure S8: HMBC spectrum of compound **1**; Figure S9: ^{1}H-^{1}H COSY spectrum of compound **1**; Figure S10: NOESY spectrum of compound **1**; Figure S11: HRESIMS spectrum of compound **2**; Figure S12: IR spectrum of compound **2**; Figure S13: ECD spectra of compound **2** in MeOH; Figure S14: ^{1}H-NMR (600 MHz, CD_3OD) spectrum of compound **2**; Figure S15: ^{13}C-NMR and DEPT (150 MHz, CD_3OD) spectrum of compound **2**; Figure S16: HSQC spectrum of compound **2**; Figure S17: HMBC spectrum of compound **2**; Figure S18: ^{1}H-^{1}H COSY spectrum of compound **2**; Figure S19: HRESIMS spectrum of compound **3**; Figure S20: IR spectrum of compound **3**; Figure S21: ECD spectra of compound **4** in MeOH; Figure S22: ^{1}H-NMR (600 MHz, CD_3OD) spectrum of compound **3**; Figure S23: ^{13}C-NMR and DEPT (150 MHz, CD_3OD) spectrum of compound **3**; Figure S24: HSQC spectrum of compound **3**; Figure S25: HMBC spectrum of compound **3**; Figure S26: ^{1}H-^{1}H COSY spectrum of compound **3**; Figure S27: HRESIMS spectrum of compound **4**; Figure S28: IR spectrum of compound **4**; Figure S29: ^{1}H-NMR (600 MHz, CD_3OD) spectrum of compound **4**; Figure S30: ^{13}C-NMR and DEPT (150 MHz, CD_3OD) spectrum of compound **4**; Figure S31: HSQC spectrum of compound **4**; Figure S32: HMBC spectrum of compound **4**; Figure S33: ^{1}H-^{1}H COSY spectrum of compound **4**; Figure S34: HRESIMS spectrum of compound **5**; Figure S35: IR spectrum of compound **5**; Figure S36: ^{1}H-NMR (600 MHz, $CDCl_3$) spectrum of compound **5**; Figure S37: ^{13}C-NMR and DEPT (150 MHz, $CDCl_3$) spectrum of compound **5**; Figure S38: HSQC spectrum of compound **5**; Figure S39: HMBC spectrum of compound **5**; Figure S40: ^{1}H-^{1}H COSY spectrum of compound **5**; Figure S41: HRESIMS spectrum of compound **6**; Figure S42: IR spectrum of compound **6**; Figure S43: ^{1}H-NMR (600 MHz, CD_3OD) spectrum of compound **6**; Figure S44: ^{13}C-NMR and DEPT (150 MHz, CD_3OD) spectrum of compound **6**; Figure S45: HSQC spectrum of compound **6**; Figure S46: HMBC spectrum of compound **6**; Figure S47: ^{1}H-^{1}H COSY spectrum of compound **6**; Figure S48: HRESIMS spectrum of compound **7**; Figure S49: IR spectrum of compound **7**; Figure S50: ^{1}H-NMR (600 MHz, CD_3OD) spectrum of compound **7**; Figure S51: ^{13}C-NMR and DEPT (150 MHz, CD_3OD) spectrum of compound **7**; Figure S52: HSQC spectrum of compound **7**; Figure S53: HMBC spectrum of compound **7**; Figure S54: 1H-1H COSY spectrum of compound **7**; Figure S55: HRESIMS spectrum of compound **8**; Figure S56: IR spectrum of compound **8**; Figure S57: ^{1}H-NMR (600 MHz, CD_3OD) spectrum of compound **8**; Figure S58: ^{13}C-NMR and DEPT (150 MHz, CD_3OD) spectrum of compound **8**; Figure S59: HSQC spectrum of compound **8**; Figure S60: HMBC spectrum of compound **8**; Figure S61: ^{1}H-^{1}H COSY spectrum of compound **8**.

Author Contributions: S.Y.: supervision, writing—review & editing, and project administration. J.Z.: Investigation, data collection, isolation of the compounds, and writing of the original draft. Y.W.: isolation of the compounds. X.W.: methodology and validation. M.Y.: Resources and software. H.W.: data curation and visualization. T.Y.: conceptualization, funding acquisition, writing—original draft, and supervision. All authors have read and agreed to the published version of the manuscript.

https://doi.org/10.3390/jof12030161

Funding: This work was financially supported by the Science and Technology Fund Project of the Guizhou Provincial Health Commission (No. gzwkj2025-140), the Youth Science and Technology Talent Development Project of the Guizhou Provincial Department of Education (No. Qian Jiao Ji [2024] 135), the Guizhou Provincial Science and Technology Project (Qian Ke He Jichu-[2024] Yiban 263), the National Natural Science Foundation of China (No. 32460017), the Zunyi Medical University Zhuhai Campus Key Construction Discipline (No. ZHGY2024-1), and the Science and Technology Innovation Team of Zhuhai Campus of Zunyi Medical University (Nos. ZHTD2024-2 and ZHTD2024-3).

Data Availability Statement: The datasets used and/or analyzed during the current study are available from the corresponding author on reasonable request.

Conflicts of Interest: The authors declare no conflicts of interest.

eferences

Nazir, A.; Puthuveettil, A.R.; Hussain, F.H.N.; Hamed, K.E.; Munawar, N. Endophytic fungi: Nature's solution for antimicrobial resistance and sustainable agriculture. *Front. Microbiol.* **2024**, *15*, 1461504. [CrossRef] [PubMed]

Islam, M.M.; Rakib, A.R.; Mahmood, M.R.; Singha, A.; Jahan, S.; Sultana, N.; Rafi, M.T.H.; Islam, S.; Alamin, M.; Kuri, C.; et al. Bioactive metabolites from endophytes: Focus on anticancer and antimicrobial potential. *Microbe* **2025**, *9*, 100553. [CrossRef]

Fan, Y.; Shi, B. Endophytic fungi from the four staple crops and their secondary metabolites. *Int. J. Mol. Sci.* **2024**, *25*, 6057. [CrossRef]

Xu, T.C.; Lu, Y.H.; Wang, J.F.; Song, Z.Q.; Hou, Y.G.; Liu, S.S.; Liu, C.S.; Wu, S.H. Bioactive secondary metabolites of the genus *Diaporthe* and anamorph *Phomopsis* from terrestrial and marine habitats and endophytes: 2010–2019. *Microorganisms* **2021**, *9*, 217. [CrossRef]

Jiang, L.; Ma, Q.; Li, A.; Sun, R.; Tang, G.; Huang, X.; Pu, H. Bioactive secondary metabolites produced by fungi of the genus *Diaporthe* (Phomopsis): Structures, biological activities, and biosynthesis. *Arab. J. Chem.* **2023**, *16*, 105062. [CrossRef]

Charria-Girón, E.; Marin-Felix, Y.; Beutling, U.; Franke, R.; Brönstrup, M.; Vasco-Palacios, A.M.; Caicedo, N.H.; Surup, F. Metabolomics insights into the polyketide-lactones produced by *Diaporthe caliensis* sp. nov. an endophyte of the medicinal plant *Otoba gracilipes*. *Microbiol. Spectr.* **2023**, *11*, e02743-23. [CrossRef]

Yuan, S.; Zhuang, S.; Qiao, Y.; Wu, Q.; Ma, Y.; Chen, S.; Liu, L.; Yan, Y.; Gao, Z. Metabolic shunting in marine-derived *Diaporthe* sp. SYSU-MS4722 yields diverse γ-butyrolactone derivatives with anti-inflammatory activity. *J. Nat. Prod.* **2025**, *88*, 1605–1615. [CrossRef]

Wang, Z.; Ma, Q.; Wu, G.; Zhong, Y.; Feng, B.; Huang, P.; Li, A.; Tang, G.; Huang, X.; Pu, H. Bioactive alpha-Pyrone analogs from the endophytic fungus *Diaporthe* sp. CB10100: Alpha-glucosidase inhibitory activity, molecular docking, and molecular dynamics studies. *Molecules* **2024**, *29*, 1768. [CrossRef]

Pu, H.; Liu, J.; Wang, Y.; Peng, Y.; Zheng, W.; Tang, Y.; Hui, B.; Nie, C.; Huang, X.; Duan, Y.; et al. Bioactive alpha-Pyrone derivatives from the endophytic fungus *Diaporthe* sp. CB10100 as inducible nitric oxide synthase inhibitors. *Front. Chem.* **2021**, *9*, 679592. [CrossRef] [PubMed]

Zhang, D.; Liu, J.m.; Chen, R.; Zhang, M.; Yu, L.y.; Wu, J.; Dai, J. A new lactone derivative from plant endophytic fungus *Periconia* sp. F-31. *China J. Chin. Mater. Medica* **2015**, *40*, 2349–2351.

Kim, H.W.; Zhang, C.; Reher, R.; Wang, M.; Alexander, K.L.; Nothias, L.-F.; Han, Y.K.; Shin, H.; Lee, K.Y.; Lee, K.H.; et al. DeepSAT: Learning molecular structures from nuclear magnetic resonance data. *J. Cheminform.* **2023**, *15*, 71. [CrossRef] [PubMed]

Zheng, J.; Wang, H.; Wang, X.; Zeng, S.; Yuan, S.; Yin, T. Genome mining and OSMAC strategies unveil diverse secondary metabolites from the endophytic fungus *Diaporthe kyushuensis* ZMU-48-1 with antifungal activity. *Front. Microbiol.* **2025**, *16*, 1604639. [CrossRef]

Schulz, B.; Wanke, U.; Draeger, S.; Aust, H.J. Endophytes from herbaceous plants and shrubs: Effectiveness of surface sterilization methods. *Mycolo. Res.* **1993**, *97*, 1447–1450. [CrossRef]

Dalinova, A.; Chisty, L.; Kochura, D.; Garnyuk, V.; Petrova, M.; Prokofieva, D.; Yurchenko, A.; Dubovik, V.; Ivanov, A.; Smirnov, S.; et al. Isolation and bioactivity of secondary metabolites from solid culture of the fungus, *Alternaria sonchi*. *Biomolecules* **2020**, *10*, 81. [CrossRef] [PubMed]

Yan, F.; Fang, J.; Ding, W.; Tang, X.; Chen, X.; Ma, Z.; Wang, J. Structurally diverse metabolites from the marine-derived *Streptomyces* sp. DS-27 based on two different culture conditions. *Chem. Biodivers.* **2023**, *20*, e202301017. [CrossRef] [PubMed]

Qiu, Y.; Chen, Y.; Pei, Y.; Matsuda, H.; Yoshikawa, M. Constituents with radical scavenging effect from *Opuntia dillenii*: Structures of new alpha-pyrones and flavonol glycoside. *Chem. Pharm. Bull.* **2002**, *50*, 1507–1510. [CrossRef]

17. Morimitsu, Y.; Hirota, A. Two oxazolyl compounds and a monosubstituted α-pyrone as free-radical scavengers isolated from a fungus. *Biosci. Biotechnol. Biochem.* **2014**, *61*, 1428–1433. [CrossRef]
18. Kirihara, M.; Suzuki, S.; Ishihara, N.; Yamazaki, K.; Akiyama, T.; Ishizuka, Y. Synthesis of acylsilanes via catalytic dedithioacetalization of 2-Silylated 1,3-Dithianes with 30% hydrogen peroxide. *Synthesis* **2017**, *49*, 2009–2014. [CrossRef]
19. Guo, Z.; Chen, B.; Chen, D.; Deng, X.; Yuan, J.; Zhang, S.; Xiong, Z.; Xu, J. New isocoumarin and pyrone derivatives from the chinese mangrove plant *Rhizophora* mangle-associated fungus *Phomopsis* sp. DHS-11. *Molecules* **2023**, *28*, 3756. [CrossRef]
20. Dong, J.-Y.; Tang, M.-C.; Liu, L. α-pyrone derivatives from *Calcarisporium* arbuscula discovered by genome mining. *J. Nat. Prod.* **2023**, *86*, 2496–2501. [CrossRef]
21. Bruguière, A.; Derbré, S.; Dietsch, J.; Leguy, J.; Rahier, V.; Pottier, Q.; Bréard, D.; Suor-Cherer, S.; Viault, G.; Le Ray, A.-M.; et al. MixONat, a software for the dereplication of mixtures based on ^{13}C NMR Spectroscopy. *Analyt. Chem.* **2020**, *92*, 8793–8801. [CrossRef]
22. Ryu, B.; Glukhov, E.; Teixeira, T.R.; Caffrey, C.R.; Madiyan, S.; Joseph, V.; Avalon, N.E.; Leber, C.A.; Naman, C.B.; Gerwick, W.H. The Kavaratamides: Discovery of linear lipodepsipeptides from the marine *Cyanobacterium Moorena* bouillonii using a comparative chemogeographic analysis. *J. Nat. Prod.* **2024**, *87*, 1601–1610. [CrossRef]
23. Huo, R.; Ji, M.; Liu, G.; Liu, L. Discovery of new bianthrones and chlorinated bianthrones with cytotoxic activity against cancer cells from *Penicillium hispanicum* guided by HSQC-based DeepSAT. *Mycology* **2025**, 1–12. [CrossRef]
24. Wieder, C.; Künzer, M.; Wiechert, R.; Seipp, K.; Andresen, K.; Stark, P.; Schüffler, A.; Opatz, T.; Thines, E. Biosynthesis of the antifungal polyhydroxy-polyketide acrophialocinol. *Org. Lett.* **2025**, *27*, 1036–1041. [CrossRef] [PubMed]
25. Khan, R.A.A.; Najeeb, S.; Hussain, S.; Xie, B.; Li, Y. Bioactive secondary metabolites from *Trichoderma* spp. against phytopathogenic fungi. *Microorganisms* **2020**, *8*, 817. [CrossRef] [PubMed]
26. Yuan, C.; Ding, G.; Wang, H.Y.; Guo, Y.H.; Shang, H.; Ma, X.J.; Zou, Z.M. Polyketide-terpene hybrid metabolites from an endolichenic fungus *Pestalotiopsis* sp. *Biomed. Res. Int.* **2017**, *2017*, 6961928. [CrossRef]

https://doi.org/10.3390/jof1203016

Journal of
Fungi

MDPI

·ticle

·ranscriptome Analysis and Its Application in Screening Genes ·elated to the Growth and Development of *Sarcomyxa edulis*

·anzhu Jiang [1,†], Xiao Huang [2,†], Peng Wang [2], Bilal Ahmad [3], Ting Yang [1], Ziyuan Wang [1], Tianyu Ren [1] ·d Jize Xu [1,*]

1 Agricultural College, Jilin Agricultural Science and Technology College, Jilin 132101, China; jiangwanzhu@jlnku.edu.cn (W.J.)
2 Economic Plants Institute, Jilin Academy of Agricultural Sciences (Northeast Agricultural Research Center of China), Changchun 130033, China; huangxiaoxiwang@sina.com (X.H.)
3 College of Plant Protection, Jilin Agricultural University, Changchun 130118, China
* Correspondence: xujize@163.com
† These authors contributed equally to this work.

Abstract: *Sarcomyxa edulis* is a characteristic edible and medicinal mushroom found in Northeast China that is highly valued by consumers for its tender texture, pleasant flavor, and high nutritional value. To gain a deeper understanding of the molecular mechanisms underlying the development of *S. edulis* fruiting bodies, this study utilized the Illumina NovaSeq platform to perform transcriptome sequencing at three growth and development stages of *S. edulis* strain SE8, namely primordia (SE8–P), fruiting body differentiation (SE8–F), and mature fruiting body (SE8–M). A total of 54.67 Gb of clean data was obtained, with a GC content of around 51%. After assembly, 36,423 Unigenes were obtained. Functional annotation was performed on the Unigenes, resulting in 21,206 Unigene annotation results. Differential expression gene analysis showed that 79,606 and 523 DEGs were annotated in at least one database during the SE8–P vs. SE8–F, SE8–F vs. SE8–M, and SE8–P vs. SE8–M processes, respectively. Among these, the genes encoding aldehyde dehydrogenase and fungal hydrophobins were consistently downregulated, playing a negative regulatory role in the growth and development of *S. edulis.* The genes encoding glycoside hydrolase and AB hydrolase superfamily proteins were consistently upregulated, playing a positive regulatory role in growth and development. Among these, the genes encoding aldehyde dehydrogenase were annotated to the Tryptophan metabolism (ko00380) pathway through KEGG, suggesting that aldehyde dehydrogenase regulates indoacetate formation in the fruiting body of *S. edulis*. The accuracy of RNA–Seq and DEG analysis was validated using quantitative PCR. This study enriches our knowledge of the genetic information and provides a theoretical basis for the molecular mechanisms of fruiting body development of *S. edulis*.

Keywords: *Sarcomyxa edulis*; transcriptome sequencing; growth and development; tryptophan metabolism

1. Introduction

Sarcomyxa edulis are widely distributed in the Northern Hemisphere, especially in the forests of Northeast China, where they are most abundant [1]. Their fruiting bodies are fleshy and delicious, with extremely high nutritional and medicinal value [2–4]. Currently, a lot of work has been carried out in the breeding of excellent *S. edulis* strains [5–8] and genetic mechanism research [9–11].

The fruiting body is the main edible part of *S. edulis* and an important structure for growth and development. Its developmental regulation mechanisms have important research value. Currently, there is relatively little research on the formation mechanisms of fruiting bodies in *S. edulis*, and only Duan [12] has conducted a systematic analysis of different growth and development stages. The development of fruiting bodies is an extremely complex process, which is regulated not only by environmental factors but also by genetic factors [13]. Although growth and development mechanisms have been studied in various fungi, their molecular mechanisms are still not fully understood. However, through the unremitting efforts of researchers in various countries, a large number of functional genes related to fruiting body development have been discovered and identified [14]. Pelkmans [15] found in a study on *Schizophyllum commune* that *Bri1* and *Hom1* play a role in the later stages of development, while *Wc–2*, *Hom2*, and *Fst4* play a role in the early stages of development. Ohm [16] also found that, in *S. commune*, seven genes, including *hom1*, *hom2*, *c2h2*, *gat1*, *bri1*, *fst3*, and *fst4*, play a role in the development of primordia and fruiting bodies. Boulianne [17] found that *cgl1* and *cgl2* genes regulate the formation of fruiting bodies during the development of *Coprinus cinereus*. *cgl2* was expressed during hyphal formation, while *cgl1* was expressed during primordia and subsequent development. Wagemaker [18] found that hydrophobic protein genes, urease genes, and mannitol dehydrogenase genes were involved in the formation process of *Agaricus bisporus* fruiting bodies at different developmental stages. Meanwhile, Wu [19] discovered various amino acid synthases, transcription factors, GTP-binding proteins, etc., in *A. bisporus*, which were key genes related to fruiting body development. In addition, Almási [20] and Krizsán [21] found that heat shock proteins, small secreted proteins, kinases, F-Box proteins, and genes related to cell wall remodeling may be involved in the development of fruiting bodies in other basidiomycetes. Pei [22] discovered and identified a gene, g13394, in *Pleurotus ostreatus* that can promote hyphal growth and accelerate fruiting body development.

S. edulis, as a characteristic low-temperature edible and medicinal mushroom found in Northeast China, has great development prospects. Our team has been engaged in genetic breeding and cultivation research. We have collected and evaluated the germplasm resources of *S. edulis*, optimized cultivation formulas and techniques, analyzed the biological characteristics of core germplasm, optimized the SRAP–PCR reaction system, and obtained and analyzed the whole genome sequence (to be published). On this basis, this study performed transcriptome sequencing on samples from three growth and development stages: primordia (SE8–P), fruiting body differentiation (SE8–F), and mature fruiting body (SE8–M). The transcriptome sequence of the *S. edulis* strain was obtained using Illumina NovaSeq platform technology. By comparing different growth and development stages, DEGs were obtained and enriched for analysis to screen genes and metabolic pathways that may be involved in the growth and development of *S. edulis*. The discovery of these genes and metabolic pathways can not only improve and enrich the transcriptome data of *S. edulis* but also provide some assistance for the study of the growth and development mechanism of *S. edulis* and other edible mushrooms and provide a theoretical basis for genetic breeding research of edible mushrooms.

2. Materials and Methods

2.1. Strain and Collection of Samples at Different Developmental Stages

The strain used for transcriptome sequencing of *S. edulis* was SE8, characterized by a yellow–brown cap. It was collected from Baishan City, Jilin Province, and preserved at Jilin Agricultural Science and Technology College. The strain was inoculated into a sawdust culture medium (sawdust 78%, wheat bran 20%, lime 1%, gypsum 1%, pH 6.0), with 250 g of dry material per bag, cultivation temperature of 15 °C, humidity of 90%, and natural light. Samples were collected during the three stages of primordia (the 5th day of fruiting), fruiting body differentiation (the 10th day of fruiting), and mature fruiting body (the 15th day of fruiting) (Figure 1), with three biological replicates per stage. The samples were mixed evenly according to the same location and stored in a −80 °C freezer for transcriptome sequencing.

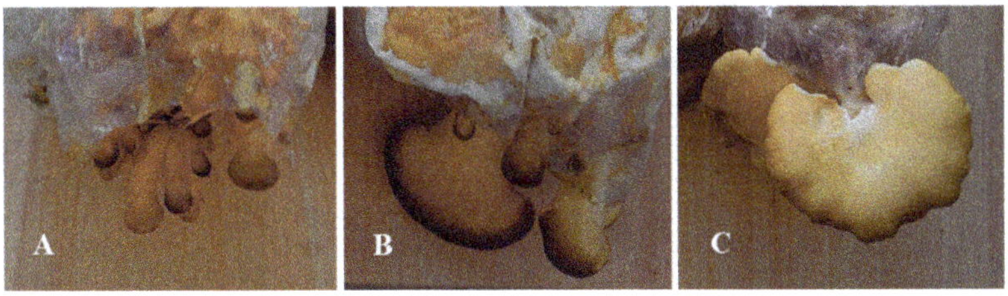

Figure 1. Development stage of *S. edulis* strain SE8. (**A**): Primordia (SE8–P). (**B**) Fruiting body differentiation (SE8–F). (**C**) Mature fruiting body (SE8–M).

2.2. RNA Isolation, Library Construction, and Sequencing

To summarize the gene expression profile of *S. edulis* at three developmental stages, cDNA samples were prepared from the primordia (SE8–P), fruiting body differentiation (SE8–F), and mature fruiting body (SE8–M). The libraries were sequenced on an Illumina NovaSeq platform to generate 150 bp paired-end reads, according to the manufacturer's instructions. Total RNA was extracted from the sample using a Tiangen DP441 assay kit (Tiangen Biotechnology, Beijing, China). Nanodrop 2000 (Thermo Fisher Scientific, Wilmington, DE, USA) and Agilent 2100 (Agilent Technologies, Santa Clara, CA, USA) were used to detect the concentration and integrity of extracted RNA. Each stage was sampled in triplicate, resulting in a total of 9 libraries. The raw data has been submitted to NCBI with the login number PRJNA1267801.

2.3. Analysis of Differentially Expressed Genes (DGEs)

Using Trinity software (2.14.0) [23], clean data were assembled into sequences to obtain the Unigene library of *S. edulis* and obtain transcriptome data. DIAMOND (v2.0.4) [24] software was used to combine the Unigene sequence with Swiss–Prot [25], NR [26], COG [27], KOG, GO [28], eggNOG4.5 [29], and KEGG [30,31] to compare with the database to obtain annotation information for Unigene. Using Bowtie [32], the reads obtained from sequencing were compared with the Unigene library. Based on the comparison results, RSEM (v1.2.19) was used to estimate the expression level, and FPKM values were used to represent the expression abundance of corresponding Unigenes [33]. Differential analysis was performed using DESeq2 software (1.30.1), and the screening criteria for differential genes were FDR < 0.01 and $|\log 2\ FC| \geq 2$. Functional annotation of DEGs in the database was performed using the database [34].

2.4. Validation of DEGs by qRT–PCR

Primer validation was a good practice for confirming the efficiency of qRT–PCR primers. To ensure the accuracy of RNA–Seq data, six representative DEGs were selected for qRT–PCR. The RNA used for qRT–PCR validation was the same RNA aliquots used for transcriptome sequencing. The FPKM values of these genes were available and showed significant changes in the transcriptional expression data of 9 samples. According to RNA seq data, the expression of the Actin gene was stable at three developmental stages and was therefore used as an internal reference gene (Table S1). Primer v5.0 was also used to design specific primers for internal reference genes and DEGs (Table S2). RNA extraction was performed using the RNAprep Pure polysaccharide polyphenol plant total RNA extraction kit (DP441) manufactured by Tengen (Wenzhou, China), and PrimeScript provided by Takara was used. The RT reagent kit with gDNA Eraser (Perfect Real Time) synthesis kit was used to synthesize cDNA. Specific primer pairs were obtained from transcriptome data and subsequently used for qRT–PCR. qRT–PCR was performed using the C1000 real-time PCR detection system (Bio–Rad, Hercules, CA, USA) in triplicate. The reaction mixture consisted of 10 μL, which included 5 μL of 1 × ChamQ Blue Universal SYBR qPCR Master Mix (Novozymes, Bagsværd, Denmark), 3 μL of diluted cDNA, and 1 μL of specific forward and reverse primers each. The $2^{-\Delta\Delta Ct}$ method was used to analyze gene expression data [35].

3. Results

3.1. Global Transcriptomic Analysis

Transcriptome sequencing was performed on nine samples from three distinct growth and development stages of *S. edulis* strains, yielding a total of 54.67 Gb of clean data. The clean data for each sample reached 5.91 Gb, with a GC content of approximately 51% and a Q30 base percentage of 95.22% or higher. The data indicators performed well, providing a reliable data basis for subsequent analysis (Table S3). Following assembly, a total of 36,423 Unigenes were obtained, with an N50 of 2335. Of these, 8986 Unigenes were longer than 1 kb (Figure 2a). Functional annotation of the Unigene sequences was conducted, resulting in 21,206 annotated Unigenes. Among these, 3277 genes were shared across annotations, with the distribution of genes in each annotation category illustrated in the petals (Table 1, Figure 2b)

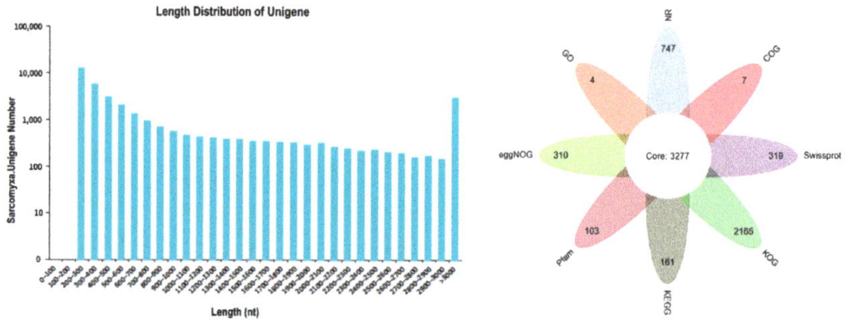

(a) (b)

Figure 2. Transcriptomic analysis of SE8. (**a**) Unigene length distribution map. The horizontal axis represents the different length intervals of Unigene; the vertical axis represents the number of Unigenes within a certain length interval. (**b**) Unigene annotation petal diagram.

Table 1. Unigene annotation statistics table.

Annotation Database	Annotated Number	300 ≤ Length < 1000	Length ≥ 1000
COG annotation	7457	2791	3156
GO annotation	15,203	5495	6029
KEGG annotation	10,853	3595	4831
KOG annotation	7728	2435	3854
Pfam annotation	12,045	4091	5879
Swissprot annotation	7717	2327	4167
TrEMBL annotation	18,204	6402	7696
eggNOG annotation	12,883	4622	5472
NR annotation	18,261	6421	7712
All annotated Unigenes	21,206	7960	7837

3.2. Gene Expression Level Analysis

The analysis of gene expression levels showed that the density of the nine samples was relatively concentrated, with log10 (FPKM) values ranging from −1.09 to 3.17 (Figure 3a). Principal component analysis and Spearman correlation coefficient rho analysis showed good sample repeatability, significant inter-group differences among samples at different developmental stages, and clear clustering among samples at the same developmental stage. These results demonstrate the stability and reliability of the test data. A stable data foundation for subsequent differential gene screening and enrichment analysis is provided (Figure 3b,c, Table S4).

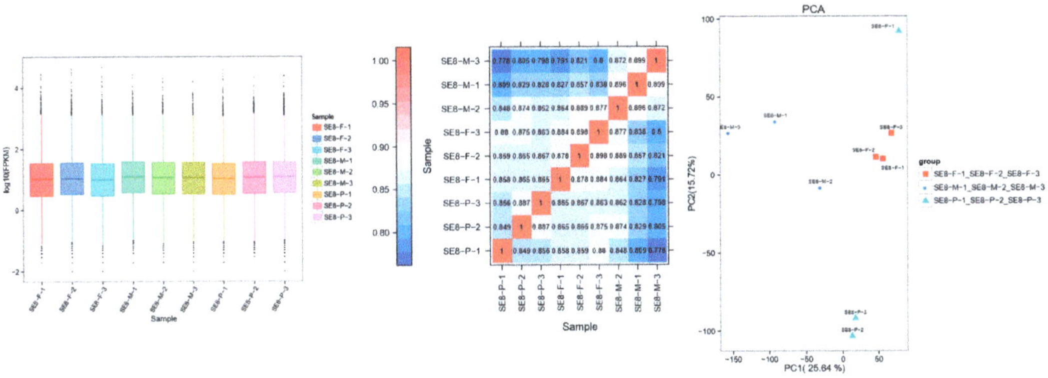

(**a**) (**b**) (**c**)

Figure 3. Gene expression level analysis of SE8. (**a**) Box plot of gene expression level distribution (FPKM) in different samples. The horizontal axis in the figure represents different samples; the vertical axis represents the logarithmic value of the sample expression level FPKM. This graph measures the expression levels of each sample from the perspective of the overall dispersion of expression levels. The vertical axis of the point represents the probability density. The peak of the distribution curve represents the region with the highest concentration of gene expression in the entire sample. (**b**) Correlation heatmap of different sample types. The horizontal and vertical axes in the figure are sample numbers, and the colors reflect the correlation between the samples. The corresponding relationship is shown in the legend on the left. (**c**) Principal Component Analysis Diagram. The coordinates PC1 and PC2 in the figure represent the principal components of different samples, and percentages represent the contribution values of corresponding principal components to sample differences. Each point represents a sample, and samples from different groups were represented by different colors and shapes.

3.3. Identification of DEGs Across Various Developmental Stages

This study identified DEGs in the SE8 strain at three stages: primordia (SE8–P), fruiting body differentiation (SE8–F), and mature fruiting body (SE8–M). There were 87 DEGs between SE8–P and SE8–F, with 44 genes upregulated and 43 genes downregulated. There were 715 DEGs between SE8–F and SE8–M, with 532 genes upregulated and 183 genes downregulated. There were 636 DEGs between SE8–P and SE8–M, with 444 genes upregulated and 192 genes downregulated (Figure 4a). To visually demonstrate the overlap of DEGs between different comparison groups, this study conducted a Venn diagram analysis on the DEGs of different comparison combinations. Among them, there were 6 common genes, 28 unique DEGs in SE8–P vs. SE8–F, 298 unique DEGs in SE8–F vs. SE8–M, and 198 unique DEGs in SE8–P vs. SE8–M (Table 2, Figure 4b).

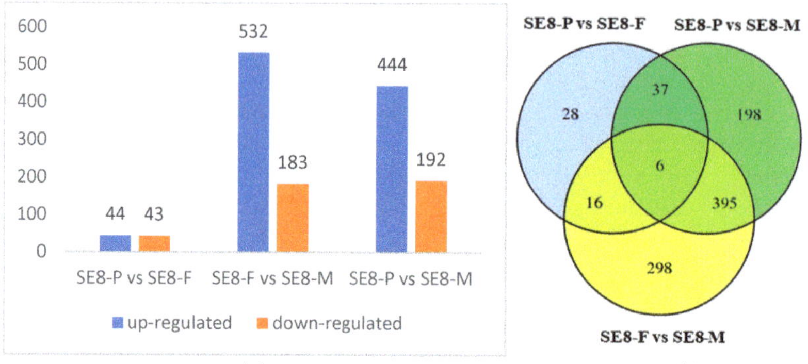

(a) (b)

Figure 4. Differentially expressed genes of SE8. (**a**) Statistics of differentially expressed genes. (**b**) Venn diagram of a differentially expressed gene.

Table 2. Information on differentially expressed genes in three stages.

Gene ID	Swissprot Annotation	Nr Annotation
TRINITY_DN327_c0_g1	6,7–dimethyl–8–ribityllumazine synthase	hypothetical protein
TRINITY_DN1035_c0_g1	Nitronate monooxygenase	2–nitropropane dioxygenase
TRINITY_DN3926_c0_g1	-	hypothetical protein
TRINITY_DN5751_c0_g1	Trimethyllysine dioxygenase	hypothetical protein
TRINITY_DN579_c0_g1	3,4–dihydroxy–2–butanone 4–phosphate synthase	3,4–dihydroxy–2–butanone 4–phosphate synthase–domain–containing protein
TRINITY_DN1480_c1_g1	Aldehyde dehydrogenase	aldehyde dehydrogenase

3.4. KOG Functional Classification of DEGs

This study performed KOG functional annotation on DEGs from the SE8 strain at various developmental stages. General function predictions only were made according to KOG annotations, post-translational modification, protein turnover, chaperones, energy production and conversion, coenzyme transport and metabolism, lipid transport and metabolism, secondary metabolite biosynthesis, transport, and catabolism. There was a high enrichment in the three stages of SE8–P vs. SE8–F, SE8–F vs. SE8–M, and SE8–P vs. SE8–M, which were mainly related to the synthesis, transportation, and metabolism of organic matter (Figure 5).

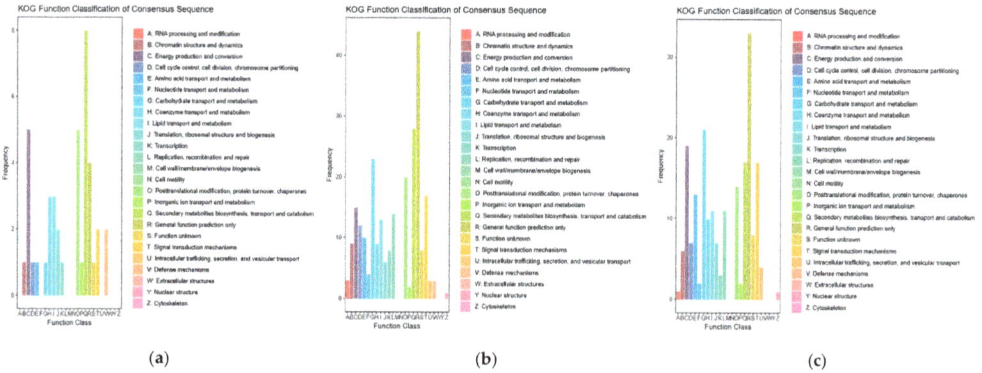

(a) (b) (c)

Figure 5. Statistical chart of KOG annotation classification for DEGs. (**a**) SE8–P vs. SE8–F, (**b**) SE8–F vs. SE8–M, (**c**) SE8–P vs. SE8–M.

3.5. GO Annotation Analysis of DEGs

The GO annotation includes three categories: Molecular Function (MF), Cellular Component (CC), and Biological Process (BP). The GO analysis during the growth process of the *S. edulis* strain is shown in Figure 6, and the results indicate that DEGs were involved in various biologically significant processes. In the comparison of SE8–P vs. SE8–F, SE8–F vs. SE8–M, and SE8–P vs. SE8–M, the significantly enriched terms were almost the same. The differential gene numbers enriched in the metabolic process and cellular process in BP, cellular anatomical entity and intracellular in CC, and catalytic activity and binding entries in MF were significantly higher than those in the other entries.

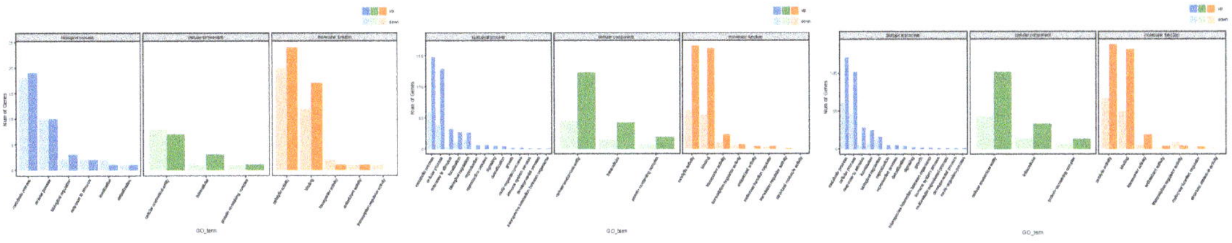

(**a**) (**b**) (**c**)

Figure 6. GO annotation of DEGs. (**a**) SE8–P vs. SE8–F, (**b**) SE8–F vs. SE8–M, (**c**) SE8–P vs. SE8–M.

During the differentiation stage of primordia and fruiting bodies (SE8–P vs. SE8–F), 60 DEGs were annotated, of which 29 genes were upregulated and 31 genes were downregulated. These DEGs were distributed among 34 BP, 7 CC, and 54 MF components. The enriched BP terms include Sphingomyelin catabolic process, Polyphosphate metabolic process, plasma membrane selenite transport, Cyanamide metabolic process, Carnitine biosynthetic process, and Nitrile metabolic process. This was related to the decomposition, synthesis, and metabolism of compounds. The enriched CC terms include Integral component of plasma membrane, Integral component of membrane, Host cell nucleus, Nucleus, and Cytoplasm. This was related to the formation of the nucleus, cytoplasm, and membrane. The enriched MF terms include Oxidoreductase activity, Monooxygenase activity, Heme binding, and Iron ion binding, this was related to oxidative stress response and reactant binding.

During the differentiation and maturation stages of the fruiting body (SE8–F vs. SE8–M), 425 DEGs were annotated, with 312 genes upregulated and 113 genes downregulated. These DEGs were distributed among 741 BP, 122 CC, and 260 MF components. The enriched BP terms include Vitamin biosynthetic process, Water-soluble vitamin metabolic process, Regulation of protein serine/threonine kinase activity, and Phospholipid catabolic process. This was related to vitamin biosynthesis and metabolism, as well as amino acid kinase activity. The enriched CC terms include Integral component of membrane, Nucleosome, Host cell nucleus, Extracellular space, and Cyclin-dependent protein kinase holoenzyme complex. This was related to the formation of the cell nucleus, membrane, and protein kinase complexes. Enriched MF is a term that includes Oxidoreductase activity, 5′–flap endonuclease activity, FAD binding, Carboxy–lyase activity, Zinc ion binding, monooxygenase activity, and Iron ion binding. This was related to oxidative stress response and reactant binding.

In total, 372 DEGs were annotated during the maturation stages of primordia and fruiting bodies (SE8–P vs. SE8–M), with 252 genes upregulated and 120 genes downregulated. These DEGs were distributed among 712 BP, 112 CC, and 240 MF components. The enriched BP terms include Vitamin biosynthetic process, Water-soluble vitamin biosynthetic process, Vitamin metabolic process, Xenobiotic metabolic process, Response to xenobiotic stimulus, and Nitrile metabolic process. This was related to compound metabolism, vitamin biosynthesis metabolism, and stress response. The enriched CC terms include Protein kinase complex, Cyclin-dependent protein kinase holoenzyme complex, Serine/threonine protein kinase complex, Host cell nucleus, and Incipient cellular bud site; this was related to protein kinase complexes and cell formation. The enriched MF terms include Cyanamide hydratase activity, Monooxygenase activity, Iron ion binding, Peroxidase activity, Oxidoreductase activity, and Glucan 1,4–alpha–glucosidase activity. This was related to oxidative stress response and reactant binding (Figure 6).

3.6. KEGG Annotation Analysis of DEGs

In order to further elucidate the biological pathways associated with DEGs, the top 20 enriched pathways were selected for KEGG enrichment analysis of DEGs in the three stages (Figure 7); 53 DEGs were annotated between SE8–P and SE8–F, and these DEGs were annotated into 29 KEGG entries. The most significantly enriched pathways included Tryptophan metabolism, Histidine metabolism, Lysine degradation, Riboflavin metabolism, Ascorbate and aldarate metabolism, Pyruvate metabolism, Fatty acid degradation, and beta–Alanine metabolism. These enriched pathways indicate that during the process from the primordium to the cap formation stage, the formation of the cap may be promoted by regulating amino acid metabolism activity. A total of 330 DEGs were annotated between SE8–F and SE8–M, and these DEGs were annotated into 88 KEGG entries. The most significantly enriched pathways included Riboflavin metabolism, Tryptophan metabolism, Lysine degradation, Pyruvate metabolism, Pentose and glucuronate interconversions, Starch and sucrose metabolism, Biotin metabolism, and Propanoate metabolism. The enrichment of these pathways suggests that during the formation of fruiting bodies, cells may accumulate substances and promote fruiting body maturation by participating in the synthesis of carbohydrate compounds and various amino acid metabolism activities. A total of 288 DEGs were annotated between SE8–P and SE8–M, and these DEGs were annotated into 87 KEGG entries. The most significantly enriched pathways included Riboflavin metabolism, Lysine degradation, Tryptophan metabolism, Fatty acid degradation, Pyruvate metabolism, beta–Alanine metabolism, Starch and sucrose metabolism, and Pantothenate and CoA biosynthesis. The enrichment of these pathways indicates that during the formation stage of fruiting bodies, substance accumulation and fruiting body maturation can be promoted by coordinating the synthesis and metabolism of various secondary metabolites, carbohydrate compounds, amino acid biosynthesis, and degradation.

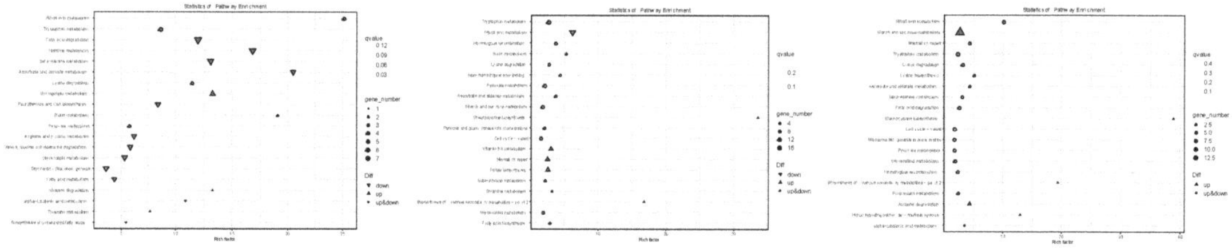

(a) (b) (c)

Figure 7. Classification of differentially expressed genes KEGG. (**a**) SE8–P vs. SE8–F, (**b**) SE8–F vs. SE8–M, (**c**) SE8–P vs. SE8–M.

This study conducted KEGG enrichment analysis on three stages: primordia (SE8–P vs. SE8–F), mature fruiting body (SE8–F vs. SE8–M), and primordia towards mature fruiting body (SE8–P vs. SE8–M). The statistical results showed that a total of 28 pathways were enriched in all three stages (Table 3). Among them, Tryptophan metabolism (ko00380), Starch and sucrose metabolism (ko00500), Carbon metabolism (ko01200), Riboflavin metabolism (ko00740), Lysine degradation (ko00310), Pyruvate metabolism (ko00620), and Glycerolipid metabolism (ko00561), which were eight pathways related to substance metabolism and degradation, have a relatively large number of enriched genes, with upregulation and downregulation of DEGs in each stage. In Pantothenate and CoA biosynthesis (ko00770), DEGs showed a downregulation trend in all three stages, while in Atrazine degradation (ko00791), DEGs showed an upregulation trend in all three stages. In addition, Basal transcription factors (ko03022), Ubiquitin-mediated proteolysis (ko04120), Ribosome (ko03010), and Phosphatidylinositol signaling system (ko04070) were the four pathways that only appeared in the SE8–F vs. SE8–M stage. Selenocompound metabolism (ko00450) and Mitophagy yeast (ko04139) pathways only appear in the SE8–P vs. SE8–M stage.

Table 3. Statistical analysis of common pathways in the three developmental stages of SE8.

ko ID	Pathway	SE8–P vs. SE8–F			SE8–F vs. SE8–M			SE8–P vs. SE8–M		
		Total	Up	Down	Total	Up	Down	Total	Up	Down
ko00010	Glycolysis/Gluconeogenesis	3	0	3	6	3	3	7	3	4
ko00053	Ascorbate and aldarate metabolism	3	0	3	4	1	3	4	1	3
ko00061	Fatty acid biosynthesis	1	0	1	4	2	2	2	0	2
ko00071	Fatty acid degradation	5	0	5	5	2	3	7	1	6
ko00280	Valine, leucine, and isoleucine degradation	3	0	3	5	1	4	5	1	4
ko00310	Lysine degradation	4	1	3	7	3	4	7	3	4
ko00330	Arginine and proline metabolism	3	0	3	5	1	4	4	0	4
ko00340	Histidine metabolism	4	0	4	4	1	3	3	0	3
ko00380	Tryptophan metabolism	7	2	5	16	6	10	12	5	7
ko00410	beta–Alanine metabolism	4	0	4	5	1	4	6	1	5
ko00500	Starch and sucrose metabolism	1	1	0	11	10	1	13	13	0
ko00511	Other glycan degradation	1	1	0	2	0	2	1	1	0
ko00561	Glycerolipid metabolism	3	0	3	8	5	3	7	3	4
ko00592	alpha–Linolenic acid metabolism	1	0	1	1	1	0	2	1	1
ko00620	Pyruvate metabolism	4	1	3	11	5	6	8	3	5
ko00640	Propanoate metabolism	1	0	1	5	4	1	5	2	3
ko00730	Thiamine metabolism	1	1	0	3	2	1	1	0	1
ko00740	Riboflavin metabolism	4	1	3	7	0	7	10	1	9
ko00770	Pantothenate and CoA biosynthesis	3	0	3	3	0	3	4	0	4
ko00780	Biotin metabolism	2	1	1	4	2	2	1	0	1
ko00791	Atrazine degradation	1	1	0	1	1	0	2	2	0
ko00910	Nitrogen metabolism	1	0	1	3	1	2	2	1	1
ko01040	Biosynthesis of unsaturated fatty acids	1	0	1	1	1	0	2	0	2
ko01200	Carbon metabolism	2	1	1	10	6	4	9	3	6
ko01212	Fatty acid metabolism	2	0	2	5	3	2	4	0	4
ko02010	ABC transporters	1	0	1	2	1	1	1	0	1
ko03440	Homologous recombination	1	0	1	9	6	3	5	3	2
ko04146	Peroxisome	1	0	1	3	2	1	4	2	2

3.7. *Tryptophan Metabolism*

KEGG enrichment analysis showed that Tryptophan metabolism (ko00380) contained the highest number of DEGs across the three developmental stages, with more genes downregulated than upregulated during growth and development (Table S5, Figure 8). During primordia-to-fruiting body differentiation (SE8–P vs. SE8–F), a total of seven DEGs were detected, including two upregulated genes, mainly encoding cytochrome P450 monooxygenases, and five downregulated genes, primarily encoding tryptamine 4–monooxygenase, NADPH–cytochrome P450 reductase, and aldehyde dehydrogenases. In the transition from fruiting body differentiation to the mature fruiting body (SE8–F vs. SE8–M), 16 DEGs were identified, with 6 upregulated genes, mainly encoding acetamidase, MFS-type transporter OryC, FAD-linked oxidoreductase, L–tyrosine:2–oxoglutarate aminotransferase (Amt1), and cytochrome P450 monooxygenases, while 10 downregulated genes mainly encoded multifunctional cytochrome P450 monooxygenases, cytochrome P450 monooxygenases, aldehyde dehydrogenases, aspirochlorine biosynthesis protein N, tryptamine 4–monooxygenase, and β–apo–4′–carotenal oxidase. During the primordia-to-mature fruiting body transition (SE8–P vs. SE8–M), 12 DEGs were detected, including 5 upregulated genes, mainly encoding cytochrome P450 monooxygenases, MFS-type transporter OryC, and cyanide hydratase, and seven downregulated genes, primarily encoding cytochrome P450 monooxygenases, bifunctional cytochromes, tryptamine 4–monooxygenase, aldehyde dehydrogenases, and aspirochlorine biosynthesis protein N. According to Table 4 and Figure 8, it can be seen that in Tryptophan metabolism, aldehyde dehydrogenase coordinates with Aldehyde oxidase to convert 5–hydroxyindoleacetate to 5–hydroxyindoleacetate, and it coordinates with indole–3–acetaldehyde oxidase to convert Indole3–acetaldehyde to Indooleacetate. In this study, the gene encoding aldehyde dehydrogenase showed a downregulation trend with the growth and development of the fruiting body of the *S. edulis*, and the two were negatively correlated. It was preliminarily inferred that aldehyde dehydrogenase plays a negative regulatory role in the growth of the *S. edulis*. Related studies have shown that indoacetate was involved in regulating cell division and elongation in different parts of the fruiting body, affecting the length, thickness, and normal expansion and size of the stem cap. Its mechanism of action was similar to that of promoting stem node elongation in plants [36–38].

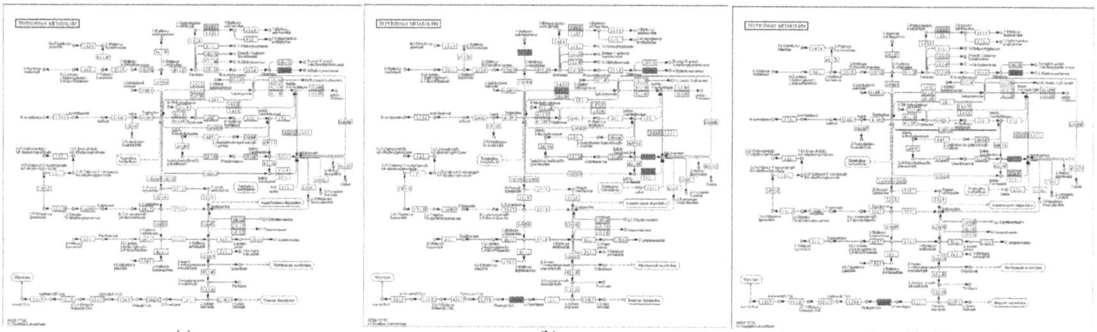

(a) (b) (c)

Figure 8. The DEGs in SE8–P vs. SE8–F, SE8–F vs. SE8–M, and SE8–P vs. SE8–M were mapped to Tryptophan metabolism (ko00380). The red represents upregulation, the green represents downregulation, and the purple represents upregulation and downregulation. (**a**) SE8–P vs. SE8–F, (**b**) SE8–F vs. SE8–M, (**c**) SE8–P vs. SE8–M.

Table 4. Genes and proteins are potentially critical for the growth and development of *S. edulis*.

Gene ID	FDR	$\log_2$ FC	Regulated	Pfam Annotation	Swiss–Prot Annotation
			SE8–P vs. SE8–F		
TRINITY_DN281_c2_g1	0.0091	1.1308	up	alpha/beta hydrolase fold	AB hydrolase superfamily protein
TRINITY_DN5420_c0_g1	0.0043	1.0802	up	alpha/beta hydrolase fold	AB hydrolase superfamily protein
TRINITY_DN1480_c1_g1	0.0009	−1.2947	down	Aldehyde dehydrogenase family	Aldehyde dehydrogenase
TRINITY_DN17839_c0_g1	0.0000	−2.1396	down	Aldehyde dehydrogenase family	Aldehyde dehydrogenase
TRINITY_DN32179_c0_g1	0.0001	−2.1490	down	Aldehyde dehydrogenase family	Aldehyde dehydrogenase
TRINITY_DN327_c0_g1	0.0000	−1.1550	down	6,7–dimethyl-8–ribityllumazine synthase	6,7–dimethyl-8–ribityllumazine synthase
			SE8–F vs. SE8–M		
TRINITY_DN281_c2_g1	0.0000	−1.4898	down	alpha/beta hydrolase fold	AB hydrolase superfamily protein
TRINITY_DN1890_c0_g1	0.0004	−1.0374	down	alpha/beta hydrolase fold	-
TRINITY_DN4014_c0_g1	0.0001	−1.0902	down	Aldehyde dehydrogenase family	Aldehyde dehydrogenase 5, mitochondrial
TRINITY_DN679_c0_g1	0.0010	−1.2214	down	Aldehyde dehydrogenase family	Beta–apo-4′–carotenal oxygenase
TRINITY_DN1480_c1_g1	0.0059	−1.0289	down	Aldehyde dehydrogenase family	Aldehyde dehydrogenase

Table 4. *Cont.*

Gene ID	FDR	$\log_2$ FC	Regulated	Pfam Annotation	Swiss–Prot Annotation
TRINITY_DN3228_c0_g1	0.0000	1.1984	up	Glycosyl hydrolases family 16	Short–chain dehydrogenase/reductase VdtF
TRINITY_DN2724_c0_g1	0.0001	1.1198	up	Glycosyl hydrolases family 5	Glucan endo–1,6–beta–glucosidase B
TRINITY_DN6981_c0_g1	0.0053	1.5132	up	Glycosyl hydrolase family 71	Glucan endo–1,3–alpha–glucosidase agn1
TRINITY_DN5133_c0_g1	0.0000	1.7507	up	Glycosyl Hydrolase Family 88	-
TRINITY_DN6759_c0_g1	0.0004	1.5236	up	Glycosyl hydrolase family 76	-
TRINITY_DN9190_c0_g1	0.0000	−1.0044	down	Thaumatin family	-
TRINITY_DN1190_c2_g1	0.0000	−1.1813	down	Thaumatin family	-
TRINITY_DN711_c10_g1	0.0007	1.0551	up	Eukaryotic aspartyl protease	Aspartic protease
TRINITY_DN1448_c0_g1	0.0059	2.2227	up	Eukaryotic aspartyl protease	Aspartic protease
TRINITY_DN1508_c1_g1	0.0017	1.2047	up	Aldo/keto reductase family	Alcohol oxidase
TRINITY_DN8842_c0_g1	0.0044	1.3520	up	Aldo/keto reductase family	NADPH–dependent conjugated polyketone reductase
			SE8–P vs. SE8–M		
TRINITY_DN2724_c0_g1	0.0000	1.2785	up	Glycosyl hydrolases family 5	Glucan endo–1,6–beta–glucosidase B
TRINITY_DN6759_c0_g1	0.0009	1.4974	up	Glycosyl hydrolases family 76	-
TRINITY_DN49_c2_g1	0.0000	−2.2034	down	Fungal hydrophobin	Hydrophobin–B
TRINITY_DN270_c0_g1	0.0014	−1.3033	down	Fungal hydrophobin	Lipase 1

3.8. qRT–PCR Validation of DEGs at Different Growth and Development Stages

In order to confirm the transcriptome data, this study screened six highly expressed genes for qRT–PCR to examine the expression patterns of genes related to the growth and development of *S. edulis*. The qRT–PCR results were consistent with the RNA–Seq results, demonstrating the reliability of the RNA–Seq data through the consistency between the two methods (Figure 9). The TRINITYDN1480uc1_g1 gene was a common gene among the three growth and development stages and shows significant expression. According to the annotation library, the gene mainly encodes aldehyde dehydrogenase, which shows a decreasing trend in FPKM value during the three growth and development processes of *S. edulis*.

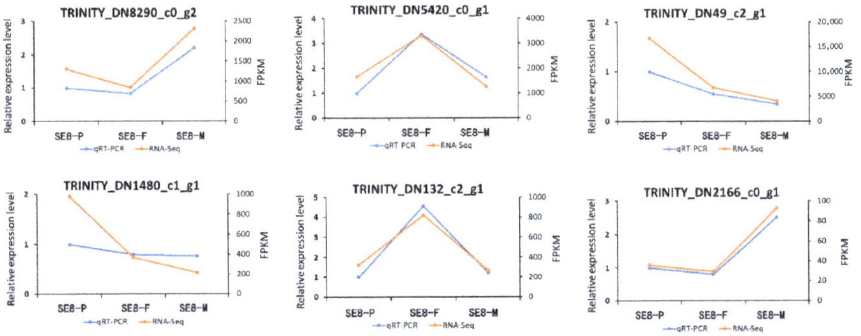

Figure 9. Verifies the expression profile of genes related to the growth and development of *S. edulis* through qRT–PCR analysis.

3.9. Genes and Proteins Potentially Critical for Growth and Development of S. edulis

The analysis of DEGs during the growth and development of *S. edulis* (Table 4) showed that, compared with the fruiting body formation stage (SE8–P vs. SE8–F), the gene encoding AB hydrolase superfamily protein was upregulated in the primordium stage, while the gene encoding aldehyde dehydrogenase family was downregulated. Compared with the mature stage of the fruiting body (SE8–F vs. SE8–M), the genes encoding AB hydrolase superfamily protein, aldehyde dehydrogenase family, and Thaumatin family were downregulated during the formation stage of the fruiting body, while the genes encoding glycoside hydrolase family, Aldo/keto reductase family, and Eukaryotic aspartyl protease were upregulated. Upregulation of the Amidohydrolase gene was also noted. In the DEGs between the primordium stage and the fruiting body maturation stage (SE8–P vs. SE8–M), the genes encoding the glycoside hydrolase family were upregulated, while the genes encoding the fungal hydrophobin were downregulated. These proteins can promote the growth and development of *S. edulis* [39].

4. Discussion

4.1. Comparative Analysis of the Transcriptome of Edible Fungi

Transcriptome sequencing technology can comprehensively and quickly obtain mRNA sequence information transcribed by specific tissues or organs of a species in a certain state and study gene expression levels and structures at the overall level, revealing molecular mechanisms in specific biological processes. Non-parametric transcriptome sequencing technology has the characteristics of high throughput, high resolution, wide applicability, high sensitivity, and quantitative dynamics and was the foundation and starting point for gene function and structure research. In recent years, with the popularization of high-throughput sequencing technology, transcriptome research on edible fungi has shifted from single-species functional annotation to cross-species comparative analysis. At present, relevant research has been conducted on typical edible mushrooms such as *Auricularia auricula, Ganoderma lucidum, Hericium erinaceus, Dictyophora indusiata, Lentinula edodes, Pleurotus ostreatus*, and *Flammulina velutipes*. The main research focuses on the gene expression differences in the growth and development, environmental adaptation, secondary metabolism, and morphogenesis of edible fungi under different developmental stages, strains, and environmental stresses [40,41]. These findings provide valuable insights into the evolutionary strategies and functional differentiation of edible fungi. Wen [42] conducted multi-omics sequencing on *D. indusiata* balls and mature *D. indusiata*, and the results showed that the key to the morphological development of *D. indusiata* balls after harvesting was stem elongation, with the fastest elongation rate in the middle of the stem. Fu [43] conducted comparative transcriptomic analysis on the mycelium, primordia, and fruiting bodies of *Pleurotus Nebrodensi*, identifying DEGs involved in morphogenesis, primary carbohydrate metabolism, cold stimulation, and blue light response. Tang [44] conducted transcriptomic studies on three treatments of *L. edodes* hyphae (light avoidance culture for 30 days, light avoidance culture for 80 days, and light avoidance for 30 days plus 50 days), revealing the molecular mechanism of light-induced *L. edodes* color transformation. It was found that photosensitive genes, light signal transduction pathways, and pigment-forming genes were involved in *L. edodes* color transformation. Wang [45] identified a total of 8495 Unigenes, 4047 proteins, and 30 metabolites in the de novo sequencing analysis of *Morchella importuna*. These results provide a basis for understanding the taste formation, texture regulation, and color changes in fruiting bodies. Wang [46] subjected the hyphae, primordia, fruiting bodies, and spores of *P. ostreatus* to high-temperature stress, H_2O_2 stress, NaCl stress, and pH stress to detect the expression levels of peroxidase (CAT). The results showed that CAT responded most significantly to high-temperature stress; however, the response mechanisms of different CAT to high temperatures were different. Zhang [47] found through transcriptome KEGG enrichment analysis of the growth and development stages of *Hypsizygus marmoreus* fruiting bodies that the color transition stage was associated with MAPK, cAMP, and blue light signaling pathways. Light affects the expression of genes related to the initiation of *H. marmoreus* fruiting bodies, and nitrogen stress may enhance fruiting body maturation. Zhou [48] analyzed the DEGs between the hyphae and fruiting bodies of *Auricularia polytricha* and found that tyrosinase had the highest expression level in the fruiting bodies, which may be an important regulatory gene in pigment synthesis in *A. polytricha* fruiting bodies. Ma [49,50] analyzed the pigment synthesis pathway of *Auricularia cornea* through multiple omics studies and identified the main pigments (γ–glutamyl–3,4–dihydroxybenzoate), major intermediates (α–D–glucose–1P, citrate, 2–oxogluconate, and glutamate), and key enzymes, such as phenoloxidase, in the fruiting body of *A. cornea*. Li [51] used corn cob and tobacco straw as two different substrates to cultivate *P. ostreatus*. Based on this, the DEGs and metabolic pathways of *P. ostreatus* at different developmental stages were analyzed. The results revealed the transcriptome adaptation of *P. ostreatus* to tobacco straw and provided new insights for the molecular mechanism of biomass conversion by edible bacteria using straw. Duan [12] studied *S. edulis* mycelium growing to a half bag (B1), mycelium in cold stimulation after a full bag (B2), mycelium in primordia appearing (B3) and primordia (B4), mycelium at the harvest stage (B5), and mature fruiting body (B6). They performed transcriptome sequencing for the six growth and development stages of *S. edulis* to screen for genes related to lignin degradation. The materials used in this study were *S. edulis* strains collected from the field, and transcriptome sequencing was performed by sampling tissues from three typical stages: primordia (SE8–P), fruiting body differentiation (SE8–F), and mature fruiting body (SE8–M). The results of this study showed significant differences in DEGs among different tissue types, especially in the SE8–F vs. SE8–M stage, where the number of DEGs was significantly higher than in the SE8–P vs. SE8–F stage and SE8–P vs. SE8–M stage. GO and KEGG enrichment analysis of DEGs between different tissue types showed the same functional distribution trend of DEGs (Figures 5–7), confirming that mature fruiting bodies were the main gene expression tissues in the formation process. This study preliminarily speculated that the genes encoding aldehyde dehydrogenase and fungal hydrophobin play a negative regulatory role in the growth and development of *S. edulis* through GEG screening and KEGG enrichment analysis between different growth and development stages. Related studies have shown that the aldehyde dehydrogenase gene was an important gene discovered in recent years that can regulate plant branching (tillering). It metabolizes endogenous and exogenous aliphatic and aromatic aldehyde molecules into corresponding carboxylic acids [52,53]. Fungal hydrophobin was crucial for the formation of fungal aerial hyphae, but it has a negative regulatory effect on the development of fruiting bodies [54,55], which was consistent with the results of this study. The genes encoding glycoside hydrolase and AB hydrolase superfamily proteins play a positive regulatory role. Related studies have shown that the GH16 family contains multiple glycoside hydrolases, and its members have activity towards various glycosidic bonds of pectin; these hydrolases play a role in the degradation of lignocellulose [56]. The AB hydrolase superfamily protein assists in disrupting the structure of plant cell walls, creating conditions for glycoside hydrolase to degrade cellulose and hemicellulose, allowing hyphae to absorb and utilize nutrients and promote the growth and development of fruiting bodies. Although both this study and Duan's [12] used *S. edulis* strains, there were differences in the source of the strains and the transcriptome sequencing techniques used, as well as different research purposes, resulting in certain differences in the research results. This also provides a data reference for the genetic diversity analysis of *S. edulis* and enriches its genetic information.

4.2. Signal Pathways Involved in the Development of Fruiting Bodies in S. edulis

The development of fruiting bodies was one of the most critical stages in the life cycle of edible fungi, involving complex metabolic regulatory networks. This process was regulated by multiple metabolic pathways, including various carbohydrate metabolism, signal transduction, synthesis of secondary metabolites, and cell wall remodeling. During the developmental stage of the fruiting body, a large amount of carbon sources and energy was required, among which glycolysis and the tricarboxylic acid cycle were core energy metabolism pathways [57,58]. The genes encoding hexokinase [59], phosphofructokinase (PFK) [60], and trehalose–6–phosphate synthase (TPS) [61] were crucial for hyphal development and fruiting body formation. In addition, the development of fruiting bodies was also regulated by environmental signals (such as light and temperature) and intracellular signaling pathways, including cAMP and MAPK [62]. In *Agaricus bisporus*, MAPK signaling promotes the aggregation of fungal filaments into primordia by inhibiting hyphal branching-related genes. In *A. bisporus*, MAPK signaling promotes the aggregation of fungal filaments into primordia by inhibiting hyphal branching-related genes [63]. During the development of fruiting bodies, a large number of secondary metabolites were synthesized, including terpenes [64–66], polyphenols [67], and sterols [68], which can promote cap differentiation and may also affect spore release. The morphological construction of sub-entities relies on the dynamic remodeling of cell wall polysaccharides. Chitin synthase (CHS) and β–β–glucan synthase (FKS) genes promote the formation of the cell wall cytoskeleton [69]. Lignin-degrading enzymes (such as laccase and manganese peroxidase MnP) enhance the mechanical strength of organelles by modifying the cross-linked structure of cell walls [70]. In this study, a total of 28 pathways appeared in all three stages of *S. edulis* growth and development. The pathways with more enriched genes were mostly related to substance metabolism and degradation, including tryptophan metabolism, starch and sucrose metabolism, carbon metabolism, riboflavin metabolism, lysine degradation, pyruvate metabolism, glycerolipid metabolism, and glycolysis/gluconeogenesis. In addition, the Pantothenate and CoA biosynthesis pathways showed a downregulation trend of DEGs in all three growth and development stages, while the Atrazine degradation pathway showed an upregulation trend of DEGs in all three growth and development stages. The discovery of these pathways provides a valuable reference for research on the development of *S. edulis* fruiting bodies. At present, although relevant pathways have been discovered, further analysis of the genes and metabolites related to the pathways has not been carried out. Further in-depth analysis should be conducted in combination with metabolomics and proteomics.

4.3. The Role of Tryptophan in Edible Fungi

Tryptophan was a precursor of bioactive substances such as niacin, serotonin, melatonin, etc. It not only participates in the synthesis and metabolic regulation of proteins in the body but also affects the metabolism of nutrients such as proteins, sugars, and fats. It can also participate in the growth and development of plants, thereby affecting the development of edible mushroom fruiting bodies [71,72]. Tryptophan plays multiple roles in the growth and development of edible fungi. It not only participates in protein synthesis as an essential amino acid but also acts as a precursor for secondary metabolites. It affects hyphal growth, fruiting body differentiation, and stress resistance by regulating signaling pathways [73,74]. Yan [75] conducted comparative transcriptome analysis on different forms of *L. edodes* fruiting bodies and found that amino acid metabolism pathways, such as trypsin metabolism, were significantly enriched in malformed *L. edodes* fruiting bodies. Genes encoding heat shock proteins, G proteins, and β –1,3–glucanase in the GH5 family were involved in the development of *L. edodes* fruiting bodies. Duan [76] conducted transcriptome sequencing on the five growth and development stages of *D. indusiata*, and the results showed that the cap was the main source of indole–3–acetic acid synthesis, confirming the importance of tryptophan metabolism in the differentiation of *D. indusiata* fruiting bodies. At the same time, three new genes related to the tryptophan metabolism pathway, IAA synthesis, were identified in the cap. In this study, five genes encoding aldehyde dehydrogenase were screened in Tryptophan metabolism. These genes were negatively correlated with the growth and development of the fruiting body of the *S. edulis* and showed significant expression of aldehyde dehydrogenase in the primary stage, which may promote the formation of hyphal tangles. As the fruiting body matures, the expression level of Aldehyde dehydrogenase was downregulated, suggesting that it may promote stem elongation and cap formation by regulating the formation of indoacetate, leading to a shift in the development of the fruiting body from the cap growth stage to the fruiting body maturity stage of *S. edulis*. Aldehyde dehydrogenase was an important enzyme class in the process of clearing reactive oxygen species in organisms, which can catalyze the oxidation of toxic aldehydes and the rapid dehydrogenation of toxic aldehydes, further enhancing the cell's aldehyde detoxification defense mechanism [77,78]. It can also regulate the growth and development of microorganisms and has been reported in various microorganisms such as *Escherichia coli* [79], *Saccharomyces cerevisiae* [80], *Tricholoma vaccinum* [81], *P. ostreatus* [82], and *Pleurotus geesteranu* [83]. At present, this study has only preliminarily screened the gene and has not yet verified its function. In the future, the gene can be accurately identified and further analyzed based on metabolomics and proteomics data to elucidate its mechanism of action in the growth and development of *S. edulis*.

5. Conclusions

The completion of transcriptome sequencing provides an effective means for studying the genetic mechanism and molecular breeding of *S. edulis*. This study obtained 54.67 GB of clean data and 21,206 Unigene annotation results. Functional annotation and enrichment of DEGs between the three growth and development stages of primordia (SE8–P), fruiting body differentiation (SE8–F), and mature fruiting body (SE8–M) were performed using GO and KEGG. At the same time, highly differentially expressed aldehyde dehydrogenase, fungal hydrophobin, glycoside hydrolase, and AB hydrolase superfamily protein genes were identified. Among them, the gene encoding aldehyde dehydrogenase was annotated into the trypsin metabolism (ko00380) pathway through KEGG, and it was preliminarily speculated that aldehyde dehydrogenase forms *S. edulis* fruiting bodies by regulating indoacetate. The accuracy of RNA–Seq and DEG analysis was validated using quantitative PCR. The above research results provide valuable information on the molecular mechanisms of *S. edulis* fruiting body development. This study aims to provide accurate genetic data on the molecular mechanisms of growth and development and enrich the genetic database of *S. edulis*. Although this study identified differential genes that may be related to the growth and development of *S. edulis*, only qRT–PCR was performed. Further functional validation and cloning of the screened differential genes were needed to clarify their mechanisms of action in growth and development. At the same time, this study used non-parametric transcriptome sequencing, which has the characteristics of wide applicability, high sensitivity, and quantitative dynamics. However, compared with parametric transcriptome sequencing, its accuracy was poorer. Future studies will integrate and analyze transcriptome data based on whole-genome data to improve the accuracy of the results.

Supplementary Materials: The following supporting information can be downloaded at https://www.mdpi.com/article/10.3390/jof11100750/s1. Table S1: Internal reference gene sequence information; Table S2: Primer sequences for internal reference genes and DEGs genes; Table S3: Statistical able of sequencing data evaluation of three strains at different development stages; Table S4: Comparison of correlation between transcriptome samples; Table S5: DEGs encoding annotation results of Tryptophan metabolism (ko00380) at three growth and development stages.

Author Contributions: Conceptualization, W.J. and J.X.; methodology, X.H. and P.W.; formal analysis, B.A. and T.Y.; resources, P.W.; data curation, Z.W. and T.R.; writing—original draft preparation, W.J.;

writing—review and editing, W.J., X.H. and J.X. All authors have read and agreed to the published version of the manuscript.

Funding: This research was funded by the Science and Technology Department of Jilin Province, grant number 20230202117NC.

Institutional Review Board Statement: Not applicable.

Informed Consent Statement: Not applicable.

Data Availability Statement: The transcriptome data presented in this study have been deposited in the National Center for Biotechnology Information Sequence Read Archive under accession number PRJNA1267801: https://dataview.ncbi.nlm.nih.gov/object/PRJNA1267801?reviewer=mjt1amaerauoilo1h6f60ic9or (accessed on 27 May 2025).

Conflicts of Interest: The authors declare no conflicts of interest.

Abbreviations

COG	Cluster of Orthologous Groups of proteins
GO	Gene Ontology
KEGG	Kyoto Encyclopedia of Genes and Genomes
KOG	Clusters of orthologous groups for eukaryotic complete genomes
Pfam	Protein Families Database
TrEMBL	Translation of EMBL nucleotide sequence database
eggNOG	evolutionary genealogy of genes: Non-supervised Orthologous Groups
NR	Non-Redundant Protein Database
DEG	Differentially expressed gene
SE8–P	Primordia
SE8–F	Fruiting body differentiation
SE8–M	Mature fruiting body
MF	Molecular Function
CC	Cellular Component
BP	Biological Process
A. bisporus	*Agaricus bisporus*
C. cinereus	*Coprinus cinereus*
S. edulis	*Sarcomyxa edulis*
S. commune	*Schizophyllum commune*
P. ostreatus	*Pleurotus ostreatus*
D. indusiata	*Dictyophora indusiata*
L. edodes	*Lentinula edodes*
M. importuna	*Morchella importuna*
H. marmoreus	*Hypsizygus marmoreus*
A. polytricha	*Auricularia polytricha*

References

1. Wang, G.S.; Cai, Q.; Hao, Y.J.; Bau, T.; Chen, Z.H.; Li, M.X.; David, N.; Kraisitudomsook, N.; Yang, Z.L. Phylogenetic and taxonomic updates of Agaricales, with an emphasis on *Tricholomopsis*. *Mycology* **2023**, *15*, 180–209. [CrossRef]
2. Sui, Y.; Shu, F.; Cui, S.; Yang, S.R.; Liu, Y.; Wang, Q. Characteristics and antioxidant activities of polysaccharides from *Sarcomyxa edulis*. *Acta Edulis Fungi* **2021**, *28*, 48–56.
3. Zhang, R.; Zhao, L.Y.; Wang, H.X.; Ng, T.B. A novel ribonuclease with antiproliferative activity toward leukemia and lymphoma cells and HIV-1 reverse transcriptase inhibitory activity from the mushroom, *Hohenbuehelia serotina*. *Int. J. Mol. Med.* **2014**, *33*, 209–214. [CrossRef]
4. Li, X.Y.; Wang, L.; Wang, Z.Y. Radioprotective activity of neutral polysaccharides isolated from the fruiting bodies of *Hohenbuehelia serotina*. *Phys. Medica* **2015**, *31*, 352–359. [CrossRef]
5. Qiu, T.M.; Wang, H.T.; Yu, Y.N.; Xu, L.L.; Lu, X.H. Domestication and nutritional composition analysis of a wild *Panellus edulis* strain. *Edible Fungi China* **2022**, *41*, 19–29.

6. Jiang, W.Z.; Miao, L.; Wang, P. Biological characteristics and domestication cultivation analysis of wild *Sarcomyxa edulis* strain. *J. Jilin Agric. Sci. Technol. Univ.* **2024**, *33*, 1–8.
7. Wang, P.; Yao, F.J.; Fang, M.; Jiu, L.D.; Zhai, Y.; Yuan, W.D. A new *Panellus edulis* cultivar 'Qidong 1'. *Acta Hortic. Sin.* **2015**, *42*, 605–606.
8. Jiu, L.D.; Yao, F.J.; Zhang, Y.M.; Zhao, N.; Zhu, M.Q. Study on the basidiospore germination characteristics and mycelia growth characteristics of *Panellus edulis*. *Edible Fungi China* **2015**, *34*, 10–12.
9. Tian, M.X.; Rong, C.B.; Niu, Y.R.; Song, S.; Liu, Y.; Chen, Q.J. ISSR and SRAP analysis of genetic diversity of *Hohenbuehelia serotina*. *Jiangsu Agric. Sci.* **2018**, *46*, 43–47.
10. Tian, F.H.; Li, C.T.; Li, Y. Genomic analysis of *Sarcomyxa edulis* reveals the basis of its medicinal properties and evolutionary relationships. *Front. Microbiol.* **2021**, *12*, 1614. [CrossRef]
11. Duan, C.; Tian, F.H.; Yao, L.; Lv, J.H.; Jia, C.W.; Li, C.T. Comparative transcriptome and WGCNA reveal key genes involved in lignocellulose degradation in *Sarcomyxa edulis*. *Sci. Rep.* **2022**, *12*, 18379. [CrossRef]
12. Duan, C.; Yao, L.; Lv, J.H.; Jia, C.W.; Tian, F.H.; Li, C.T. Systematic analysis of changes across different developmental stages of the mushroom *Sarcomyxa edulis*. *Gene* **2022**, *824*, 146450. [CrossRef] [PubMed]
13. Song, L.L.; Xing, X.K.; Guo, S.X. Morphological process and regulation mechanisms of fruiting body differentiation in macro-fungi:A review. *Mycosystema* **2018**, *37*, 671–684.
14. Shen, N.; Zhang, J.C.; Wang, C.C.; Bian, Y.B.; Xiao, Y. Studies on transcriptome during fruiting body development of *Lentinula edodes*. *Acta Hortic. Sin.* **2022**, *49*, 801–815.
15. Pelkmans, J.F.; Patil, M.B.; Gehrmann, T.; Reinders, M.J.T.; Wösten, H.A.B.; Lugones, L.G. Transcription factors of *Schizophyllum commune* involved in mushroom formation and modulation of vegetative growth. *Sci. Rep.* **2017**, *7*, 310. [CrossRef] [PubMed]
16. Ohm, R.A.; Jong, J.F.D.; Bekker, C.D.; Han, A.B.W.; Lugones, L.G. Transcription factor genes of *Schizophyllum commune* involved in regulation of mushroom formation. *Mol. Microbiol.* **2011**, *81*, 1433–1445. [CrossRef]
17. Boulianne, R.P.; Liu, Y.; Aebi, M.; Lu, B.C.; Kües, U. Fruiting body development in *Coprinus cinereus*: Regulated expression of two galectins secreted by a nonclassical pathway. *Microbiology* **2000**, *146*, 1841–1853. [CrossRef] [PubMed]
18. Wagemaker, M.J.; Eastwood, D.C.; Van, D.D.C.; Jetten, M.S.; Burton, K.; Van Griensven, L.J.; Hj, O.D.C. Expression of the urease gene of *Agaricus bisporus*: A tool for studying fruit body formation and post-harvest development. *Appl. Microbiol. Biotechnol.* **2006**, *71*, 486–492.
19. Wu, X.M.; Zhang, X.; Li, N.Y. Transcriptome analysis of *Agaricus bisporus* fruiting at different stages. *Mycosystema* **2017**, *36*, 193–203.
20. Almási, E.; Sahu, N.; Krizsán, K.; Bálint, B.; Kovács, G.M.; Kiss, B.; Cseklye, J.; Drula, E.; Henrissat, B.; Nagy, I.; et al. Comparative genomics reveals unique wood-decay strategies and fruiting body development in the Schizophyllaceae. *New Phytol.* **2019**, *224*, 902–915. [CrossRef]
21. Krizsánk, K.; Almási, E.; Merényi, Z.; Sahu, N.; Virágh, M.; Kószó, T.; Mondo, S.; Kiss, B.; Bálint, B.; Kües, U.; et al. Transcriptomic atlas of mushroom development reveals conserved genes behind complex multicellularity in fungi. *Proc. Natl. Acad. Sci. USA* **2019**, *116*, 7409–7418. [CrossRef] [PubMed]
22. Pei, J.Q.; Zhao, M.R.; Huang, C.Y.; Wu, X.L. Discovery and verification of a functional gene influencing the growth and development of *Pleurotus ostreatus*. *Biotechnol. Bull.* **2025**, *41*, 327–334.
23. Grabherr, M.G.; Haas, B.J.; Yassour, M.; Levin, J.Z.; Thompson, D.A.; Amit, I.; Adiconis, X.; Fan, L.; Raychowdhury, R.; Zeng, Q.D.; et al. Full-length transcriptome assembly from RNA Seq data without a reference genome. *Nat. Biotechnol.* **2011**, *29*, 644–652. [CrossRef]
24. Buchfink, B.; Xie, C.; Huson, D.H. Fast and Sensitive Protein Alignment Using Diamond. *Nat. Methods* **2015**, *12*, 59–60. [CrossRef]
25. Boeckmann, B.; Bairoch, A.M.; Apweiler, R.; Blatter, M.; Estreicher, A.; Gasteiger, E.; Martin, M.; Michoud, K.; Odonovan, C.; Phan, I.; et al. The SWISS- PROT protein knowledgebase and its supplement TrEMBL in 2003. *Nucleic Acids Res.* **2003**, *31*, 365–370. [CrossRef] [PubMed]
26. Deng, Y.Y.; Li, J.Q.; Wu, S.F.; Zhu, Y.P.; Chen, Y.W.; He, F.C. Integrated nr Database in Protein Annotation System and Its Localization. *Comput. Eng.* **2006**, *32*, 71–72.
27. Tatusov, R.L.; Galperin, M.Y.; Natale, D.A. The COG database: A tool for genome scale analysis of protein functions and evolution. *Nucleic Acids Res.* **2000**, *28*, 33–36. [CrossRef]
28. Ashburner, M.; Ball, C.A.; Blake, J.A.; Botstein, D.; Butler, H.L.; Cherry, J.M.; Davis, A.P.; Dolinski, K.; Dwight, S.S.; Eppig, J.T.; et al. Gene ontology: Tool for the unification of biology. *Nat. Genet.* **2000**, *25*, 25–29. [CrossRef]
29. Huerta-Cepas, J.; Szklarczyk, D.; Forslund, K.; Cook, H.; Heller, D.; Walter, M.C.; Rattei, T.; Mende, D.R.; Sunagawa, S.; Kuhn, M.; et al. eggNOG 4.5: A hierarchical orthology framework with improved functional annotations for eukaryotic, prokaryotic and viral sequences. *Nucleic Acids Res.* **2016**, *44*, 286–293. [CrossRef]
30. Kanehisa, M.; Goto, S.; Kawashima, S.; Okuno, Y.; Hattori, M. The KEGG resource for deciphering the genome. *Nucleic Acids Res.* **2004**, *32*, 277–280. [CrossRef]

31. Kanehisa, M.; Goto, S.; Hattori, M.; Aokikinoshita, K.F.; Itoh, M.; Kawashima, S.; Katayama, T.; Araki, M.; Hirakawa, M. From genomics to chemical genomics: New developments in KEGG. *Nucleic Acids Res.* **2006**, *34*, 354–357. [CrossRef] [PubMed]
32. Langmead, B.; Trapnell, C.; Pop, M.; Salzberg, S. Ultrafast and memory-efficient alignment of short DNA sequences to the human genome. *Genome Biol.* **2009**, *10*, R25. [CrossRef] [PubMed]
33. Li, B.; Colin, N.D. RSEM: Accurate transcript quantification from RNA Seq data with or without a reference genome. *BMC Bioinform.* **2011**, *4*, 323. [CrossRef]
34. Love, M.I.; Huber, W.; Anders, S. Moderated estimation of fold change and dispersion for RNA-seq data with DESeq2. *Genome Biol.* **2014**, *15*, 550. [CrossRef]
35. Li, X.; Wang, F.; Liu, Q.; Li, Q.P.; Qian, Z.M.; Zhang, X.L.; Li, K.; Li, W.J.; Dong, C.H. Developmental transcriptomics of Chinese cordyceps reveals gene regulatory network and expression pro les of sexual development-related genes. *BMC Genom.* **2019**, *20*, 337.
36. Müller, D.; Leyser, O. Auxin, cytokinin and the control of shoot branching. *Ann. Bot.* **2011**, *107*, 1203–1212. [CrossRef]
37. Dhale, D.A.; Ankita, P.; Rathod, M.C. Production of IAA by endophytic fungi as *Phyllanthus emblica* Linn. *Curr. Agric. Res. J.* **2023**, *11*, 881–889.
38. Cui, X.; Kong, W.L. Research progress of auxin IAA on the growth and development of edible fungi. *China Cucurbits Veg.* **2024**, *37*, 8–14.
39. Jiang, X.L.; Huang, Q.X.; Li, H.; Zhao, J.P. A review of advances in plant thaumatin-like proteins. *J. Zhejiang A&F Univ.* **2012**, *29*, 279–287.
40. Yang, Y.; Hu, X. A chromosome-scale genome of *Trametes versicolor* and transcriptome-based screening for light-induced genes that promote triterpene biosynthesis. *J. Fungi* **2025**, *11*, 81. [CrossRef] [PubMed]
41. Zhang, J.X.; Li, Y.X.; Mao, Y.F.; Zhang, Y.S.; Zhou, B.T.; Liu, W.; Wang, W.; Zhang, C. Integrated transcriptomic and proteomic analyses reveal molecular mechanism of response to heat shock in *Morchella sextelata*. *J. Fungi* **2025**, *11*, 76. [CrossRef]
42. Wen, X.F. Molecular Mechanism of Post-Harvest Morphological Development of *Dictyophora indusiate*. Master's Thesis, Chengdu University, Chendu, China, 2022.
43. Fu, Y.P.; Dai, Y.T.; Yang, C.T.; Wei, P.; Song, B.; Yang, Y.; Sun, L.; Zhang, Z.W.; Li, Y. Comparative transcriptome analysis identified candidate genes related to Bailinggu mushroom formation and genetic markers for genetic analyses and breeding. *Sci. Rep.* **2017**, *7*, 9266. [CrossRef]
44. Tang, L.H.; Jian, H.H.; Song, C.Y.; Bao, D.P.; Shang, X.D.; Wu, D.Q.; Tan, Q.; Zhang, X.H. Transcriptome analysis of candidate genes and signaling pathways associated with light-induced brown film formation in *Lentinula edodes*. *Appl. Microbiol. Biotechnol* **2013**, *97*, 4977–4989. [CrossRef]
45. Wang, J.Q.; Ye, H.L.; Liang, D.W.; Liu, D.Y.; Geng, F.; Li, X. De novo sequencing and transcriptome analysis of *Morchella importuna* fruiting body. *Food Sci.* **2018**, *39*, 81–87.
46. Wang, L.N. Characterization and Function Analysis of Catalase Genes in *Pleurotus ostreatus*. Ph.D. Thesis, Chinese Academy of Agricultural Sciences, Beijing, China, 2019.
47. Zhang, J.J.; Ren, A.; Chen, H.; Zhao, M.W.; Shi, L.; Chen, M.G.; Wang, H.; Feng, Z.Y. Transcriptome analysis and its application in identifying genes associated with fruiting body development in basidiomycete *Hypsizygus marmoreus*. *PLoS ONE* **2015**, *10*, e0123025. [CrossRef] [PubMed]
48. Zhou, Y.; Chen, L.F.; Fan, X.Z.; Bian, Y.B. De novo assembly of *Auricularia polytricha* transcriptome using Illumina sequencing for gene discovery and SSR marker identification. *PLoS ONE* **2014**, *9*, e91740. [CrossRef] [PubMed]
49. Ma, X.X.; Lu, L.S.; Zhang, Y.M.; Fang, M.; Shao, K.S.; Sun, X.; Yao, F.J. Velvet family members regulate pigment synthesis of the fruiting bodies of *Auricularia cornea*. *J. Fungi* **2023**, *9*, 412. [CrossRef]
50. Ma, X.X.; Lu, L.X.; Yao, F.J.; Fang, M.; Wang, P.; Meng, J.J.; Shao, K.S.; Sun, X.; Zhang, Y.M. High-quality genome assembly and multi-omics analysis of pigment synthesis pathway in *Auricularia cornea*. *Front. Microbiol.* **2023**, *14*, 1211795. [CrossRef] [PubMed]
51. Li, X.R.; Luo, L.; Wang, X.Y.; Zhu, M. Further insights into the molecular mechanisms underlying tobacco straw cultivation of *Pleurotus ostreatus* by comparative transcriptome analyses. *Genomics* **2025**, *117*, 110992. [CrossRef]
52. Xu, D.; Liu, Q.; Chen, G.; Yan, Z.; Hu, H. Aldehyde dehydrogenase ALDH3F1 involvement in flowering time regulation through histone acetylation modulation on *FLOWERING LOCUS C*. *J. Integr. Plant Biol.* **2020**, *62*, 1080–1092. [CrossRef]
53. Tola, A.J.; Jaballi, A.; Germain, H.; Missihoun, T.D. Recent development on plant aldehyde dehydrogenase enzymes and their functions in plant development and stress signaling. *Genes* **2020**, *12*, 51. [CrossRef]
54. Li, X.B.; Qin, Y.; Kong, Y.F.; Karunarathna, S.C.; Liang, Y.; Xu, J.Z. Optimization of protoplast preparation conditions in *Lyophyllum decastes* and transcriptomic aunctional analysis of hydrophobin genes in sexual development of Botrytis cinerea. *J. Fungi* **2024**, *10*, 886. [CrossRef] [PubMed]
55. Terhem, R.B.; van Kan, J.A. Functional analysis of hydrophobin genes in sexual development of *Botrytis cinerea*. *Fungal Genet. Biol.* **2014**, *71*, 42–51. [CrossRef]

6. Li, X.; Liu, M.; Dong, C. Hydrophobin Gene Cmhyd4 negatively regulates fruiting body development in edible fungi *Cordyceps militaris*. *Int. J. Mol. Sci.* **2023**, *24*, 13. [CrossRef]
7. Arnold, P.K.; Finley, L.W.S. Regulation and function of the mammalian tricarboxylic acid cycle. *J. Biol. Chem.* **2023**, *299*, 102838. [CrossRef]
8. Fernie, A.R.; Carrari, F.; Sweetlove, L.J. Respiratory metabolism: Glycolysis, the TCA cycle and mitochondrial electron transport. *Curr. Opin. Plant Biol.* **2004**, *7*, 254–261. [CrossRef]
9. Qian, L.; Zhang, Z.J.; Chen, X.M.; Li, S.F.; Luo, Y.; Liu, J.H. Polysaccharides content and activity of enzymes related to sugar metabolism from the fruiting bodies of *Ganoderma lucidum* at different growth stages. *Storage Process* **2019**, *19*, 128–131.
0. Liu, P.H.; Xie, B.G.; Deng, Y.J.; Jiang, Y.J. Cloning, structural analyses and expression levels of phosphofructokinase gene in different strains of *Volvariella volvacea*. *Mycosystema* **2013**, *32*, 253–260.
1. Xu, C.T.; Li, Z.H.; Pan, J.L.; Li, H.K.; Hu, Q.X.; Zou, Y.J.; Chen, X.H.; Mi, C.X. Cloning and expression analysis of Trehalose-6-Phosphate Synthase from *Sanghuangporus vaninii*. *Acta Edulis Fungi* **2024**, *31*, 45–52.
2. Bhadoriya, P.; Jain, M.; Kaicker, G.; Saidullah, B.; Saran, S. Deletion of Htt cause alterations in cAMP signaling and spatial patterning in *Dictyostelium discoideum*. *J. Cell. Physiol.* **2019**, *234*, 18858–18871. [CrossRef]
3. Bai, S.; Zhou, Z.C.; Mariga, A.M.; Shang, X.L.; Ma, N.; Fang, D.L.; Yang, W.J.; Hu, Q.H.; Gao, H.Y.; Chen, H.J.; et al. CA-g-CS/PLA film packaging improved storage stability through the MAPK signaling pathway of postharvest *Agaricus bisporus*. *Postharvest Biol. Technol.* **2024**, *209*, 112688. [CrossRef]
4. Agger, S.; Gallego, F.L.; Dannert, C.S. Diversity of sesquiterpene synthases in the basidiomycete *Coprinus cinereus*. *Mol. Microbiol.* **2009**, *72*, 1181–1195. [CrossRef]
5. Bezmenova, A.V.; Zvyagina, E.A.; Fedotova, A.V.; Kasianov, A.S.; Neretina, T.V.; Penin, A.A.; Bazykin, G.A.; Kondrashov, A.S. Rapid accumulation of mutations in growing mycelia of a hypervariable fungus *Schizophyllum commune*. *Mol. Biol. Evol.* **2020**, *37*, 2279–2286. [CrossRef]
6. Masunaga, N.; Kitaoka, T.; Ichinose, H. Biocatalyst collection and heterologous expression of sesquiterpene synthases from basidiomycetous fungi: Discovery of a novel. *Microb. Biotechnol.* **2023**, *16*, 632–644. [CrossRef]
7. Abdelshafy, A.M.; Belwal, T.; Liang, Z.; Wang, L.; Li, D.; Luo, Z.S.; Li, L. A comprehensive review on phenolic compounds from edible mushrooms: Occurrence, biological activity, application and future prospective. *Crit. Rev. Food Sci. Nutr.* **2022**, *62*, 6204–6224. [CrossRef]
8. Hu, C.C.; Yu, P.J.; Li, S.J. Regulatory mechanisms of ergosterol biosynthesis in fungi. *J. Fungal Res.* **2019**, *17*, 138–146.
9. Tartar, A.; Shapiro, A.M.; Scharf, D.W.; Boucias, D.G. Differential expression of chitin synthase (CHS) and glucan synthase (FKS) genes correlates with the formation of a modified, thinner cell wall in in vivo-produced beauveria bassiana cells. *Mycopathologia* **2005**, *160*, 303–314. [CrossRef]
0. Janusz, G.; Kucharzyk, K.H.; Pawlik, A.; Staszczak, M.; Paszczynski, A.J. Fungal laccase, manganese peroxidase and lignin peroxidase: Gene expression and regulation. *Enzym. Microb. Technol.* **2013**, *52*, 1–12. [CrossRef] [PubMed]
1. Li, W.P.; Zhang, Z.; Wu, R.Y.; Mao, M.Y.; Ji, Y.K.; Wang, X.N.; Dou, S.G.; Yan, M.; Chen, W.T. Fusobacterium nucleatum-derived outer membrane vesicles promote immunotherapy resistance via changes in tryptophan metabolism in tumour-associated Macrophages. *J. Extracell. Vesicles* **2025**, *14*, e70070. [CrossRef] [PubMed]
2. Wang, S.; Schooten, F.J.V.; Jin, H.; Jonkers, D.; Godschalk, R. The involvement of intestinal tryptophan metabolism in inflammatory bowel disease identified by a meta-analysis of the transcriptome and a systematic review of the metabolome. *Nutrients* **2023**, *15*, 19. [CrossRef]
3. Nutaratat, P.; Srisuk, N.; Arunrattiyakorn, P.; Limtong, S. Indole-3-acetic acid biosynthetic pathways in the basidiomycetous yeast *Rhodosporidium paludigenum*. *Arch. Microbiol.* **2016**, *198*, 429–437. [CrossRef]
4. Sun, S.L.; Yang, W.L.; Fang, W.W.; Zhao, Y.X.; Guo, L.; Dai, Y.J. The plant growth-promoting rhizobacterium variovorax boronicumulans CGMCC 4969 regulates the level of Indole-3-acetic acid synthesized from Indole-3-acetonitrile. *Appl. Environ. Microbiol.* **2018**, *84*, e00298-18. [CrossRef] [PubMed]
5. Yan, D.; Gao, Q.; Rong, C.B.; Liu, Y.; Song, S.; Yu, Q.Y.; Zhou, K.X.; Liao, Y.L. Comparative transcriptome analysis of abnormal cap and healthy fruiting bodies of the edible mushroom *Lentinula edodes*. *Fungal Genet. Biol.* **2021**, *156*, 103614. [CrossRef]
6. Duan, M.Z.; Long, S.F.; Wu, X.J.; Feng, B.; Qin, S.Q.; Li, Y.J.; Li, X.; Li, C.N.; Zhao, C.G.; Wang, L.Q.; et al. Genome, transcriptome, and metabolome analyses provide new insights into the resource development in an edible fungus *Dictyophora indusiate*. *Front. Microbiol.* **2023**, *14*, 1137159. [CrossRef]
7. Skibbe, D.S.; Liu, F.; Wen, T.J.; Yandeau, M.D.; Cui, X.Q.; Cao, J.; Simmons, C.R.; Schnable, P.S. Characterization of the aldehyde dehydrogenase gene families of *Zea mays* and *Arabidopsis*. *Plant Mol. Biol.* **2002**, *48*, 751–764. [CrossRef]
8. Singh, S.; Brocker, C.; Koppaka, V.; Chen, Y.; Jackson, B.C.; Matsumoto, A.; Thompson, D.C.; Vasiliou, V. Aldehyde dehydrogenases in cellular responses to oxidative/electrophilic stress. *Free. Radic. Biol. Med.* **2012**, *56*, 89–101. [CrossRef]
9. Jozefczuk, S.; Klie, S.; Catchpole, G.; Szymanski, J.; CuadrosInostroza, A.; Steinhauser, D.; Selbig, J.; Willmitzer, L. Metabolomic and transcriptomic stress response of *Escherichia coli*. *Mol. Syst. Biol.* **2010**, *6*, 364. [CrossRef] [PubMed]

80. Jiang, S.; Wang, H.; He, Y.H.; Chen, Y.F.; Deng, Q.B.; Guo, X.W.; Xiao, D.G. Effect of *Saccharomyces cerevisiae* overexpressing acetaldehyde dehydrogenase genes on reducing higher alcohols in Huangjiu. *China Brew.* **2020**, *39*, 153–159.
81. Asiimwe, T.; Krause, K.; Schlunk, I.; Kothe, E. Modulation of ethanol stress tolerance by aldehyde dehydrogenase in the mycorrhizal fungus *Tricholoma vaccinum*. *Mycorrhiza* **2012**, *22*, 471–484. [CrossRef]
82. Qi, Y.C.; Liu, Y.D.; Sun, X.K.; Zhang, M.L.; Zhang, Q.; Wen, Q.; Qiu, L.Y.; Shen, J.W. Cloning and expression analysis of *Pleurotus ostreatus* aldehyde dehydrogenase gene PoALDH1 under abiotic stresses. *Mycosystema* **2017**, *36*, 1121–1131.
83. Song, X.L. Characterizations of Mycelium Polysaccharides from *Pleurotus geesteranus* and the Mechanisms on Improving the Alcoholic Liver Disease in Mice. Ph.D. Thesis, Shandong Agricultural University, Tai'an, China, 2022.

Journal of
Fungi

MDPI

rticle

3iodegradation of Chloroquine by a Fungus from Amazonian 5oil, *Penicillium guaibinense* CBMAI 2758

'atrícia de Almeida Nóbrega [1,2], Samuel Q. Lopes [1], Lucas S. Sá [1], Ryan da Silva Ramos [2], Fabrício H. e Holanda [3], nana F. de Araújo [1], André Luiz M. Porto [4], Willian G. Birolli [5] and Irlon M. Ferreira [1,*]

1 Biocatalysis and Applied Organic Synthesis Laboratory, Federal University of Amapá, Macapá 68903-419, AP, Brazil; doutoradobiotecnologia21@gmail.com (P.d.A.N.); samuql12@gmail.com (S.Q.L.); lucassa720@gmail.com (L.S.S.); inanafauro24@gmail.com (I.F.d.A.)

2 Post-Graduate Program in Biotechnology and Biodiversity—BIONORTE Network, Federal University of Amapá, Macapá 68902-280, AP, Brazil; ryan.ramos@ueap.edu.br

3 Federal Institute of Science and Technology Education of Amapá, Rua Nilo Perçanha, Bairro Cajari, Laranjal do Jari 68920-000, AP, Brazil; holanda_fabricio@yahoo.com.br

4 Laboratory of Organic Chemistry and Biocatalysis, São Carlos Institute of Chemistry, University of São Paulo, Ed. Prof. Douglas Wagner Franco, Av. João Dagnone 1100, Santa Angelina, São Carlos 13563-120, SP, Brazil; almporto@iqsc.usp.br

5 Molecular Oncology Research Center, Educational and Research Institute, Barretos Cancer Hospital, Av. Ébano 165-1, Dr. Paulo Prata, Barretos 14784-384, SP, Brazil; willianbirolli@gmail.com

* Correspondence: irlon.ferreira@gmail.com

Abstract: Concern over the presence of pharmaceutical waste in the environment has prompted research into the management of emerging organic micropollutants (EOMs). In response, sustainable technologies have been applied as alternatives to reduce the effects of these contaminants. This study investigated the capacity of filamentous fungi isolated from iron mine soil in the Amazon region to biodegrade the drug chloroquine diphosphate. An initial screening assessed the growth of four fungal strains on solid media containing chloroquine diphosphate: *Trichoderma pseudoasperelloides* CBMAI 2752, *Penicillium rolfsii* CBMAI 2753, *Talaromyces verruculosus* CBMAI 2754, and *Penicillium* sp. cf. *guaibinense* CBMAI 2758. Among them, *Penicillium* sp. cf. *guaibinense* CBMAI 2758 was selected for further testing in liquid media. A Box–Behnken factorial design was applied with three variables, pH (5, 7, and 9), incubation time (5, 10, and 15 days), and chloroquine diphosphate concentration (50, 75, and 100 mg·L^{-1}), totaling 15 experiments. The samples were analyzed by gas chromatography–mass spectrometry (GC-MS). The most effective conditions for chloroquine biodegradation were pH 7, 100 mg·L^{-1} concentration, and 10 days of incubation. Four metabolites were identified: one resulting from *N*-deethylation M1 (*N*4-(7-chloroquinolin-4-yl)-*N*1-ethylpentane-1,4-diamine), two from carbon–carbon bond cleavage M2 (7-chloro-*N*-ethylquinolin-4-amine) and M3 (*N*1,*N*1-diethylpentane-1,4-diamine), and one from aromatic deamination M4 (*N*1-ethylbutane-1,4-diamine) by enzymatic reactions. The toxicity analysis showed that the products obtained from the biodegradation of chloroquine were less toxic than the commercial formulation of this compound. These findings highlight the biotechnological potential of Amazonian fungi for drug biodegradation and decontamination.

Keywords: Amazonian fungi; biodegradation; pharmaceutical compounds; environmental pollution; COVID-19

1. Introduction

The COVID-19 pandemic, caused by the SARS-CoV-2 virus, triggered a global public health crisis that demanded rapid and innovative responses. In this context, repositioning of drugs, also known as repurposing, has gained prominence as a strategy for accelerating the development of treatments, given that the safety profiles for these drugs have already been established. As a result, repositioning became an essential tool in the search for effective therapies against COVID-19 [1].

During the pandemic, numerous drugs used to treat other diseases were tested as potential therapies for COVID-19 [2], in some cases without robust scientific evidence. Among the most widely used were hydroxychloroquine [3], chloroquine [4], ivermectin [5], azithromycin [6], favipiravir [7], remdesivir [8], lopinavir–ritonavir, ribavirin, and dexamethasone [9] (Figure 1).

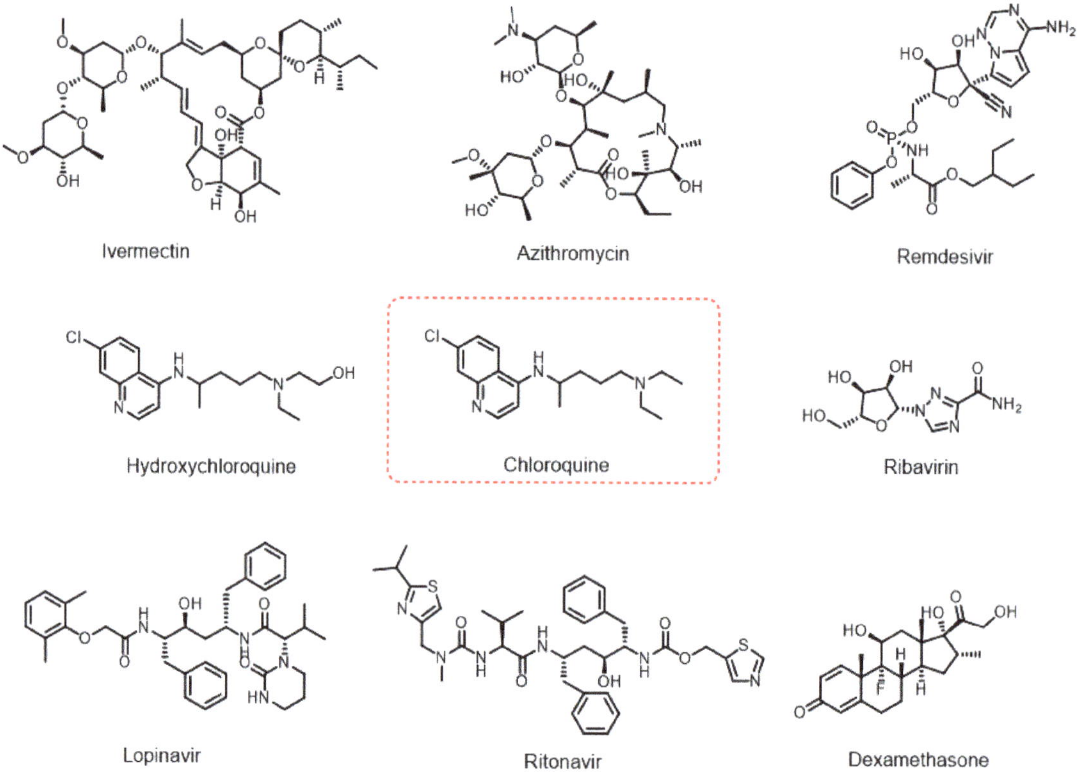

Figure 1. Chemical structures of selected drugs investigated as potential treatments for COVID-19 during the pandemic.

Increased consumption and incorrect disposal of these drugs during the COVID-19 pandemic generated worrying impacts on public health and the environment, with an increased risk of environmental contamination [10,11]. This risk mainly arises from the potential adverse effects that pharmaceutical waste can have on humans and on aquatic and terrestrial ecosystems [11–13]. Much of this waste enters the environment through domestic, industrial, and hospital effluents, as most drugs are not fully metabolized in the human body and are excreted in urine or feces, either unchanged or as active metabolites [14,15].

A study reported significant increases in the concentrations of drugs, including those used against COVID-19, in aquatic environments during the pandemic in comparison with pre-pandemic levels [16]. This highlights how the intensified use of pharmaceuticals during the health crisis exacerbated the issue of environmental pollution by these compounds as micropollutants. In particular, drugs such as chloroquine present a concern in wastewater treatment plants and natural water bodies, as they may not be fully removed during these processes and can enter the environment due to their moderate persistence [10]. Furthermore, certain pharmaceutical compounds can become highly toxic when improperly disposed, posing significant risks to human health and ecosystems, even at low concentrations [17,18].

The removal of pharmaceutical compounds from the environment using conventional chemical and physical treatments is often inefficient, unsustainable, or economically unfeasible. In recent years, a number of sustainable technologies have been developed to manage pharmaceutical organic micropollutant (OMP) waste. These methods include the use of oxidation–reduction techniques, membranes [19], adsorption [20], photodegradation [21], and biodegradation [22–24].

Although still limited in number, some studies have demonstrated the degradation of chloroquine by physical and chemical methods. For example, Midasi et al. [25] compared oxidation methods and observed that the electro–Fenton–BDD process was the most efficient, achieving complete elimination of chloroquine and 92% removal of total organic carbon (TOC). Dai et al. [10] evaluated the degradation of chloroquine phosphate (CQP) and found that the UV/H_2O_2 process outperformed the isolated use of UV or H_2O_2. Similarly, Han Qi et al. [26] tested ozonation for the degradation of low concentrations of chloroquine phosphate (0.1 mg/L), and under ideal conditions (pH 7, 30 °C, and 0.192 mg/L ozone), complete removal was achieved within 40 min.

Biodegradation is the process by which organic substances are broken down by microorganisms into simpler natural compounds (e.g., CO_2, H_2O, and NH_3) that can be integrated into natural biogeochemical cycles. The biodegradation of pharmaceuticals by microorganisms such as fungi and bacteria has emerged as an effective, environmentally sustainable, and cost-efficient alternative for the removal of organic micropollutants [24,27,28].

Several studies have documented the use of microorganisms in the process of biodegradation and biotransformation of pharmaceuticals. For example, a recent study demonstrated the effectiveness of *Aspergillus* sp., isolated from activated sludge, in degrading chlortetracycline, achieving a biodegradation rate of 95% after 72 h under controlled conditions [29–31]. Likewise, a strain of the filamentous fungus *Cunninghamella echinulata* var. *elegans* (ATCC 8688a) grown in a medium containing hydroxychloroquine biotransformed this compound after 96 h of incubation, producing various metabolites. The main metabolite identified was 4-(1,2,3,4-tetrahydroquinoline-4-yl amino)pentan-1-ol [32].

Among the promising fungi in biotechnological processes, those found in the Amazon Rainforest, though still little studied, stand out for their potential [33]. The forest's ecological richness is sustained by its microbial diversity, which plays vital roles in recycling organic matter and supporting both plant and animal life. In particular, metabolically active soil fungi produce valuable compounds that can be harnessed for processes such as biodegradation [34] and biotransformation [24].

In this context, and with the aim of advancing the knowledge of the enzymatic potential of fungi for the biodegradation of chloroquine, this study investigated the ability of filamentous fungi isolated from Brazilian Amazonian soil to degrade this pharmaceutical product.

2. Results and Discussion

2.1. Identification of Filamentous Fungi

The fungi used in the biodegradation process were previously isolated and identified [35]. Species identification was performed using macromorphological, micromorphological, and molecular analyses, resulting in the identification of *Trichoderma pseudoasperelloides* CBMAI 2752, *Penicillium rolfsii* CBMAI 2753, *Talaromyces verruculosus* CBMAI 2754, and *Penicillium* sp. cf. *guaibinense* CBMAI 2758. Figure S1 (Supplementary Materials) shows the fungal isolates grown on Sabouraud dextrose agar and 2% malt extract agar (pH 5) in Petri dishes, incubated at approximately 28 °C for 7 days. All fungi were identified at the Multidisciplinary Center for Chemical, Biological, and Agricultural Research (CPQBA) from Unicamp (https://site.cpqba.unicamp.br/).

The soil of the Amazon region stands out due to its remarkable microbiological diversity, hosting a wide variety of microorganisms with biotechnological potential [36]. Numerous studies have explored this rich microbial reservoir, focusing on the characterization and isolation of species, especially filamentous fungi, that play key roles in ecological processes and offer promising applications in biodegradation, biocontrol, and the production of industrially relevant metabolites.

Radial growth assays were conducted to preliminarily evaluate the effects on filamentous fungal strains isolated from iron mine soil. Chloroquine diphosphate at the concentrations of 25, 50, and 100 $mg{\cdot}L^{-1}$ was added to the strains *T. pseudoasperelloides* CBMAI 2752, *P. rolfsii* CBMAI 2753, *T. verruculosus* CBMAI 2754, and *P.* sp cf. *guaibinense* CBMAI 2758 on solid culture media (2% Sabouraud dextrose agar and 2% malt). Colony growth was measured every 24 h over a seven-day period. Plates of solid media without chloroquine diphosphate were used as fungal control (biotic). The experiments were carried out in triplicate [24].

The results for mycelial growth, based on the average diameter (cm) in the Petri dish, are shown in Figure 2. Radial growth rates were determined by applying linear regression to the data for all tested species. Three strains (*T. verruculosus* CBMAI 2754, *P. rolfsii* CBMAI 2753, and *P.* sp cf. *guaibinense* CBMAI 2758) showed adaptive growth behavior in the presence of chloroquine diphosphate. Among these, *P.* sp cf. *guaibinense* CBMAI 2758 was particularly notable, displaying enhanced growth when cultivated in medium enriched with this substance at 25 $mg{\cdot}L^{-1}$ and 50 $mg{\cdot}L^{-1}$, as evidenced by larger colony diameters in comparison with the control group.

The radial growth velocity or radial growth rate (RGR) corresponds to the slope of the line obtained from linear regression of the colony radius over time. A greater slope indicates a higher RGR and, consequently, greater fungal growth potential [37]. Based on the RGR results, *Talaromyces verruculosus* CBMAI 2754, *Penicillium* cf. *guaibinense* CBMAI 2758, and *Penicillium rolfsii* CBMAI 2753 exhibited higher growth rates in medium containing 25 $mg{\cdot}L^{-1}$ chloroquine diphosphate compared to the control medium (absence of chloroquine diphosphate).

In contrast, *T. pseudoasperelloides* CBMAI 2752 showed an RGR of 2.22 cm/d for the medium enriched with chloroquine and 3.10 cm/d for the medium in the absence of chloroquine. Therefore, the presence of chloroquine in the medium led to a reduction in the radial growth rate at all tested concentrations. In contrast, *Penicillium* sp. cf. *guaibinense* CBMAI 2758 appeared capable of utilizing chloroquine as a carbon source to support its mycelial growth. These findings suggest that this microorganism holds significant potential as a biotechnological agent for the biodegradation of chloroquine. Detailed growth data are provided in Table S1 (Supplementary Material).

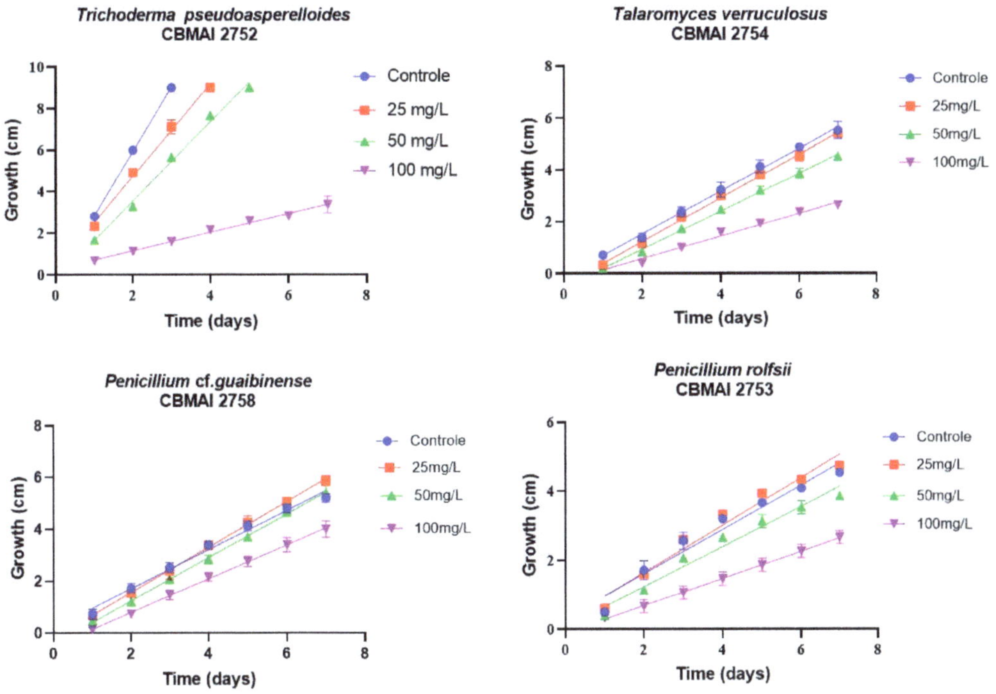

Figure 2. Radial growth of fungal strains isolated from iron mine soil cultured on solid media (Sabouraud dextrose agar and 2% malt, pH 5, at ± 28 °C for 7 days).

Pearson's correlation analysis comparing the growth rates of the fungal strains with varying concentrations of chloroquine diphosphate yielded a correlation coefficient of −0.520. This moderate negative correlation indicates that as the concentration of chloroquine diphosphate increases, fungal growth tends to decrease, demonstrating the inhibitory effect of the drug on fungal development. In the experiments with *T. verruculosus* CBMAI 2754 and *P. rolfsii* CBMAI 2753, the diameter of the strains treated with 25 $mg{\cdot}L^{-1}$ did not differ significantly from those of the control. This suggests that chloroquine diphosphate does not interfere with the growth of these strains at this concentration.

However, it is important to emphasize that further studies are needed to assess the effects of longer exposure periods. *T. verruculosus* CBMAI 2754, *P. rolfsii* CBMAI 2753, and *T. pseudoasperelloides* CBMAI 2752 showed differences in colony coloration when grown in the presence of chloroquine diphosphate, as shown in Table 1. Notably, *T. pseudoasperelloides* CBMAI 2752 showed significant inhibition of growth with 100 $mg{\cdot}L^{-1}$ chloroquine diphosphate in comparison with the control experiment (Table 1). This inhibition may have been the result of a harmful effect caused by the drug itself or from metabolites produced during its biodegradation.

2.2. Evaluation of the Biodegradation of Chloroquine Diphosphate Through Experimental Design

The biodegradation process was evaluated based on the percentage reduction of chloroquine diphosphate, as determined by the experimental design using *P.* cf. *guaibinense* CBMAI 2758 as biodegradation agent. The variables' values are summarized in Table 2. A Pareto diagram of standardized effects was generated for the tests with this microorganism (Figure 3) and, although the diagram presents all tested effects, only pH was statistically significant, as indicated by the reference significance line.

In the Pareto diagram, the length of each bar represents the magnitude of the estimated effect for each parameter, while the vertical line indicates the threshold for statistical significance at the 5% level. Within the experimental range tested, the variables that exceeded this threshold were considered statistically significant at a 95% confidence level ($p < 0.05$), as shown in Figure 3 [24].

Table 1. Radial growth of *T. pseudoasperelloides* CBMAI 2752, *P. rolfsii* CBMAI 2753, *T. verruculosus* CBMAI 2754, and *P.* sp cf. *guaibinense* CBMAI 2758 in the presence of chloroquine diphosphate.

Fungus	**Growth Medium**	**Time (Days)**		**Growth Speed (cm/d)**
		3	**6**	
Trichoderma pseudoasperelloides CBMAI 2752	Control			3.10
	Chloroquine 25 mg·L^{-1}			2.22
	Chloroquine 50 mg·L^{-1}			1.9
	Chloroquine 100 mg·L^{-1}			0.44
Penicillium rolfsii CBMAI 2753	Control		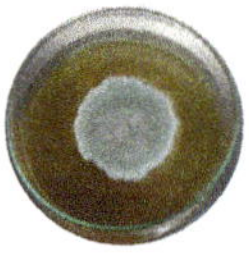	0.82
	Chloroquine 25 mg·L^{-1}			0.84
	Chloroquine 50 mg·L^{-1}			0.733
	Chloroquine 100 mg·L^{-1}			0.435

Table 1. *Cont.*

Fungus	Growth Medium	Time (Days) 3	6	Growth Speed (cm/d)
Talaromyces verruculosus CBMAI 2754	Control			0.75
	Chloroquine 25 mg·L^{-1}			0.87
	Chloroquine 50 mg·L^{-1}			0.83
	Chloroquine 100 mg·L^{-1}			0.65
Penicillium cf. *guaibinense* CBMAI 2758	Control			0.64
	Chloroquine 25 mg·L^{-1}			0.68
	Chloroquine 50 mg·L^{-1}			0.58
	Chloroquine 100 mg·L^{-1}			0.39

Table 2. The three independent variables used in the Box–Behnken factorial design.

Factors	Variables	Levels		
		−1	0	+1
X1	Time (days)	5	10	15
X2	pH	5	7	9
X3	Chloroquine diphosphate concentration (mg·L^{-1})	50	75	100

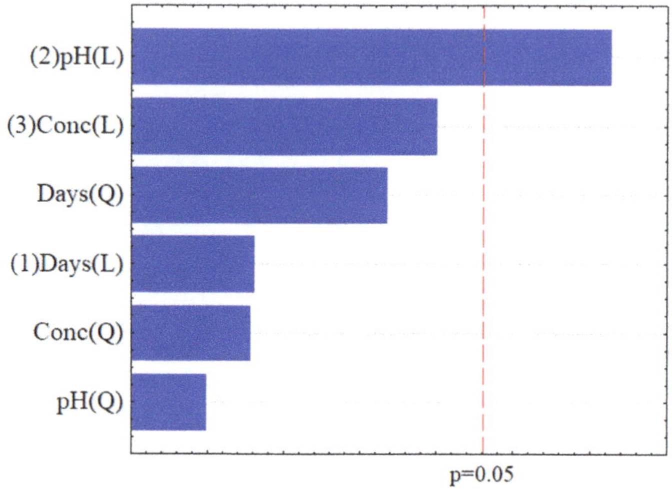

Figure 3. Pareto diagram of the linear (L) and quadratic (Q) effects and order of factors (1, time; 2, pH; 3, and concentration) on the biodegradation of chloroquine (%) by the fungal strain *Penicillium* cf. *guaibinense* CBMAI 2758.

According to Figure 3, pH was the only variable with a statistically significant effect on the biodegradation of chloroquine diphosphate by *P.* cf *guaibinense* CBMAI 2758. Notably, pH emerged as an important factor, with neutral conditions (pH 7) yielding the highest degradation percentages.

Fungi have different nutritional and environmental requirements, which include their preference for an optimal pH range for growth. This pH preference can vary among species and depend on the specific environmental conditions in which the fungi develop. However, most fungi thrive in slightly acidic environments, typically around pH 5 to 6. This range is favorable to many filamentous fungi and yeasts, enabling adequate growth and development [38,39].

According to Rigo et al. [40], variables such as pH, temperature, humidity, and nutrient availability play a decisive role in biotechnological processes, affecting microbial growth and reaction efficiency. The author also emphasizes that factors such as medium composition, incubation time, and substrate concentration are key determinants for optimizing these processes.

Figure 4 presents the projection of three response surface graphs used to optimize the biodegradation conditions. The quadratic model demonstrated a significant correlation between the response and the factors time, pH, and concentration. These graphs allow an analysis of the interactions between pairs of variables, providing a detailed evaluation of their combined effects on the biodegradation efficiency [41].

The biodegradation of chloroquine diphosphate by the strain *P.* sp. cf. *guaibinense* CBMAI 2758 was influenced by the following variables: time and pH of the reaction medium (Figure 4A), the interaction between time and concentration (Figure 4B), and the interaction between pH and concentration (Figure 4C). Among these variables, the analysis of the incubation time (evaluated at 5, 10, and 15 days) showed that 10 days had the most significant positive effect on biodegradation. Thus, extending the incubation period beyond 10 days did not lead to further meaningful increases in the degradation rate, indicating that longer durations offered no added benefit. Furthermore, an analysis of the interaction between time and pH revealed that decreasing the pH below 7 significantly reduced the biodegradation rate, whereas increasing it above 7 did not increase the biodegradation rate significantly. Therefore, the response surface graph confirmed that the optimal conditions, considering the combined effects of these variables, were 10 days of incubation at pH 7.

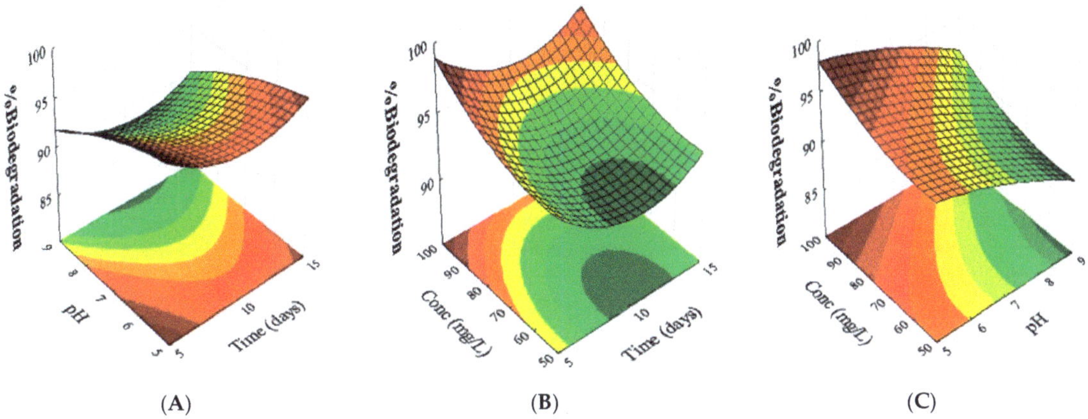

Figure 4. Response surface plots for the biodegradation of chloroquine diphosphate by the fungus *Penicillium* sp. cf. *guaibinense* CBMAI 2758 as a function of (**A**) the reaction time (X1) and pH of the culture medium (X2); (**B**) reaction time (X1) and chloroquine diphosphate concentration (X3); and (**C**) pH of the culture medium (X2) and chloroquine diphosphate concentration (X3).

A comparison of the variables concentration versus time showed that increasing the concentration of chloroquine diphosphate led to higher biodegradation rates, reaching up to 100%, which indicates that this was the most significant variable alongside time. An increase or decrease in time above or below 10 days did not greatly influence the biodegradation rate. Thus, 10 days remained the optimal time, as observed in graph A, which shows the interaction between pH and time.

In the analysis of pH versus concentration ($mg \cdot L^{-1}$), it was observed again that concentration was the variable that most influenced the biodegradation rate, reaching the optimal performance at 100 $mg \cdot L^{-1}$. Deviations from pH 7, either higher or lower, led to a reduction in biodegradation efficiency, confirming that pH 7 was the optimal condition for the process studied.

It is important to highlight that the solubility of chloroquine, especially in water at an adequate temperature and pH, can facilitate its availability for microbial activity, enhancing the biodegradation process. According to Daneshfar and Vafafard [42], the solubility of chloroquine diphosphate increases with temperature, reaching its maximum between 298.2 and 333.2 K. Thus, environments where chloroquine is more soluble are more favorable for efficient biodegradation, reinforcing the importance of optimized conditions for the treatment of this compound.

This study demonstrated that the chloroquine diphosphate biodegradation level achieved in most experiments at pH 7 was higher than 90%, as shown in Table 3. Several studies have reported on the use of microorganisms for the biodegradation of micropollutants, offering several advantages in comparison with other physical and chemical methods [43]. The remarkable effectiveness of fungi in bioremediation is largely due to the production of a broad spectrum of intracellular enzymes, with an emphasis on the cytochrome P450 complex. This superfamily of monooxygenases, which uses a heme group as a cofactor, plays an essential role in the degradation of recalcitrant compounds. Through reactions such as deamination, dealkylation, dehalogenation, and hydroxylation, these enzymes promote structural modifications in pollutants, making them more susceptible to degradation and eventual mineralization [44,45].

Table 3. Experimental design matrix and biodegradation responses of *Penicillium* cf. *guaibinense* CBMAI 2758.

Run	Levels of Uncoded and Coded Variables						Response (%BD)
	X1		X2		X3		
T1	5	−1	5	−1	75	0	>99.0
T2	15	1	5	−1	75	0	98.2
T3	5	−1	9	1	75	0	88.2
T4	15	1	9	1	75	0	91.2
T5	5	−1	7	0	50	−1	97.0
T6	15	1	7	0	50	−1	89.0
T7	5	−1	7	0	100	1	>99.4
T8	15	1	7	0	100	1	98.6
T9	10	0	5	−1	50	−1	95.5
T10	10	0	9	1	50	−1	87.9
T11	10	0	5	−1	100	1	93.9
T12	10	0	9	1	100	1	93.8
T13	10	0	7	0	75	0	92.8
T14	10	0	7	0	75	0	92.5
T15	10	0	7	0	75	0	91.7

In addition to intracellular enzymes, fungi also produce extracellular enzymes that contribute to the breakdown of complex organic molecules. Laccases and peroxidases, in particular, have been associated with the degradation of various xenobiotics [44,46]. However, specific evidence of their activity against chloroquine is not yet available in the literature. Nonetheless, the enzymatic diversity of fungi remains a key factor in their potential application for environmental bioremediation [45].

This metabolic versatility gives fungi a distinct advantage over other organisms in the remediation of contaminated environments, underscoring their biotechnological potential for use in sustainable decontamination processes. Furthermore, the ability of fungi to act under extreme pH and temperature conditions expands their applicability for bioremediation compared with other microorganisms such as bacteria, whose enzyme systems are often more restricted [43,45].

A study conducted to investigate the efficacy of the strain *Trichoderma pubescens* DAOM 166162 for treating amoxicillin-containing wastewater showed that this strain removed more than 98% of the antibiotic within 24 h, with a biodegradation efficiency of 70%. Degradation occurs through hydroxyl groups in extracellular polymeric substances (EPS), which help to disrupt the structure of amoxicillin in alkaline environments, thus highlighting the potential of *Trichoderma* for the bioremediation of pharmaceutical wastewater [47].

A study by Li et al. [48] revealed that the fungal strains *Trichoderma harzianum*-LJ245, *Penicillium oxilacum*-LJ302, and *Penicillium citrinum*-LJ318 were effective at the biodegradation of chlortetracycline (CTC) in crude residue. The degradation rate reached 96% with LJ245, 99% with LJ302, and 98% with LJ318 over a 40-day period, in addition to increasing the pH of CTC crude residue from 2.30 to 8.32. These findings emphasize the potential of these strains as effective treatments for pharmaceutical waste.

Another study investigated the biodegradation of a mixture of nonsteroidal anti-inflammatory drugs (NSAIDs), including celecoxib, diclofenac, and ibuprofen, using the fungal consortium *Ganoderma applanatum* and *Laetiporus sulphureus*. The results showed that this consortium achieved a degradation efficiency exceeding 99%. Enzymatic analysis revealed a significant increase in ligninolytic enzyme production during the degradation process, indicating the importance of these enzymes for the effective removal of NSAIDs from the environment [49].

2.3. Identification of Biodegradation Metabolites by GC-MS

In the process of chloroquine diphosphate biodegradation by filamentous fungus from Amazonian soil, four metabolites were qualitatively identified. After 10 days of biodegradation by *Penicillium* sp. cf. *guaibinense* CBMAI 2758, the following metabolites were identified by GC-MS (mass spectra are presented in Table S2, Supplementary Material): M1, *N*4-(7-chloroquinolin-4-yl)-*N*1-ethylpentane-1,4-diamine, is a product of the action of *N*-deethylation with a molecular ion of m/z 290 generated from the loss of the m/z 29 ion (-CH_2-CH_3). The *N*-deethylation reaction is recognized as a detoxification pathway for environmental pollutants and the drug quetiapine [50]; M2, 7-chloro-*N*-ethylquinolin-4-amine, possibly results a carbon–carbon bond cleavage, producing a metabolite with m/z 205; whereas M3, *N*1,*N*1-diethylpentane-1,4-diamine, and M4, *N*1-ethylbutane-1,4-diamine, result from aromatic deamination (Figure 5).

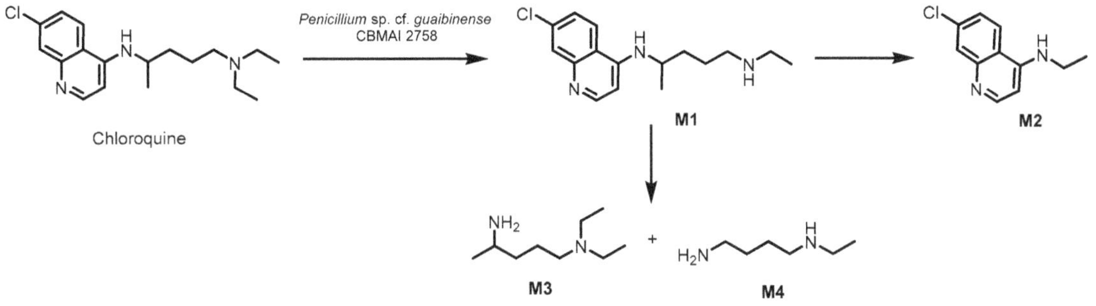

Figure 5. Proposal for the biodegradation of chloroquine by *Penicillium* sp. cf. *guaibinense* CBMAI 2758.

2.4. Toxicological Analysis

Although chloroquine is widely employed for treating malaria and autoimmune diseases, it is associated with significant toxic effects on both human and rodent models, especially with prolonged exposure or at high doses. In humans, the main adverse effects include retinal toxicity, characterized by degeneration of the retinal pigment epithelium, which can lead to irreversible visual loss [51]. In addition, cardiac disorders have been reported, such as prolongation of the QT interval that can evolve into fatal arrhythmias [52]. In murine and rat models, studies have shown that hepatotoxicity, nephrotoxicity, and myocardial damage are associated with the accumulation of the drug in target tissues, as well as the induction of oxidative stress and cell apoptosis [53,54]. These findings emphasize the need for close monitoring during the clinical use of chloroquine and the importance of detailed preclinical evaluations using animal models.

The toxicological evaluation of the products derived from the biodegradation of chloroquine revealed a safety profile significantly superior to that of the parent drug. The results indicated a low risk of toxicity in relation to the carcinogenic potential, with TD_{50} values (median tumorigenic dose) $\geq$ 0.798 mg.kg^{-1} for rat models and $\geq$16.00 mg.kg^{-1} for mice. In addition, the maximum tolerated doses (MTD) in rats for the biodegradation products were higher than that of chloroquine, for which the value recorded was 0.016 g.kg^{-1}, as shown in Table 4. These data suggest that the metabolites formed during the degradation process are less toxic, thus helping to reduce risks to human health and minimize the environmental impacts associated with the disposal of this pharmaceutical.

Table 4. Computational toxicity parameters in animal models, with predictions of toxicity risks and tolerated doses of the potential carcinogen TopHit8.

	Computational Toxicological Parameters						Prediction of Toxicity Risks					Tolerated Doses of Potential Carcinogen		
							Pimephales promelas	*Daphnia magna*	Rat			Rat	Mouse	Rat
Molecules	Female/Male		Ames	Skin		Ocular			Oral	Chronic	Inhalation	(Body Weight/Day)		
	Mouse	Rat	Mutagenicity	Irritancy	Sensitization	Irritation	LC_{50} (g/L)	EC_{50} (mg/L)	LD_{50} (g/kg)	LOAEL (g/kg)	LC_{50} ($mg/m^3/h$)	TD_{50} (mg/kg)		RMTD (g/kg)
Chloroquine	C−	C−	M+	−	−	++++	0.009	11.40	1.068	0.062	24,400.2	0.376	12.73	0.016
M1	C−	C−	M−	−	−	+++	0.001	0.22	1.465	0.092	1511.87	3.394	28.74	0.093
M2	C	C−	M−	++	−	+++	2.56×10^{-5}	0.32	0.853	0.007	2230.09	94.806	16.00	0.103
M3	C−	C−	M−	++	−	−	0.001	0.72	0.138	0.065	8395.07	3.020	27.90	0.392
M4	C−	C−	M−	−	+	+++	0.001	0.33	6.877	0.070	771.57	0.798	45.80	0.736

C− (non-carcinogen); M− (non-mutagen); M+ (mutagen); + (weak); ++ (mild); +++ (moderate); ++++ (severe).

The *Penicillium* genus has been identified as a prolific producer of extracellular oxidative and hydrolytic enzymes, such as laccases, manganese peroxidases, lipases, proteases, esterases, and cellulases, that exhibit significant potential for the biodegradation of persistent organic compounds via molecular docking simulations [55,56]. For example, *Penicillium simplicissimum* secretes laccase (~66 kDa) and manganese peroxidase (~60 kDa), demonstrating effective activity in the degradation of polyethylene, as evidenced in spectrophotometric and FTIR/NMR studies. Tests with *Penicillium chrysogenum* demonstrated the production of laccase, lipase, and esterase under halophilic conditions, facilitating the decolorization of oily effluents [57].

Furthermore, molecular docking studies confirmed the interaction of fungal laccase and peroxidase with mycotoxins using structures with PDB IDs 1HFU and 1MNP, suggesting their potential in the degradation of aromatic-structured drugs [58–60]. Application of these enzymes, especially laccases and peroxidases, in chloroquine docking allows interactions in the active site to be simulated, identifying key residues (His, Asp, Cu, Ser, and Tyr) and anticipating possible mechanisms of enzymatic manipulation [55]. This approach supports the development of bioremediation strategies for pharmaceutical waste, as in silico evidence helps optimize the design of sustainable processes for environmental applications.

The results from the validation of the molecular docking simulation protocols were considered satisfactory, since the relative overlap (pose + orientation + torsion) of the crystallographic ligand (experimental control model) and the docking ligand (docking pose theoretical model) were considered similar. Recovering the pose of the complexes in Figure S2 (Supplementary Material), enabled validation of the docking protocols through the calculation of the root mean square deviation (RMSD). The average distances between atoms in the control molecules were 1.365, 1.433, and 0.642 Å, respectively.

Extracellular phytases are generally absent in plants and animals, but are commonly isolated from bacteria and fungi. Protein overlap similarity analysis of 3-phytase from *Aspergillus niger* with *Penicillium* sp. indicated 63.68% shared identity in expressing the enzyme, and the three-dimensional structures of the proteins are virtually identical. According to Oakley [61], the $1'$, $3'$, $4'$, $5'$, and $6'$-sulphate groups of the IHS bound to PhyA are in the equatorial conformation, while the $2'$-sulphate group is in the axial position. This sulfate group also has the lowest β-factors among all six groups and is the most deeply inserted into the active site cavity. In addition, Arg58, Arg62, Arg142, and Asp339 interact with this group through hydrogen bonds and salt bridges. The molecular docking analysis of the chloroquine interaction showed a binding affinity of −7097 kcal/mol (Table 5), and similarity with the Arg62 amino acid residue (salt bridge type, Figure 6) located in the β-sheet.

Table 5. Binding affinity values of enzymes expressed by fungi with chloroquine.

Ligand	Binding Affinity (kcal/mol)		
	Phytase	Protease	Penicillopepsin-1
HIS	−7.208	--	--
PMS	--	−7.475	--
PP7	--	--	−9.547
Chloroquine	−7.097	−7.262	−7.923

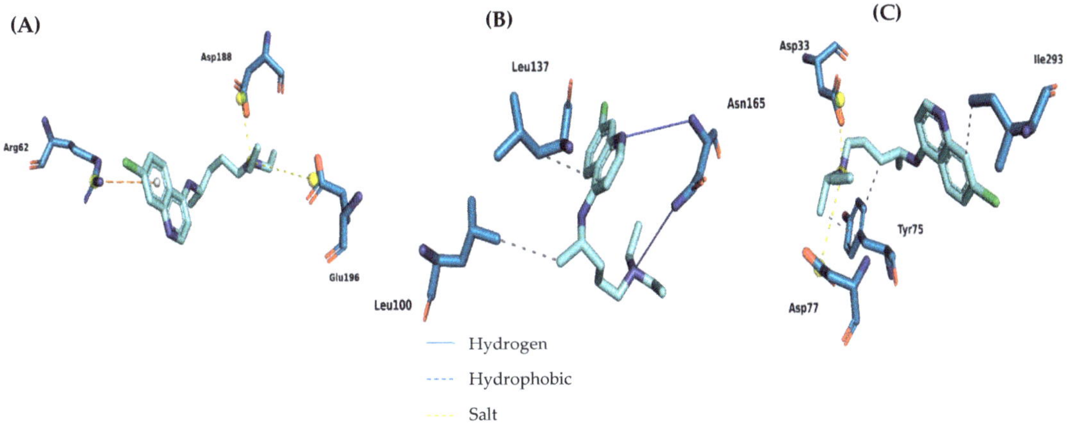

Figure 6. Map of the interactions of chloroquine with the receptors phytase (**A**), protease (**B**) and penicillopepsin-1 (**C**).

The aspartic protease secreted by fungi of the *Penicillium* genus, notably *penicillopepsin*, has wide industrial and biotechnological use due to its high thermal stability, catalytic efficiency at acidic pH, and specificity for peptide bonds between hydrophobic residues [57]. In the environmental context, this enzyme contributes significantly to the biodegradation of nitrogen-containing organic residues, such as those present in industrial effluents, being considered an important tool for bioremediation processes and reuse of protein biomass. Its three-dimensional structure was one of the first to be elucidated among aspartic proteases, and it has become a model for mechanistic and enzyme engineering studies. In the study by Koszelak et al. [62], an ordered PMSF was found in the groove formed by the segments Leu137-Gly 138-Gly 139 and Gly 164-Asn 165-Glu 166 and had Ser173 and Asn198 below it. The molecular docking analysis in the present study showed that chloroquine interacted with the active site of the protease with a binding affinity of −7475 kcal/mol (Figure 6 and Table 5), with similarity to the Leu100 and Asn165 residues.

In penicillopepsin-1, the two catalytic aspartate residues (Asp33 and Asp213) reside in two similarly folded loops at the base of the hydrophobic substrate-binding groove and in front of the central 'sheet'. These active enzyme structures contain catalytic water between the aspartates that deprotonate to initiate acid–base-catalyzed hydrolysis. The tip of the flap makes two hydrogen bonds in the main chain via the nitrogen atoms (NH) of the main structure of Gly76 and Asp 77 with PPi3, further anchoring the inhibitor to the active site. The molecular docking analysis in the present study showed interactions with similar amino acid residues in chloroquine with Asp33, Asp77 and Tyr 75 (Figure 5). The binding affinity value for the penicillopepsin-1 site with chloroquine was −7923 kcal/mol (Table 5).

3. Materials and Methods

3.1. Reagents and Solvents

Malt extract was purchased from Kasvi (São Paulo, Brazil), Sabouraud dextrose agar from Himedia (São Paulo, Brazil), and chloroquine diphosphate from Fundação Oswaldo Cruz (Rio de Janeiro, Brazil) (batch 19090737). Chloroquine of analytical grade (99%, CAS no. 50-63-5) was from Sigma-Aldrich (São Paulo, Brazil). Ethyl acetate (99.5%), dimethyl sulfoxide (97%), and methanol (99.8%) of HPLC grade were purchased from Sigma-Aldrich (São Paulo, Brazil), along with alpha-naphthol (99%) and diethylamine (99%). Na_2SO_4 (99%) was purchased from Reagen (São Paulo, Brazil).

3.2. Isolation and Identification of Filamentous Fungi

Four filamentous fungi (*Trichoderma pseudoasperelloides* CBMAI 2752, *Penicillium rolfsii* CBMAI 2753, *Talaromyces verruculosus* CBMAI 2754, and *Penicillium* cf. *guaibinense* CBMAI 2758) were isolated from soil on land owned by the company UNAMGEN Mining and Metallurgy, located in the municipality of Mazagão (state of Amapá, Brazil), as described in a previous study [35]. These strains were identified under a microscope in accordance with previously described methodology [41] and were used for the biodegradation of chloroquine diphosphate.

Molecular approaches were also applied to confirm these identifications at CPQBA-Unicamp. Genomic DNA from cultures was purified using the phenol DNA extraction protocol. Amplification of gene markers was performed using PCR methodology with the extracted genomic DNA as template. The primers (synthetic oligonucleotides) used for the PCR reaction were as follows: 728f and TEFIr targeting the TEF (translation elongation factor) region for CBMAI 2752; and Bt2a and Bt2b targeting the beta-tubulin region for isolates CBMAI 2753, CBMAI 2754, CBMAI 2756 and CBMAI 2758. The amplification product was column-purified (GFX PCR DNA and gel band purification kit, GE Healthcare, São Paulo, Brazil) and was directly subjected to sequencing using the ABI 3500XL series automated sequencer (Applied Biosystems, São Paulo, Brazil).

To perform the genetic distance analysis, the partial gene sequences obtained through the primers were assembled into a consensus (single consensus sequence combining the different fragments obtained), and then compared with the sequences of organisms from the GenBank (http://www.ncbi.nlm.nih.gov/) and CBS (https://wi.knaw.nl, accessed on 20 May 2022) databases. Sequences of microorganisms related to the unknown sample were subsequently selected for construction of a dendrogram. The DNA sequences were aligned using the CLUSTAL X software within BioEdit 7.2.6 and the genetic distance analyses were performed using the MEGA software, version 6.0. A distance matrix was calculated using the model and the dendrogram was constructed from the genetic distances using the neighbor-joining method, with bootstrap values calculated from 1000 resamples, in the MEGA 6.0 package.

3.3. Radial Growth of Strains in the Presence of Chloroquine Diphosphate (Solid)

Each strain was cultivated in solid culture media (2% Sabouraud dextrose agar and 2% malt) in pH 5 adjusted with HCl (10%) or KOH (10%) solutions. The solutions, flasks, and all material were sterilized in an autoclave (Phoenix, model AV-75, São Paulo, Brazil) at 121 °C for 20 min. Then, chloroquine diphosphate solutions at different concentrations (25, 50, and 100 $mg{\cdot}L^{-1}$) were solubilized in 600 μL of DMSO. The drug was transferred to an Erlenmeyer flask with the culture medium at 45 ± 5 °C to avoid thermal degradation [63]. The plates were inoculated at a central point with a nickel–chromium inoculation needle. The plates were then incubated at 28 °C (B.O.D. LUCADEMA® model LUCA 161/03, Grupo Lucadema, São Paulo, Brazil). Radial growth was monitored every 24 h for 7 days using calipers to measure the diameter in perpendicular positions. Solid-medium plates without chloroquine diphosphate were used as fungal (biotic) controls. The experiments were performed in triplicate.

3.4. Experimental Design and Statistical Model

A three-level, three-variable Box–Behnken factorial design was applied to determine the optimal conditions for the biodegradation of chloroquine diphosphate using the filamentous fungus *Penicillium* sp. cf. *guaibinense* CBMAI 2758. The variables were reaction time (days), pH, and chloroquine diphosphate concentration (mg·L^{-1}). The independent variables were designated as X1, X2, and X3, and their values are shown in Table 2.

The biodegradation rate (BD%) was employed as the response variable for determining the ideal conditions. STATISTICA® software (version 10, Statesoft Inc., Tulsa, OK, USA, trial version, 2011) was used for experimental design and data analysis. Analysis of variance (ANOVA) was used to evaluate the effects of the independent variables and the significance of their interactions. A Pareto chart was used to assess the significance of the effects of the tested variables on the biodegradation rate.

3.5. Biodegradation Reaction of Chloroquine Diphosphate by the Fungus in Liquid Medium

In the screening on solid medium, *Penicillium* cf. *guaibinense* CBMAI 2758 exhibited superior growth, and was selected for quantitative biodegradation experiments in liquid medium, in accordance with the aforementioned design. Erlenmeyer flasks (125 mL) containing 50 mL of liquid medium (2% malt) at different pH values (5, 7, and 9), adjusted using a 10% HCl or KOH solution, were sterilized in an autoclave (20 min at 121 °C and 1 atm) and used for fungal cultivation. Inoculations were performed with six circular fragments (0.5 cm in diameter) taken from the solid culture. These cultures were then incubated in an orbital shaker (LUCADEMA®, model LUCA-222) for 5 days (28 °C, 130 rpm). The experiments were conducted as previously described, with slight modifications [27].

Then, solutions of chloroquine diphosphate with different final concentrations (50, 75 and 100 mg·L^{-1}) were added, and the reactions were incubated in an orbital shaker for 5, 10, and 15 days at 28 °C and 130 rpm. All the biodegradation experiments were performed in triplicate. An abiotic control was performed with the medium containing chloroquine diphosphate at 50 mg·L^{-1} with pH 5, 7, and 9, without the presence of the fungus.

3.6. Extraction of Chloroquine Diphosphate and Its Metabolites

To extract chloroquine diphosphate, 0.025 mg of alpha-naphthol (internal standard) dissolved in 400 µL of ethanol and 50 mL of ethyl acetate was added to each experiment, resulting in a solution of ethyl acetate and culture medium (1:1), which was kept under vigorous magnetic stirring for 30 min. The fungal cells were then filtered. Subsequently, the mixture was extracted by liquid–liquid partitioning in three steps with ethyl acetate (3 × 30 mL). Soon afterwards, anhydrous sodium sulfate (Na_2SO_4) was added, and the solution was filtered again. The excess solvent was removed under reduced pressure using a rotary evaporator (Quimis Q344B, Diadema, Brazil), followed by drying in an oven for 40 h at 40 °C. The samples were then lyophilized [24].

3.7. Method Validation

The method was validated in accordance with the Collegiate Board Resolution no. 166 (RDC, 2017) of the Brazilian Health Surveillance Agency (ANVISA). Selectivity, sensitivity, linearity, precision, detection, and quantification limit were assessed in the validation of the chromatographic method. In the analysis of method selectivity, it was confirmed that the substances or metabolites present in the samples did not co-elute with chloroquine diphosphate in the GC-MS analysis. For method validation, an abiotic control experiment was performed with chloroquine in 2% malt culture medium.

The reactions were prepared in a sterile environment and extracted as described in Section 3.6. Extraction of Chloroquine Diphosphate and Its Metabolites. Lastly, the solutions were prepared for chromatographic analysis in a 2% diethylamine methanolic solution (GC-MS) [24]. Linearity was verified through a calibration curve composed of five chloroquine diphosphate solutions (50, 350, 650, 950, and 1250 mg·L^{-1}). To determine the chloroquine diphosphate concentration, an internal standard method was performed, employing the following equation: c = ax + B, where c corresponds to the analyte concentration in mg·L^{-1}, x refers to the analyte chloroquine area over the internal standard alpha-naphtol area, a is the angular coefficient, and B is the linear coefficient.

Accuracy was defined as the degree of agreement between the results of the employed method and the value accepted as true (RDC, 2017). Precision was expressed as the relative standard deviation (RSD), also known as coefficient of variation (CV), which was calculated according to the equation below:

$$CV = [SD/DMC] \times 100$$

where SD refers to the standard deviation and DMC is the average concentration determined.

The relative precision was calculated using the accuracy percentage equation as follows:

$$\text{Precision\%} = [\text{Mean detected concentration}/\text{Nominal concentration}]$$

The detection limit (LD) and quantification limit (LQ) were determined based on the obtained analytical curve, which took into consideration the confidence intervals of the linear regression.

3.8. Detection of Chloroquine Diphosphate and Metabolites Through Gas Chromatography–Mass Spectrometry (GC-MS) Analysis

The extracts obtained from the biodegradation process were analyzed in a Shimadzu/GC-2010 gas chromatograph (Kyoto, Japan) equipped with a Shimadzu/AOC-5000 autoinjector and coupled to Shimadzu MS-2010 plus mass spectrometer (GC-MS) in SCAN mode at 70 eV. The furnace was equipped with a DB-5 fused silica column (J&W Scientific, 30 mm × 0.25 mm × 0.25 μm, Levittown, PA, USA) and the carrier gas was helium (He) at 65 kPa. The analyses were performed using a flow rate of 3 mL/min and a split of 5 with the injector at 250 °C and the detector at 270 °C. The initial oven temperature was 90 °C for 2 min, and then increased to 280 °C at a heating rate of 6 °C/min. This temperature was maintained for 26 min, resulting in a total analysis time of 60 min [64]. The scan mode was m/z 40–500, and metabolite identification was performed using the Wiley 8, NIST 5, NIST 21, and NIST 107 spectral libraries [4].

3.9. Toxicity Predictions

Toxicity prediction tests were carried out using Discovery Studio 2015 software (Accelrys, Inc., San Diego, CA, USA) via the Toxicity Prediction by Komputer Assisted Technology (TOPKAT) function (BIOVIA, San Diego, CA, USA, 2024). The toxicological properties analyzed were the following: carcinogenicity in rodents (female mice and female rats), Ames mutagenicity, skin irritation, and skin sensitization. Toxicity risk prediction calculations were carried out via TOPKAT and measured the following parameters: oral LD_{50} (g/kg body weight), *Daphnia magna* EC_{50} (mg/L), lowest dose at which a toxic or adverse effect is observed (LOAEL, g/kg body weight), and *Pimephales promelas* LC_{50} (g/L) [65]. In addition, the carcinogenic potential was also predicted using the parameters TD_{50} (mg/kg body weight/mouse-day) and maximum tolerated dose in rats (MRTD) [66].

3.10. Molecular Docking Simulations

Crystal structures were downloaded from the Protein Data Bank (PDB) with the ID codes 3K4Q (*Aspergillus niger*-phytase), 5Z6O (*Penicillium cyclopium*-protease), and 1BXO (penicillopepsin-1) with resolutions of 2.20, 1.70, and 0.50 Å, respectively [61,62,67]. The crystal structures were elucidated using the X-ray diffraction method and the controls myo-inositol (IHS), phenyl metasulfonic acid (PMS), and phosphonate inhibitor (PP7) were used to validate the molecular docking protocols.

Molecular docking simulations were carried out using the DockThor server (https://dockthor.lncc.br/v2/, accessed on 22 April 2025) [68]. The parameters of the algorithm used were set as follows: (1) 12 runs of docking conformations, (2) 500,000 evaluations per docking run, and (3) a population of 750 individuals. The quality of the protein–ligand docking score was evaluated based on the root mean square deviation (RMSD) between the best score of the docking pose and the experimental binding mode of the crystallographic ligand.

4. Conclusions

Among the four fungal strains investigated, *Talaromyces verruculosus* CBMAI 2754, *Penicillium* sp. cf. *guaibinense* CBMAI 2758, and *Penicillium rolfsii* CBMAI 2753 demonstrated better adaptation to the presence of chloroquine diphosphate. At a concentration of 25 $mg \cdot L^{-1}$, these strains exhibited higher radial growth rates (RGRs) than those observed in the control medium without this compound. *Penicillium* sp. cf. *guaibinense* CBMAI 2758 stood out by demonstrating superior growth at both 25 and 50 $mg \cdot L^{-1}$ chloroquine diphosphate compared to the control, highlighting its potential for application in biodegradation processes. Based on these results, this strain was selected for the optimization experiments involving time, pH, and xenobiotic concentration using a Box–Behnken factorial design. Analyses of the Pareto diagram and response surface graphs indicated that the pH and chloroquine diphosphate concentration were the most significant variables influencing the biodegradation process. The optimal conditions for maximizing biodegradation efficiency were identified as pH 7, an incubation time of 10 days, and concentration of 100 $mg \cdot L^{-1}$.

The GC-MS analyses identified four main metabolites of chloroquine diphosphate formed through *N*-deethylation M1 (*N*4-(7-chloroquinolin-4-yl)-*N*1-ethylpentane-1,4-diamine), carbon-carbon bond cleavage M2 (7-chloro-*N*-ethylquinolin-4-amine) and M3 (*N*1,*N*1-diethylpentane-1,4-diamine), and aromatic deamination M4 (*N*1-ethylbutane-1,4-diamine) by enzymatic reactions, thus confirming the biodegradation process.

The toxicological assessment revealed that the biodegradation of chloroquine produces compounds with a significantly improved safety profile than the parent drug. These results indicate lower toxicity, underscoring the potential of the biodegradation process as an effective and environmentally safer approach for treatment and disposal of chloroquine.

The filamentous fungi evaluated in this study demonstrated effective biodegradation of the drug chloroquine diphosphate, highlighting the potential of Amazonian biodiversity as a valuable source of microorganisms for developing sustainable technologies to treat organic micropollutants.

Supplementary Materials: The following supporting information can be downloaded at https://www.mdpi.com/article/10.3390/jof11080579/s1: Figure S1: Morphology and molecular identification of filamentous fungi isolated from iron mine soil; Figure S2: RMSD representation of the crystallographic ligand (green) and best docking pose (gray) in the (A) Phytase; (B) Protease and (C) Penicillopepsin-1; Figure S3: Chromatographic analysis coupled to mass spectrometry (GC/MS) of the biodegradation reaction of chloroquine by the fungus *Penicillium* cf. *guaibinense:* CBMAI 2758; Table S1: Growth rates of the strains *Trichoderma pseudoasperelloides*, *T. verruculosus*, *Penicillium* cf. *guaibinense* and *Penicillium rolfsii* in the presence of chloroquine diphosphate, on 2% Sabouraud dextrose agar, 2% malt, pH 5, at 28 °C for 7 days; Table S2: Main ions observed in the mass spectra of metabolites identified from the biodegradation of chloroquine diphosphate (DCQ) by *Penicillium* cf. *guaibinense* CBMAI 2758.

Author Contributions: Conceptualization, P.d.A.N. and I.M.F.; methodology, P.d.A.N.; software, F.H.e.H.; validation, F.H.e.H., A.L.M.P., I.M.F. and W.G.B.; formal analysis, I.M.F., S.Q.L., I.F.d.A.; investigation, P.d.A.N.; resources, I.M.F.; data curation, P.d.A.N. and L.S.S.; writing—original draft preparation, P.d.A.N. and I.F.d.A.; writing—review and editing, P.d.A.N., A.L.M.P., W.G.B. and I.M.F.; toxicological and docking analysis, R.d.S.R.; visualization, I.M.F. and P.d.A.N.; supervision, I.M.F.; project administration, P.d.A.N.; funding acquisition, I.M.F. All authors have read and agreed to the published version of the manuscript.

Funding: This research was funded by the National Council for Scientific and Technological Development CNPQ, grant number 141512/2024-5; 402635/2023-0; 443823/2024-3 and 444009/2024-8 and the project was carried out using the infrastructure of the Federal University of Amapá UNIFAP.

Institutional Review Board Statement: Not applicable.

Informed Consent Statement: Not applicable.

Data Availability Statement: The data are contained in the article or Supplementary Material.

Conflicts of Interest: The authors declare no conflict of interest.

Abbreviations

The following abbreviations are used in this manuscript:

CBMAI	Brazilian Collection of Environmental and Industrial Microorganisms
pH	hydrogen of potential
EOMs	Emerging organic contaminants
GC-MS	Gas chromatography–mass spectrometry
SARS-CoV-2	Severe acute respiratory syndrome coronavirus 2
COVID-19	Coronavirus disease 2019
RGR	Radial growth rate
EPS	Extracellular polymeric substances
CTC	Chlortetracycline
NSAIDs	Non-steroidal anti-inflammatory drugs
CAS	Chemical Abstracts Service
HPLC	High-performance liquid chromatography
CPQBA	Multidisciplinary Center for Chemical, Biological, and Agricultural Research
DNA	Deoxyribonucleic acid
PCR	Polymerase chain reaction
HCl	Hydrochloric acid
KOH	Sodium hydroxide
DMSO	Dimethyl sulfoxide
BD	Biodegradation rate
RDC	Collegiate Board Resolution

ANVISA	National Health Surveillance Agency
RSD	Relative standard deviation
CV	Coefficient of variation
SD	Standard deviation
DMC	Determined mean concentration
LD	Limit of detection
LQ	Limit of quantification

References

1. Serafin, M.B.; Bottega, A.; Foletto, V.S.; da Rosa, T.F.; Hörner, A.; Hörner, R. Drug Repositioning Is an Alternative for the Treatment of Coronavirus COVID-19. *Int. J. Antimicrob. Agents* **2020**, *55*, 105969. [CrossRef]
2. Furlan, L.; Caramelli, B. The Regrettable Story of the "Covid Kit" and the "Early Treatment of COVID-19" in Brazil. *Lancet Reg. Health Am.* **2021**, *4*, 100089. [CrossRef] [PubMed]
3. Liu, J.; Cao, R.; Xu, M.; Wang, X.; Zhang, H.; Hu, H.; Li, Y.; Hu, Z.; Zhong, W.; Wang, M. Hydroxychloroquine, a Less Toxic Derivative of Chloroquine, Is Effective in Inhibiting SARS-CoV-2 Infection in Vitro. *Cell Discov.* **2020**, *6*, 6–9. [CrossRef] [PubMed]
4. Cortegiani, A.; Ingoglia, G.; Ippolito, M.; Giarratano, A.; Einav, S. A systematic review on the efficacy and safety of chloroquine for the treatment of COVID-19. *J. Crit. Care* **2020**, *57*, 279–283. [CrossRef] [PubMed]
5. Caly, L.; Druce, J.D.; Catton, M.G.; Jans, D.A.; Wagstaff, K.M. The FDA-Approved Drug Ivermectin Inhibits the Replication of SARS-CoV-2 in Vitro. *Antivir. Res.* **2020**, *178*, 3–6. [CrossRef] [PubMed]
6. Mirtaleb, M.S.; Mirtaleb, A.H.; Nosrati, H.; Heshmatnia, J.; Falak, R.; Zolfaghari Emameh, R. Potenciais agentes terapêuticos para COVID-19: Uma revisão atualizada sobre terapia antiviral, imunoterapia e terapia celular. *Biomed. Pharmacother.* **2021**, *138*, 111518. [CrossRef]
7. Lou, Y.; Liu, L.; Yao, H.; Hu, X.; Su, J.; Xu, K.; Luo, R.; Yang, X.; He, L.; Lu, X.; et al. Clinical Outcomes and Plasma Concentrations of Baloxavir Marboxil and Favipiravir in COVID-19 Patients: An Exploratory Randomized, Controlled Trial. *Eur. J. Pharm. Sci.* **2021**, *157*, 105631. [CrossRef]
8. Antinori, S.; Cossu, M.V.; Ridolfo, A.L.; Rech, R.; Bonazzetti, C.; Pagani, G.; Gubertini, G.; Coen, M.; Magni, C.; Castelli, A.; et al. Compassionate Remdesivir Treatment of Severe COVID-19 Pneumonia in Intensive Care Unit (ICU) and Non-ICU Patients: Clinical Outcome and Differences in Post-Treatment Hospitalisation Status. *Pharmacol. Res.* **2020**, *158*, 104899. [CrossRef]
9. Khalili, J.S.; Zhu, H.; Mak, N.S.A.; Yan, Y.; Zhu, Y. Novel Coronavirus Treatment with Ribavirin: Groundwork for an Evaluation Concerning COVID-19. *J. Med. Virol.* **2020**, *92*, 740–746. [CrossRef]
10. Dai, H.; Wang, C.; Yu, W.; Han, J. Tracing COVID-19 Drugs in the Environment: Are We Focusing on the Right Environmental Compartment? *Environ. Pollut.* **2023**, *339*, 122732. [CrossRef]
11. Saadat, D.; Rawtani, C.M. Hussain. Environmental perspective of COVID-19. *Sci. Total Environ.* **2020**, *728*, 1138870. [CrossRef] [PubMed]
12. Ahmed, B.; Tscharke, P.M.; Bertsch, K.; Bibby, A.; Bivins, P.; Choi, L.; Clarke, J.; Dwyer, J.; Edson, T.M.H.; Nguyen, J.W.O.; et al. SARS-CoV-2 RNA monitoring in wastewater as a potential early warning system for COVID-19 transmission in the community. *Sci. Total Environ.* **2020**, *761*, 144216. [CrossRef] [PubMed]
13. Picó, Y.; Barceló, D. Microplastics and other emerging contaminants in the environment after COVID 19 pandemic: The need of global reconnaissance studies. *Curr. Opin. Environ. Sci. Health* **2023**, *33*, 100468. [CrossRef] [PubMed]
14. Aus der Beek, T.; Weber, F.A.; Bergmann, A.; Hickmann, S.; Ebert, I.; Hein, A.; Küster, A. Pharmaceuticals in the Environment-Global Occurrences and Perspectives. *Environ. Toxicol. Chem.* **2016**, *35*, 823–835. [CrossRef] [PubMed]
15. Nannou, C.; Ofrydopoulou, A.; Evgenidou, E.; Heath, D.; Heath, E.; Lambropoulou, D. Antiviral Drugs in Aquatic Environment and Wastewater Treatment Plants: A Review on Occurrence, Fate, Removal and Ecotoxicity. *Sci. Total Environ.* **2020**, *699*, 134322. [CrossRef] [PubMed]
16. Morales-Paredes, C.A.; Rodríguez-Díaz, J.M.; Boluda-Botella, N. Pharmaceutical Compounds Used in the COVID-19 Pandemic: A Review of Their Presence in Water and Treatment Techniques for Their Elimination. *Sci. Total Environ.* **2022**, *814*, 152691. [CrossRef]
17. Asif, M.B.; Li, C.; Ren, B.; Maqbool, T.; Zhang, X.; Zhang, Z. Elucidating the Impacts of Intermittent In-Situ Ozonation in a Ceramic Membrane Bioreactor: Micropollutant Removal, Microbial Community Evolution and Fouling Mechanisms. *J. Hazard. Mater.* **2021**, *402*, 123730. [CrossRef]
18. Mohd Hanafiah, Z.; Wan Mohtar, W.H.M.; Wan-Mohtar, W.A.A.Q.I.; Bithi, A.S.; Rohani, R.; Indarto, A.; Yaseen, Z.M.; Sharil, S.; Binti Abdul Manan, T.S. Removal of Pharmaceutical Compounds and Toxicology Study in Wastewater Using Malaysian Fungal Ganoderma Lucidum. *Chemosphere* **2024**, *358*, 142209. [CrossRef]

19. Williams, N.S.; Ray, M.B.; Gomaa, H.G. Removal of Ibuprofen and 4-Isobutylacetophenone by Non-Dispersive Solvent Extraction Using a Hollow Fibre Membrane Contactor. *Sep. Purif. Technol.* **2012**, *88*, 61–69. [CrossRef]
20. Stachowiak, M.; Cegłowski, M.; Kurczewska, J. Hybrid Chitosan/Molecularly Imprinted Polymer Hydrogel Beads Doped with Iron for Selective Ibuprofen Adsorption. *Int. J. Biol. Macromol.* **2023**, *251*, 126356. [CrossRef]
21. Magaela, B.N.; Ndlovu, K.S.; Tshangana, C.S.; Muleja, A.A.; Mamba, B.B.; Nyokong, T.; Managa, M. Photodegradation of Ibuprofen Using 5-10-15-20-Tetrakis(4-Bromophenyl) Porphyrin Conjugated to Graphene Quantum Dots. *Opt. Mater.* **2022**, *134*, 113147. [CrossRef]
22. Danner, M.C.; Robertson, A.; Behrends, V.; Reiss, J. Antibiotic Pollution in Surface Fresh Waters: Occurrence and Effects. *Sci. Total Environ.* **2019**, *664*, 793–804. [CrossRef]
23. Hasan, M.K.; Alfredo, S.; Murthy, R. Riffat. Biodegradation of salicylic acid, acetaminophen and ibuprofen by bacteria collected from a full-scale drinking water biofilter. *J. Environ. Manag.* **2021**, *295*, 113071. [CrossRef] [PubMed]
24. Holanda, F.H.e.; Birolli, W.G.; Morais, E.d.S.; Sena, I.S.; Ferreira, A.M.; Faustino, S.M.M.; Grace, d.S.; Solon, L.; Porto, A.M. Ferreira, I.M. Study of Biodegradation of Chloramphenicol by Endophytic Fungi Isolated from *Bertholletia excelsa* (Brazil Nuts) *Biocatal. Agric. Biotechnol.* **2019**, *20*, 101200. [CrossRef]
25. Midassi, S.; Bedoui, A.; Bensalah, N. Efficient Degradation of Chloroquine Drug by Electro-Fenton Oxidation: Effects of Operating Conditions and Degradation Mechanism. *Chemosphere* **2020**, *260*, 127558. [CrossRef] [PubMed]
26. Han, Q.; Huang, Y.; Wang, X.; Liu, Y.; Lin, W.; Zhang, F.; Chen, X. Degradation of Low Concentration Chloroquine Phosphate by Ozone Oxidation. *Proc. J. Phys. Conf. Ser. Inst. Phys.* **2024**, *2679*, 012050. [CrossRef]
27. Birolli, W.G.; de A. Santos, D.; Alvarenga, N.; Garcia, A.C.F.S.; Romão, L.P.C.; Porto, A.L.M. Biodegradation of Anthracene and Several PAHs by the Marine-Derived Fungus *Cladosporium* Sp. *CBMAI* 1237. *Mar. Pollut. Bull.* **2018**, *129*, 525–533. [CrossRef]
28. Alvarenga, N.; Birolli, W.G.; Nitschke, M.; Rezende, M.O.O.; Seleghim, M.H.R.; Porto, A.L.M. Biodegradation of Chlorpyrifos by Whole Cells of Marine-Derived Fungi *Aspergillus sydowii* and *Trichoderma* sp. *J. Microb. Biochem. Technol.* **2015**, *7*, 133–139 [CrossRef]
29. Li, Z.; Wang, J.; Gu, C.; Guo, Y.; Wu, S. Marine Bacteria-Mediated Abiotic-Biotic Coupling Degradation Mechanism of Ibuprofen *J. Hazard. Mater.* **2022**, *435*, 128960. [CrossRef]
30. Murshid, S.; Dhakshinamoorthy, G.P. Biodegradation of Sodium Diclofenac and Mefenamic Acid: Kinetic Studies, Identification of Metabolites and Analysis of Enzyme Activity. *Int. Biodeterior. Biodegrad.* **2019**, *144*, 104756. [CrossRef]
31. Arun, K.B.; Madhavan, A.; Tarafdar, A.; Sirohi, R.; Anoopkumar, A.N.; Kuriakose, L.L.; Awasthi, M.K.; Binod, P.; Varjani, S. Sindhu, R. Filamentous fungi for pharmaceutical compounds degradation in the environment: A sustainable approach. *Environ Technol. Innov.* **2023**, *31*, 103182. [CrossRef]
32. dos Santos, V.H.P.; dos Santos, W.T.; Ionta, M.; de Paula, A.C.C.; Silva, E.O. Biotransformation of hydroxychloroquine to evaluate the cytotoxicity of its metabolites and mimic mammalian metabolism. *Results Chem.* **2023**, *5*, 100761. [CrossRef] [PubMed]
33. Correa, L.O.; Bezerra, A.F.M.; Honorato, L.R.S.; Cortez, A.C.A.; Souza, J.V.B.; Souza, E.S. Amazonian soil fungi are efficient degraders of glyphosate herbicide; novel isolates of *Penicillium*, *Aspergillus*, and *Trichoderma*. *Braz. J. Biol.* **2023**, *83*, e242830 [CrossRef] [PubMed]
34. Ramos, M.M.; Dos SMorais, E.; da SSena, I.; Lima, A.L.; de Oliveira, F.R.; de Freitas, C.M.; Fernandes, C.P.; de Carvalho, J.C.T. Ferreira, I.M. Silver nanoparticle from whole cells of the fungi *Trichoderma* spp. isolated from Brazilian Amazon. *Biotechnol. Lett* **2020**, *42*, 833–844. [CrossRef] [PubMed]
35. Nóbrega, P.A.; Ferreira, B.L.; Sá, L.S.; Braga, F.C.; Bezerra, R.M.; Ferreira, M.I. Isolation and Identification of Iron, Chromium, and Manganese Tolerant Filamentous Fungi from Mining Soil. *Curr. Biotechnol.* **2024**, *13*, 174–184. [CrossRef]
36. Oliveira, A.N.; Oliveira, L.A.; Andrade, J.S. Chagas júnior, A.F. Atividade enzimática de isolados de rizóbia nativos da amazônia central crescendo em diferentes níveis de acidez. *Ciênc. Tecnol. Aliment. Camp.* **2006**, *26*, 204–210. [CrossRef]
37. Colla, L.M.; Primaz, A.L.; de Lima, M.; Bertolin, T.E.; Costa, J.A.V. Isolation and Screening of Fungi to Bioremediation from Triazine Herbicide Contaminated Soil. *Cienc. Agrotecnologia* **2008**, *32*, 809–813. [CrossRef]
38. Tortora, G.J.; Case, C.L.; Funke, B.R. *Microbiologia*; Artmed: Porto Alegre, Brazil, 2016; pp. 1–964. ISBN 8582713541/9788582713549
39. Palanivel, T.M.; Pracejus, B.; Novo, L.A. Bioremediation of copper using indigenous fungi *Aspergillus* species isolated from an abandoned copper mine soil. *Chemosphere* **2023**, *314*, 37688. [CrossRef]
40. Rigo, D.; Luana, G.; Tres, G.A.; Camera, F.D.; Zeni, J. Microbiological Production of Enzymes: A Review. *Braz. J. Dev.* **2021**, *7* 9232–9254. [CrossRef]
41. Sena, I.S.; Ferreira, A.M.; Marinho, V.H.; e Holanda, F.H.; Borges, S.F.; de Souza, A.A.; de Carvalho, R.; Koga, R.; Lima, A.L. Florentino, A.C.; et al. *Euterpe oleracea* Mart (Açaizeiro) from the Brazilian Amazon: A Novel Font of Fungi for Lipase Production *Microorganisms* **2022**, *10*, 2394. [CrossRef]
42. Daneshfar, A.; Vafafard, S. Solubility of Chloroquine Diphosphate and 4,7-Dichloroquinoline in Water, Ethanol, Tetrahydrofuran Acetonitrile, and Acetone from (298.2 to 333.2) K. *J. Chem. Eng. Data* **2009**, *54*, 2170–2174. [CrossRef]

3. Tomasini, A.; Leon-Santiesteban, H.H. *Fungal Bioremediation: Fundamentals and Applications*; Tomasini, A., Leâon-Santiesteban, H.H., Eds.; CRC Press: Boca Raton, FL, USA, 2019.
4. Rodriguez-Rodrıíguez, C.E.; Caminal, G.; Vicent, T.; Diaz-Cruz, M.S.; Eljarrat, E.; Farre, M.; López de Alda, M.J.; Petrovic, M.; Barcelo, D. Fungal-Mediated Degradation of Emerging Pollutants in Sewage Sludge. *Handb. Environ. Chem.* **2013**, *24*, 137–164. [CrossRef]
5. Akerman-Sanchez, G.; Rojas-Jimenez, K. Fungi for the Bioremediation of Pharmaceutical-Derived Pollutants: A Bioengineering Approach to Water Treatment. *J. Fungi* **2021**, *4*, 100071. [CrossRef]
6. Park, H.; Choi, I.G. Genomic and transcriptomic perspectives on mycoremediation of polycyclic. aromatic hydrocarbons. *Appl. Microbiol. Biotechnol.* **2020**, *104*, 6919–6928. [CrossRef]
7. Cai, Y.; Yu, H.; Ren, L.; Ou, Y.; Jiang, S.; Chai, Y.; Chen, A.; Yan, B. Treatment of amoxicillin-containing wastewater by *Trichoderma* strains selected from activated sludge. *Sci. Total Environ.* **2023**, *867*, 161565. [CrossRef]
8. Li, Y.; Chen, H.; Wang, Y.; Yang, Z.; Zhang, H. Efficient biodegradation of chlortetracycline in high concentration from strong-acidity pharmaceutical residue with degrading fungi. *J. Hazard. Mater.* **2022**, *424*, 127671. [CrossRef] [PubMed]
9. Bankole, P.O.; Adekunle, A.A.; Jeon, B.H.; Govindwar, S.P. Novel cobiomass degradation of NSAIDs by two wood rot fungi, Ganoderma applanatum and Laetiporus sulphureus: Ligninolytic enzymes induction, isotherm and kinetic studies. *Ecotoxicol. Environ. Saf. Ecotoxicol.* **2020**, *203*, 110997. [CrossRef] [PubMed]
0. Wang, Z.; Wang, Z.; Wang, G.; Zhang, G.; Wang, Q.; Wang, W. New insight into biodegradation mechanism of phenylurea herbicides by cytochrome P450 enzymes: Successive N-demethylation. *Environ. Int.* **2023**, *182*, 108332. [CrossRef] [PubMed]
1. Marmor, M.F.; Kellner, U.; Lai, T.Y.Y.; Melles, R.B.; Mieler, W.F. Recommendations on Screening for Chloroquine and Hydroxy-chloroquine Retinopathy (2016 Revision). *Ophthalmology* **2016**, *123*, 1386–1394. [CrossRef]
2. Chorin, E.; Dai, M.; Shulman, E.; Wadhwani, L.; Bar-Cohen, R.; Barbhaiya, C.; Aizer, A.; Holmes, D.; Bernstein, S.; Spinelli, M.; et al. The QT Interval in Patients with COVID-19 Treated with Hydroxychloroquine and Azithromycin. *Nat. Med.* **2020**, *26*, 808–809. [CrossRef]
3. Van Eaton, K.M.; Gustafson, D.L. Pharmacokinetic and Pharmacodynamic Assessment of Hydroxychloroquine in Breast Cancer. *J. Pharmacol. Exp. Ther.* **2021**, *379*, 331–342. [CrossRef] [PubMed]
4. Rynes, R.I. Antimalarial drugs in the treatment of rheumatological diseases. *Rheumatology* **1997**, *36*, 799–805. [CrossRef] [PubMed]
5. Matúš, P.; Littera, P.; Farkas, B.; Urík, M. Review on Performance of *Aspergillus* and *Penicillium* Species in Biodegradation of Organochlorine and Organophosphorus Pesticides. *Microorganisms* **2023**, *11*, 1485. [CrossRef] [PubMed]
6. Sowmya, H.V.; Ramalingappa; Krishnappa, M.; Thippeswamy, B. Degradation of Polyethylene by *Penicillium simplicissimum* Isolated from Local Dumpsite of Shivamogga District. *Environ. Dev. Sustain.* **2015**, *17*, 731–745. [CrossRef]
7. Boucherit, Z.; Flahaut, S.; Djoudi, B.; Mouas, T.N.; Mechakra, A.; Ameddah, S. Potential of Halophilic *Penicillium chrysogenum* Isolated from Algerian Saline Soil to Produce Laccase on Olive Oil Wastes. *Curr. Microbiol.* **2022**, *79*, 178. [CrossRef] [PubMed]
8. Selim, S.; Al-Sanea, M.M.; Alhejely, A.; Moawad, H.; Masmali, I.; Hendawy, O.M. Degradative potential of laccase and manganese peroxidase to mycotoxins on infected maize grains by fungi with docking interaction studies. *Bioresources* **2024**, *19*, 9773–9787. [CrossRef]
9. Sundaramoorthy, M.; Kishi, K.; Gold, M.H.; Poulos, T.L. Preliminary crystallographic analysis of manganese peroxidase from *Phanerochaete chrysosporium*. *J. Mol. Biol.* **1994**, *238*, 845–848. [CrossRef]
0. Ducros, V.; Brzozowski, A.M.; Wilson, K.S.; Østergaard, P.; Schneider, P.; Svendson, A.; Davies, G.J. Structure of the laccase from *Coprinus cinereus* at 1.68 Å resolution: Evidence for different 'type 2 Cu-depleted' isoforms. *Acta Crystallogr. D Biol. Crystallogr.* **2001**, *57 Pt 3*, 333–336. [CrossRef]
1. Oakley, A.J. The Structure of *Aspergillus niger* Phytase PhyA in Complex with a Phytate Mimetic. *Biochem. Biophys. Res. Commun.* **2010**, *397*, 745–749. [CrossRef]
2. Koszelak, S.; Ng, J.D.; Day, J.; Ko, T.P.; Greenwood, A.; McPherson, A. The Crystallographic Structure of the Subtilisin Protease from *Penicillium cyclopium*. *Biochemistry* **1997**, *36*, 6597–6604. [CrossRef]
3. Vacondio, B.W.G.; Birolli, I.M.; Ferreira, M.H.; Seleghim, S.; Gonçalves, S.P.; Vasconcellos, A.L. Biodegradation of pentachlorophe-nol by marine-derived fungus *Trichoderma harzianum* CBMAI 1677 isolated from ascidian *Didemnun ligulum*. *Biocatal. Agric. Biotechnol.* **2015**, *4*, 266–275. [CrossRef]
4. Soares, P.R.S.; Birolli, W.G.; Ferreira, I.M.; Porto, A.L.M. Biodegradation pathway of the organophosphate pesticides chlorpyrifos, methyl parathion and profenofos by the marine-derived fungus *Aspergillus sydowii* CBMAI 935 and its potential for methylation reactions of phenolic compounds. *Mar. Pollut. Bull.* **2021**, *166*, 2–12. [CrossRef] [PubMed]
5. Ramos, R.S.; Borges, R.S.; de Souza, J.S.N.; Araujo, I.F.; Chaves, M.H.; Santos, C.B.R. Identification of Potential Antiviral Inhibitors from Hydroxychloroquine and 1,2,4,5-Tetraoxanes Analogues and Investigation of the Mechanism of Action in SARS-CoV-2. *Int. J. Mol. Sci.* **2022**, *23*, 1781. [CrossRef] [PubMed]

66. Rodrigues, E.S.R.; De Faveri Moreira, R.; da Silva Ramos, R.; de Souza, S.A.; Filho, J.W.M.S.; da Silva, B.J.P.; Jumbo, L.O.V.; de Oliveira, E.E.; Lima, E.S.; de Souza Aguiar, R.W. Safety Assessment of the Ethanolic Extract of *Siparuna guianensis*: Cell Viability, Molecular Risk Predictions and Toxicity Risk for Acute and Sub-Chronic Oral Ingestion. *J. Ethnopharmacol.* **2025**, *347*, 119751. [CrossRef] [PubMed]
67. Khan, A.R.; Parrish, J.C.; Fraser, M.E.; Smith, W.W.; Bartlett, P.A.; James, M.N.G. Lowering the Entropic Barrier for Binding Conformationally Flexible Inhibitors to Enzymes. *Biochemistry* **1998**, *37*, 16839–16845. [CrossRef]
68. Guedes, I.A.; Pereira da Silva, M.M.; Galheigo, M.; Krempser, E.; de Magalhães, C.S.; Correa Barbosa, H.J.; Dardenne, L.E. DockThor-VS: A Free Platform for Receptor-Ligand Virtual Screening. *J. Mol. Biol.* **2024**, *436*, 168548. [CrossRef]

Journal of
Fungi

MDPI

rticle

Development and Characterization of Mycelium-Based Composite Using Agro-Industrial Waste and *Ganoderma lucidum* as Insulating Material

ustavo Jiménez-Obando [1,2], Juan Sebastian Arcila [3], Ricardo Augusto Tolosa-Correa [1,4], enny Leandra Valencia-Cardona [3] and Sandra Montoya [3,*]

[1] Escuela de Arquitectura y Urbanismo, Universidad Nacional de Colombia, Calle 65 # 23-29, Manizales 170004, Colombia; gjimenezob@unal.edu.co (G.J.-O.); ratolosac@unal.edu.co (R.A.T.-C.)
[2] GESA Grupo de Estudios en Sostenibilidad Ambiental, Universidad Nacional de Colombia, Carrera 45 # 26-85, Bogotá 11001, Colombia
[3] Center for Technological Development—Bioprocess and Agro-Industry Plant, Department of Engineering, Universidad de Caldas, Calle 65 # 26-10, Manizales 170004, Colombia; juan.arcila@ucaldas.edu.co (J.S.A.); leandra.valenciacardona@udecaldas.edu.co (Y.L.V.-C.)
[4] Grupo de Procesos Químicos, Catalíticos y Biotecnológicos, Universidad Nacional de Colombia, Campus La Nubia, Bloque T, Km 7 vía al Aeropuerto, Manizales 170003, Colombia
* Correspondence: sandra.montoya@ucaldas.edu.co; Tel.: +57-(606)-8781500 (ext. 11677)

Abstract: Mycelium-based composites (MBCs) have emerged as eco-friendly alternatives, utilizing fungal mycelium as a natural binder for agro-industrial residues. This study focuses on developing an MBC based on abundant waste in Colombia, pith Arboloco (A) (*Montanoa quadrangularis*), a plant endemic to the Colombian–Venezuelan Andes with outstanding insulating properties, and natural fiber of Kikuyu grass (G) (*Cenchrus clandestinus*), utilizing *Ganoderma lucidum* as an agent to form a mycelium network in the MBC. Three formulations, T (100% A), F1 (70% A/30% G), and F2 (30% A/70% G), were evaluated under two different Arboloco particle size ranges (1.0 to 5.6 mm) for their physical, mechanical, and thermal properties. The Arboloco particle sizes did not show significant differences in the MBC properties. An increase in Kikuyu grass proportion (F2) demonstrated superior density (60.4 ± 4.5 kg/m^3), lower water absorption (56.6 ± 18.4%), and better compressive strength (0.1686 MPa at 50% deformation). Both mixing formulations (F1–F2) achieved promising average thermal conductivity and specific heat capacity values of 0.047 ± 0.002 W m^{-1} K^{-1} and 1714 ± 105 J kg^{-1} K^{-1}, comparable to commercial insulation materials. However, significant shrinkage (up to 53.6%) and high water absorption limit their scalability for broader applications. These findings enhance the understanding of MBC's potential for non-structural building materials made of regional lignocellulosic waste, promoting a circular economy in waste management for developing countries.

Keywords: mycelium-based composite; *Ganoderma lucidum*; thermal insulation; *Montanoa quadrangularis* Shultz Bip

1. Introduction

The construction industry represents one of the sectors with the highest energy consumption globally. It utilizes approximately 40% of the energy generated annually during the various stages of its life cycle, including manufacturing, on-site construction, and final disposal of materials. The demand for construction materials has grown 23 times, and it is estimated that between 2017 and 2026, the number of square meters built in buildings worldwide will increase from 163 to 184 billion [1]. This trend presents a challenging outlook from both economic and environmental perspectives, where eco-design and engineering play a crucial role in developing sustainable building alternatives that minimize the environmental impact associated with greenhouse gas emissions, energy consumption, water pollution, and waste inherent in the production and transportation of building materials [2]. The evolution of sustainable building materials traces a path from traditional practices to cutting-edge innovation. The pioneering adoption of renewable resources, such as bamboo [3] and wood [4,5], has been characterized by their low environmental impact and structural adaptability. This foundation shifted toward circular economy principles, incorporating recycled materials such as crushed concrete, plastic, and glass as aggregates in concrete [6–8]; the development of novel materials based on the inclusion of steel fibers as a reinforced agent to form High-Toughness Recycled Aggregate Concrete (HTRAC) [9]; or the use of recycled brick powder-based geopolymer brick powder as a precursor to prepare alkali-activated based geopolymers (AABGs) [10], which led to the reduction of waste and conserving resources without substantially affecting the strength, deformability, and toughness properties of concrete.

Concomitantly with global warming changes and its sharp and long outdoor temperature fluctuation, the external envelope of a building has become a crucial aspect in eco-design projects due to the necessity of controlling the surrounding microclimate, which is directly associated with the thermal comfort of inhabitants and energy loss, which decreases energy consumption due to heating and cooling demands [11]. Mycelium-based composites (MBCs) are emerging as promising insulating materials due to their sustainable and effective thermal properties. White-rot fungi (WRT), a primary fungal species in MBC production, preferentially utilize agro-industrial residues due to their lignocellulosic matrix composition [12]. The activation of the enzymatic machinery of white-rot fungi through the production of laccase, manganese peroxidases (MnP), and lignin peroxidases (LiP) breaks down lignin while preserving cellulose, which favors the formation of dense mycelia that bind particle residues into cohesive networks [13]. Fungal species such as *Pleurotus* spp., *Ganoderma* spp., and *Trametes* spp. are commonly used due to their ability to bind various lignocellulosic residues, forming solid MBC materials with outstanding characteristics such as fire resistance, thermal stability, low thermal conductivity, and acoustic absorption capacity [13–17]. Multiple results reported in the review literature have demonstrated the versatility of the thermal and mechanical properties of MBCs, as well as the dependence of strength on fungal species and substrate type. Agro-industrial residues, such as sawdust, wood, and rye berries, that exhibit a high affinity for colonization by *Pleurotus* spp. and *Ganoderma* spp. demonstrate low thermal conductivity, making them effective in the production of thermal insulator MBCs [18–21]. In contrast, MBC production from *Trametes* spp. utilizing fiber materials such as hemp, flax waste, and wood residues supports the enhancement of mechanical properties [18,22,23]. Despite the various results reported above, the mechanical properties, dimensional stability, and water adsorption are considered factors that affect the structural integrity and durability of MBCs during an extended period, limiting the standardization of processes for large-scale building applications [24]. To overcome these constraints, natural alternatives such as fiber-reinforced materials, including jute, hemp, bamboo, and flax, in MBCs have been shown to significantly improve the mechanical properties of MBCs [18,25]. Additionally, pineapple fiber residues reduce water adsorption and enhance the dimensional stability of lignocellulosic substrates made from rice bran and coffee grounds [26]. In this context, the limited knowledge about the impact of lignocellulosic waste on the formation and physical, mechanical, and thermal characteristics of MBCs emphasizes the need to explore alternatives of new lignocellulosic substrates to optimize MBC properties, which can lead to building applications that align with a region's waste management demands.

In the present research, we focus on developing MBC materials for constructive applications by combining two regionally available lignocellulosic residues: grass clippings (Kikuyu grass) from urban pruning activities and the pith of *Montanoa quadrangularis* (commonly known as Arboloco), a plant endemic to the Colombian–Venezuelan Andes. We use the fungal species *Ganoderma lucidum* as a binding agent to form the bio-composite's mycelium network structure. Unlike conventional substrates, these agro-industrial residues do not compete with the agri-food or forestry sectors. Arboloco holds significant cultural and ecological value in Andean communities, where it has historically been used for firewood and vernacular construction during the Antioquian colonization [27], contributing to the insulating characteristics of the MBC. At the same time, Kikuyu grass serves as a fiber-reinforced material, which, according to the literature mentioned above, enhances the mechanical properties of the MBC. The experimental design assessed the effects of two particle size ranges of Arboloco pith (1.0–2.36 mm and 2.37–5.66 mm) and three variations of substrate formulation (Arboloco/Kikuyu grass) on the properties of mycelium-based composition. The physical properties (apparent density, shrinkage, and water absorption), mechanical properties (compressive strength, modulus of rupture, and elasticity), and thermal properties (diffusivity, conductivity, and specific heat capacity) of the mycelium-based composite, as well as its morphological and chemical properties (as observed using SEM and FTIR), were evaluated. This study contributes to a broader understanding of the potential applications of local lignocellulosic residues in non-structural building systems, providing a sustainable alternative to improve regional waste management challenges.

2. Materials and Methods

2.1. Organism and Spawn Preparation

Figure 1 illustrates the schematic process for preparing the organism and spawn. A strain of *Ganoderma lucidum* WC806 obtained from the Pennsylvania State University Mushroom Culture Collection, USA, was used. It is available at the Technological Development Center of the Bioprocess and Agroindustry Plant of the University of Caldas and maintained on potato dextrose agar (PDA) at 4 °C with periodic transfers according to the protocols described by Montoya et al. [28]. The spawn for inoculating the solid substrate was prepared on wheat and hydrated with hot water up to 35–40% humidity. The hydrated wheat was packed in 200 g units in bioriented polypropylene bags, provided with a cotton filter on the bag top for gas exchange. The grain was sterilized at 121 °C for 2 h and inoculated with four 1 cm side pieces of mycelium extended on agar in a laminar flow cabinet. Inoculated bags were incubated in penumbra at 25 °C until complete colonization. The second stage continued with preparing the seed or inoculum packed with 1000 g of hydrated grain and inoculated with 4% (*w*/*w* fresh).

2.2. Culture Medium to Produce Mycelium-Based Composite (MBC)

To prepare culture media (Figure 1), lignocellulosic materials were selected: pith from the Arboloco tree (*Montanoa quadrangularis* Schultz Bip.) and Kikuyu grass (*Cenchrus clandestinus*). Both materials were collected from various locations in Manizales, Colombia (2280 masl, average temperature 17 °C, and 70% relative humidity). Arboloco (abbreviated A) and Kikuyu grass (abbreviated G) were dried at 60 °C and under ambient conditions, respectively. The dried Arboloco was mechanically crushed and sieved to separate particles into two size ranges: 1.0–2.36 mm (AL: Arboloco low particle size) and 2.37–5.66 mm (AH: Arboloco high particle size). Six culture medium formulations (Table 1) with an average moisture content of 94% were prepared to produce *Ganoderma lucidum* mycelium blocks. Each formulation was packaged in 2000 g bioriented polypropylene bags and sterilized at 121 °C for 2 h using the Tyndall method.

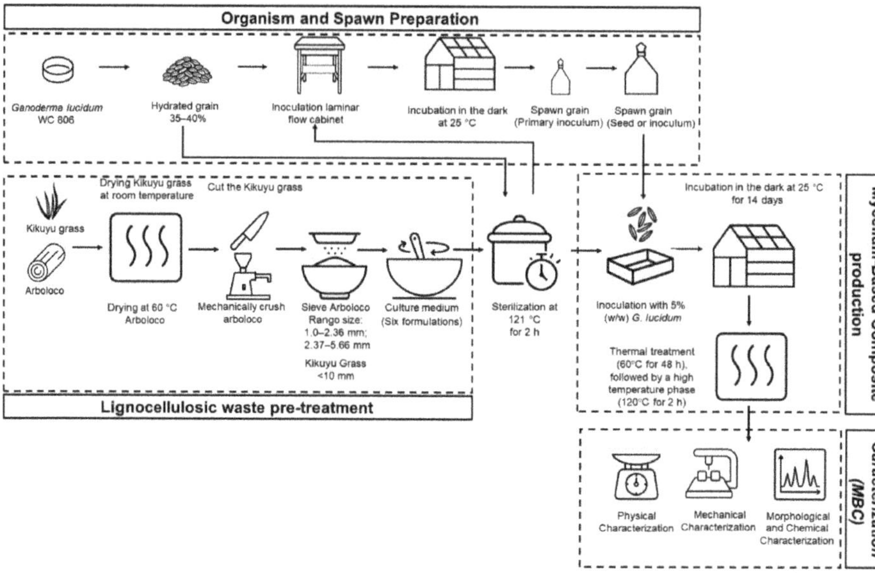

Figure 1. Schematic of the production and characterization process of a mycelium-based material (MBC) from lignocellulosic waste and *Ganoderma lucidum*. The diagram illustrates four main stages of the process: (1) Organism and spawn preparation involves cultivating Ganoderma lucidum WC 806 on sterilized grain until complete colonization. This process involves two phases: the preparation of a primary inoculum, followed by the production of fungal spawn intended for substrate inoculation. (2) Lignocellulosic waste pre-treatment occurs in parallel with the first stage. Both wastes are crushed, and after that, Arboloco is sieved into two particle ranges. Finally, the lignocellulosic waste is mixed based on the six formulations proposed in this work. (3) Mycelium-based composite production uses mixtures of Kikuyu grass (*Pennisetum clandestinum*) and Arboloco (*Montanoa quadrangularis*). This stage includes the preliminary conditioning of materials, substrate formulation, sterilization, and inoculation, followed by incubation until complete colonization of the material. (4) Material characterization involves thermal treatment for drying and inactivation, followed by physical, mechanical, and chemical analyses to evaluate the potential of the composite as an alternative biomaterial.

Table 1. Substrate formulation conditions based on the *w/w* % of the substrate on a dry basis and Arboloco particle size distribution.

Particle Size, mm	F1 (70%A/30% G)	F2 (30% A/70% G)	T (100% A/0% G)
L: 1.0–2.36	70	30	100
H: 2.37–5.66	30	70	0

Six formulations: TH, TL, F1L, F1H, F2L, F2H, considering A: Arboloco, G: Kikuyo grass, L: low particle sizes, H: high particle sizes.

After sterilization, the substrate was transferred to acrylic molds disinfected with a 7% (*v/v*) formaldehyde solution. Four mold sizes were employed: M1 (6 cm × 6 cm × 4 cm; 15-cell mold), M2 (31 cm × 8.5 cm × 4 cm; 3-cell mold), M3 (12 cm × 6 cm × 7 cm; 6-cell mold), and M4 (33 cm × 33 cm × 4 cm). Twelve replicates were prepared for each combination of formulation and particle size. Each mold cell was inoculated with 5% (*w/w*) *G. lucidum* mycelium (wet basis relative to substrate mass) and introduced into a polypropylene bag with a cotton plug to enable gas exchange. The molds were incubated in the dark at 25 °C for 14 days to achieve complete mycelial colonization. At the midpoint of the incubation period, molds were inverted to ensure homogeneous mycelial coverage. A stepped thermal treatment was applied to inactivate the *G. lucidum* mycelium: an initial low-temperature phase (60 °C for 48 h) followed by a high-temperature phase (120 °C for 2 h).

2.3. MBC Physical Characterization

2.3.1. Apparent Density

The MBC samples were previously dried at 50 °C for 96 h. The apparent density was calculated by weighing the dried samples and calculating their volume according to the ASTM D3573-20 standard protocol [29]. The formula for dry density was $\rho = m/V$, where ρ is the dry density in kg/m^3, m is the mass of the composites in kg, and v denotes the sample volume in m^3. At least four replicates were recorded for each experimental condition.

2.3.2. Water Absorption

The water absorption test was developed based on the standard method ASTM D570-22 [30]. Before the composite was dried at 50 °C for 24 h, the initial composite mass was determined after cooling. Subsequently, the samples were submerged in deionized water for 48 h and weighed. Due to the low density of the MBC material, it was forced to remain submerged using wire mesh and crystal balls. The weight change was calculated using the following equation:

$$\text{Mass Change (\%)} = (W - D)/D \times 100, \quad (1)$$

where W = wet mass (kg) and D = dry mass. Each experimental condition was carried out five times.

2.3.3. Shrinkage

The shrinkage of the sample was determined and calculated based on the wet and dry volumes using the following expression:

$$\text{Shrinkage percentage (\%)} = [(V_1 - V_2)/V_1] \times 100, \quad (2)$$

where V_1 = the wet volume of the sample and V_2 = the dry volume of the sample. Thirty-six replications evaluated each experimental condition.

2.4. MBC Morphological and Chemical Characterization

The functional groups for each MBC were assessed by Fourier Transform Infrared Spectroscopy (FTIR) using the Thermo Scientific model Nicolet iS5 (Thermo Fisher Scientific, Waltham, MA, USA), covering a range of 400–4000 cm^{-1}. The microscopic morphology of the interior of the composite (approximately 3 mm from the surface) was photographed and examined with a scanning electron microscope (SEM), TENSCAN VEGA 4th generation, (TESCAN, Brno, Czech Republic) using an accelerating voltage of 15 kV.

2.5. MBC Mechanical Characterization

2.5.1. Compression Strength

The compressive strength tests were determined according to procedure B of the standard norma ASTM C165-23 [31] utilizing a HUMBOLT HM-5030 F instrument (Humboldt Mfg, Elgin, IL, USA) with a 5 kN load capacity. Each sample (5 × 5 × 2.2 cm) was compressed until it achieved 50% of its original thickness at 12.5 mm/min. The contact surface on the plates was not perfect due to the rough surface of the samples. The loads and tenses were recorded periodically. An initial pre-load of 2% of the final load was considered to ensure the homogeneity of the samples.

2.5.2. Flexural Strength

The flexural strength was measured based on procedure B of the standard norma AST C203-22 [32]. A rectangular cross-section (27 × 6 × 2.9) was exposed to apply force through the 3-point bend test at 36.55 mm/min speed using a SHIMADZU AG-X instrument (Shimadzu Corporation, Kyoto, Japan) with a load cell of 10 kN. The force was applied to the sample until it reached fracture or a maximum deflection of 22 mm. The Modulus of Rupture (MOR) and Modulus of Elasticity (MOE) were calculated using the following equation.

Maximum deflection was calculated by:

$$D = (\varepsilon L^2)/6d, \quad (3)$$

where

D: deflection at the center of the sample (mm);

ε: strain (mm/mm);

L: support span (mm);

d: depth of the beam (mm).

The modulus of rupture (MOR) and modulus of elasticity (MOE) were calculated using the following equations:

$$MOR = 3PL/(2bd^2), \quad (4)$$

where

P: load at a given point on the load–deflection curve (N);

b: width of the beam tested (mm);

and

$$MOE = (PL^3)/48ID, \quad (5)$$

where I: moment of inertia (mm^4).

2.6. MBC Thermal Characterization

The thermal properties, conductivity, diffusivity, and specific heat were recorded using the ISOMET 2114 instrument (APPLIED PRECISION Measurement and test solution, Bratislava, Slovakia) following the standard ASTM-D5334 [33]. The cylindrical samples (10 × 4.5 × 4.5 cm) were dried at 50 °C until they reached a constant weight. Subsequently, the samples were wrapped with an ultra-thin sheet of low-density polyethylene to preserve their anhydrous state. They were then kept at a constant temperature of 22 °C for 4 h before the thermal measurement. The needle probe was inserted into each prismatic sample to a minimum depth of 8 cm, ensuring proper contact. Finally, the thermal data were recorded.

3. Results and Discussion

3.1. Morphological Characteristics

Figures 2 and 3 illustrate the morphological characteristics of mycelium growth. The analysis of the top surface reveals a dense mycelial coverage with pronounced surface irregularities, particularly in conditions containing some Kikuyu grass (Figure 2). Cross-sectional images show that high Arboloco particle sizes under conditions TH and F1H favor dense internal hyphal colonization, occupying mainly the inner air voids of the biomaterial, while conditions of low particle size (TL and F1L) evidence a complete hyphal coverage (both inner and outer). The F2 formulation shows an inverse trend, where a tightly packed structure associated with a higher mycelium yield was associated with larger particle sizes (F2H).

Figure 2. Photographs of the mycelium-based composite at the top surface and cross-section under six different formulations—TH, TL, F1L, F1H, F2L, and F2H—where A: Arboloco pith; G: Kikuyu grass; F1: 70% A/30% G; F2: 30% A/70% G; and Arboloco pith particle sizes are L: 1.0–2.36 mm and H: 2.37–5.66 mm.

The analysis of SEM images (Figure 3A–I) demonstrates noticeable elongated, highly branched, and ramified hyphae with randomly arranged and oriented growth. The T1 formulation (100% A/0% G) exhibits a low-density mycelium surface layer around the substrate, with no hyphal adherence (Figure 3A–C). At the same time, a higher number of air voids is observed in the material. Nevertheless, increasing the proportion of Kikuyu grass favors the densification of the mycelium. In the case of F1, hyphae growing around the substrate without adhering to the material (Arboloco pith and Kikuyu grass) were detected (Figure 3D–F). In contrast, Figure 3G–I show a higher densification of hyphae and firm adherence to the substrate, developing a hyphal bundle embedded in a polysaccharide matrix (Figure 3G,H) as observed by Pesciaroli et al. [34] These characteristics seem to be correlated with the complete and tight mycelium structure observed with the F2 formulation (Figure 2).

Fourier transform infrared spectroscopy (FTIR) analysis of the MBC colonized by *G. lucidum* (Figure 4) revealed distinct structural modifications compared to non-colonized raw material (uncrushed Arboloco pith) obtained in previous studies in our laboratory [35]. The peaks in the fingerprint and their functional group interactions are assigned based on previous studies FTIR studies on the changes in wood chemistry following decay by brown-rot and white-rot fungi [18,36]. A progressive decrease (T1 < F1 < F2) in intensity at 1042 cm^{-1} (C-O-C and C-O stretching vibrations in cellulose and hemicellulose) and 890 cm^{-1} (C–H deformation of cellulose) demonstrates fungal hydrolysis of polysaccharides in cellulose, while the decline of the 1735 cm^{-1} band (ester C=O stretching) indicates hemicellulose deacetylation [18,37]. A notable change was observed at 1247 cm^{-1}, where the strong peak in non-colonized Arboloco material declines after fungal colonization, indicating the degradation of aryl-ether and ester linkages in lignin and hemicellulose, respectively. In concordance with the enzymatic white-rot fungi, the complete disappearance of peaks at 1425 cm^{-1} (CH_2 bending in crystalline cellulose), 1596 cm^{-1}, and 1505 cm^{-1} (aromatic C=C vibrations in lignin) in the MBC samples provided further evidence of the structural breakdown of lignin and complex carbohydrate bonds. On the other hand, the progressive broadening and elongation of the 1634 cm^{-1} peak (C=O stretching) shows the accumulation of oxidized lignin derivatives (carbonyl and quinoid structure) associated to components such as chitin and glucans, while the broadening and intensification of the 3317 cm^{-1} (O-H/N-H stretching) band with respect to the control (uncrushed Arboloco pith) suggests increased hydrophilicity resulting from both the exposure of cellulose hydroxyl groups after lignin removal and the water-binding capacity of fungal mycelium. The inverse relationship between peaks at 1042 cm^{-1} and 1634 cm^{-1} (Figure 4) suggests the accumulation of exopolysaccharides, as observed in Figure 3G–I under the F2 formulation. Previous studies support this behavior, in which a decline in the peak at 1042 cm^{-1} is attributed to the depolymerization of cellulose into monomeric sugars, which fungi metabolize to synthesize the EPS as β-glucans [18], while an increase in the peak at 1634 cm^{-1} is linked to oxidized lignin derivatives and the simultaneous production of carbonyls and glucuronic acid (a component of fungal β-glucans) [38,39]. These spectroscopic changes collectively illustrate the unique ability of white-rot fungi to degrade all major lignocellulose components through a combination of hydrolytic and oxidative mechanisms while simultaneously introducing fungal exopolysaccharides into the substrate matrix.

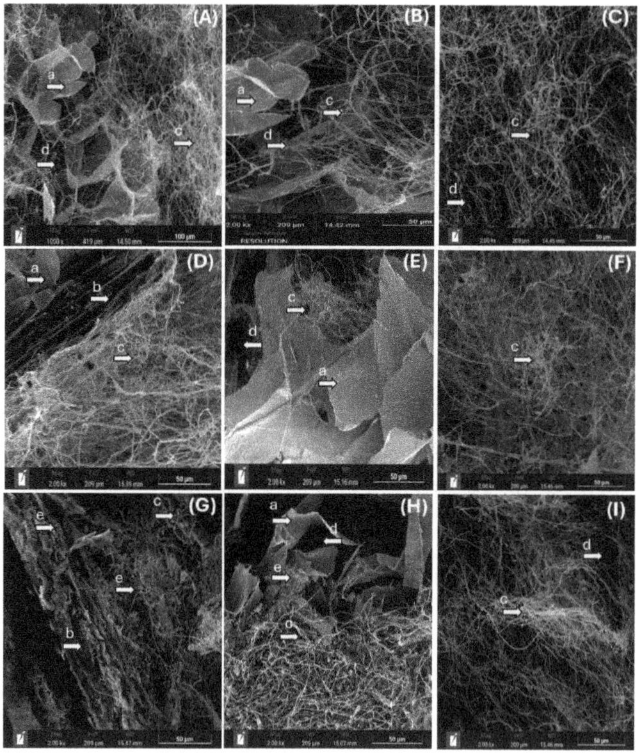

Figure 3. SEM images of the mycelium-based composite (MBC) considering the following formulations: (**A–C**) 100% A/0 G; (**D–F**) 70% A/30% G; (**G–I**) 30% A/70% G. Characteristics of the MBC are based on the following aspects: (a) Arboloco pith, (b) Kikuyu grass, (c) hyphae structure, (d) air void, and (e) exopolysaccharides (EPSs).

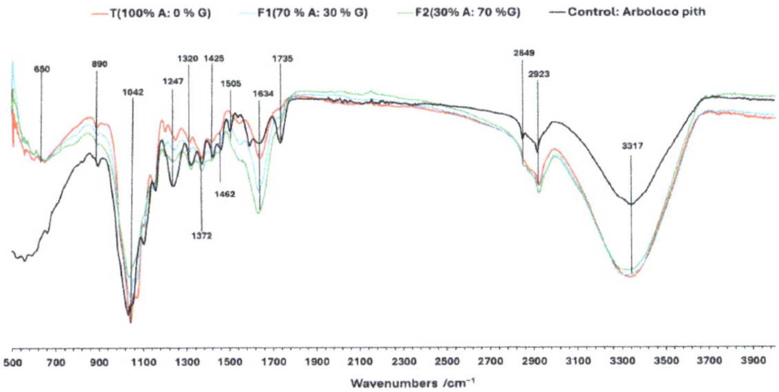

Figure 4. FTIR spectra of MBC under different formulations: 100% A/0%G, 70% A/30% G, 30% A/70% G, and control composed of uncrushed Arboloco pith without colonization. A: Arboloco pith; G: Kikuyu grass. Previously, the transmittance signals were standardized.

3.2. Physical Properties

The analysis of the physical properties of the MBC considered three parameters—apparent density, shrinkage, and water adsorption—which were recorded throughout the experiment. Figure 5A shows that the biomaterial composed of 100% Arboloco achieved the lowest average density of 31 $\pm$ 1.6 kg m^{-3}. These trends are reversed as the percentage of Arboloco decreases from 100 to 30%, while the Kikuyu grass content increases from 0 to 70%, causing a rise in the average density up to 60.4 $\pm$ 4.5 kg m^{-3}. Comparatively, the MBC density values were 3.3-fold lower than those reported for lignocellulosic substrates such as beech sawdust [12], wheat bran [40], and rice and wheat straw [41] used in *G. lucidum* mycelium composite (260 to 370 kg m^{-3}). The low apparent density of the Arboloco–Kikuyu grass MBC (31–60.4 kg m^{-3}) positions it as a lightweight biomaterial, comparable to natural insulators like sheep's wool (30–114 kg m^{-3}) [42] and synthetic alternatives such as glass wool (10–100 kg m^{-3}) [43]. This lightweight property suggests its suitability for applications in non-structural and building envelopes, being typically used for thermal insulation materials. The inverse correlation between the Arboloco pith content and apparent density is probably linked to low-density mycelium growth and the high presence of air voids (Figure 3), causing an increase in the MBC porosity, a typical characteristic of lightweight biomaterials. A similar trend was observed in the water absorption (Figure 5B), achieving values above 1700% after 48 h of water immersion under the formulations TL and TH, surpassing raw uncrushed Arboloco pith without mycelium colonization (425%) previously evaluated in our laboratory [35]. This trend could be promoted by the increase in the porosity of the bio-composite, a consequence of the Arboloco particle size distribution and the hygroscopic properties of the mycelium, which accelerate the hydraulic penetration and lower drag force, increasing the water intake of the MBC [44]. However, increasing the fiber content in the substrate triggers a sharp decline in the average water absorption, reaching 556 $\pm$ 184% in the F2 conditions and achieving consistent values as reported in the MBC literature, which ranges from 40 to 580 [45]. Regarding shrinkage post-drying, the presence of fiber in the substrate (F1–F2), independent of particle size, leads to a maximum shrinkage percentage of 53.6 $\pm$ 5.6%, exceeding that of substrates with only Arboloco (T1–T2) by 10%. This shrinkage percentage was markedly higher than other studies of MBCs utilizing sawdust (6.8–14.5%) [46], hemp and flax (7–15%) [26] as a mycelium growth support. According to Islam et al. [47], the random direction of hyphal growth in the mycelium microstructures, as observed in our MBC (Figure 3), is evidence of intense heterogeneous deformation, resulting in an increase in the shrinkage properties. These high shrinking and water absorption properties make it difficult to predict the geometric and dimensional properties after drying treatment, becoming a challenge in this novel biomaterial. Hot pressing post-treatment on the *Trametes versicolor* MBC has demonstrated a decrease in the hydrophobicity of the mycelium and porosity of the bio-composite [22], which favors the diminishment of water uptake and shrinking properties. This alternative arises as an option to implement in our material that can contribute to overcoming the constraint associated with forming regular constructive structures for implementation in massive-scale production.

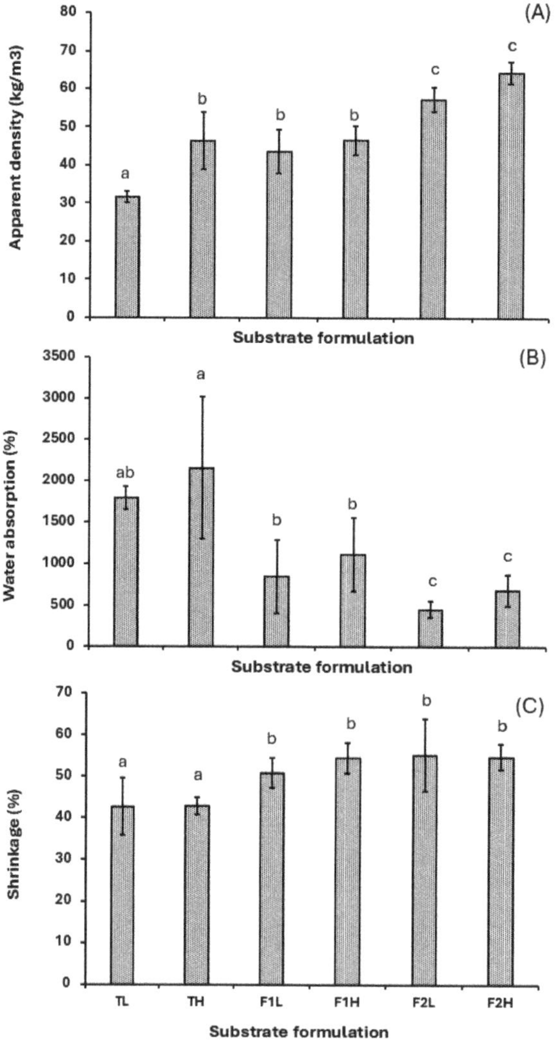

Figure 5. Physicochemical properties of MBC: (**A**) apparent density, (**B**) water absorption, and (**C**) shrinkage under different substrates. A (Arboloco pith), G (Kikuyu grass), and particle sizes L (1.0–2.36 mm) and H (2.37–5.66 mm). Substrate formulations T (100% A/0% G), F1 (70% A/30% G), and F2 (30% A/70% G). The results are shown as the mean ± SD. The lowercase letters above bars indicate significant differences between substrate formulations (Tukey, $p \leq 0.05$, $n \geq 10$).

3.3. Mechanical Properties

The mechanical properties of the developed MBC showed no significant variations in the Young's modulus of compression due to the composition of the raw materials used (Figure 6A). The average value obtained was 0.264 ± 0.045 MPa, indicating low stiffness but notable elastic capacity. This behavior suggests that the material can absorb energy and deform without compromising its structural integrity. Such behavior aligns with prior findings highlighting the potential of lightweight biomaterials with low stiffness but high energy absorption capacity, such as cork [48]. The compressive strength analysis showed a progressive increase with the deformation percentage, with average values ranging from 0.017 MPa at 10% to 0.146 MPa at 50% deformation (Figure 6B). This trend reflects a nonlinear strengthening response under higher compressive loads. Regarding compressive strength, the formulations exhibited consistent performance, with formulation F1H achieving strengths of 0.0031, 0.0589, 0.0934, 0.1465, and 0.154 MPa at deformation levels of 10%, 20%, 30%, 40%, and 50%, respectively. On the other hand, increasing the Kikuyu grass content (formulation F2) improved the material's performance at higher deformation levels, reaching an average compressive strength of 0.1686 MPa at 50% deformation. Although these values are consistent with those reported for low-density mycelium-based materials, such as the 0.024 MPa reported by Vidholdová et al. [49] at 20% deformation using *Trametes versicolor* and spruce wood, they are notably lower than those observed in densified composites. For instance, Sisti et al. [50] achieved compressive strengths ranging from 0.04 to 0.08 MPa at 25% deformation to 0.53 to 1.00 MPa at 75% deformation using wheat bran and industrial fibers inoculated with *Basidiomycetes*, while Etinosa et al. [51] reported strengths of 0.9 MPa at 10% deformation using *Pleurotus ostreatus* and a substrate enriched with calcium and starch. In contrast, recent research suggests that the relationship between density and compressive strength in mycelium-based materials is not always direct. Peng et al. [52] observed a positive correlation between these properties. In contrast, the results of our study did not show significant differences in the compressive strength despite an increase in the apparent density due to substrate composition. This indicates that the mechanical properties remain unchanged under low-density conditions (<0.1 g/cm^3). This aligns with the results of Vidholdová et al. [49], who attributed the low compressive strength of porous mycelium-based materials to increased air voids, which contribute to early-stage deformation. Compared to other studies, composites with higher densities and base materials such as beech sawdust or natural fibers have demonstrated significantly higher strengths. For example, Sivaprasad et al. [53] reported compressive strengths of up to 2.49 MPa at 20% deformation using *Pleurotus ostreatus* and *Ganoderma lucidum* grown on beech sawdust and other agro-industrial substrates. At higher deformations (>50%), these strengths can be up to five times higher than those observed in this study, highlighting the critical role of substrate selection and densification processes in mechanical performance. Regarding flexural properties (Figure 6C,D), the MBCs exhibited a modulus of rupture (MOR) of 0.0339 ± 0.0145 MPa and a modulus of elasticity (MOE) of 0.75 ± 0.3237 MPa. Although these values fall within the range reported for non-pressed mycelium-based composites with a MOR from 0.05 to 0.29 MPa [54], these are relatively low compared to other studies. For instance, Rigobello & Ayres [55] achieved a MOE of 3.32 MPa and a MOR of 0.306 MPa using *Ganoderma lucidum* and European beech particles. Additionally, some formulations (F1 and F2) outperformed samples TH, suggesting that the substrate composition can influence flexural properties. However, variations between different particle sizes were not significant. The combination of low stiffness and high resilience observed in this study highlights the potential of mycelium-based composites for applications requiring flexibility and energy absorption, such as impact-resistant panels and safety components in the automotive industry. While the mechanical performance observed is relatively low compared to other MBCs reported in the literature, the biodegradable and lightweight nature of these materials positions them as sustainable alternatives for various industrial applications. Additionally, the limited dependence of mechanical properties on material density under low-density configurations could be advantageous for developing lightweight materials for non-structural applications.

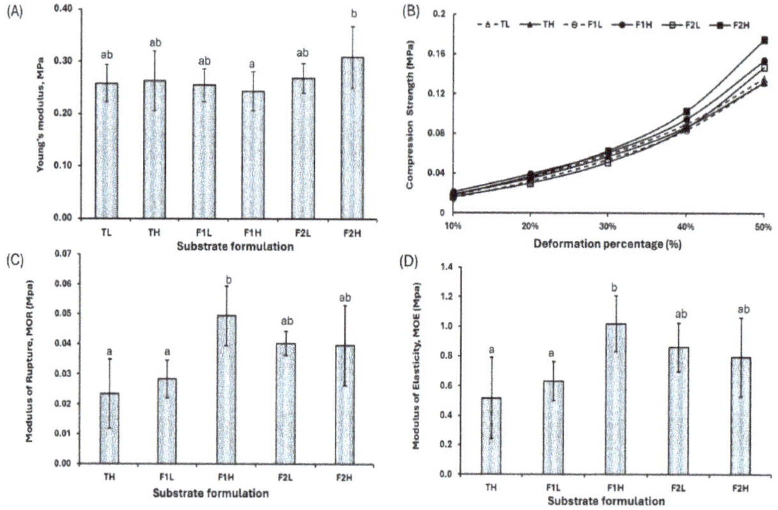

Figure 6. Mechanical properties of MBC: (**A**) Young's modulus, (**B**) compressive strength, (**C**) modulus of rupture (MOR), and (**D**) modulus of elasticity (MOE) under different substrates A (Arboloco pith) and G (Kikuyu grass) and particle sizes L (1.0–2.36 mm) and H (2.37–5.66 mm). Substrate formulations T (100% A/0% G), F1 (70% A/30% G), and F2 (30% A/70% G). The results are shown as the mean ± SD. The lowercase letters above bars indicate significant differences between substrate formulations (Tukey, $p \leq 0.05$, $n \geq 6$).

3.4. Thermal Properties

Previous studies in our laboratory on the raw substrate of uncrushed Arboloco pith have demonstrated its promising use as a building insulation material due to its low thermal conductivity of 0.044 W m^{-2} K [35]. Based on the results observed in Figure 7A, this thermal conductivity was conserved in the MBC, reaching an average value of 0.047 ± 0.002 W m^{-1} K^{-1} throughout the experiment. A comparative analysis of the experimental conditions shows that the particle size of the crushed Arboloco pith did not significantly affect the thermal conductivity across formulations (T, F1, and F2). However, conditions with a high proportion of Kikuyu grass F2 (30% A/70% G) exhibit the lowest thermal conductivity at 0.045 W m^{-1} K^{-1}. These outstanding thermal performances are typically attributed to materials with a low density and high porosity, as observed in our experiments (<80 kg m^{-3}), being comparable with the properties of commercial lightweight insulation boards (density below 150 kg m^{-3}) such as expanded polystyrene and cork, which exhibit thermal conductivity values of 0.03–0.05 W m^{-1} K^{-1} [7,56]. For MBCs, prior studies emphasize that thermal properties are strongly influenced by the substrate composition more than the fungal species. For instance, lignocellulosic substrates such as birch wood [23], bamboo fiber [57], sawdust [58], and coffee silver skin [59] bonded by the mycelium of *Ganoderma lucidum*, *Pleurotus ostreatus*, and *Trametes versicolor* have reported wide thermal conductivity ranges from 0.04 to 0.18 W m^{-1} K^{-1}. In this sense, our MCB demonstrates competitive thermal insulation properties, positioning it as a sustainable, low-cost biomaterial for construction applications.

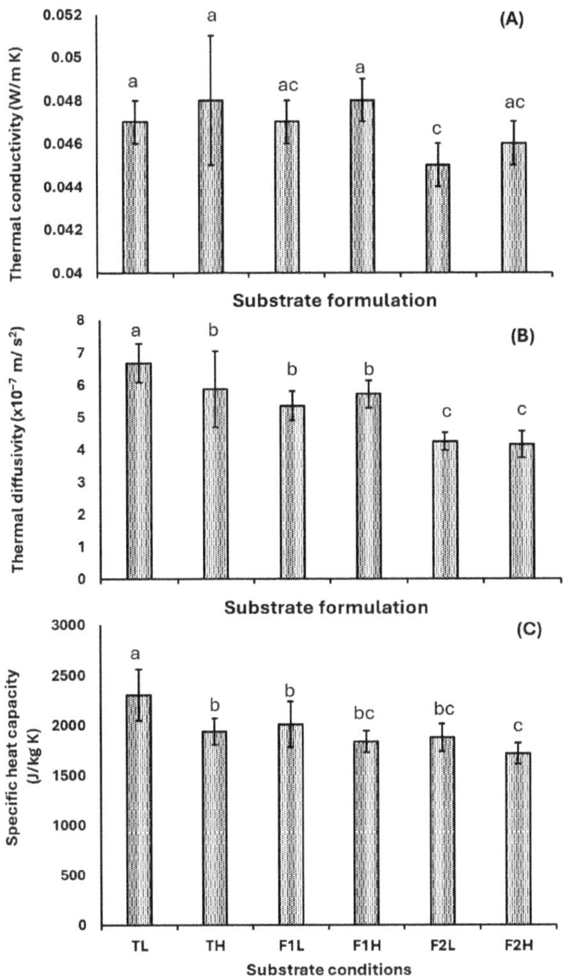

Figure 7. Thermal properties of the MBC: (**A**) thermal conductivity, (**B**) thermal diffusivity, and (**C**) specific heat capacity under different substrates T (Arboloco) and C (Kikuyu grass) and particle sizes L (1.0–2.36 mm) and H (2.37–5.66 mm). Substrate formulations T (100% A/0% G), F1 (70% A/30% G), and F2 (30% A/70% G). The results are shown as the mean ± SD. The lowercase letters above bars indicate significant differences between substrate formulations (Tukey, $p \leq 0.05$, $n \geq 9$).

Thermal diffusivity and specific heat capacity show a decline in the average values from $6.6 \pm 0.6 \times 10^{-7}$ to $4.1 \pm 0.4\ ^{3}\ 10^{-7}$ $m^2\ s^{-1}$ and 2303 ± 254 to 1714 ± 105 J/kg K, respectively, as the proportion of Kikuyu grass increases in the substrate (Figure 7B,C). Prior studies on plaster and high-density polyethylene (HDPE) material demonstrated similar trends, where the addition of wood, wheat, and flax fiber reduced the thermal diffusivity of the bio-composite [60,61]. Furthermore, the highest thermal diffusivity patterns under TL and TH conditions suggest a direct association with the low mycelium densification (Figure 3A–C) and apparent density properties (Figure 5), leading to highly porous structures and air-entrapped voids. These microstructural features enhance heat retention while lowering the thermal diffusivity. The specific heat capacity (Figure 7C) shows a higher performance than commercial insulating materials such as expanded polystyrene, polyurethane foam, and wood fiber (800–1500 J/kg K) [62,63], even reaching competitive properties as observed with natural composites of bamboo fibers, cotton waste, or epoxy–natural fiber (1300–1800 J/kg K) [64,65]. These characteristics suggest that our MBC performs better in absorbing and storing heat, reducing temperature fluctuations, and enhancing indoor thermal stability and comfort, which is a critical advantage for green building design. In contrast, mycelium-based composites (MBCs) fabricated from *Ganoderma resinaceum* [66], *Pleurotus ostreatus* [19], and *Ganoderma Lucidum* (previous studies in our laboratory), using rye berries and wheat straw as substrates, exhibited a specific heat capacity ranging from 4100 to 6500 J kg^{-1} K^{-1}, reaching values threefold higher than those observed across our study. These findings suggest that thermal properties are primarily determined by substrate composition rather than fungal species. Future studies should prioritize optimizing substrate formulations (e.g., blends of lignocellulosic agro-industrial wastes) to improve thermal performance while aligning with circular economy principles and carbon neutrality objectives.

4. Conclusions

- The development of mycelium-based composite materials (MBCs) using *Ganoderma lucidum* and Colombian agro-industrial wastes (Arboloco and Kikuyu grass) illustrates significant potential as a lightweight thermal insulation material with a low environmental impact, consistent with the principles of sustainable construction.
- Formulation F2, consisting of 30% Arboloco and 70% Kikuyu grass, exhibited an outstanding thermal performance, with a thermal conductivity of 0.045 W m^{-1} K^{-1} and a specific heat capacity of 1714 $\pm$ 105 J kg^{-1} K^{-1}. Moreover, the low apparent density (60.4 $\pm$ 4.5 kg·m^{-3}) positions it as a promising lightweight insulating material that could serve as an alternative to synthetic insulation materials like expanded polystyrene. These results establish our MBC as a viable energy-efficient option.
- Despite the thermal and ecological advantages of these biomaterials, significant challenges, including high volumetric shrinkage (53.6%) and elevated water absorption (556 $\pm$ 184%), must be addressed, as they limit the materials' direct applicability without additional treatment. These findings highlight the necessity of incorporating post-processing techniques, such as hot pressing, to enhance dimensional stability and developing strategies aimed at reducing water absorption capacity.
- Future research should focus on substrate optimization by exploring hybrid combinations of lignocellulosic biomass to enhance mechanical performance without compromising eco-design principles. Additionally, the standardization of growth and treatment protocols will be essential for achieving controlled and reproducible scaling of MBC production.

Author Contributions: G.J.-O.: Conceptualization, methodology, writing—original draft preparation, and formal analysis; J.S.A.: writing—review and editing, data curation and analysis, and supervision; Y.L.V.-C.: conceptualization and methodology; R.A.T.-C.: project administration, funding acquisition, and review and editing; S.M.: project administration, funding acquisition, and review and editing. All authors have read and agreed to the published version of the manuscript.

Funding: This work was supported by project HERMES No. 57919 under the Joint Call of Universidad de Caldas and Universidad Nacional de Colombia–Sede Manizales for Research Projects 2021 (Second Cohort).

Institutional Review Board Statement: Not applicable for studies not involving humans or animals.

Informed Consent Statement: Not applicable for studies not involving humans.

Data Availability Statement: The original contributions presented in the study are included in the article, further inquiries can be directed to the corresponding author.

Acknowledgments: The authors thank the Bioprocess and Agroindustry Plant at Universidad de Caldas and the laboratories of Processes, Materials, Chemistry, Biopolymers, and Nanoparticulate Materials at Universidad Nacional de Colombia–Sede Manizales for their invaluable support.

Conflicts of Interest: The authors declare no conflicts of interest.

References

1. Yadav, M.; Agarwal, M. Biobased building materials for sustainable future: An overview. *Mater. Today Proc.* **2021**, *43*, 2895–2902. [CrossRef]
2. Cintura, E.; Faria, P.; Duarte, M.; Nunes, L. Eco-efficient boards with agro-industrial wastes—Assessment of different adhesives. *Constr. Build. Mater.* **2023**, *404*, 132665. [CrossRef]
3. Xu, P.; Zhu, J.; Li, H.; Wei, Y.; Xiong, Z.; Xu, X. Are Bamboo Construction Materials Environmentally Friendly? A Life Cycle Environmental Impact Analysis. *Environ. Impact Assess. Rev.* **2022**, *96*, 106853. [CrossRef]
4. Wang, R.; Haller, P. Applications of Wood Ash as a Construction Material in Civil Engineering: A Review. *Biomass Convers. Biorefin* **2022**. [CrossRef]
5. Sun, X.; He, M.; Li, Z. Novel Engineered Wood and Bamboo Composites for Structural Applications: State-of-Art of Manufacturing Technology and Mechanical Performance Evaluation. *Constr. Build. Mater.* **2020**, *249*, 118751. [CrossRef]
6. Ginga, C.P.; Ongpeng, J.M.C.; Daly, M.K.M. Circular Economy on Construction and Demolition Waste: A Literature Review on Material Recovery and Production. *Materials* **2020**, *13*, 2970. [CrossRef]
7. Haque, S.E.; Nahar, N.; Chowdhury, N.N.; Gazi-Khan, L.; Sayanno, T.K.; Muktadir, M.A.; Haque, M.S. Identification of Recycling Potential of Construction and Demolition Waste: Challenges and Opportunities in the Greater Dhaka Area. *Environ. Monit. Assess.* **2025**, *197*, 646. [CrossRef]
8. Nyika, J.; Dinka, M. Recycling Plastic Waste Materials for Building and Construction Materials: A Minireview. *Mater. Today Proc.* **2022**, *62*, 3257–3262. [CrossRef]
9. Wang, C.; Liu, J.; Lu, B.; Zhang, Y.; Ma, Z. Stiffness Degradation and Mechanical Behavior of Microfiber-Modified High-Toughness Recycled Aggregate Concrete under Constant Load Cycling. *Eng. Fract. Mech.* **2024**, *312*, 110608. [CrossRef]
10. Ma, Z.; Wang, B.; Zhang, Z.; Zhang, Y.; Wang, C. New Insights into the Effects of Silicate Modulus, Alkali Content and Modification on Multi-Properties of Recycled Brick Powder-Based Geopolymer. *J. Build. Eng.* **2024**, *97*, 110989. [CrossRef]
11. Abu-Jdayil, B.; Mourad, A.H.; Hittini, W.; Hassan, M.; Hameedi, S. Traditional, state-of-the-art and renewable thermal building insulation materials: An overview. *Constr. Build. Mater.* **2019**, *214*, 709–735. [CrossRef]
12. Vašatko, H.; Gosch, L.; Jauk, J.; Stavric, M. Basic Research of Material Properties of Mycelium-Based Composites. *Biomimetics* **2022**, *7*, 51. [CrossRef]
13. Contreras, E.; Flores, R.; Gutiérrez, A.; Cerro, D.; Sepúlveda, L.A. Agro-Industrial Wastes Revalorization as Feedstock: Production of Lignin-Modifying Enzymes Extracts by Solid-State Fermentation Using White Rot Fungi. *Prep. Biochem. Biotechnol.* **2023**, *53*, 488–499. [CrossRef] [PubMed]
14. Ekblad, A.; Wallander, H.; Godbold, D.L.; Cruz, C.; Johnson, D.; Baldrian, P.; Björk, R.G.; Epron, D.; Kieliszewska-Rokicka, B.; Kjøller, R.; et al. The production and turnover of extramatrical mycelium of ectomycorrhizal fungi in forest soils: Role in carbon cycling. *Plant Soil.* **2013**, *366*, 1–27. [CrossRef]
15. Aiduang, W.; Suwannarach, N.; Kumla, J.; Thamjaree, W.; Lumyong, S. Valorization of agricultural waste to produce myco-composite materials from mushroom mycelia and their physical properties. *Agric. Nat. Resour.* **2022**, *56*, 1083–1090.
16. Chulikavit, N.; Wang, C.; Huynh, T.; Yuen, A.C.Y.; Khatibi, A.; Kandare, E. Fireproofing Flammable Composites Using Mycelium: Investigating the Effect of Deacetylation on the Thermal Stability and Fire Reaction Properties of Mycelium. *Polym. Degrad. Stab.* **2023**, *215*, 110419. [CrossRef]
17. Irbe, I.; Kirpluks, M.; Kampuss, M.; Andze, L.; Milbreta, U.; Filipova, I. Assessing the Conformity of Mycelium Biocomposites for Ecological Insulation Solutions. *Materials* **2024**, *17*, 6111. [CrossRef]
18. Elsacker, E.; Vandelook, S.; Brancart, J.; Peeters, E.; De Laet, L. Mechanical, physical and chemical characterisation of mycelium-based composites with different types of lignocellulosic substrates. *PLoS ONE* **2019**, *14*, 7. [CrossRef]
19. Zhang, X.; Hu, J.; Fan, X.; Yu, X. Naturally grown mycelium-composite as sustainable building insulation materials. *J. Clean. Prod.* **2022**, *342*, 130784. [CrossRef]
20. Camilleri, E.; Narayan, S.; Lingam, D.; Blundell, R. Mycelium-Based Composites: An Updated Comprehensive—Overview. *Biotechnol. Adv.* **2025**, *79*, 108517. [CrossRef]
21. Jin, Q.; Zhang, Z.; Chen, J. A Study on the Thermal Performance of *Pleurotus ostreatus*/Straw Mycelium Composites and Its Application in Building Envelopes. *J. Build. Eng.* **2024**, *92*, 109646. [CrossRef]
22. Nussbaumer, M.; Van Opdenbosch, D.; Engelhardt, M.; Briesen, H.; Benz, J.P.; Karl, T. Material characterization of pressed and unpressed wood–mycelium composites derived from two *Trametes species*. *Environ. Technol. Innov.* **2023**, *30*, 103063. [CrossRef]

23. Sun, W.; Tajvidi, M.; Howell, C.; Hunt, C.G. Insight into mycelium-lignocellulosic bio-composites: Essential factors and properties. *Compos. Part. A Appl. Sci. Manuf.* **2022**, *161*, 107125. [CrossRef]
24. Yang, L.; Park, D.; Qin, Z. Material Function of Mycelium-Based Bio-Composite: A Review. *Front. Mater.* **2021**, *8*, 737377. [CrossRef]
25. Voutetaki, M.E.; Mpalaskas, A.C. Natural Fiber-Reinforced Mycelium Composite for Innovative and Sustainable Construction Materials. *Fibers* **2024**, *12*, 57. [CrossRef]
26. Kohphaisansombat, C.; Jongpipitaporn, Y.; Laoratanakul, P.; Tantipaibulvut, S.; Euanorasetr, J.; Rungjindamai, N.; Chuaseeharonnachai, C.; Kwantong, P.; Somrithipol, S.; Boonyuen, N. Fabrication of Mycelium (*Oyster mushroom*)-Based Composites Derived from Spent Coffee Grounds with Pineapple Fibre Reinforcement. *Mycology* **2024**, *15*, 665–682. [CrossRef] [PubMed]
27. Álvarez Mejía, L.M. *Biología, uso y Manejo del Arboloco (Montanoa quadrangularis)*, 1st ed.; Editorial Universidad de Caldas: Manizales, Colombia, 2006; pp. 223–305.
28. Montoya, S.; López, D.M.; Segura, B. Influencia de la luz azul sobre la productividad del cultivo sólido de *Ganoderma lucidum*. *Rev. Colomb Biotecnol.* **2018**, *1*, 51–58. [CrossRef]
29. *ASTM, D3575-20*; Standard Test Methods for Flexible Cellular Materials Made From Olefin Polymers. ASTM International: West Conshohocken, PA, USA, 2020; 15, pp. 1–11.
30. *ASTM D570*; Test Method for Water Absorption of Plastics. ASTM International: West Conshohocken, PA, USA, 2022.
31. *ASTM C165-23*; Test Method for Measuring Compressive Properties of Thermal Insulations. ASTM International: West Conshohocken, PA, USA, 2023.
32. *ASTM C203-22*; Test Methods for Breaking Load and Flexural Properties of Block-Type Thermal. ASTM International: West Conshohocken, PA, USA, 2022.
33. *ASTM D5334-22*; Test Method for Determination of Thermal Conductivity of Soil and Rock by Thermal Needle Probe Procedure. ASTM International: West Conshohocken, PA, USA, 2022.
34. Pesciaroli, L.; Petruccioli, M.; Federici, F.; D'Annibale, A. *Pleurotus ostreatus* Biofilm-Forming Ability and Ultrastructure Are Significantly Influenced by Growth Medium and Support Type. *J. Appl. Microbiol.* **2013**, *114*, 1750–1762. [CrossRef]
35. Jimenes Obando, G.; Tolosa Correa, R.; Arcila Henao, J.S.; Montoya Barreto, S. Isolamento Térmico Natural: Médula de *Montanoa quadrangularis* Shultz BIP. In Proceedings of the 5 Congresso Luso-Brasileiro de Materiais de Construcao sustentáveis: Congresso Construcao 2024, Lisboa, Portugal, 6–8 November 2024.
36. Pandey, K.K.; Pitman, A.J. FTIR studies of the changes in wood chemistry following decay by brown-rot and white-rot fungi. *Int. Biodeterior. Biodegrad.* **2003**, *52*, 151–160. [CrossRef]
37. Qi, J.; Li, F.; Jia, L.; Zhang, X.; Deng, S.; Luo, B.; Zhou, Y.; Fan, M.; Xia, Y. Fungal Selectivity and Biodegradation Effects by White and Brown Rot Fungi for Wood Biomass Pretreatment. *Polymers* **2023**, *15*, 1957. [CrossRef]
38. Supramani, S.; Rejab, N.A.; Ilham, Z.; Ahmad, R.; Show, P.L.; Ibrahim, M.F.; Wan-Mohtar, W.A.A.Q.I. Performance of Biomass and Exopolysaccharide Production from the Medicinal Mushroom *Ganoderma lucidum* in a New Fabricated Air-L-Shaped Bioreactor (ALSB). *Processes* **2023**, *11*, 670. [CrossRef]
39. Hamza, A.; Khalad, A.; Kumar, D.S. Enhanced Production of Mycelium Biomass and Exopolysaccharides of *Pleurotus ostreatus* by Integrating Response Surface Methodology and Artificial Neural Network. *Bioresour. Technol.* **2024**, *399*, 130577. [CrossRef]
40. Brazkova, M.; Angelova, G.; Goranov, B. Food Science and Applied Biotechnology Effect of the lignocellulose substrate type on mycelium growth and biocomposite formation by *Ganoderma lucidum* GA3P. *Food Sci. Appl. Biotechnol.* **2022**, *5*, 211–218.
41. Cai, J.; Han, J.; Ge, F.; Lin, Y.; Pan, J.; Ren, A. Development of impact-resistant mycelium-based composites (MBCs) with agricultural waste straws. *Constr. Build. Mater.* **2023**, *389*, 131730. [CrossRef]
42. Dénes, T.O.; Iştoan, R.; Tămaş-Gavrea, D.R.; Manea, D.L.; Hegyi, A.; Popa, F.; Vasile, O. Analysis of Sheep Wool-Based Composites for Building Insulation. *Polymers* **2022**, *14*, 2109. [CrossRef] [PubMed]
43. Neila González, F.J. *Arquitectura Bioclimática en un Entorno Sostenible*, 1st ed.; Editorial Munilla-Lería: Madrid, Spain, 2004; pp. 89–99.
44. Alaneme, K.K.; Anaele, J.U.; Oke, T.M.; Kareem, S.A.; Adediran, M.; Ajibuwa, O.A.; Anabaranze, Y.O. Mycelium based composites: A review of their bio-fabrication procedures, material properties and potential for green building and construction applications. *Alex. Eng. J.* **2023**, *83*, 234–250. [CrossRef]
45. Jones, M.; Mautner, A.; Luenco, S.; Bismarck, A.; John, S. Engineered mycelium composite construction materials from fungal biorefineries: A critical review. *Mat. Des.* **2020**, *187*, 108397. [CrossRef]
46. Aiduang, W.; Kumla, J.; Srinuanpan, S.; Thamjaree, W.; Lumyong, S.; Suwannarach, N. Mechanical, Physical, and Chemical Properties of Mycelium-Based Composites Produced from Various Lignocellulosic Residues and Fungal Species. *J. Fungi* **2022**, *8*, 1125. [CrossRef]
47. Islam, M.R.; Tudryn, G.; Bucinell, R.; Schadler, L.; Picu, R.C. Morphology and mechanics of fungal mycelium. *Sci. Rep.* **2017**, *7*, 13070. [CrossRef]

8. Walsh, J.; Kim, H.I.; Suhr, J. Low velocity impact resistance and energy absorption of environmentally friendly expanded cork core-carbon fiber sandwich composites. *Compos. Part. A Appl. Sci. Manuf.* **2017**, *101*, 290–296. [CrossRef]
9. Vidholdová, Z.; Kormúthová, D.; Ždinsky, J.I.; Lagaña, R. Compressive Resistance of the Mycelium Composite. *Ann. WULS For. Wood Technol.* **2020**, *107*, 31–36. [CrossRef]
0. Sisti, L.; Gioia, C.; Totaro, G.; Verstichel, S.; Cartabia, M.; Camere, S.; Celli, A. Valorization of wheat bran agro-industrial byproduct as an upgrading filler for mycelium-based composite materials. *Ind. Crops Prod.* **2021**, *170*, 113742. [CrossRef]
1. Etinosa, P.O.; Salifu, A.A.; Salifu, T.A.; Obayemi, J.D.; Onche, E.O.; Aina, T.; Soboyejo, W.O. Self-organized mycelium biocomposites: Effects of geometry and laterite composition on compressive behavior. *J. Mech. Behav. Biomed. Mater.* **2023**, *142*, 105831. [CrossRef] [PubMed]
2. Peng, L.; Yi, J.; Yang, X.; Xie, J.; Chen, C. Development and characterization of mycelium bio-composites by utilization of different agricultural residual byproducts. *J. Bioresour. Bioprod.* **2022**, *8*, 78–89. [CrossRef]
3. Sivaprasad, S.; Byju, S.K.; Prajith, C.; Shaju, J.; Rejeesh, C.R. Development of a novel mycelium bio-composite material to substitute for polystyrene in packaging applications. *Mater. Today Proc.* **2021**, *47*, 5038–5044. [CrossRef]
4. Appels, F.V.W.; Camere, S.; Montalti, M.; Karana, E.; Jansen, K.M.B.; Dijksterhuis, J.; Krijgsheld, P.; Wöstenet, H.A. Fabrication factors influencing mechanical, moisture- and water-related properties of mycelium-based composites. *Mater. Des.* **2019**, *161*, 64–71. [CrossRef]
5. Rigobello, A.; Ayres, P. Compressive behaviour of anisotropic mycelium-based composites. *Sci. Rep.* **2022**, *12*, 1–13. [CrossRef]
6. Fu, H.; Ding, Y.; Li, M.; Li, H.; Huang, X.; Wang, Z. Research on thermal performance and hygrothermal behavior of timber-framed walls with different external insulation layer: Insulation Cork Board and anti-corrosion pine plate. *J. Build. Eng.* **2020**, *28*, 101069. [CrossRef]
7. Hadini, M.H.; Susanto, D.; Chalid, M.; Alkadri, M.F. Investigating the mechanical performance of mycelium biocomposite using Oil Palm Empty Fruit Bunch (OPEFB) fiber and pine sap bioresin as sandwich insulation panels. *Constr. Build. Mater.* **2024**, *448*, 138173. [CrossRef]
8. Zhang, M.; Zhang, Z.; Zhang, R.; Peng, Y.; Wang, M.; Cao, J. Lightweight, thermal insulation, hydrophobic mycelium composites with hierarchical porous structure: Design, manufacture and applications. *Compos. B Eng.* **2023**, *266*, 111003. [CrossRef]
9. Bonga, K.B.; Bertolacci, L.; Contardi, M.; Paul, U.C.; Zafar, M.S.; Mancini, G.; Marini, L.; Ceseracciu, L.; Fragouli, D.; Athanassiou, A. Mycelium Agrowaste-Bound Biocomposites as Thermal and Acoustic Insulation Materials in Building Construction. *Macromol. Mater. Eng.* **2024**, *309*, 202300449. [CrossRef]
0. Mehrez, I.; Hachem, H.; Jemni, A. Thermal insulation potential of wood-cereal straws/plaster composite. *Case Studies in Constr. Mat.* **2022**, *17*, e01353. [CrossRef]
1. Li, X.; Tabil, L.G.; Oguocha, I.N.; Panigrahi, S. Thermal diffusivity, thermal conductivity, and specific heat of flax fiber-HDPE biocomposites at processing temperatures. *Compos. Sci. Technol.* **2008**, *68*, 1753–1758. [CrossRef]
2. Ghazi Wakili, K.; Rädle, W.; Rohner, T. Measuring the Specific Heat Capacity of Thermal Insulation Materials Used in Buildings by Means of a Guarded Hot Plate Apparatus. *Int. J. Thermophys.* **2023**, *44*, 15. [CrossRef]
3. Yousefi, Y.; Tariku, F. Thermal conductivity and specific heat capacity of insulation materials at different mean temperatures. *J. Phys. Conf. Ser.* **2021**, *2069*, 012090. [CrossRef]
4. Shah, D.U.; Bock, M.C.D.; Mulligan, H.; Ramage, M.H. Thermal conductivity of engineered bamboo composites. *J. Mater. Sci.* **2016**, *51*, 2991–3002. [CrossRef]
5. Alazzawi, S.; Mahmood, W.A.; Shihab, S.K. Comparative study of natural fiber-Reinforced composites for sustainable thermal insulation in construction. *Int. J. Thermofluids* **2024**, *24*, 100839. [CrossRef]
6. Xing, Y.; Brewer, M.; El-Gharabawy, H.; Griffith, G.; Jones, P. Growing and testing mycelium bricks as building insulation materials. *IOP Conf. Ser. Earth Environ. Sci.* **2018**, *121*, 022032. [CrossRef]

Journal of
Fungi

MDPI

Article

Trichoderma brevicompactum 6311: Prevention and Control of *Phytophthora capsici* and Its Growth-Promoting Effect

Jien Zhou [1,2,†], Junfeng Liang [2,†], Xueyan Zhang [2], Feng Wang [2], Zheng Qu [2], Tongguo Gao [3], Yanpo Yao [2,*] and Yanli Luo [1,*]

1 College of Resources and Environment, Xinjiang Agricultural University, Urumqi 830052, China; zjientongxue@163.com
2 Agro-Environmental Protection Institute, Ministry of Agriculture and Rural Affairs, Tianjin 300191, China; liangjunfeng@caas.cn (J.L.); 13571732795@163.com (X.Z.); wangfeng_530@163.com (F.W.); quzhengwf@163.com (Z.Q.)
3 College of Life Sciences, Hebei Agricultural University, Baoding 071002, China; gtgrxf@163.com
* Correspondence: yao7707@126.com (Y.Y.); luoyanlimail@sina.com (Y.L.)
† These authors contributed equally to this work.

Abstract: Pepper Phytophthora blight caused by *Phytophthora capsici* results in substantial losses in global pepper cultivation. The use of biocontrol agents with the dual functions of disease suppression and crop growth promotion is a green and sustainable way of managing this pathogen. In this study, six biocontrol strains of *Trichoderma* with high antagonistic activity against *P. capsici* were isolated and screened from the rhizosphere soil of healthy peppers undergoing long-term continuous cultivation. Morphological identification and molecular biological identification revealed that strains 2213 and 2221 were *T. harzianum*, strains 5111, 6311, and 6321 were *T. brevicompactum*, and strain 7111 was *T. virens*. The results showed that *T. brevicompactum* 6311 had the greatest inhibitory effect against *P. capsici*. The inhibition rate of 6311 on the mycelial growth of *P. capsici* was 82.22% in a double-culture test, whereas it reached 100% in a fermentation liquid culture test. Meanwhile, the pepper fruit tests showed that 6311 was 29% effective against *P. capsici* on pepper, and a potting test demonstrated that the preventive and controlling effect of 6311 on pepper epidemics triggered by *P. capsici* was 55.56%. The growth-promoting effect, germination potential, germination rate, radicle-embryonic axis length, germination index, and fresh weight of peppers cultured in the 6311 fermentation broth were significantly increased compared with the results for the control group. Scanning electron microscopy revealed that 6311 achieved the parasitism of *P. capsici*, producing siderophores and the growth hormone indoleacetic acid (IAA) to achieve disease-suppressive and growth-promoting functions. Transcriptomic results indicated that genes encoding proteins involved in plant disease resistance, namely flavanone 3-hydroxylase (F3H) and growth transcription factor (AUX22), were generally upregulated after the application of 6311. This study demonstrated that 6311 exhibits significant bioprotective and growth-promoting functions.

Keywords: *T. brevicompactum*; biocontrol strains; soil-borne pathogens; pepper Phytophthora blight; disease inhibition; growth promotion

1. Introduction

Peppers (*Capsicum annuum* L.) are vegetables that are well known and widely consumed globally, and they are prized for their culinary versatility and unique flavor. Yearly accumulation of pathogenic fungi in the soil due to continuous cropping can lead to outbreaks of soil-borne diseases [1,2]. As a result of intensive and protected cultivation, soil-borne diseases are occurring extensively. Pepper yields have been reduced by more than 60%, causing significant economic losses to farmers and seriously hampering the development of the pepper industry [3,4].

Cultivated peppers are highly susceptible to *P. capsici*. Pepper Phytophthora blight caused by this pathogen tends to develop rapidly, resulting in severe yield loss or even crop failure. *P. capsici* is a fungal pathogen capable of causing rot and wilt in a wide range of crops. The sporangia of *P. capsici* can germinate and invade plant cells directly or produce zoospores, which are released into water, attaching to plant roots and producing germ tubes that penetrate plant cells to cause pepper Phytophthora blight. This disease is one of the most common soil-borne fungal diseases globally and can occur at all reproductive stages of peppers [5]; thus, soil contaminated by *P. capsici* is the main cause of pepper Phytophthora blight. The application of chemical pesticides is currently the most common method for managing pepper Phytophthora blight; however, the long-term application of pesticides not only destroys the health of the soil environment, but also tends to cause pepper Phytophthora blight to become mold resistant. For this reason, adopting biocontrol measures that are both environmentally friendly and crop safe would be a very promising strategy. Therefore, the application of functional microorganisms and their metabolites is proposed as an alternative to conventional chemical pesticides.

Soil microbial communities are essential components of soil ecosystems and are important indicators of soil microbiological health [6]. The fungi *Trichoderma* spp., living in soil, can inhibit the growth of *P. capsici*, improve plant resistance to pathogens, and promote plant growth, regulating the microbiological structure of the soil. It has been successfully developed for use as a commercial biocontrol agent [7–10], such as in the form of *T. harzianum*, *T. hamatum*, *T. virens*, and *T. asperellum* [11]. These species are effective against a wide range of pathogens, such as *Phytophthora*, *Fusarium*, and *Pythium* [12]. *Trichoderma* spp. antagonizes pathogens through a variety of direct and indirect mechanisms. The direct mechanisms of their biocontrol include competition for space and nutrients [12], as well as secretion of lytic enzymes [13] and antibiotics [14]. For example, *T. asperellum* can surround and penetrate *P. capsici* hyphae, leading to mycelial collapse, and can also use spores to enter *P. capsici* oospores, develop mycelium, and produce conidia, thereby leading to the decomposition of *P. capsici* oospores [15]. Meanwhile, the antibiotic gliadin produced by *T. virens* has been shown to inhibit the growth of *Mycosphaerella* spp., with a significant reduction (by 62.64%) in disease incidence [16]. The main indirect mechanisms of their biocontrol include the induction of plant resistance, increased resistance to stress, and root growth. For example, lilac seedlings treated with *T. afroharzianum* T52 biofertilizer showed a significant increase in the number of lateral roots, and the incidence of the plant infection was reduced by 12.28% compared with that in the control [17]. Meanwhile, *T. ghanense* and *T. citrinoviride* increased the shoot length, root length, leaf length, leaf width, and dry weight of cucumbers, while suppressing disease caused by *P. infestans* [18]. The application of *T. viride* not only significantly reduced the disease index of soybeans, but also favored the maintenance of network stability and increased network complexity, as well as being more friendly to soil microhabitats [19]. *T. brevicompactum* controlled root rot of *Atractylodes macrocephala* caused by *Fusarium oxysporum*, reducing the severity of root rot by 48% to 58% and significantly increasing plant growth and biomass compared to those of the control (healthy) group [20].

Peppers from Jize County, Handan City, Hebei Province, China, were used as court tributes during the Ming and Qing dynasties, while Wangdu County in Baoding also has a 600-year history of pepper cultivation. In this study, healthy pepper plants were collected from soil with long-term continuous pepper cropping in Hebei Province, China, to perform targeted cultivation and screening of rhizosphere soil strains, along with the isolation and identification of functional strains. The biocontrol and biopromotion performance and mechanism of action of biocontrol microorganisms against *P. capsici* were verified through in vitro tests, such as the double-culture test, the fermentation liquid culture test, and the isolated peppers fruit test, along with in vivo tests, such as the seed germination test and the pot planting test. The goals of this study will provide insights into the effective prevention and control of pepper Phytophthora blight and promote the sustainable production of peppers.

2. Materials and Methods

2.1. Source, Isolation, and Identification of Biocontrol Fungi

Source of biocontrol fungi and pathogenic strains: From July to September 2023, blight occurred in some long-term continuous pepper cropping plots in Handan City and Baoding City, Hebei Province, P.R. China, resulting in massive plant deaths. Two batches of inter-root soil of healthy pepper plants were collected from long-term continuous cropping plots during the blight outbreaks and stored at 4 °C. The collected soil was used to cultivate potential biocontrol fungi and isolate and purify potential biocontrol fungi. *P. capsici* was provided by the Institute of Plant Protection, Hebei Academy of Agricultural and Forestry Sciences, Hebei Province, China.

Isolation and screening of biocontrol fungi: A total of 5 g of fresh inter-root soil was placed in 45 mL of sterile water and shaken on a shaker at 28 °C and 180 rpm for 1 h. The soil samples were diluted to 10^{-1}, 10^{-2}, 10^{-3}, 10^{-4}, 10^{-5}, and 10^{-6} using the stepwise dilution method, and 100 μL of each soil suspension was aspirated on potato dextrose agar (PDA) medium plates for spreading; this procedure was repeated three times. After incubation at 28 °C for 3 days, single colonies were picked up with an inoculation ring and transferred to a PDA plate for further purification. This step was repeated five times until pure colonies were obtained, which were numbered and stored in a refrigerator at 4 °C.

Biocultural testing of biocontrol fungi and *P. capsici*: A 5 mm radius sterilized hole punch was used to punch holes at the edge of the *P. capsici* colonies cultured for 5 days. The *P. capsici* section was inoculated on one side of a 9 cm diameter PDA plate, and the symmetrical position of the *P. capsici* block was inoculated with the six potential biomicrobial blocks to create a 5 cm distance between the blocks. The plate was then sealed with sealing film and subjected to inverted incubation for 7 days in the dark at 28 °C. Only the PDA plate inoculated with the *P. capsici* discs was used as a control, and each treatment was replicated three times. The percentage of fungi inhibition was calculated using the crossover method (see Equation (1)).

$$\text{Inhibition rate} = \frac{(\text{D} - \text{d})}{\text{D}} \times 100\% \quad (1)$$

D: diameter of *P. capsici* in the control group; d: diameter of *P. capsici* in the treatment group

Identification of biocontrol microbial strains: (1) The strains to be identified were inoculated onto PDA plates and cultured at a constant temperature of 28 °C. After 5 days, the morphology and color of the colonies were observed, and the morphological characteristics of conidia and conidiophores were observed under a light microscope.

Molecular biological identification: The cultured strains were ground to fine powder in liquid nitrogen, and total genomic DNA was extracted from mycelium using a modified version of the CTAB method described by Tiwari [21]. The following primers were used: rDNA internal transcribed spacer region ITS4: (5′-TCCTCCGCTTATTGATATATGC-3′); ITS5: (5′-GGAAGTAAAAGTCGTAACAAGG-3′). PCR was performed in accordance with the procedure of Korkom [22]. The ITS sequences of the biocontrol microorganisms were constructed, along with the sequences of other *Trichoderma* from the GenBank database of NCBI, and the phylogenetic tree was constructed using the neighbor-joining method in the MEGA11 program. The ITS sequences of the biocontrol microorganisms have been submitted to the GenBank database of NCBI. Further confirmation was made through *tef1* sequencing, which can distinguish different strains of Trichoderma, using the primer pair EF1-728F and TEFl rev and PCR conditions as described by Samuels [23].

2.2. Inhibition of the Disease and Enhancement of Growth Performance by the Biocontrol Organisms

In vitro inhibitory activity testing of fermentation solution against *P. capsici*: Activated PDA plates of six *Trichoderma* strains developed on potato dextrose liquid medium (PDB) were inoculated conical and cultured by oscillation at 180 rpm and 28 °C for 7 days. The fermentation broth was obtained by filtration through eight layers of sterile gauze and then a 0.22-μm sterile filter membrane. The plates were made by mixing the fermentation broth with PDA at a ratio of 1:3 (by volume), and the center was inoculated with fresh cakes of *P. capsici*, while equal-volume PDA plates, without the addition of the fermentation broth, were used as a control. Each treatment was repeated three times. The other procedures were the same as those described in Section 2.1.

Fermentation solution for the pepper promotion test: Healthy, uniform-sized pepper seeds were disinfected with 5% NaClO for 5 min, rinsed well with sterile water, drained, and placed in a five-fold dilution of fermentation solution of six *Trichoderma* strains for 24 h. The seeds were evenly placed in a germination box with a double layer of aseptic germination paper, 50 seeds per dish, and each treatment was replicated three times, followed by culture in a humidified atmosphere at 28 °C for 10 days. Sterile water was used for a control group to determine the germination potential of the seeds, germination rate, fresh weight, embryonic axes' radicle length, and 5-day germination index.

$$\text{Germination rate (\%)} = \text{number of seeds germinated at 10 days/total number of seeds} \quad (2)$$

$$\text{Germination potential (\%)} = \text{number of seeds germinated at 5 days/total number of seeds} \times 100\% \quad (3)$$

$$\text{Germination index (GI)} = \sum \text{Gt}\frac{\text{Gt}}{Dt} \quad (4)$$

Gt is the number of germinations per day, and *Dt* is the number of germination days.

Pepper fruit blight antagonism test: Peppers that were disease-free, unwounded, mature, and basically the same size were selected and immersed in 2% sodium hypochlorite solution for 5 min, followed by rinsing with sterile water three times, and then left to dry. A wound of 5 mm in diameter and 1 mm in depth was created at the waist of the pepper using a sterile perforator, and six discs of six strains of biocontrol microorganisms were inoculated into the wound. The wounds were covered with blotting paper that absorbed a sufficient amount of sterile water, and the fruits were placed in sterilized plastic pots covered with film to keep them warm and moisturized. After 24 h of incubation at room temperature, the fungal cake was removed, and then blocks of *P. capsici* were inoculated into the wounds to continue incubation. The diseased area was observed and photographed every day, and the area was determined, with each treatment being repeated three times. Treatment involving inoculation with *P. capsici* alone was used as a positive control, while treatment with sterile water was used as a negative CK. The inhibition rate was calculated as described in Section 2.1.

Potting experiment: Six treatments were designed for potting. Treatment 1 was the sterile water control (CK), treatment 2 was *Trichoderma* 2213 + *P. capsici*, treatment 3 was

Trichoderma 6311 + *P. capsici*, treatment 4 was *P. capsici*, treatment 5 was *Trichoderma* 2213, and treatment 6 was *Trichoderma* 6311.

Experimental method: Three peppers of TianYu No. 5 were planted in each pot, cultivated for 90 days, followed by wounding of the roots, and in treatments 2, 3, 5, and 6, respectively, 10 mL of *Trichoderma* spore solution with a spore count of 1×10^6 was added to the pepper' root systems. Five days after inoculation with the *Trichoderma* suspensions, a suspension of 1×10^5 conidia of *P. capsici* was added in treatments 2, 3, and 4, with each treatment being replicated three times. All roots of the plants were collected on the 5th day of inoculation of peppers with *P. capsici*. Disease severity was rated as follows: level 0: the whole plant was free of disease; level 1: less than one-fifth of the lateral branches were affected, or watery spots appeared at the base of the stems; level 2: one-fifth to one-half of the lateral branches were affected, or decay spots appeared at the base of the stems, but the plants had not wilted; level 3: one-half to three-quarters of the lateral branches were affected, or decay spots appeared at the base of the stems, and the plants showed recoverable wilting; level 4: more than three-quarters of the whole plant's lateral branches were affected, some branches were dead, or the plant showed irrecoverable wilting phenomenon; and level 5: the whole plant was dead.

$$\text{Blight disease condition index} = \frac{\Sigma\text{Disease severity of diseased plants} \times \text{number of diseased plants}}{\text{Highest incidence severity } \times \text{ total number of plants surveyed}} \quad (5)$$

$$\text{Prevention and control effects} = \frac{\text{Comparison Condition Index} - \text{Treatment Condition Index}}{\text{Comparison of condition indices}} \quad (6)$$

2.3. Analysis of Disease-Suppressing and Growth-Promoting Mechanisms of Biocontrol Microorganisms

Comparison of antagonistic properties between strains: Scanning electron microscopy (SEM) was used to examine the role of biocontrol fungi on the plates in the double-culture experiment. The above junction isolated the strains of biocontrol fungi, and *P. capsici* was cut off with a sterile blade, observed, and recorded using a light microscope. The junction was then collected and fixed in a 1.5 mL centrifuge tube, 1 mL of 2.5% glutaraldehyde solution was added, and the samples were rinsed with 0.1 M, pH 7.0 phosphate buffer three times for 15 min each. The samples were fixed with 1% osmioglycolic acid solution for 1–2 h and rinsed with 0.1 M, pH 7.0 phosphate buffer three times for 15 min each. The samples were then dehydrated with five solutions containing increasing concentrations of ethanol. The samples were rinsed three times with 0.1 M, pH 7.0 phosphate buffer solution for 15 min each. Next, the samples were dehydrated with five solutions containing increasing concentrations of ethanol for 15 min and then treated with 100% ethanol twice for 20 min each. The samples were treated with a mixture of ethanol and isoamyl acetate (1:1 by volume) for 30 min and then treated with pure isoamyl acetate for 1 h. Finally, the samples were subjected to critical point drying, coated, and then observed via SEM.

Siderophores performance test: Six strains of the potential biocontrol microbial were cultured on PDA plates for 7 days to achieve colony growth. Then, the cooled Chrome azurol S (CAS) test medium was poured onto the PDA plates and left to stand for 24 h to observe the plate color. Each biocontrol microbial strain comprised one treatment, and each treatment was repeated three times. The spore liquid was inoculated into PDB with 5% (v/v) inoculum, incubated at 28 °C and 180 rpm for 5 days, and then centrifuged at 8000 rpm for 15 min to obtain the supernatant. A total of 3 mL of CAS assay solution was added to the polycarbonate tube, and then 3 mL of supernatant was added, followed by mixing, and the mixture was allowed to stand for 1 h. Absorbance was detected using a spectrophotometer at 630 nm (A1). A total of 3 mL of uninoculated PDB and 3 mL of supernatant were removed to observe the color of the plate, with each treatment being repeated three times. The inoculated PDB and 3 mL of CAS assay solution were mixed well, and the absorbance was measured in the same way as for the reference value (A0). The rate of synthesis of the siderophores was determined using Equation (7).

$$\text{Relative concentration of siderophores } (\%) = \frac{(\text{A0} - \text{A1})}{\text{A0}} \tag{7}$$

IAA growth-promoting properties: We referred to the method used in the research of Ma et al. and improved upon it [24]. The bioprophylactic microorganisms were activated on PDA plates at 28 °C for 5 days. Holes were punched at the edge of the colony with a 5 mm-radius punch, and five clusters were placed in 100 mL of PDB and shaken at 180 rpm for 7 days at 28 °C before removal, and then filtered through eight layers of sterile gauze. This yielded an IAA concentration that was determined using Salkowski's reagent [25], which was then compared with the IAA standard curve.

Transcriptomic analysis: Whole roots from treatments 1, 5, and 6 were sampled from pepper plants, snap-frozen in liquid nitrogen, and stored in a freezer at −80 °C. Total RNA was isolated and extracted from the tissues using Invitrogen TRIzol reagent, which was outsourced to Shanghai Personal Bio-technology Co., Ltd., Shanghai, China. Total RNAs of ≥1 μg were selected and enriched for poly (A) RNAs with Poly (A) RNA using Oligo (dT) magnetic beads, and cDNA synthesis was performed using RNA for cDNA synthesis. Purified size-selected and adapter-conjugated cDNA fragments were used for library construction. The mixed libraries were sequenced on the Illumina NovaSeq 6000 platform. Both raw and clean data generated from each library were sequenced to a depth of no less than 5 G. Raw reads obtained from the NovaSeq sequencing platform were filtered to remove shorter reads, junctions, and low-quality reads. All clean reads were localized to the pepper reference genome (http://peppergenome.snu.ac.kr/) using HISAT2 (v2.1.0) and statistically matched to each gene using HTSeq (v0.9.1). The differentially expressed genes (DEGs) were identified using the edgeR package, with DEGs defined as follows: |log2FoldChange|>1 and *p*-value <0.05. Kyoto Encyclopedia of Genes and Genomes (KEGG) enrichment analysis was performed using clusterProfiler (v4.6.0), and the threshold was set as a *p*-value <0.05.

2.4. Data Analysis and Visualization

Excel 2019 was used to handle the statistical data; SAS9.4 was used for one-way analysis of variance (ANOVA); and Duncan's multiple comparison test was used to analyze the significance of differences, with $p < 0.05$ indicating a significant difference. Graphs were prepared using Origin 9.1 software.

3. Results

3.1. Antagonism and Identification of Biocontrol Microorganisms Against *P. capsici*

The antagonistic activity of the isolates against *P. capsici* was assessed using a double-culture test. Six biocontrol microorganisms from strains isolated from the inter-root soil of peppers demonstrated high antagonistic ability in the dual-culture test. The control strain of *P. capsici* grew uniformly in a radial direction on PDA (Figure 1a). In contrast, *P. capsici* mycelia were inhibited under the *Trichoderma* strain culture conditions, and colony expansion stopped, whereas *Trichoderma* strains continued to expand until they completely covered the *P. capsici* colonies and the whole plate (Figure 1b,c). All six biocontrol microorganism strains inhibited conidia of *P. capsici* by more than 70% (Figure 1d), with strain 6311 showing the strongest antagonistic effect on *P. capsici* filaments (reaching 82.22%). In addition, all six strains of biocontrol microorganisms covered the entire PDA plate with colonies of *P. capsici* after 5 days of incubation at 28 °C, and the production of *Trichoderma* strain conidia was observed in the medium.

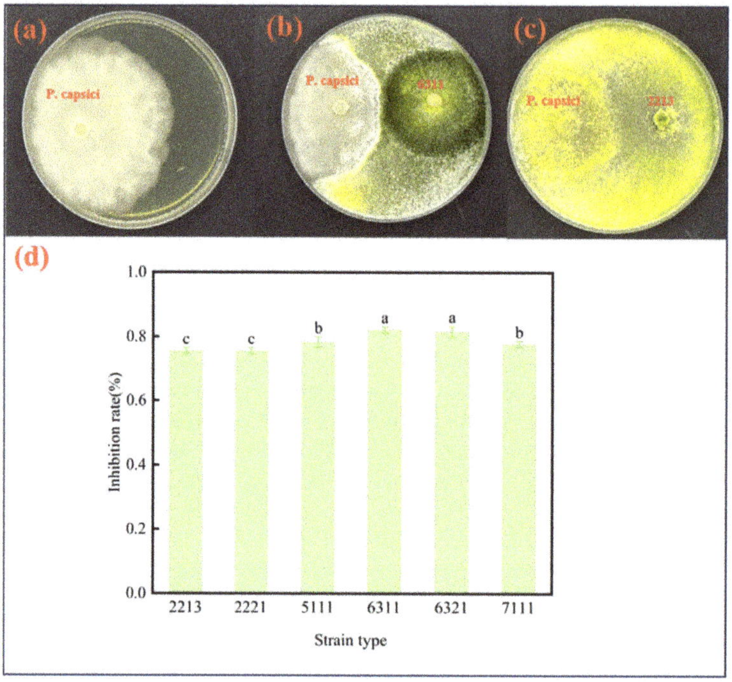

Figure 1. Biocontrol microorganisms and *P. capsici* dual-culture test and developmental tree of strains after 7 days of inoculation. (**a**) Inoculation with *P. capsici* only; (**b**) inoculation with *Trichoderma* 6311 and *P. capsici*; (**c**) inoculation with *Trichoderma* 2213 and *P. capsici*; (**d**) inhibition of *P. capsici* by the biocontrol microorganisms. Identical letters denote no significant differences between the respective groups ($p > 0.05$), whereas differing letters indicate a significant difference ($p < 0.05$).

Fungal isolates 2213 and 2221 produced smooth and spherical-to-subspherical conidia on the long main axis (Figures 2a,b and S1). Meanwhile, 5111, 6311, and 6321 showed well-developed aerial mycelia on the PDA plate, and the colonies were fluffy to felted, white protruding, thick and fluffy, and formed clear concentric circles. After 5 days, the plates were covered with a large number of dark-green or yellowish-green spores, with a yellowish-brown backside (Figures 2c,d and S1). Moreover, 7111 colonies were observed in the medium covering the plate. The fungal colonies of 7111 were white and light green, which turned dark green after 5 days (Figure S1). The ITS sequences of fungal isolates 2213, 2221, 5111, 6311, 6321, and 7111 were deposited in the GenBank database under the accession numbers SUB144343272213PP770682, SUB144343272211PP770683, SUB144343275111PP770684, SUB144343276311PP770685, SUB144343276321PP770686, and SUB144343277111PP770687. By using primers ITS4, ITS5, EF1-728F, and TEFl rev, the isolates 2213 and 2221 were molecularly identified as *T. harzianum*; 5111, 6311, and 6321 were identified as *T. brevicompactum*; and 7111 was identified as *T. virens*. These *Trichoderma* strains showed a greater than 90% homology with *T. harzianum* (MF871551.1), *T. brevicompactum* (KR094463.1), and *T. virens* (EU280080.1), respectively (Figures 2e,f, S2 and S3).

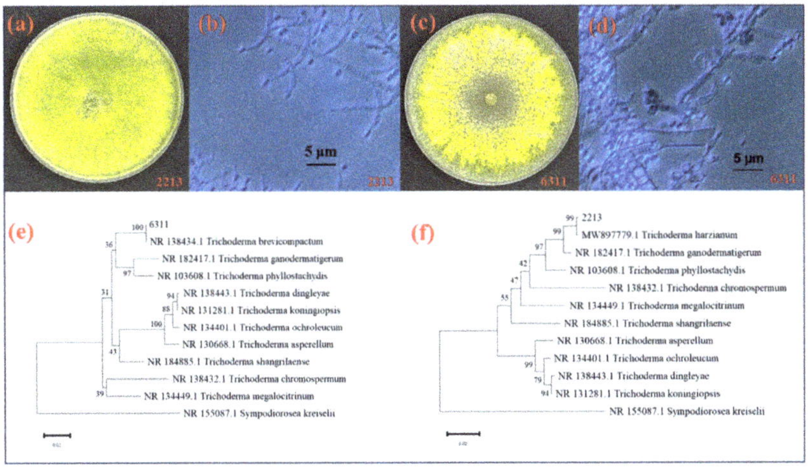

Figure 2. Morphological and molecular identification of biocontrol microorganisms. (**a**,**b**) *Trichoderma* 2213 colony morphology and spore morphology; (**c**,**d**) *Trichoderma* 6311 colony morphology and spore morphology; (**e**,**f**) 2213 and 6311 developmental trees.

3.2. Analysis of Disease-Suppressing and Growth-Promoting Properties of Biocontrol Microorganisms

3.2.1. Analysis of Disease-Suppressing Properties of Biocontrol Microorganisms

The effect of the *Trichoderma* spp. on the biocontrol of *P. capsici* on isolated pepper fruits is shown in Figure 3. The uninoculated *P. capsici* treatment displayed no lesions, and no pathogenic mycelial growth was seen (Figure 3a), whereas the *P. capsici* treatment group inoculated with *P. capsici* alone exhibted visible lesions, with white mycelial growth. The diameters of the diseased spots on pepper fruits treated with the biocontrol microorganisms were all significantly smaller than those of the *P. capsici* treatment group (Figure 3d). Figure 3b,c indicate that the isolated strains of *Trichoderma* inhibited *P. capsici* growth in isolated pepper fruits. Among them, the 2213 and 6311 treatment groups displayed the smallest diameter of *P. capsici*-infested spots on the pepper fruits, with an average of 4.47 ± 0.03 cm and 5.23 ± 0.13 cm, and the inhibition rates were 40% and 29%, respectively (Figure 3n). The inhibition test of the *Trichoderma* fermentation solution showed that the fermentation solutions of six strains of *Trichoderma* exhibited different abilities to inhibit the mycelial growth of *P. capsici* (Figure 3e–g). Compared with the control, the fermentation solutions of strains 5111 and 6311 inhibited the mycelial growth of *P. capsici* by 100%. Meanwhile, the fermentation solutions of 6321 and 7111 also significantly reduced the growth of *P. capsici* mycelium, with inhibition rates of 69.72% and 27.52%, respectively. Meanwhile, the 2213 and 2221 fermentation broths achieved inhibition rates of only 4.59% and 8.26%, respectively (Figure 3o).

In the pot experiment, the *Trichoderma* strains also showed excellent pepper blight-suppressive effects (Figure 3h–m). Among the three treatments involving *P. capsici* inoculation, the disease indices of the 6311-added treatment groups were lower than those of the other two treatment groups (Figure 3p). Notably, the roots of peppers treated with 6311 displayed the smallest lesions (Figure 3m), and the control effect reached 55.56% (Figure 3q). This was followed by the appearance of watery spots on pepper roots inoculated with 2213 (Figure 3l). In the inoculation of blight alone treatment, the pepper stems showed greater shrinkage at the base (Figure 3k), and the leaves wilted and fell off, indicating entry into the necrotrophic stage.

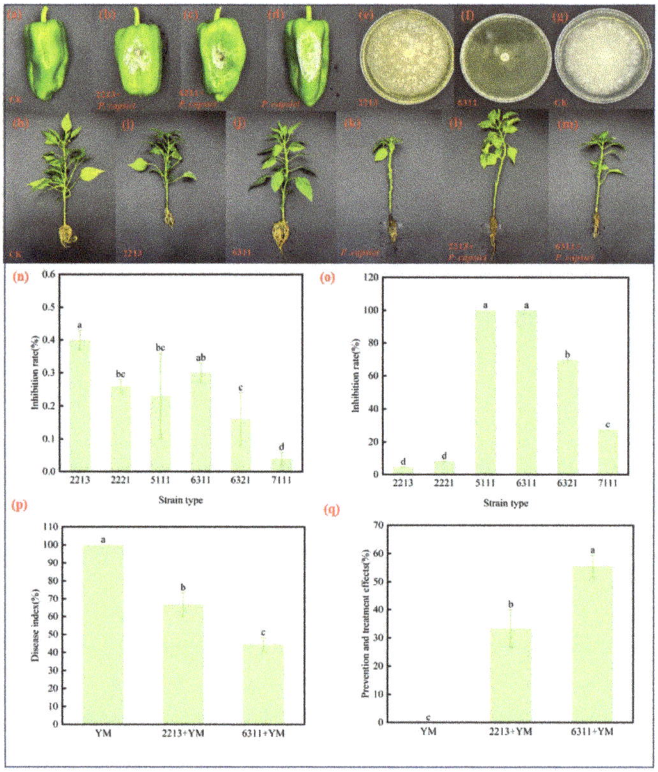

Figure 3. Results of pepper fruit or plant potting experiment. (**a**) CK group without any added microorganisms; (**b**) 2213+ *P. capsici* treatment group; (**c**) 6311+ *P. capsici* treatment group; (**d**) *P. capsici* treatment group; (**e**) no fermentation solution inhibition of *P. capsici*; (**f**) 2213 fermentation solution inhibition of *P. capsici*; (**g**) 6311 fermentation solution inhibition of *P. capsici*; (H) normal growth of *P. capsici* without fermentation solution; (**h–m**) growth of pepper after 5 days of each treatment; (h) CK; (i) 2213; (j) 6311; (k) *P. capsici*; (l) 2213+ *P. capsici*; (m) 6311+ *P. capsici*; (**n**) inhibition of *P. capsici* in pepper fruits; (**o**) *Trichoderma* fermentation broth inhibition; (**p**) disease indices for each treatment in the potting experiment; and (**q**) preventive and control effects of each treatment. Identical letters denote no significant differences between the respective groups ($p > 0.05$), whereas differing letters indicate a significant difference ($p < 0.05$).

3.2.2. Analysis of the Growth-Promoting Properties of Biocontrol Microorganisms

The growth indexes of the effects of bioprophylactic microbial fermentation solution on pepper seeds are shown in Figure 4. Following the treatments of pepper seeds with six bioprophylactic microbial fermentation solutions, the seeds achieved a germination potential and a germination rate of 40.67–57.33% and 90.67–98.67%, respectively, which were significantly greater than those in the control group (Figure 4a). Among the six strains, 2213 and 6311 displayed the best growth-promoting effects, with a germination potential and a germination rate reaching 57.33% and 55.33% and 98.67% and 94.67%, respectively. In addition, the fermentation broths of the six biocontrol microorganisms exhibited significant germination-promoting effects on the embryonic roots and embryonic axes of pepper seeds (Figure 4c). The best germination-promoting effects were observed for the fermentation broths of 2213 and 6311, with the embryonic roots and embryonic axes in treatment 2213 increasing by 1.66-fold and 2.45-fold and those in 6311 by 1.50-fold and 2.21-fold compared with those of the control, respectively. The germination indices of the seeds treated with the six biocontrol microorganisms were significantly increased compared with those of the control treatment (Figure 4d), with 46.01% and 42.62% for 2213 and 6311, respectively. Meanwhile, the fresh weight of pepper seedlings increased by 47.62% and 42.86% compared with that of the control, respectively (Figure 4b). These results indicated that the *Trichoderma* fermentation solution improved the growth indexes of all pepper seedlings, with *T. harzianum* 2213 and *T. brevicompactum* 6311 exhibiting significant growth-promoting effects.

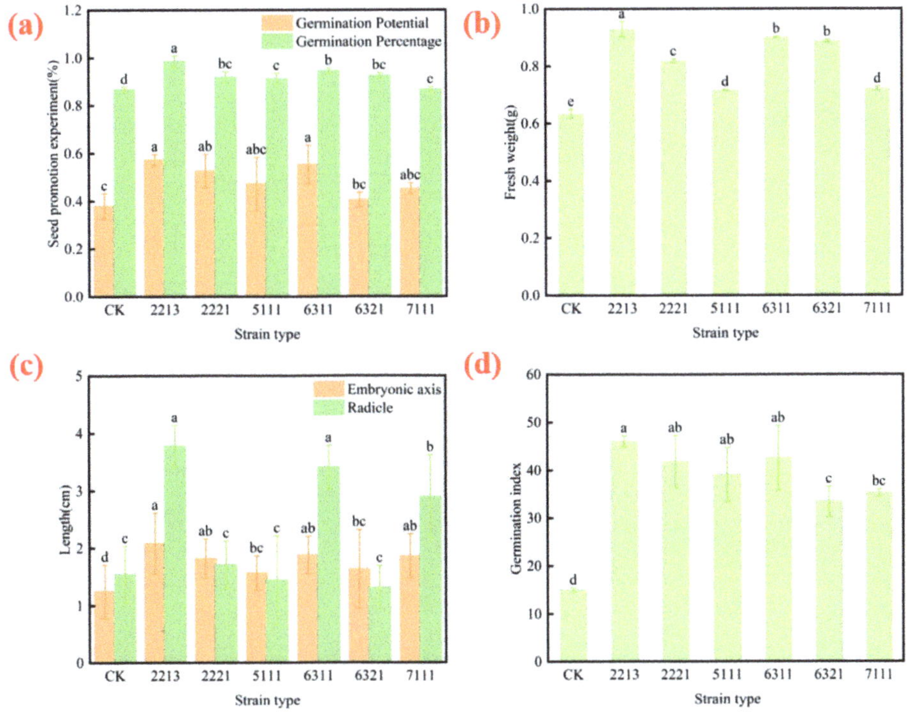

Figure 4. Effect of biocontrol microbial fermentation solution on the growth of pepper seedlings, where CK stands for no fermentation solution applied. (**a**) Seed germination potential and germination rate; (**b**) fresh weight; (**c**) lengths of embryonic axis and radicle; (**d**) germination index. Identical letters denote no significant differences between the respective groups ($p > 0.05$), whereas differing letters indicate a significant difference ($p < 0.05$).

3.3. *Mechanisms of Disease Suppression and Promotion by Biocontrol Microorganisms*

3.3.1. Physiological Evidence

Siderophores are capable of chelating iron to prevent the uptake of siderophores by plant pathogens [26], thus protecting plants by reducing the number of pathogens [27]. In this study, compared to unspliced fungi (Figure 5b), the *Trichoderma* 6311 treatment group showed a bright purple color on a double-layer chromogenic plate (Figure 5a), indicating the production of siderophores. The relative concentration of siderophores produced by different *Trichoderma* strains was determined (Figure 5c). The highest relative concentration of siderophores reached 72.88% for *T. brevicompactum* 6311, while 5111 and 6321 produced only small amounts of siderophores, and no siderophores were detected for *T. harzianum* 2213 and 2221. Light microscopy and SEM revealed the entangled parasitism of the biocontrol microbial mycelium on *P. capsici* hyphae and growth along the *P. capsici* mycelium, with a large number of branches entangled around the *P. capsici* filaments. In addition, the *P. capsici* idem were deformed or enlarged, indicating the infestation of its mycelium by 2213 (Figure 5f,g) and 6311 (Figure 5e,h–j). IAA, as an endogenous growth hormone that promotes plant growth and facilitates the formation of plant roots, plays an important role in plant growth and development [28]. Concentrations of IAA produced by biocontrol microorganisms are shown in Figure 5d. The findings showed that 2213, 2221, and 6311 produced high concentrations of IAA, namely, 30.91, 26.07, and 14.56 mg/L, respectively, whereas 5111, 6321, and 7111 produced only small amounts. High levels of IAA are known to promote lateral root formation [29] and increase root length [30].

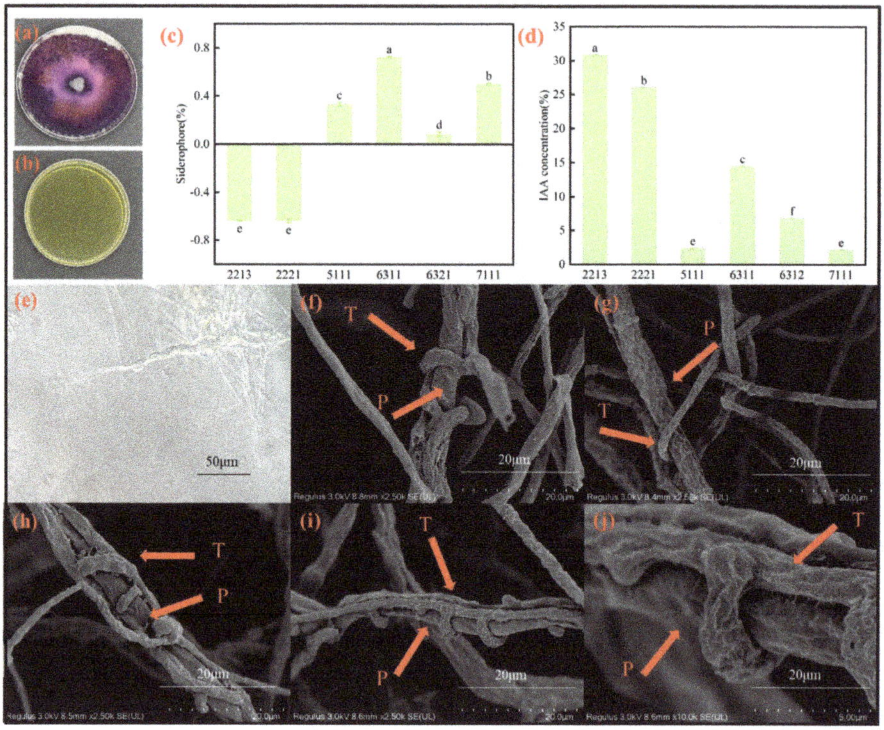

Figure 5. (**a**,**b**) Bilayer plates inoculated with 6311 and not inoculated with fungi; (**c**,**d**) relative concentration of siderophores and concentration of IAA upon *Trichoderma* treatment, respectively; (**e**) 6311 (T) hyphae entangled in *P. capsici* (P) hyphae under a 400× optical microscope; (**f**,**g**) SEM images of 2213 (T) hyphae entangled in *P. capsica* (P) hyphae; (**h**–**j**) SEM images of 6311 (T) hyphae entangled in *P. capsica* (P) hyphae. Identical letters denote no significant differences between the respective groups ($p > 0.05$), whereas differing letters indicate a significant difference ($p < 0.05$).

3.3.2. Transcriptomic Evidence

Among all the treatments, the proportion of Q20 and Q30 bases in the transcriptome of the pepper root was the lowest, with 98.87% and 96.69%, respectively (Table S1). In the alignment results, at least 78.04% of the high-quality reads mapped successfully to the reference genome of CM334 (Table S2). Principal component analysis of all transcripts showed that the first two principal components explained 82.9% of the total variation, and the three biological triplicates of each group clustered together (Figure 6a). Overall, 1346 pepper genes exhibited upregulated expression, and 829 genes displayed downregulated expression after treatment with strain 2213 (Figure 6b). A total of 930 pepper genes showed upregulated expression and 1363 genes displayed downregulated expression after treatment with strain 6311 (Figure 6b). Those treated with the 6311 strain exhibited

1107 upregulated genes and 2469 downregulated genes compared with the results for the treatment with strain 2213, indicating that the two types of *Trichoderma* had different effects on the pepper root system. To further understand the transcriptomic response of peppers to the two *Trichoderma* species, a Venn diagram of DEGs was produced. Overall, 7 genes (Figure 6c) and 16 genes (Figure 6d) were commonly upregulated and downregulated in the 2213 and 6311 strain treatments, suggesting that these DEGs are the key genes for the responses to the two *Trichoderma* species.

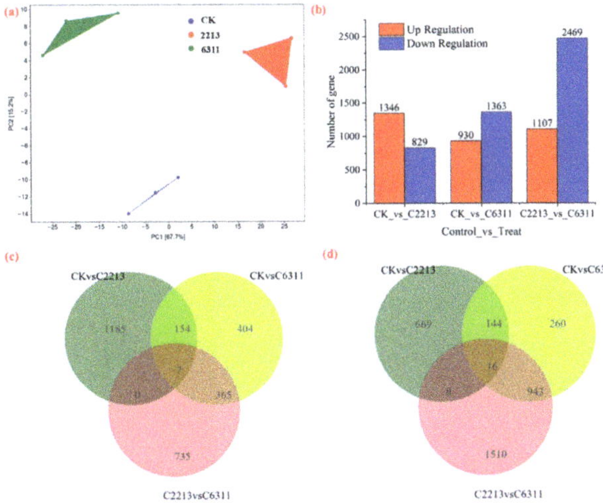

Figure 6. Sample correlations, and Venn diagrams of differentially expressed genes (DEGs). (**a**) The PCA for all samples; (**b**) the DEGs of all samples; (**c**,**d**) Venn diagrams of upregulated and downregulated DEGs, respectively.

Among the 1346 upregulated genes in the strain 2213 treatment, there were 12 significantly enriched KEGG pathways, 11 of which were related to metabolism, including phenylpropane biosynthesis and zeatin biosynthesis (Figure 7a). Among the 930 upregulated genes in the strain 6311 treatment, there were 5 significantly enriched pathways, including phytohormone signal transduction and zeatin biosynthesis (Figure 7b). The seven co-upregulated genes in the strain 2213 and 6311 treatments are thought to be potentially involved in the defense response, suggesting that strains 2213 and 6311 could activate the defensive responses of peppers. Among the downregulated genes in the 2213 treatment group, phenylacetone biosynthesis, terpene skeleton biosynthesis, isoquinoline alkaloid biosynthesis, valine and leucine, and isoleucine degradation were particularly enriched. Meanwhile, phenylacetone biosynthesis, interconversion of pentose and glucuronic acid, and cyanoamino acid metabolism were enriched pathways associated with the 6311 downregulated genes (Figure 7c,d).

In both the strain 2213 and 6311 treatments, the growth hormone transporter protein (AUX22) genes (*CA06g16770* and *CA08g08660*) were upregulated (Tables S3 and S4); AUX22 genes are members of the Aux/IAA family of proteins. AUX22 is a short-lived protein localized in the nucleus, playing a crucial role in the IAA signal transduction pathway [31]. This suggests that treatments with strains 2213 and 6311 might promote pepper plant growth. In addition, two cellulose synthase (CesA) genes (*CA07g01680* and *CA02g13540*) and two xylem cysteine protease 1 (XCP1) genes (*CA12g18250* and *CA04g23350*) were uniquely upregulated in the strain 2213 treatment (Table S3). CesA is usually associated with cell wall reinforcement to enhance structural resistance and improve disease defense [32]. Meanwhile, XCP1 is an important enzyme in plant plastid ectodomains and belongs to the class of xylose-like cysteine proteases [33]. It can activate defense responses by producing pathogen-associated molecular pattern (PAMP)-like peptides recognized by pattern-recognition receptors. The flavanone 3-hydroxylase (F3H) controlling genes (*CA08g16150* and *CA08g00130*) were upregulated in the stain 6311 treatments (Table S4). Flavanone 3-hydroxylase (F3H) is a key enzyme that directs carbon flow to produce 3-hydroxylated flavonoids, including flavonols and anthocyanins. These substances can be used against pathogens such as those of *Colletotrichum sublineola* [34].

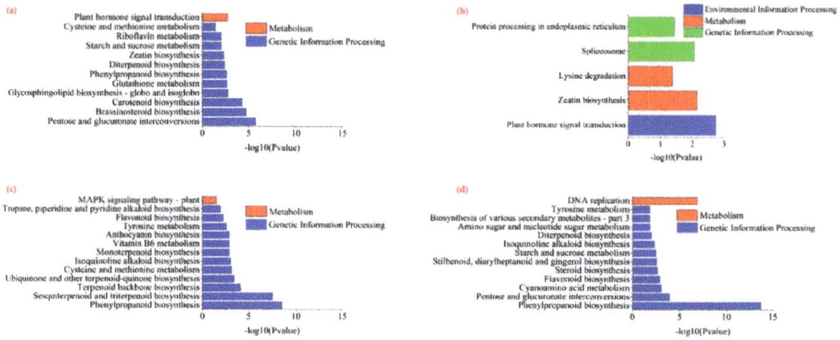

Figure 7. KEGG enrichment analysis of DEGs. (**a**,**b**) KEGG enrichment analyses of upregulated and downregulated genes in strain 2213 treatment, respectively; (**c**,**d**) KEGG enrichment analysis of upregulated and downregulated genes in strain 6311 treatment, respectively. Identical letters denote no significant differences between the respective groups ($p > 0.05$), whereas differing letters indicate a significant difference ($p < 0.05$).

4. Discussion

P. capsici is one of the best known fungal pathogens and has serious adverse effects on crop yields globally, especially on peppers. Although chemical fungicides are available to counter this pathogen, they pose serious threats to human health and the environment [35]. With growing demands for sustainable agricultural development, biological control is considered to be an environmentally friendly and economical alternative to fungicides. *Trichoderma* strains are important representatives in the field of biological control, with significant practical value and potential for the biocontrol of plant diseases; thus, these have been studied worldwide. Indeed, the studies showed that *Trichoderma* strains can effectively inhibit approximately 30 plant pathogenic fungi, including *P. capsici*. In the present study, six strains of biocontrol *Trichoderma* with strong antagonistic activity against *P. capsici* were isolated from the rhizosphere soil of pepper plants collected from plots with long-term continuous cropping in Heibei Province, China. Morphological and molecular analyses led to the identification of two strains of *T. harzianum*, three strains of *T. brevicompactum*, and one strain of *T. virens*, with *T. brevicompactum* 6311 being identified as the strongest inhibitor of the mycelial growth of *P. capsici*. Fungal parasitism is a key means of biocontrol in species of *Trichoderma*, which can parasitize pathogenic microorganisms such as *Aspergillus rotundus* [18], *Mycosphaerella epidermidis* [15], and *Fusarium oxysporum* [36] and can invade or damage the mycelium, resulting in the enlargement, distortion, and rupture of the pathogen's cells. Light microscopy and SEM of pathogenic fungal mycelium from the boundary of the zone of inhibition of dual-culture plates showed the mycelial parasitism of *T. brevicompactum* 6311 on *P. capsici*, suggesting that *T. brevicompactum* may attack the *P. capsici* cell wall/cell membranes and inhibit the growth of the *P. capsici* mycelium through the production of an iron-carrying antifungal substance, as well as through competition for space and nutrients [37,38].

It was previously reported that *Trichoderma* produces a variety of compounds with antagonistic properties, including a range of enzymes and metabolites such as cellulases, xylanases, pectinases, and chitinases [39]. Moreover, it has been shown that fermentation broths can be used to study the antagonistic capacity of *Trichoderma* metabolites. *Trichoderma* fermentation broth causes the malformation of pathogenic mycelium, with swollen, fragmented, and twisted hyphae. This study showed that the fermentation broth of isolate 6311 inhibited the growth of *P. capsici* mycelium by 100%, which was the first experimental result revealing that *T. brevicompactum* exhibited the highest inhibitory efficiency. Meanwhile, the fermentation broth of isolate 2213 showed little or no inhibitory effect. The reduction of the mycelial growth diameter in the isolated pepper fruit inhibition assay also demonstrated the ability of species of *Trichoderma* to inhibit the *P. capsici* mycelium, with isolates 2213 and 6311 achieving the most significant inhibition rates of 40% and 29%, respectively. Meanwhile, *Trichoderma* produces the growth hormone IAA [40], iron carriers [41], and other plant growth-promoting substances [42]. In this study, isolate 6311 was found to display the greatest ability to produce iron carriers, confirming its strong potential to inhibit pathogenic fungi. Meanwhile, for the first time, it was found that *T. brevicompactum* significantly enhanced the radicle-embryonic axis, fresh weight, germination rate, and germination index of pepper seedlings (Figure 4), which, to the best of our knowledge, has not been reported previously. Moreover, the results of the pot experiment showed that isolate 6311 could achieve a 55.56% prevention and control rate of pepper Phytophthora blight (Figure 3q). High-throughput sequencing technology was used for transcriptomic analysis on roots from treatment and control groups in which *Trichoderma* was applied alone. KEGG enrichment analysis of upregulated genes showed that genes related to the transcription factors AUX22 and AHP1, which were promoted by the application of isolate 6311, were generally upregulated. Growth hormones are central hormones for plant growth and development and are transported by a variety of transporter proteins with different biochemical and structural properties. Members of the Aux/IAA family have been identified as short-lived nuclear proteins that play a key role in suppressing the expression of ARF-activated genes [43,44]. The studies showed that the Aux/IAA genes exhibit high expression levels that may play roles in pepper fruit development and ripening, and AUX22 was identified as belonging to the Aux/IAA family. In addition, the AHP gene family is important for cytokinin signaling. Cytokinins are involved in the regulation of almost all important processes in plant growth and development. Cytokinin signaling is mediated through a mechanism involving continuous phosphorylation (phosphorelay). In *Arabidopsis*, receptors activated by cytokinins autophosphorylate and deliver the phosphate group to the phosphotransfer protein (AHP) and further to the downstream response regulator (ARR), thereby mediating cytokinin-regulated responses. In the current study, both AUX22 and AHP1 were upregulated in the isolate 2213 and 6311 treatment groups compared with the levels in the control group, indicating that both *Trichoderma* species have the ability to promote pepper growth. Meanwhile, genes related to XCP1 showed a separate upregulation in the treatment of isolate 2213. XCP1 is a key protease for plant immunity that activates systemic immunity by producing the cytokine CAPE9 from the classical salicylic acid signaling marker PR1. Genes related to XCP1 were individually upregulated in the isolate 2213 treatment [33]. Thus, *Trichoderma* 2213 has the ability to inhibit pepper Phytophthora blight. The F3H gene of the phenylpropane pathway was individually upregulated in the isolate 6311 treatment, and it has been shown that the phenylpropane biosynthetic pathway plays an important role in root resistance to pepper Phytophthora blight [45]. It was also found that the phenylpropane biosynthetic pathway leads to the synthesis of secondary metabolites involved in plant defense against pathogens.

5. Conclusions

In summary, *T. brevicompactum* 6311 was isolated from the rhizosphere soil of peppers in this study. Experimentally, it was proven that in terms of disease suppression, isolate 6311 could effectively inhibit the growth of *P. capsici* mycelia on both plates and in pepper fruit, and the metabolites produced in its fermentation broth could also inhibit *P. capsici*. Moreover, in potting experiments, isolate 6311 demonstrated excellent preventive and control effects on pepper Phytophthora blight. In terms of growth promotion, the sterile fermentation broth of isolate 6311 promoted the growth of pepper seedlings. By exploring the mechanisms involved, it was found that isolate 6311 inhibited the growth of pepper Phytophthora blight mycelia and improved its controlling effect by secreting antimicrobial substances, performing parasitism, and producing iron carriers. It was also revealed that its ability to secrete IAA could effectively improve the growth indexes of pepper seedlings. In addition, the results of transcriptomic sequencing analysis on the pepper root system showed that isolate 6311 affected the gene expression in the pepper root system and regulated the promotion of pepper growth, resistance-related signal transduction, and phytohormone pathways, such as the growth hormone pathway and the phenylpropane biosynthesis pathway. Our study provides a potential biocontrol agent for the biocontrol of *P. capsici* in peppers and offers possible insights into the mechanism of interaction between biocontrol agents and pathogens. In future work, we plan to focus on the application of isolate 6311 for biocontrol purposes in agricultural production.

Supplementary Materials: The following supporting information can be downloaded at: https://www.mdpi.com/article/10.3390/jof11020105/s1, Figure S1. Colony morphology of four biocontrol microorganisms; Figure S2. Molecular identification of four biocontrol microorganisms based on ITS; Figure S3. Molecular identification of four biocontrol microorganisms based on tef1 primers; Table S1. The sequencing data; Table S2. The alignment information; Table S3. Upregulated gene information of strain 2213; Table S4. Upregulated gene information of strain 6311.

Author Contributions: J.Z.: writing—original draft and data curation. J.L.: investigation and data curation. X.Z.: visualization and conceptualization. F.W.: methodology, investigation, and funding acquisition. Z.Q.: conceptualization. T.G.: visualization. Y.Y.: writing—review and editing and Supervision. Y.L.: writing—review and editing, supervision, and methodology. All authors have read and agreed to the published version of the manuscript.

Funding: This research is financially supported by the National Key Research and Development Program of China [2022YFD1901305], Yunnan Fundamental Research Projects [202101AT070002], the Wangfeng Expert Primary Level Workstation, and the Intelligent Aid to Xinjiang Innovation and Development Talent Plan.

Institutional Review Board Statement: Not applicable.

Informed Consent Statement: Not applicable.

Data Availability Statement: The original contributions presented in this study are included in the article/Supplementary Materials. Further inquiries can be directed to the corresponding author(s).

Conflicts of Interest: The authors declare that they have no known competing financial interests or personal relationships that could have appeared to influence the work reported in this paper.

References

1. Zhou, J.; Zhang, X.; Qu, Z.; Zhang, C.; Wang, F.; Gao, T.; Yao, Y.; Liang, J. Progress in Research on Prevention and Control of Crop Fungal Diseases in the Context of Climate Change. *Agriculture* **2024**, *14*, 1108. [CrossRef]
2. Mao, T.; Jiang, X. Changes in Microbial Community and Enzyme Activity in Soil Under Continuous Pepper Cropping in Response to *Trichoderma hamatum* MHT1134 Application. *Sci. Rep.* **2021**, *11*, 21585. [CrossRef] [PubMed]

3. Zhang, D.; Yan, D.; Cheng, H.; Fang, W.; Huang, B.; Wang, X.; Wang, X.; Yan, Y.; Ouyang, C.; Li, Y.; et al. Effects of Multi-Year Biofumigation on Soil Bacterial and Fungal Communities and Strawberry Yield. *Environ. Pollut.* **2020**, *256*, 113415. [CrossRef] [PubMed]
4. Wan, L.; Zhu, W.; Dai, Y.; Zhou, G.; Chen, G.; Jiang, Y.; Zhu, M.e.; He, M. Identification of Pepper Leaf Diseases Based on TPSAO-AMWNet. *Plants* **2024**, *13*, 1581. [CrossRef] [PubMed]
5. Admassie, M.; González-Pérez, E.; Woldehawariat, Y.; Alemu, T. Screening of Potential Bacterial Isolates Against *Phytophthora capsici* and Its Plant Growth-Promoting Effect on Pepper Plants. *Physiol. Mol. Plant Pathol.* **2023**, *127*, 102028. [CrossRef]
6. Kang, H.; Chai, A.; Lin, Z.; Shi, Y.; Xie, X.; Li, L.; Fan, T.; Xiang, S.; Xie, J.; Li, B. Deciphering Differences in Microbial Community Diversity Between Clubroot-Diseased and Healthy Soils. *Microorganisms* **2024**, *12*, 251. [CrossRef]
7. Bonaterra, A.; Badosa, E.; Daranas, N.; Francés, J.; Roselló, G.; Montesinos, E. Bacteria as Biological Control Agents of Plant Diseases. *Microorganisms* **2022**, *10*, 1759. [CrossRef]
8. Weiland, J.E. *Pythium* Species and Isolate Diversity Influence Inhibition by the Biological Control Agent *Streptomyces lydicus*. *Plant Dis.* **2014**, *98*, 653–659. [CrossRef]
9. Pandit, M.A.; Kumar, J.; Gulati, S.; Bhandari, N.; Mehta, P.; Katyal, R.; Rawat, C.D.; Mishra, V.; Kaur, J. Major Biological Control Strategies for Plant Pathogens. *Pathogens* **2022**, *11*, 273. [CrossRef]
10. Liu, D.; Yan, R.; Fu, Y.; Wang, X.; Zhang, J.; Xiang, W. Antifungal, Plant Growth-Promoting, and Genomic Properties of an Endophytic Actinobacterium *Streptomyces* sp. NEAU-S7GS2. *Front. Microbiol.* **2019**, *10*, 2077. [CrossRef]
11. Asad, S.A. Mechanisms of Action and Biocontrol Potential of *Trichoderma* Against Fungal Plant Diseases—A Review. *Ecol. Complex.* **2022**, *49*, 100978. [CrossRef]
12. Oszust, K.; Cybulska, J.; Frąc, M. How Do *Trichoderma* Genus Fungi Win a Nutritional Competition Battle against Soft Fruit Pathogens? A Report on Niche Overlap Nutritional Potentiates. *Int. J. Mol. Sci.* **2020**, *21*, 4235. [CrossRef] [PubMed]
13. Tchameni, S.N.; Cotârleț, M.; Ghinea, I.O.; Bedine, M.A.B.; Sameza, M.L.; Borda, D.; Bahrim, G.; Dinică, R.M. Involvement of Lytic Enzymes and Secondary Metabolites Produced by *Trichoderma* spp. in the Biological Control of *Pythium myriotylum*. *Int. Microbiol.* **2020**, *23*, 179–188. [CrossRef] [PubMed]
14. Silva, R.N.; Monteiro, V.N.; Steindorff, A.S.; Gomes, E.V.; Noronha, E.F.; Ulhoa, C.J. Trichoderma/Pathogen/Plant Interaction in Pre-Harvest Food Security. *Fungal Biol.* **2019**, *123*, 565–583. [CrossRef]
15. Jiang, H.; Zhang, L.; Zhang, J.-Z.; Ojaghian, M.R.; Hyde, K.D. Antagonistic Interaction Between *Trichoderma asperellum* and *Phytophthora capsici* In Vitro. *J. Zhejiang Univ. Sci. B* **2016**, *17*, 271–281. [CrossRef]
16. Tomah, A.A.; Abd Alamer, I.S.; Li, B.; Zhang, J.-Z. A New Species of *Trichoderma* and Gliotoxin Role: A New Observation in Enhancing Biocontrol Potential of *T. virens* Against *Phytophthora capsici* on Chili Pepper. *Biol. Control* **2020**, *145*, 104261. [CrossRef]
17. Liu, B.; Ji, S.; Zhang, H.; Wang, Y.; Liu, Z. Isolation of *Trichoderma* in the Rhizosphere Soil of *Syringa oblata* from Harbin and Their Biocontrol and Growth Promotion Function. *Microbiol. Res.* **2020**, *235*, 126445. [CrossRef]
18. Al-Shuaibi, B.K.; Kazerooni, E.A.; Al-Maqbali, D.a.; Al-Kharousi, M.; Al-Yahya'ei, M.N.; Hussain, S.; Velazhahan, R.; Al-Sadi, A.M. Biocontrol Potential of Trichoderma Ghanense and Trichoderma Citrinoviride Toward *Pythium aphanidermatum*. *J. Fungi* **2024**, *10*, 284. [CrossRef]
19. Gao, P.; Qi, K.; Han, Y.; Ma, L.; Zhang, B.; Zhang, Y.; Guan, X.; Qi, J. Effect of *Trichoderma viride* on Rhizosphere Microbial Communities and Biocontrol of Soybean Root Rot. *Front. Microbiol.* **2023**, *14*, 1204688. [CrossRef]
20. Huang, X.G.; Li, M.Y.; Yan, X.N.; Yang, J.S.; Rao, M.C.; Yuan, X.F. The Potential of *Trichoderma brevicompactum* for Controlling Root Rot on *Atractylodes macrocephala*. *Can. J. Plant Pathol.* **2021**, *43*, 794–802. [CrossRef]
21. Tiwari, R.; Chandra, K.; Shukla, S.K.; Jaiswal, V.P.; Amaresan, N.; Srivastava, A.K.; Gaur, A.; Sahni, D.; Tiwari, R.K. Interference of Bio-Control *Trichoderma* to Enhance Physical and Physiological Strength of Sugarcane During Pokkah Boeng Infection. *World J. Microbiol. Biotechnol.* **2022**, *38*, 139. [CrossRef] [PubMed]
22. Korkom, Y.; Yildiz, A. Evaluation of Biocontrol Potential of Native *Trichoderma* Isolates Against Charcoal Rot of Strawberry. *J. Plant Pathol.* **2022**, *104*, 671–682. [CrossRef]
23. Samuels, G.J.; Dodd, S.L.; Lu, B.-S.; Petrini, O.; Schroers, H.-J.; Druzhinina, I.S. The *Trichoderma koningii* Aggregate Species. *Stud. Mycol.* **2006**, *56*, 67–133. [CrossRef] [PubMed]
24. Ma, Q.; Cong, Y.; Wang, J.; Liu, C.; Feng, L.; Chen, K. Pre-Harvest Treatment of Kiwifruit Trees with Mixed Culture Fermentation Broth of Trichoderma Pseudokoningii and Rhizopus Nigricans Prolonged the Shelf Life and Improved the Quality of Fruit. *Postharvest Biol. Technol.* **2020**, *162*, 111099. [CrossRef]
25. Zhang, S.; Gan, Y.; Xu, B. Mechanisms of the IAA and ACC-Deaminase Producing Strain of *Trichoderma longibrachiatum* T6 in Enhancing Wheat Seedling Tolerance to NaCl Stress. *BMC Plant Biol.* **2019**, *19*, 22. [CrossRef]
26. Chowdappa, S.; Jagannath, S.; Konappa, N.; Udayashankar, A.C.; Jogaiah, S. Detection and Characterization of Antibacterial Siderophores Secreted by Endophytic Fungi from *Cymbidium aloifolium*. *Biomolecules* **2020**, *10*, 1412. [CrossRef]

27. Shen, Y.; Zhao, J.; Zou, X.; Shi, Z.; Liao, Y.; He, Y.; Wang, H.; Chen, Q.; Yang, P.; Li, M. Differential Responses of Bacterial and Fungal Communities to Siderophore Supplementation in Soil Affected by Tobacco Bacterial Wilt (*Ralstonia solanacearum*). *Microorganisms* **2023**, *11*, 1535. [CrossRef]
28. Lü, Z.-W.; Liu, H.-Y.; Wang, C.-L.; Chen, X.; Huang, Y.-X.; Zhang, M.-M.; Huang, Q.-L.; Zhang, G.-F. Isolation of Endophytic Fungi from Cotoneaster multiflorus and Screening of Drought-Tolerant Fungi and Evaluation of Their Growth-Promoting Effects. *Front. Microbiol.* **2023**, *14*, 1267404. [CrossRef]
29. Gupta, S.; Pandey, S. Unravelling the Biochemistry and Genetics of ACC Deaminase-An Enzyme Alleviating the Biotic and Abiotic Stress in Plants. *Plant Gene* **2019**, *18*, 100175. [CrossRef]
30. Olanrewaju, O.S.; Glick, B.R.; Babalola, O.O. Mechanisms of Action of Plant Growth Promoting Bacteria. *World J. Microbiol. Biotechnol.* **2017**, *33*, 197. [CrossRef]
31. Chen, H.; Song, Z.; Wang, L.; Lai, X.; Chen, W.; Li, X.; Zhu, X. Auxin-Responsive Protein MaIAA17-like Modulates Fruit Ripening and Ripening Disorders Induced by Cold Stress in 'Fenjiao' Banana. *Int. J. Biol. Macromol.* **2023**, *247*, 125750. [CrossRef] [PubMed]
32. De Caroli, M.; Rampino, P.; Pecatelli, G.; Girelli, C.R.; Fanizzi, F.P.; Piro, G.; Lenucci, M.S. Expression of Exogenous *GFP-CesA6* in Tobacco Enhances Cell Wall Biosynthesis and Biomass Production. *Biology* **2022**, *11*, 1139. [CrossRef] [PubMed]
33. Chen, Y.-L.; Lin, F.-W.; Cheng, K.-T.; Chang, C.-H.; Hung, S.-C.; Efferth, T.; Chen, Y.-R. XCP1 Cleaves Pathogenesis-Related Protein 1 into CAPE9 for Systemic Immunity in *Arabidopsis*. *Nat. Commun.* **2023**, *14*, 4697. [CrossRef] [PubMed]
34. Wang, L.; Lui, A.C.W.; Lam, P.Y.; Liu, G.; Godwin, I.D.; Lo, C. Transgenic Expression of Flavanone 3-Hydroxylase Redirects Flavonoid Biosynthesis and Alleviates Anthracnose Susceptibility in Sorghum. *Plant Biotechnol. J.* **2020**, *18*, 2170–2172. [CrossRef]
35. Siegenthaler, T.B.; Hansen, Z.R. Sensitivity of *Phytophthora capsici* from Tennessee to Mefenoxam, Fluopicolide, Oxathiapiprolin, Dimethomorph, Mandipropamid, and Cyazofamid. *Plant Dis.* **2021**, *105*, 3000–3007. [CrossRef]
36. Yassin, M.T.; Mostafa, A.A.-F.; Al-Askar, A.A.; Sayed, S.R.M.; Rady, A.M. Antagonistic Activity of *Trichoderma harzianum* and *Trichoderma viride* Strains Against Some Fusarial Pathogens Causing Stalk Rot Disease of Maize, In Vitro. *J. King Saud. Univ. Sci.* **2021**, *33*, 101363. [CrossRef]
37. Mukherjee, P.K.; Mendoza-Mendoza, A.; Zeilinger, S.; Horwitz, B.A. Mycoparasitism as a Mechanism of Trichoderma-Mediated Suppression of Plant Diseases. *Fungal Biol. Rev.* **2022**, *39*, 15–33. [CrossRef]
38. Rodríguez, M.; Evans, H.C.; de Abreu, L.M.; de Macedo, D.M.; Ndacnou, M.K.; Bekele, K.B.; Barreto, R.W. New Species and Records of *Trichoderma* Isolated as Mycoparasites and Endophytes from Cultivated and Wild Coffee in Africa. *Sci. Rep.* **2021**, *11*, 5671. [CrossRef]
39. Mejía, C.; Ardila, H.D.; Espinel, C.; Brandão, P.F.B.; Villamizar, L. Use of *Trichoderma koningiopsis* Chitinase to Enhance the Insecticidal Activity of *Beauveria bassiana* Against *Diatraea saccharalis*. *J. Basic Microbiol.* **2021**, *61*, 814–824. [CrossRef]
40. Abdenaceur, R.; Farida, B.-t.; Mourad, D.; Rima, H.; Zahia, O.; Fatma, S.-H. Effective Biofertilizer *Trichoderma* spp. Isolates with Enzymatic Activity and Metabolites Enhancing Plant Growth. *Int. Microbiol.* **2022**, *25*, 817–829. [CrossRef]
41. Schalamun, M.; Molin, E.M.; Schmoll, M. RGS4 Impacts Carbohydrate and Siderophore Metabolism in *Trichoderma reesei*. *BMC Genom.* **2023**, *24*, 372. [CrossRef] [PubMed]
42. Singh, S.; Chanotiya, C.S.; Singh, A.; Vajpayee, P.; Kalra, A. Role of ACC-Deaminase Synthesizing Trichoderma Harzianum and Plant Growth-Promoting Bacteria in Reducing Salt-Stress in *Ocimum sanctum*. *Physiol. Mol. Biol. Plants* **2023**, *29*, 815–828 [CrossRef] [PubMed]
43. Abel, S.; Oeller, P.W.; Theologis, A. Early Auxin-Induced Genes Encode Short-Lived Nuclear Proteins. *Proc. Natl. Acad. Sci. USA* **1994**, *91*, 326–330. [CrossRef] [PubMed]
44. Dreher, K.A.; Brown, J.; Saw, R.E.; Callis, J. The Arabidopsis Aux/IAA Protein Family Has Diversified in Degradation and Auxin Responsiveness. *Plant Cell* **2006**, *18*, 699–714. [CrossRef]
45. Li, Y.; Yu, T.; Wu, T.; Wang, R.; Wang, H.; Du, H.; Xu, X.; Xie, D.; Xu, X. The Dynamic Transcriptome of Pepper (*Capsicum annuum*) Whole Roots Reveals an Important Role for the Phenylpropanoid Biosynthesis Pathway in Root Resistance to *Phytophthora capsici* *Gene* **2020**, *728*, 144288. [CrossRef]

Journal of
Fungi

MDPI

Article

Deconstruction of Alkali Lignin and Lignocellulosic Substrates by *Aspergillus ochraceus* DY1 Isolated from Rotten Wood

Namdol Nilza [1], Ram Prasad [2], Ajit Varma [1] and Menaka Devi Salam [1,*]

[1] Amity Institute of Microbial Technology, Amity University Uttar Pradesh, Noida 201313, India
[2] Department of Botany, Mahatma Gandhi Central University, Motihari 845401, India
* Correspondence: salammenaka@gmail.com

Abstract: The present study reports the ability of a fungal isolate *Aspergillus ochraceus* DY1, obtained from rotten wood, to degrade alkali lignin (AL) and lignocelluloses in an efficient manner. The efficiency of degradation was monitored by measuring the percentage of decolorization and utilizing GC-MS for identifying degradation products at different time intervals (10, 20, 30, and 40 days). The optimal degradation of alkali lignin (AL) was achieved at 0.01% concentration, 25 °C, and pH 7, resulting in 63.64% degradation after 40 days of incubation. A GC-MS analysis revealed significant degradation products, including n-hexadecanoic acid, octadecane, butylated hydroxytoluene, 2,6,11-trimethyl-dodecane, dibutyl phthalate, oleic acid, 3,5-dimethoxy-phenol acetate, and 2-(phenylmethylene)- cyclohexanone. Structural changes in AL were confirmed through HSQC 2D NMR and size-exclusion chromatography, indicating depolymerization and reduced molecular weight. Furthermore, *A. ochraceus* DY1 demonstrated substantial biomass loss in corn stover (62.5%) and sugarcane bagasse (50%) after 7 days of solid-state fermentation. Surface morphological depletion was observed in the bio-treated corn stover through SEM and confocal microscopy, which was not seen in the untreated one. These findings underscore the potential of *A. ochraceus* DY1 for efficient lignin degradation, with promising applications in biofuel production, waste management in the paper and pulp industry, and the synthesis of value-added bioproducts.

Keywords: lignin degradation; *Aspergillus ochraceus*; rotten wood; alkali lignin; lignin valorization

1. Introduction

Lignin makes up about 30% of the planet's organic carbon and is the largest naturally occurring aromatic polymer [1]. In addition to the existence of rich aromatic compounds, one of the primary reasons for lignin's importance is its availability, either directly from plants or as a by-product of industrial conversion operations. The major examples of these processes are the pulp and paper industries, as well as the lignocellulose feedstock biorefineries generating liquid fuels, such as bioethanol [2,3].

Several fungi and bacteria naturally break down this resistant branching polymer, and most of its products have commercial potential. These microbial lignin degradation pathways offer a wide range of alternatives to the problematic use of hazardous chemicals in the biodegradation of lignin waste in many industries [4]. Pulp and paper mills produce large amounts of wastewater containing lignin and face challenges in complying with environmental regulations. A number of treatment methods have been developed which rely on physico-chemical processes, but they cause environmental burdens and are costly. Therefore, efficient biological treatment methods are in demand [5]. Lignin recovery from the paper and pulp industries' wastewater, and their conversion to valuable low molecular weight aromatic chemicals by exploiting potential microorganisms, has been under substantial focus during recent years. The recovery process is challenging as it tends to become modified during the pulping process. However, advanced technologies are being studied to obtain lignin from the wastewater of paper and pulp industries [6].

Therefore, in spite of its recalcitrant nature, lignin valorization is considered crucial for the full implementation of cost-effective lignocellulosic biorefineries [7].

Fungi, bacteria, and actinomycetes act together to degrade and alter lignin through biodegradation. Most of the wood-degrading fungi belong to *Basidiomycetes*, and significant studies have been conducted on their lignocellulolytic activity [8]. However, *Ascomycetes* that dwell in grasslands and shrublands are important determinants in carbon and nitrogen cycling as they are dominant in plant litter and soils. They may be present in associations with plants as endophytes, pathogens, or saprophytes in soil, playing a role in plant biomass decomposition [8]. Therefore, analyzing the lignin degradation potential of *Ascomycetes* and studying the molecular mechanism of lignin breakdown is necessary. Members of *Ascomycota* cause the soft rot of wood decay. Important examples are *Daldinia concentrica* [9] and *Podospora anserina* [10]. The genome analysis reports of some representatives of *Ascomycota* have shown the presence of many enzymes involved in lignin depolymerization. Moreover, genome and secretome analyses of the genera *Embellisia, Coniochaeta, Chaetomium, Phoma,* and *Aspergillus* sp. have shown the presence of genes encoding potential lignin degradation enzymes as well as the proteins and molecules involved in the decomposition of plants [8].

Extracellular lignin is broken down and modified by four different types of ligninolytic oxidative enzymes: laccase (Lac), manganese peroxidase (MnP), lignin peroxidase (LiP), and universal peroxidase [11]. The chemical compounds produced after the degradation of lignin are phenylacetic acid, 4-hydroxy-benzoic acid, vanillic acid [12], dibutyl phthalate [13], 3,5-dimethoxy-phenol derivatives, 2, 4-Di-tert-butylphenol, β-hydroxypropiovanillone [14], β-ketoadipate, p-cresol, 4-hydroxybenzoate, glucose, ethanol [15], β-ketoadipate enol-lactone, 3-oxohexanedioic acid, etc. Many of these degradation products are industrially important, and some of them have antioxidant properties. Hexadecenoic acid can be used in the production of surfactants as a natural additive in organic products, napalm, etc. Higher alkanes, like octadecane and nonane, are produced from lignin as by-products upon degradation and are some of the most important components of fuel oil and lubricating oil [16]. Dibutyl phthalate is a frequent lignin breakdown intermediate. The fungal peroxidase degradation of lignosulfonate produced the phthalate derivative bis(2-ethylhexyl) phthalate [17]. This shows that lignocellulosic waste can be converted into value-added products in a sustainable way with the help of lignin-degrading microbes.

The present study reports the assessment of *A. ochraceus* DY1 from rotten wood samples for lignin and lignocellulosic degradation. Its ability to depolymerize Alkali lignin (AL) has been analyzed in order to evaluate its potential biotechnological applications. In recent years, ligninolytic enzymes have gained applications mainly in lignin valorization to biofuel and high-value products, in pharmaceutical and cosmetics industries, and in the biodegradation of environmental pollutants containing phenolic compounds [18]. Since fungal ligninolytic enzymes have the capability of degrading different lignocellulosic substrates and producing high product yield in lignin valorization, the exploration of robust fungal lignin degraders is of interest for the study of their potential applications.

2. Materials and Methods

2.1. Isolation and Screening of Fungi

Soil samples were collected from the surroundings of the Haripur paper mill (geolocations: 30°91′29.431″ N and 76°84′77.035″ E) in Baddi, Himachal Pradesh, India, agricultural land (geolocations: 28°50′68.60″ N and 77°39′42.85″ E), and the biodiversity parks (geolocations: 28°53′55.17″ N and 77°39′10.29″ E) of Noida, Uttar Pradesh, India. Rotten wood samples were also collected from the biodiversity park and a wood warehouse in Noida, Uttar Pradesh, India. All samples were collected in sterilized bags and processed immediately after collection. Ten grammes of the soil sample were diluted in 90 mL of sterile 0.8% (*w*/*v*) saline water in a 250 mL conical flask and kept under a shaking condition at 150 rpm to obtain a homogenized soil suspension. The serial dilution of the soil suspension was performed aseptically up to 10^{-5} dilution. Following this, 0.1 mL from each dilution was spread plated onto a potato dextrose agar (PDA) medium containing 50 mgL^{-1} of chloramphenicol for the isolation of fungi. The inoculated plates were incubated at 27 °C for 5 to 7 days. For the isolation of fungi from rotten wood samples, decaying wood strips were surface sterilized and cut into pieces of 5 mm × 5 mm. The wood pieces were then plated on PDA media containing 50 mgL^{-1} of chloramphenicol and incubated at 27 °C for 7 days. Morphologically distinct isolates were selected and inoculated on fresh solid media and further screened for their lignin degradation potential on PDA containing different lignin representatives for screening, such as (i) 4 mM guaiacol, to detect laccase activity, where the formation of a brown colour zone around the colonies indicated positive ligninolytic activity; (ii) 1.5 mM aniline blue dye, to detect lignin peroxidase and manganese peroxidase activities through the decolourization of the dye; and (iii) 4 mM tannic acid, to detect tannase activity with the formation of clear zone. The plates were then incubated at 28 °C for 5 days and observed for any degradative activity. In addition, the degradation potential of lignocellulose was checked by screening the isolates on PDA plates supplemented with 0.5% carboxymethylcellulose (CMC) and 0.5% birchwood xylan for cellulase and hemicellulase activities, respectively.

2.2. *Study of Alkali Lignin (AL) Removal Rate*

To determine the removal rate of AL by the positive fungal isolate, it was inoculated in a minimal medium broth containing glucose 10 gL^{-1}, KH_2PO_4 0.5 gL^{-1}, $CaCl_3$ 0.075 gL^{-1}, $FeCl_3$ 0.01 gL^{-1}, $MgSO_4$ 0.15 gL^{-1}, $MnSO_4$ gL^{-1}, and yeast extract 0.2 gL^{-1} with 0.02% (*w*/*v*) AL, and incubated at 27 °C for 7 days. The AL used in the experiment was from Tokyo Chemical Industry (India) Pvt. Ltd., Chennai, India (Product No. L0082) and is chemically processed and water-soluble. The content of the AL was measured at 280 nm and determined using the standard curve of AL. The removal rate of AL was determined using the formula, removal rate (%) = $(C_1 - C_2)/C_1 \times 100$, where C_1 is the concentration of AL at t_0 (i.e., before inoculation) and C_2 is the concentration of AL after 7 days of fungal culture inoculation [19].

2.3. *Identification of Most Potent Isolate*

The isolate having the maximum AL removal was identified based on an internal transcribed spacer (ITS)-5.8S rRNA gene sequencing analysis. For molecular identification, genomic DNA was isolated using the AllPrep fungal DNA isolation kit (Qiagen, Hilden, Germany) and the purity and concentration were analyzed using a NanoDrop 2000 spectrophotometer. The ~700 bp ITS rDNA fragment was amplified using high-fidelity PCR polymerase, TaKaRa ExTaq (DSS TAKARA Bio India Pvt. Ltd., New Delhi, India). The primers used were: ITS5 Forward 5′ TCCGTAGGTGAACCTGCGG 3′ and ITS4 Reverse 5′ TCCTCCGCTTATTGATATGC 3′. The PCR cycling conditions were initial denaturation at 94 °C for 3 min, 30 cycles with denaturation at 94 °C for 1 min, annealing at 50 °C for 1 min, and extension at 72 °C for 2 min. The final extension was given at 72 °C for 7 min. The sequence data were aligned with available databases through NCBI BLAST, and the maximum homology of the sequence was checked.

2.4. *Phylogenetic Analysis*

The phylogenetic tree was constructed for the isolated strain using the MEGA 11.0 software [20] based on the ITS rDNA sequences and other types of strains of *A. ochraceus*. The neighbour-joining method was used with bootstrap values based on 1000 replications, where the scale was 0.02.

2.5. *Effect of Different Parameters on Alkali Lignin Degradation*

The effect of the AL concentration, pH, temperature, and incubation period on AL degradation was checked by varying the AL concentration (0.01%, 0.05%, 0.1%, and 0.5%), pH (3, 5, 7, and 9), temperature (22 °C, 25 °C, 28 °C, and 30 °C), and incubation period (10,

20, 30, and 40 days). The extent of the degradation was, thereafter, analyzed via FTIR, and the AL removal rate was calculated for each of the different conditions.

FTIR was performed to predict the unknown compounds present in the DY1-treated AL medium. The DY1-treated medium was centrifuged at 10,000 rpm for 5 min, after which the supernatant was collected in a fresh tube for FTIR testing in the Perkin Elmer FTIR/ATR instrument.

2.6. Analysis of Degradation Products

A GC-MS spectrometry was performed to identify the compounds present in the medium under optimized conditions of 10, 20, 30, and 40 days of incubation. For this study, the DY1-treated medium was centrifuged at 10,000 rpm for 15 min in falcon tubes, and the supernatant was mixed well with a 1:1 ratio of ethyl acetate for 30 min in a shaker. The contents were poured into a separating funnel and left for 15–20 min until the organic and inorganic layers separated. The lower inorganic layer was discarded, and the above organic layer was transferred into a flat dish for evaporation in a hot air oven until all the liquid evaporated. The contents were scraped out, mixed with methanol, and used for the identification of the compounds through GC-MS spectrometry. Agilent HP-5 ms Low Bleed GC/MS (30 m long, 0.25 mm inner diameter, 0.2 μm film thickness, particular number: 19091S-433) was used as a capillary column. Helium was used as the carrier gas at a flow rate of 1.3 mL/min. The column temperature was initially 50 °C for 1 min, then it was raised to 325 °C (10 °C/min, with a holding time of 5 min). The transfer line, quadrupole, and the ion source temperatures were 325 °C, 150 °C, and 250 °C, respectively. The solvent delay time was 4.50 min; the injector and detector temperatures were both 275 °C. One microliter of the methanol-soluble portion was directly injected, and the solvent delay time was 4.50 min. Mass spectra were recorded in the range of 150–510 m/z. The degraded products of lignin were identified by comparing mass spectra to the National Institute of Standards and Technology 11 (NIST17) library, which is available in the used instruments, i.e., GCMS-QP2010 Ultra (Shimadzu Corporation, Kyoto, Japan).

A 2D NMR was performed to predict the unknown compounds present in the DY1-treated AL medium. The fungal isolate, DY1, was incubated in a minimal medium broth containing glucose 10 gL^{-1}, KH_2PO_4 0.5 gL^{-1}, $CaCl_3$ 0.075gL^{-1}, $FeCl_3$ 0.01 gL^{-1}, $MgSO_4$ 0.15 gL^{-1}, $MnSO_4$ gL^{-1}, and yeast extract 0.2 gL^{-1} supplemented with 0.01% (w/v) AL for 7 days at 25 °C and 160 rpm. The culture was filtered and then lyophilized; the content was collected in a fresh tube and dissolved with deuterated methanol for the 2D NMR testing. Two-dimensional NMR spectra were recorded at 25 °C in 5 mm tubes on a Bruker Avance II 300, Bruker, Billerica, MA, USA. Two-dimensional NMR spectra were recorded in HSQC (heteronuclear single-quantum correlation) spectroscopy.

Size exclusion chromatography (SEC) to determine the molecular weight distribution of the lignin before and after fungal treatment was performed on an Agilent 1260 Infinity GPC/SEC System (Agilent, Santa Clara, CA, USA) coupled to a UV detector and a RI detector. The control sample and the treated samples were dissolved in the solvent—tetrahydrofuran (THF) and filtered over a 0.45 μm syringe filter before injection. Five μL of the samples were injected into the SEC system with THF as eluent (0.5 mL min^{-1} at 40 °C). A standard calibration was performed with a set of narrow polydispersity pullulan standards for the molecular weight, ranging from 342 $\times$ 10^5 to 3.44 $\times$ 10^5 g mol^{-1}, and polystyrene standards for the molecular weight, ranging from 162 $\times$ 10^6 to 5 $\times$ 10^6 g mol^{-1} (Agilent Technologies, Santa Clara, CA, USA).

2.7. Analysis of Bio-Treatment of Corn Stover and Sugarcane Bagasse

- Determination of biomass loss
  Fungal treatment procedure

Corn stover and sugarcane bagasse were collected locally. They were crushed, washed, and dried in an oven. They were then ground, with the help of a laboratory grinding mill, into small particles of mesh, sized 1 to 2 mm, and stored in air-tight containers. The fungal treatment was carried out by inoculating the lignocellulosic substrate (corn stover or sugarcane bagasse), mixed with inorganic salt solution, having the compositions of 2 g L^{-1} $(NH_4)_2SO_4$, 2 g L^{-1} KH_2PO_4, 0.3 g L^{-1} $MgSO_4$, 0.3 g L^{-1} $CaCl_2$, 0.5 g L^{-1} NaCl, 0.005 g L^{-1} $FeSO_4$, 0.016 g L^{-1} $MnSO_4$, and 0.017 g L^{-1} $ZnCl_2$, in the ratio of 1:3 (*w*/*v*). A total of 3 mL of the spore suspension was inoculated into the medium under aseptic conditions and the mixture was cultivated for 21 days in static condition at 27 °C.

To check the biomass loss, the bio-treated corn stover or sugarcane bagasse was first regenerated via boiling at 100 °C for 15 min and the water was drained with the help of a filter paper. The sample was washed repeatedly with autoclaved distilled water until the water was visibly clear. Then, the corn stover was drained to get rid of excess water with the help of a muslin cloth, after which it was allowed to dry completely in an oven at 60 °C for 3 h until all the water weight was removed. After the biomass was completely dried, it was weighed to calculate the biomass loss according to the formula biomass loss = initial weight − final weight [21].

In order to check the correlation (if any) between biomass loss and the spore count, observations were conducted at regular intervals using the haemocytometer. A very tiny amount of bio-treated corn stover or sugarcane bagasse was taken in an Eppendorf tube, and 1 mL of distilled water was added to the tube. It was shaken well to transfer as many fungal spores into the liquid as possible. A total of 100 µL of the suspended liquid was placed on the haemocytometer and a cover slip was placed on top. The haemocytometer was then observed under a microscope. This observation was made in two-day intervals for a period of 23 days [22].

- A microscopic analysis of bio-treated corn stover and sugarcane bagasse

The regenerated biomass samples, i.e., the untreated and the DY1-treated, which were boiled and then washed several times with autoclaved distilled water, were finally dried in a hot air oven at 50 °C for 2 h. A few strands of the biomass samples were mounted on the coverslip and coated with poly L-lysine [23]. The sample was then dehydrated with methanol and kept at room temperature for drying. The dried samples were mounted on aluminium stubs, sputter coated with a thin layer (6 nm) of gold. The samples were analyzed through a scanning electron microscope (ZEISS EVO, Jena, Germany) with a beam voltage of 20 kV. The surface morphologies of the untreated and the bio-treated corn stover and sugarcane bagasse were compared.

The samples were also analyzed under a confocal scanning laser microscopy (CSLM). Very thin-sectioned samples were distributed on the glass microscope slide and covered with glass coverslips. Images were captured using Nikon's Confocal microscope A1 under 60× magnification at Amity Institute of Microbial Technology, Amity University, Noida, India, and analyzed using NIS Elements software (NIS-Elements AR 6.02.01) (Nikon, Tokyo, Japan).

2.8. Ligninolytic Activity

A total of 5 g of each substrate (corn stover and sugarcane bagasse) was taken in 250 mL conical flasks. Each substrate-containing flask was moistened with 5 mL of distilled water and sterilized by autoclaving at 121 °C for 20 min. A conidial suspension was prepared by mixing two 5 mm mycelial discs from a 7-day-old strain in 10 mL of sterile water achieving 2×10^{-6} spores/mL. The substrates were then inoculated in 3 mL of the spore suspension under aseptic conditions and incubated for 14 days in static conditions at 27 °C. For the extraction of the crude enzyme, the experimental flasks were filled with 50 mL of autoclaved distilled water and placed in a shaker incubator at 120 rpm for an hour. Using Whatman No. 1 filter paper, the biomass was individually filtered, and the filtrate was thereafter centrifuged at 5000 rpm for 20 min. The supernatant was used as a crude enzyme for testing ligninolytic enzyme activity. One ml of the supernatant was mixed with 1 ml of 2 mM guaiacol, along with 3 mL of the 10 mM sodium acetate buffer (pH 5), in a test tube. A blank was also prepared, which contained 1 ml of distilled water instead of the culture supernatant. The mixtures were incubated for 15 min at 30 °C and the absorbance read 450 nm using a UV-vis spectrophotometer. The laccase activity was determined by observing the absorbance at 450 nm (ε = 0.6740 μM/cm). One unit of laccase activity was expressed as the amount of enzyme required to oxidize 1 μmol of guaiacol per minute [24].

3. Results

3.1. Isolation and Screening of Microorganisms

Out of 22 morphologically distinct fungal isolates, eight isolates showed lignin degradation and/or lignocellulose degradation potential during preliminary screening. The fungal isolate, DY1, showed the best degrading activities on the PDA containing different lignin representatives for screening, i.e., guaiacol, aniline blue dye, and tannic acid (Table 1). The zone of decolourization and oxidation activity was observed around the culture within 48 h of incubation for the isolate DY1. Moreover, it showed positive results in the screening of cellulase and xylanase activities.

Table 1. Status of various fungal isolates on different indicators of lignin biodegradation.

Isolate	Sample Type	Source	Indicator			CMC Hydrolysis	Birchwood Xylan Hydrolysis
			Guaiacol	Aniline Blue Dye	Tannic Acid		
SL4	Soil	Agricultural field	0.2 mm	-	-	0.3 mm	0.5 mm
SL5	Soil	Agricultural field	0.3 mm	0.6 mm	-	1 mm	-
PS12	Soil	Paper mill	2.3 mm	0.9 mm	0.8 mm	0.6 mm	0.8 mm
PS14	Soil	Paper mill	0.8 mm	0.8 mm	0.7 mm	-	1.1 mm
SL17	Wood	Wood warehouse	0.4 mm	-	-	1.8 mm	2.1 mm
SL18	Wood	Wood warehouse	1.9 mm	-	-	1.6 mm	0.7 mm
DY1	Rotten wood	Biodiversity Park	4.2 mm	3.6 mm	2.3 mm	1.9 mm	1.6 mm
DY3	Soil	Biodiversity Park	0.2 mm	0.8 mm	-	1.9 mm	1 mm

3.2. Study of Alkali Lignin Removal Rate

The lignin degradation ability was determined by the alkali lignin removal rate (%). The DY1 isolate, which was selected based on the preliminary screening results, was grown in a medium containing AL as the sole carbon source and the remaining lignin content after incubation was determined using a UV-spectrophotometer at 280 nm. The percentage of AL removal rate was determined to be 76.65% after 7 days of incubation.

$$\text{Removal rate (\%)} = (C_1 - C_2)/C_1 \times 100$$

$$= (1.89 - 0.441)/1.89 \times 100$$

$$= 76.65\%$$

3.3. Identification of the Most Potent Isolate

The molecular identification of the fungal isolate DY1 was carried out based on ITS region sequencing. A total length of 593 bp of the ITS region was sequenced, and it was found that the strain had 99–100% homology, with various strains of *A. ochraceus*. Therefore, the fungal isolate in the present study was designated as *A. ochraceus* strain DY1. The Genbank accession number for the ITS region of strain *A. ochraceus* DY1 is OP804322.

3.4. Phylogenetic Analysis

To determine the relationship between *A. ochraceus* DY1 (Taxonomy ID: 40380) and 11 other related strains of *Aspergillus* sp. based on sequence analyses, a phylogenetic tree was obtained using MEGA11. The *Aspergillus niger* strain 18S and *Rhizopus oryzae* strain Y5-1 were included as members other than *A. ochraceus*, and as non-ascomycete strains, respectively. It was observed that the nearest neighbours with the same evolutionary branch were six different strains of *A. ochraceus* with bootstrap value 99 (Figure 1). These strains also showed 99% to 100% similarities with *A. ochraceus* DY1 in the BLAST analysis. A few strains of *A. ochraceus* chosen during the phylogenetic tree construction appeared under separate nodes from the ones under study and, therefore, have different evolutionary branches. They were: (i) *A. ochraceus* isolate VJP06, as a separate node with the bootstrap value 90, and two other strains, *A. ochraceus* strain C135N and *A. ochraceus* strain 567N, which were related to *A. candidus* ATCC 1002, as a separate node with bootstrap value 100; all of them arise from a common ancestor. *Aspergillus niger* and the *Rhizopus oryzae* strain Y5-1 appeared under separate evolutionary branches.

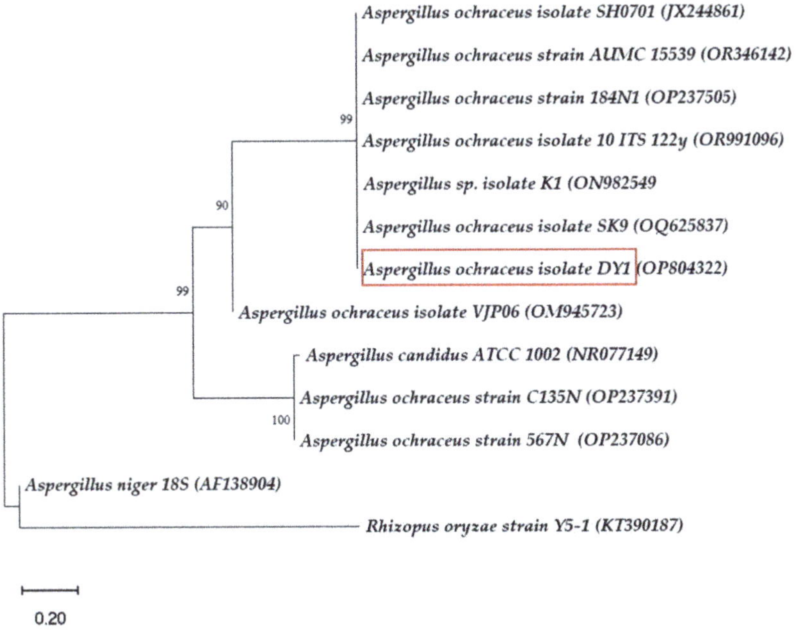

Figure 1. Phylogenetic tree of isolates based on ITS region sequencing. The isolate in the present study is highlighted with a red colour box. The NCBI accession numbers of the respective sequences are given in parentheses after the microorganism names.

3.5. Effect of Different Parameters on AL Degradation

3.5.1. Effect of Lignin Concentration on AL Degradation

Lignin degradation, using the selected fungal isolate *A. ochraceus* DY1 at different concentrations of AL, was studied. The decolourization (degradation) was not observed in the control, i.e., the uninoculated medium supplemented with AL. The test sample with 0.01% AL had the highest lignin degradation, as calculated by the percentage removal rate at 60.19%, followed by 0.05%, 0.1%, and 0.5% lignin concentrations, as shown in Figure 2.

The results of the FTIR analysis of AL treated with *A. ochraceus DY1* are shown in Figure 3. More new peaks appeared in the treated sample compared to the untreated one, indicating changes in the functional groups of the products formed during the fungal treatment of AL. A characteristic broad peak of an aliphatic alcohol with an O–H stretch at 3325 cm^{-1} and an C–O stretch at 1072 cm^{-1} was observed in the spectrum. The C–O stretching, which is a sharp peak, also supports obtaining a phenol structure. The peaks at 2982 cm^{-1} and 1635 cm^{-1} indicate the presence of the C–H stretching of the alkane group and the C=C stretching of alkene (monosubstituted), respectively. The peak at 1382 cm^{-1} shows O–H bending, which confirms the phenol structure. The peaks at 1152 and 1252 cm^{-1} show C–O stretching and the peak at 2104 cm^{-1} shows C–N and C=C stretching, as shown in Figure 3.

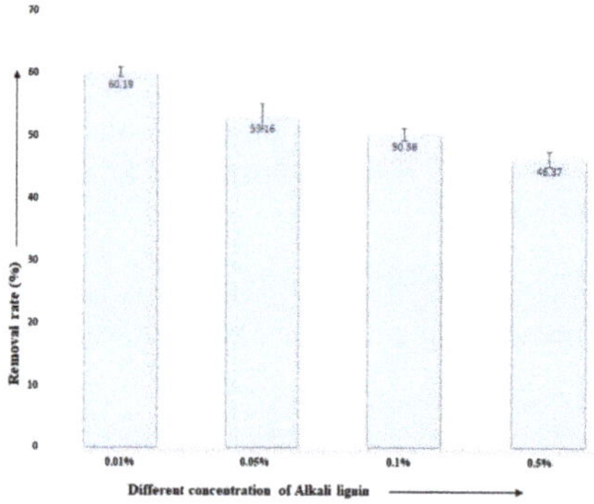

S. No.	Lignin concentration %	0hr	15th day	Removal rate (%)
1	0.01%	0.53	0.211	60.19
2	0.05%	0.901	0.422	53.16
3	0.1%	1.787	0.887	50.36
4	0.5%	4.311	2.312	46.37

Figure 2. Effect of different lignin concentrations on their degradation via *A. ochraceus* DY1. The results are the mean of three distinct replicates, with standard deviation represented by error bars.

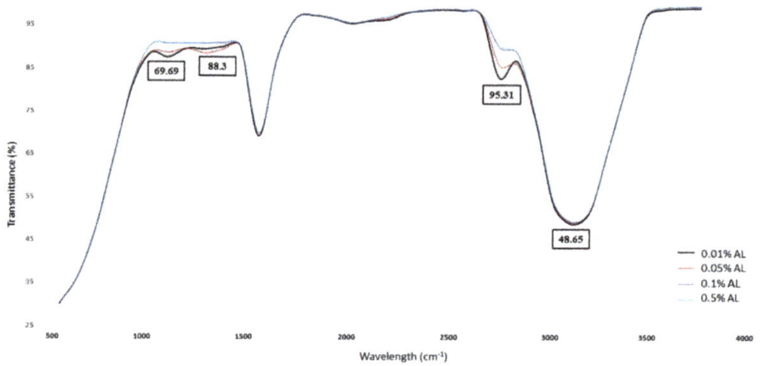

S.No.	Wavelength	Transmittance %				Functional groups
		0.01%	0.05%	0.1%	0.5%	
1	1072	88.3	88.6	90.23	90.23	C-O stretching, Primary alcohol
2	1152	89.42	88.52	90.63	90.7	C-O stretching, ester
3	1252	89.34	89.41	90.67	90.72	C-O stretching, alkyl aryl ether
4	1382	89.29	88.38	90.63	90.7	Terminal CH3
5	1635	69.69	69.8	69.72	69.71	C=C stretching, alkene
6	2104	95.31	95.48	95.48	95.47	C-N, C=C stretching bonds
7	2982	85.8	83.65	89.72	90.11	C-H stretching, alkane
8	3325	**48.65**	48.93	49.1	49.11	O-H, polyphenols

Figure 3. FTIR spectra of the effect of different AL concentrations on their degradation via *A. ochraceus* DY1. The AL concentrations used were 0.01%, 0.05%, 0.1%, and 0.5%, represented by different colour lines in the spectra. The transmittance values where major peaks were detected are indicated inside the boxes. The wavelengths where peaks were detected, and the corresponding functional groups, are shown below the spectra.

3.5.2. Effect of Temperature on AL Degradation

Lignin degradation via the selected fungal isolate *A. ochraceus* DY1 at different temperatures with 0.05% (*w*/*v*) AL was studied. The sample incubated at 25 °C had the highest percentage of lignin degradation at 54.03%. The difference in the percentage removal rate at 25 °C from the previous experiment that showed 60.19% might be due to the percentage of lignin concentrations used. At 22 °C, 28 °C, and 30 °C, the degradation percentages were 45.17%, 42.94%, and 11.92%, respectively, as shown in Figure 4.

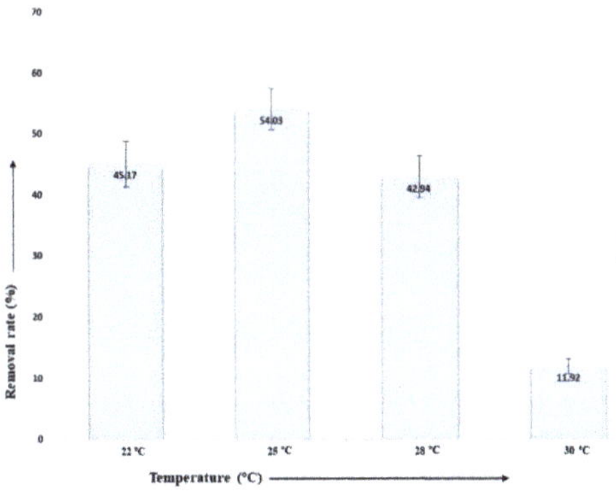

S. No.	Temperature (°C)	0hr	15th day	Removal rate (%)
1	22°C	0.901	0.494	45.17
2	25°C	0.92	**0.423**	**54.03**
3	28°C	0.913	0.521	42.94
4	30°C	0.907	0.799	11.92

Figure 4. Effect of temperature on lignin degradation via *A. ochraceus* DY1. The results are the mean of three distinct replicates, with standard deviation represented by error bars.

The result of the FTIR analysis of AL treated with *A. ochraceus* DY1 at different temperatures is as shown in Figure 5. There was no significant difference between the different temperatures, but the peaks obtained at 1892 cm^{-1} and 1982 cm^{-1}, which are of C=O stretching C–H stretching and alkane, respectively, were sharper and more prominent at 25 °C.

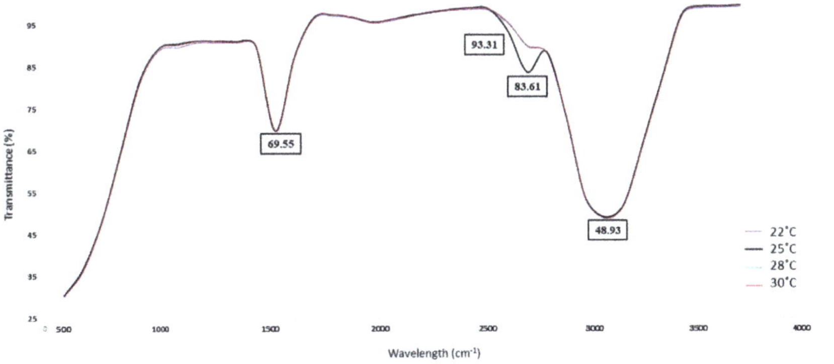

S.No.	Wavelength	Transmittance %				Functional groups
		22°C	25°C	28°C	30°C	
1	1635	69.57	69.55	69.73	69.69	C=C stretching, alkane
2	2892	95.22	93.31	95.51	95.49	C-O stretching
3	2982	89.95	83.61	89.76	89.79	C-H stretching, alkane
4	3326	49.08	**48.93**	48.76	48.71	OH, polyphenols

Figure 5. FTIR spectra of the effect of temperature on AL degradation via *A. ochraceus* DY1, shown as different colours for the temperatures 22 °C, 25 °C, 28 °C, and 30 °C. The values inside the boxes indicate transmittance values of the peaks. The wavelengths where peaks were obtained, and their corresponding functional groups are shown below the spectra.

3.5.3. Effect of pH on AL Degradation

Lignin degradation via the selected fungal isolate *A. ochraceus* DY1 at different pH levels was studied. The sample with pH7 had the highest percentage of lignin degradation at 43.07%, whereas pH3, pH5, and pH9 showed 15.27%, 25.64%, and 4%, respectively, as shown in Figure 6.

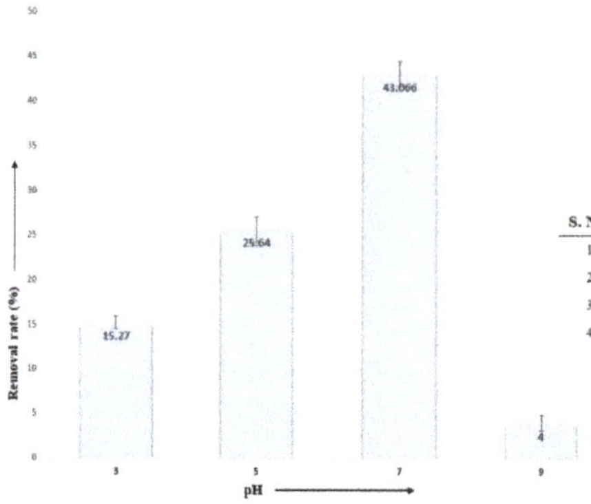

S. No.	pH	0hr	15th day	Removal rate (%)
1	3	0.071	0.06	15.27
2	5	0.077	0.0527	25.64
3	7	0.102	**0.0404**	**43.066**
4	9	0.114	0.10944	4

Figure 6. Effect of different pH on lignin degradation via *A. ochraceus* DY1. The results are the mean of triplicates, with standard deviation represented by error bars.

The results of the FTIR analysis of AL treated with *A. ochraceus* DY1 at different pH are shown in Figure 7. There was no significant difference between the peaks obtained at different pH, although the decolourization of AL was best obtained at pH7.

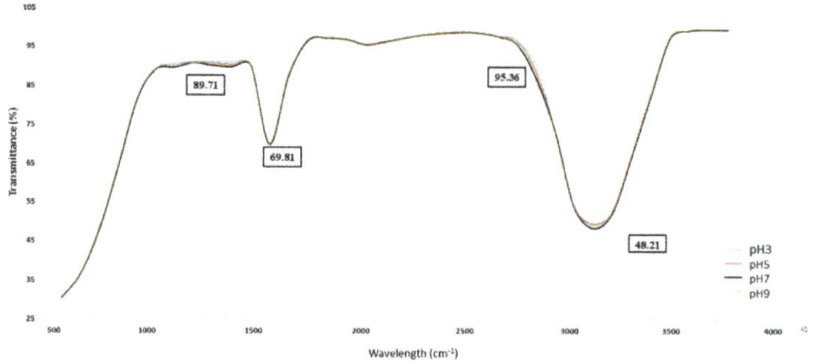

S.No.	Wavelength	Transmittance %				Functional groups
		3	5	7	9	
1	1094	90.45	90.06	89.55	90.01	C-O stretching, primary alcohol
2	1155	90.86	90.24	89.64	90.06	C-O stretching, ester
3	1267	90.91	90.58	90.04	90.45	C-O stretching, alkyl aryl ether
4	1380	90.93	90.29	89.71	90.06	Terminal CH3
5	1639	69.95	70.01	69.81	69.77	C=C stretching, alkene
6	2877	95.97	96.45	95.36	95.45	C-O stretching
7	2982	90.21	88.55	86.58	87.94	C-H stretching, alkane
8	3329	49.21	49.52	**48.21**	49.06	OH, polyphenols

Figure 7. FTIR spectra of the effect of pH on AL degradation via *Aspergillus ochraceus* DY1, as shown by different colour spectra for the pH values 3, 5, 7, and 9. The values inside the boxes indicate the transmittance values of the peaks. The absorption peaks and their corresponding functional groups are shown below the spectra.

3.5.4. Effect of Incubation Period on AL Degradation

The effect of different days of incubation on the lignin degradation in the presence of the selected fungal isolate *Aspergillus ochraceus* DY1 and 0.01% (*w*/*v*) AL was studied. The sample incubated for 40 days had the highest lignin degradation with 63.73%, whereas the sample incubated for 10 days showed an 11.77% lignin concentration; 20 days and 30 days of incubation showed 20.52% and 45.75%, respectively. This indicates that with gradual increases in the incubation period, the degradation of lignin also increases, as shown in Figure 8.

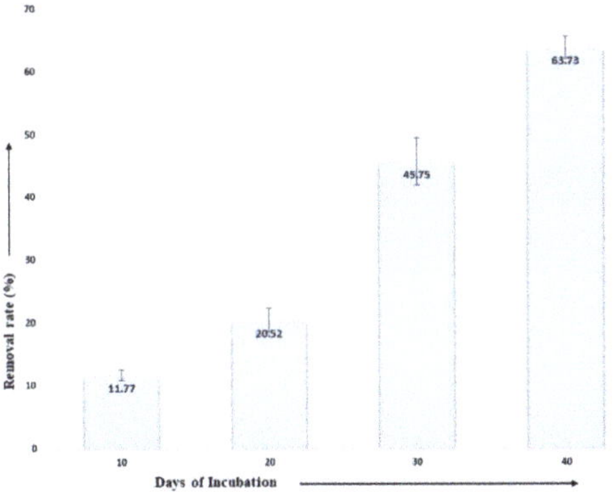

S. No.	Days	Control	Test	Removal rate (%)
1	10	0.08137	0.07178	11.77
2	20	0.0813	0.0646	20.52
3	30	0.0812	0.04403	45.75
4	40	0.08177	**0.0297**	**63.73**

Figure 8. Effect of different days of incubation on lignin degradation by *Aspergillus ochraceus* DY1. The results are the mean of triplicates with error bars representing standard deviation.

The results of the FTIR analysis of AL treated with *Aspergillus ochraceus* DY1 at different incubation periods are shown in Figure 9. The incubation period of 40 days gave peaks which were sharp and significant compared to the 10- and 20-day incubation periods. The prominent peaks were at 2981 cm^{-1} (C–H stretching, alkane), 1382 cm^{-1} (O–H bending, terminal -CH_3), 1152 cm^{-1} (C–O stretching, ester), and 953 cm^{-1} (C=C bending, alkene). These peaks were not visible during the 10- and 20-day incubation periods.

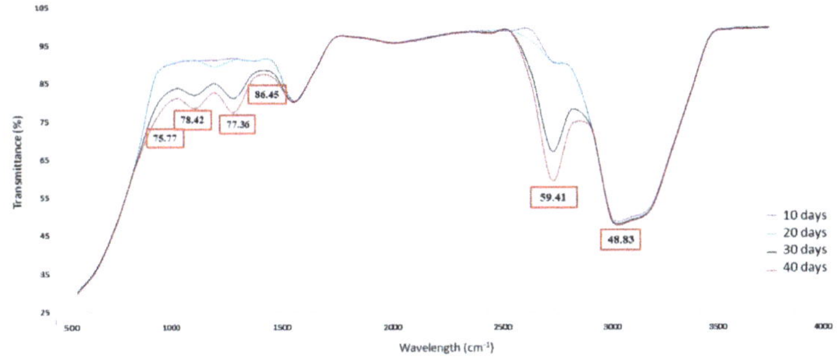

S.No.	Wavelength	Transmittance %				Functional groups
		10 days	20 days	30 days	40 days	
1	953	87.15	87.19	78.89	75.77	C=C bending, alkene
2	1073	90.36	90.55	83.59	81.04	C-O stretching, Primary alcohol
3	1152	91.08	91.21	81.84	78.42	C-O stretching, ester
4	1252	91.1	89.41	84.91	82.6	C-O stretching, alkyl aryl ether
5	1382	91.6	91.19	81.02	77.36	O-H bending, Terminal CH3
6	1463	90.84	90.89	87.56	86.32	C-H bending, alkane
7	1473	90.7	90.66	87.58	86.45	Methylene group
8	2666	98.76	98.79	98.24	98.01	S-H stretching, thiol
9	2890	98.83	96.06	87.34	84.1	C-H stretching, alkane
10	2981	90.68	90.75	67.05	59.41	C-H stretching, alkane
11	3326	49.81	49.06	49	**48.83**	O-H, polyphenols
12	3338	49.82	49.08	48.9	**48.83**	C-H stretching, alkyne

Figure 9. FTIR spectra of the effect of incubation time on AL degradation via *Aspergillus ochraceus* DY1 shown as different colour spectra for 10, 20, 30, and 40 days of incubation. The values inside the boxes indicate the transmittance values of the peaks. The absorption peaks and their corresponding functional groups are shown below the spectra.

3.6. GC MS Analysis of AL Degradation

The lignin degradation potential of *Aspegillus ochraceus* DY1 was further checked by treating the alkali lignin with the strain and analyzing the degradation products on different days using GC MS. Technical lignin may contain different functional groups, such as phosphate, sulphonate, butyl groups, etc., depending on the process and chemical reactions utilized.

The degradation products of AL by *Aspergillus ochraceus* DY1 were analyzed by GC-MS. The results of untreated and treated AL at different days of incubation are shown in Table 2. The possible degradation products are n-hexadecanoic acid, oleic acid, octadecane, butylated hydroxytoluene, 2,6,11-trimethyl-dodecane, dibutyl phthalate, 3,5-dimethoxy-phenol acetate, and 2-(phenylmethylene)-cyclohexanone, which were produced during the incubation kept up to 40 days. The larger compounds linked to the technical lignin or processed lignin structure, which gradually disappeared during the incubation period, were 13-(2-Methoxyphenyl) tricyclo[8.2.2.24,7] hexadeca-1, phenol, 2,4-bis(1,1-dimethylethyl)-phosphite, 4,4′-butylidenebis[2-(1,1-dimethylethyl)-5-methyl- phenol, and 7,9-Di-tert-butyl-1-oxaspiro (4,5) deca-6,9-diene-2,8-dione.

Table 2. GC-MS results of untreated and treated alkali lignin.

Compound	Control	10 Days	20 Days	30 Days	40 Days
13-(2Methoxyphenyl)tricyclo[8.2.2.24,7]hexadeca-1	+	−	−	−	−
Phenol, 2,4-bis(1,1-dimethylethyl)-, phosphite	+	+	+	−	−
Phenol,4,4′-butylidenebis[2-(1,1-dimethylethyl)-5-methyl-	+	−	−	−	−
7,9-Di-tert-butyl-1-oxaspiro (4,5) deca-6,9-diene-2,8-dione	+	+	+	+	+
n-hexadecanoic acid	−	+	+	+	+
oleic acid	−	−	−	+	+
Butylated hydroxytoluene	−	−	+	+	+
Dodecane, 2,6,11-trimethyl-	−	−	−	−	+
Octadecane	−	−	−	+	+
Dibutyl phthalate	−	−	+	+	+
Phenol, 3,5-dimethoxy-, acetate	−	−	+	−	−
Cyclohexanone, 2-(phenylmethylene)-	−	−	−	+	−

+ (Present), − (Absent).

The HSQC 2D NMR spectroscopy analysis was carried out to demonstrate the structural characteristics of compounds generated after the treatment of AL with *Aspergillus ochraceus* DY1, as shown in Figure 10. The structural changes in the depolymerization process were clearly demonstrated in the comparison between the fresh AL and the DY1-treated AL. It can be clearly seen from this figure that the untreated AL had δ_C/δ_H = 50.0–60.0/3.0–4.0, indicating the Cβ-Hβ in β-β′ resinol structures, the β-5′ in phenylcoumaran structures, and the C–H in -OCH3 structures. The δ_C/δ_H = 60.0–80.0/3.0–4.0 indicates the Cγ-Hγ in β-β′ resinol structures and the δ_C/δ_H = 60.0–80.0/4.0–5.0 indicates Cα=O in Hibbert ketone. The cross-peaks of Cβ–Hβ in β-O-4′substructures linked to syringyl units were indicated by the chemical shift of δ_C/δ_H = 80.0–100.0/4.0–5.0. When compared with the spectra of the untreated AL, the spent lignin after fungal treatment showed only the δ_C/δ_H 40.0–60.0/3.0–4.0, indicating the β-β resinol and β-5 phenyl coumaran structures of lignin. The difference in the ^{13}C–^{1}H correlation signals between fresh and spent AL, and also the disappearance of β-O-4′aryl ether linkages, show the process of depolymerization in the fungal-treated AL.

SEC was performed to observe the degradation of AL by comparing the changes in the molecular weight (Mw) distribution of the polymer after it was treated with *Aspergillus ochraceus* DY1. In this study, the Mw of AL, before and after 7 days of incubation with strain DY1, was evaluated. The average Mw of the sample on day 0 was approximately 15,776.2 Da. The Mw of the treated sample decreased to 12,219.2 Da on the 7th day after being treated with *Aspergillus ochraceus* DY1. In addition, the polydispersity (Mw/Mn) of lignin fragments decreased from 2.2 to 1.67 on day 7. A decrease in the molecular weight distribution of treated AL over time indicated that AL was successfully degraded after 7 days of treatment with *Aspergillus ochraceus* DY1 (Table 3).

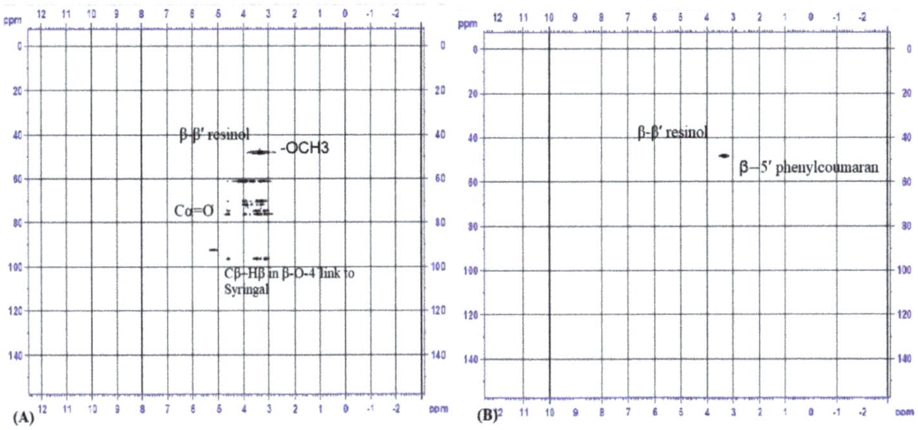

Figure 10. 2D HSQC NMR characterization of (**A**) untreated AL and (**B**) DY1-treated AL.

Table 3. SEC results of untreated and treated Alkali lignin after 7 days of inoculation with *Aspergillus ochraceus* DY1.

Sample	RT	PMw	Mw	Mn	Mw/Mn	StartMw	EndMw	Total Area	Mz
Treated	7.803	7330.6	12,219.2	7310.5	1.67145	128,513.2	631.2	1.067712	27,989
Control	7.652	9653.1	15,776.2	7165.1	2.2018	152,729.7	1055.5	1.47431	35,651

RT: retention time, **PMw**: peak molecular weight, **Mw**: weight-average molecular weight, **Mn**: number-average molecular weight, **Mw/Mn**: polydispersity index (PDI), **StartMw**: starting molecular weight, **EndMw**: ending molecular weight, **Total Area**: the total area under the GPC curve, **Mz**: Z-average molecular weight.

3.7. Analysis of Bio-Treatment of Corn Stover and Bagasse

Aspergillus ochraceus DY1 was analyzed for its potential applications in the bio-treatment of corn stover and sugarcane bagasse. In SSF using 4 g of corn stover, biomass loss of corn stover after 7 days of treatment was found to be 62.5%, indicating the consumption of corn stover during bio-treatment. Similarly, when sugarcane bagasse was used, the biomass loss was observed to be 50%. In correlation, a gradual increase in the spore count of *A. ochraceus* DY1 was observed during 23 days of incubation in corn stover and in sugarcane bagasse at 27 °C (Figure 11).

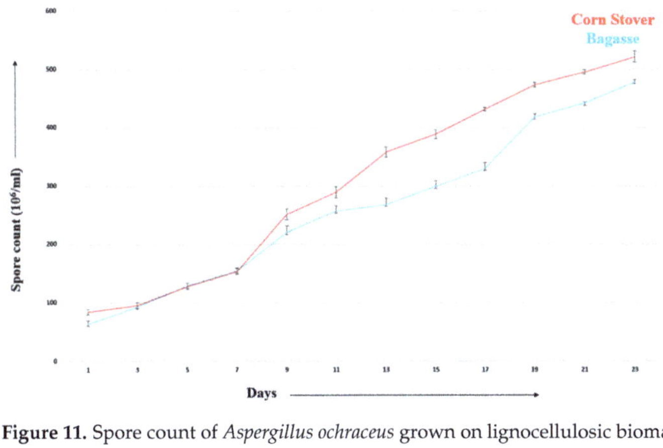

Figure 11. Spore count of *Aspergillus ochraceus* grown on lignocellulosic biomass.

3.8. Microstructure Analysis of Bio-Treated Corn Stover

Structural analyses of bio-treated corn stover and sugarcane bagasse were carried out by scanning electron microscopy (SEM), and it was found that there was a significant depletion of structural morphology (Figures 12 and 13). The surface of untreated corn stover showed smooth and uniform structural integrity, whereas DY1-treated corn stover showed various pits and disintegration, demonstrating that the corn stover surface was partially destroyed. SEM results were also supported by confocal microscopy results. The representative image of the deconstruction of the lignocellulosic raw material DY1 is shown in Figure 14, where the bio-treated corn stover is seen to have a rough surface with thin layering compared to the untreated one, which showed a smooth surface with thick layering under confocal microscopy. This is because of the degradation or removal of layers by the fungal treatment. These morphological changes indicated that degradation occurred on the surfaces of both corn stover and sugarcane bagasse after bio-treatment due to the degradation of lignocellulose by the enzymes secreted by the fungus DY1 during growth in biomass.

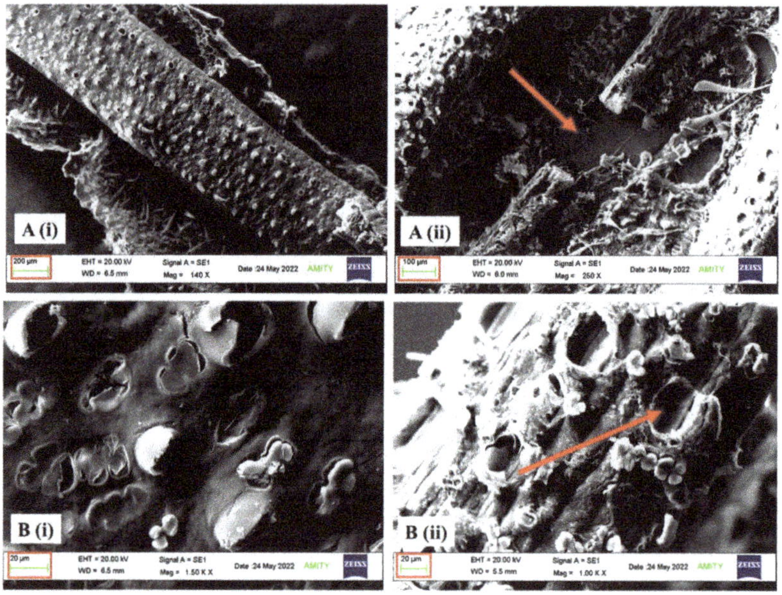

Figure 12. SEM of untreated and pre-treated corn stover. (**A**) (**i**) and (**B**) (**i**) untreated corn stover, (**A**) (**ii**) and (**B**) (**ii**) bio-treated corn stover. The red colour arrows indicate pits and disintegration of structural morphology in the bio-treated corn stover.

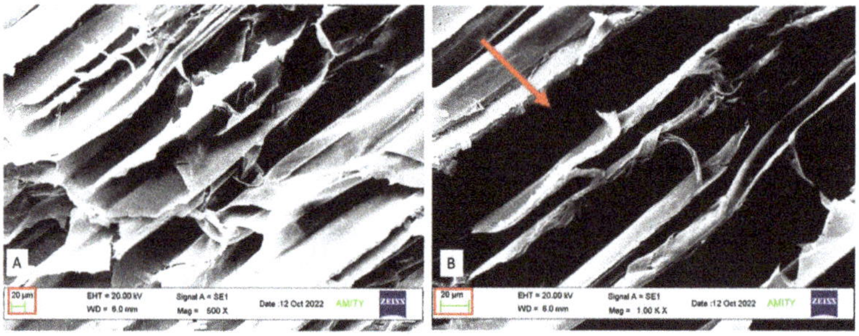

Figure 13. SEM of (**A**) untreated, and (**B**) bio-treated sugarcane bagasse. The red colour arrow indicates surface disintegration in the bio-treated sugarcane bagasse.

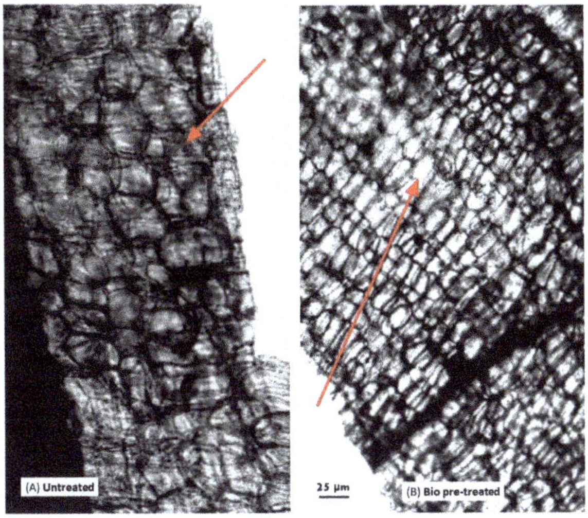

Figure 14. Surface morphologies of the (**A**) untreated (control), and (**B**) bio-treated corn stover observed by microscopy. Scale bar: 25 µm. The red colour arrows indicate smooth surface with thick layering, and rough surface with thin layering in the untreated and bio-treated cornstovers respectively.

3.9. Ligninolytic Activity

The extracellular laccase-like activity was detected through the significant colour change from brownish yellow to reddish brown in the assay tubes containing culture supernatant mixed with 2 mM guaiacol dye along with the 10 mM sodium acetate buffer. The laccase-like activity was calculated to be 0.187 UmL^{-1} and 0.168 UmL^{-1}, respectively when corn stover and sugarcane bagasse were used as raw material substrates. The control sample in which boiled culture supernatant was used, in order to denature the enzyme, did not show any colour change. Also, a reaction mixture containing 5 mM L-cysteine as a laccase inhibitor was used, which showed no colour change, indicating that the enzyme may be a laccase.

4. Discussion

The study in this paper aimed to screen potential fungi collected from rotten wood samples and wood warehouses that could possibly degrade lignin. Primary screening for ligninolytic activity through the qualitative plate assay method using aniline blue dye, guaiacol, and tannic acid showed that the fungal isolate DY1 had the maximum capability to utilize the lignin model compounds within 48 h of incubation. Furthermore, it was found that the DY1 isolate could degrade CMC and birchwood xylan when screened on a PDA medium supplemented with its respective substrates. This indicates that the isolate has cellulase and xylanase activities, which are important for the degradation of lignocellulosic biomass. The fungal isolate DY1 was identified as *A. ochraceus* DY1 through morphological and 5.8S rDNA/ITS analyses. The phylogenetic tree generated using MEGA11, with a scale bar representing 0.01 base substitutions per site and bootstrap values indicating branch reliability, shows that *A. ochraceus* DY1 (OP804322.1) clusters are closely aligned with other *A. ochraceus* strains, confirming its identification as a member of this species. Lignin-degrading microorganisms have been reported from various sources such as soil samples and effluent discharge [25], commercial coir [26], pulp and paper mill wastewater [27], forest environment [28], decayed woods [29], etc. Based on previous reports, *Pseudomonas* sp., *Ochrobactrum* sp., and *Burkholderia* sp. have high ligninolytic enzyme activity among bacteria. Whereas in the case of fungi, white-rot fungi, such as *Phanerochoete chrysosporium*, *Phlebia radiata*, *Coriolus versicolor*, *Bjerkandera adusta*, and *Panus tigrinus*, etc., have been reported as high lignin-degrading fungi [30]. Wood-decaying fungi mostly belong to *Basidiomycetes*, including some members of *Ascomycetes*. According to [31], about 10,000 species of fungi causing white rot and having lignocellulose degrading activity are estimated to exist, but they have not been fully explored. In the present study, *A. ochraceus* DY1 has been found to exhibit laccase-like activity, a typical characteristic of the *Ascomycetes* group of soft-rot fungi, according to [32]. This result is supported by the presence of their corresponding genes through the whole genome analysis of *A. ochraceus* DY1 [33]. When grown in a minimal medium and having alkali lignin as the sole carbon source, it was able to decolourize it significantly with a removal rate of 76.65% in 7 days. In previous reports, *Lentinus edodes* (UEC-2019) and *Basidiomycetes* were reported to remove 73% of AL in 5 days [34]. *Trametes hirsuta* X-13, another *Basidiomycetes* which causes white rot, was able to decolorize AL up to 50% in 7 days [35]. *Phanerochaete chrysosporium*, the white-rot lignin-degrading fungus, was reported to degrade up to 90% of AL within 7 days by performing the spectrophotometric method of lignin assay [36]. Also, *Aspergillus flavus* and *Emericella nidulans* were reported to degrade AL by 14.4% to 21% in different media by measuring the decolourization percentage and residual AL using UV-visible spectroscopy [25]. Su et al., 2018, studied 63 strains for the degradation of AL and the highest degradation ability was reported to be 36.73% in 4 days by an isolate identified as *Myrothecium verrucaria*. The AL degradation was assayed as a percentage removal rate of AL using the UV spectroscopy method [37]. The FTIR analysis of the degraded sample showed peaks corresponding to C–H bending at 1463 cm^{-1} (alkane), C=C bending at 953 cm^{-1} (alkene), C–O stretching at 1252 cm^{-1} and at 1152 cm^{-1} (alkyl-aryl ether, ester), O–H bending at 1382 cm^{-1} (terminal -CH3) as new peaks apart from the peak at 3326 cm^{-1}, which is of polyphenols. This indicated the deconstruction of AL by the strain when checked up to 40 days of incubation in the AL medium.

In the present study, many low molecular weight acidic and aromatic compounds were identified in the extract of the degraded sample. The compounds, such as hexadecenoic acid, dibutyl phthalate, butylated hydroxytoluene, 3,5-dimethoxy-phenol acetate, and 2-(phenylmethylene)-cyclohexanone, were detected after 20 days of incubation, indicating that they might be the compounds produced through AL degradation. The compounds, such as 13-(2Methoxyphenyl) tricyclo[8.2.2.24,7]hexadeca-1, Phenol, 2,4-bis(1,1-dimethylethyl)-, phosphite, Phenol,4,4′-butylidenebis[2-(1,1-dimethylethyl)-5-methyl, appeared in the initial period of incubation, i.e., up to 20 days, and then disappeared during the later stages, indicating that they belong to the structure of the technical lignin used in the experiment. The compound 7,9-Di-tert-butyl-1-oxaspiro (4,5) deca-6,9-diene-2,8-dione was detected throughout the experiment, which shows the recalcitrant part of the structure. The study by Chandra and Bhargava (2013) reported derivates, such as tetradecanoic acid, hexadecanoic acid, 1-phenathrene carboxylic acid, and Bis (2-ethylhexyl) phthalate, after the efficient degradation of AL via mixed bacterial cultures *Citrobacter freundii* (FJ581026) and *Citrobacter* sp. (FJ581023) at pH 8, temperature 35 °C, 600 ppm AL, and 6 days of incubation. The cultures were reported to produce manganese peroxidase [38]. Contrary to the result in the present study, Riyadi and his team reported the production of 3-methyl-butanoic acid, guaiacol derivatives, and 4,6-dimethyl-dodecane from 2.5 g/L Kraft lignin treated with *Streptomyces* sp. strain S6 at 30 °C for 7 days. The ligninolytic enzymes detected were lignin peroxidase, laccase, dye-decolorizing peroxidase, and aryl-alcohol oxidase activities [17]. This variation in the result may be due to the difference in the type of lignin substrates used in the experiments. A study carried out by Cui and his team showed 23 lignin degradation products as detected via GC-MS, which included n-hexadecanoic acid, oleic acid, octadecane, butylated hydroxytoluene and 2,6,11-trimethyl-dodecane [39]. The lignin degradation percentage was reported to be 50%. In another study, lignin monomer compounds like 2-methoxy phenol, 2,6-di-methoxy phenol and its alkylated derivatives were found in the degraded alkali lignin which showed 19–41.6% degradation in 21 days [25]. Xu et al., 2022, reported vanillic acid and 3,5-dimethoxy-4-hydroxycinnamaldehyde through a GC-MS analysis of alkali lignin degradation, which only showed 8 to 14% degradation [40]. The production of 15 elements (guaiacol, phenylacetic acid, benzene, benzoic acid,

4-hydroxyacetophenone, 4-allyl-2-methoxyphenol, (3-ethoxy-4-hydroxyphenyl) (hydroxy) acetic acid, vanillin) has been reported from AL, having 40% degradation in 7 days [41]. The difference in the lignin degradation profile may be due to the difference in the degradation rate. The present strain, *A. ochraceus* DY1, has an efficient degradation capability of 76.65% degradation in 7 days. Confirmation of the depolymerization process was obtained by the results of the SEC and 2D HSQC NMR analysis. In SEC, the polydispersity of lignin fragments decreased from 2.2 to 1.67 on day 7, and a decrease in the molecular weight distribution of treated AL over time indicated that AL was successfully degraded after 7 days of treatment with *A. ochraceus* DY1. In 2D HSQC NMR, the spent AL sample showed an absence of important linkages that are supposed to be present in the lignin structure. The 2D HSQC of untreated AL showed the signal characteristics of technical lignin (lignin obtained as byproducts from the pulping industry), in which phenylglycerol structures, such as guaiacylglycerol (G-gly) and syringylglycerol (S-gly), were detected at $\delta C/\delta H$ = 73.2–75.2/4.3–4.5, 74.7–76.9/3.4–3.6, and 61.2–65.6/3.1–3.8. Similar results were observed in a study on the structural analysis of technical lignins obtained after the pulping process. The other signals detected in the 2D HSQC of untreated AL were Cβ-Hβ in β-β′ resinol structures, the Cβ–Hβ in β-O-4′substructures linked to syringyl units, Cγ-Hγ in β-β′ resinol structures, the β-5′ in phenylcoumaran structures, the C–H in -OCH3 structures, and the Cα=O in Hibbert ketone. When compared with the spectra of the untreated AL, the spent lignin after fungal treatment showed only the $\delta C/\delta H$ 40.0–60.0/3.0–4.0, indicating the β-β resinol and β-5 phenyl coumaran structures of lignin. Depolymerization in the fungal-treated AL was confirmed by the disappearance of β-O-4′aryl ether linkages and by most of the other signals detected in the untreated AL. Similar results were obtained in previous studies involving SEC and 2D HSQC NMR studies, which contained comparisons between depolymerized and non-depolymerized lignin samples [21].

A. ochraceus DY1 was also observed to utilize lignocellulosic biomass, such as corn stover and sugarcane bagasse, efficiently, as shown by the lignocellulose biomass loss of 62.5% and 50%, respectively, within 7 days of incubation. There was a corresponding increase in the spore count of DY1, which correlates with the biomass utilization. Further microscopic analyses of the treated biomass, i.e., corn stover and the sugarcane bagasse, showed depletion in the structural integrity of the biomasses compared with the untreated control. A similar observation was reported in a study of the treatment of corn stover by *Ceriporiopsis subvermispora* and *Phanerochaete chrysosporium*, in which small but significant changes in the cross sections of the parenchyma cells and vascular bundles were observed under SEM after 2 weeks of incubation [42]. Su et al., 2018 also reported similar results while studying the bio-pretreatment of corn stover by *Myrothecium verrucaria* [37]. This shows the potential of the fungal strain in the present study for its application in the biotreatment of agricultural waste biomass for sustainable waste management.

5. Conclusions

The present study reports the deconstruction of AL by *A. ochraceus* DY1 and its ability to consume lignocellulosic substrates, corn stover, and sugarcane bagasse, with the production of laccase-like activity. The conditions for the maximum removal of AL in a minimal medium were found to be 0.01% AL, 25 °C, pH 7, and an incubation period of 40 days. Important compounds, such as n-hexadecanoic acid, oleic acid, octadecane, butylated hydroxytoluene, 2,6,11-trimethyl-dodecane, dibutyl phthalate, 3,5-dimethoxy-phenol acetate, and 2-(phenylmethylene)-cyclohexanone were detected when the AL containing medium was inoculated with the strain under optimized conditions. This shows the ability of *A. ochraceus* DY1 to utilize AL and transform it into valuable compounds. AL is a by-product of the pulp and paper industry, and the use of microorganisms to treat the wastewater and to convert the recovered lignin into value-added chemicals has been of interest in recent years. *A. ochraceus* DY1 can also be further used to study the large-scale production of ligninolytic enzymes using lignocellulosic waste. The ability of this strain to grow efficiently in a medium containing AL as the sole carbon source and in lignocellulosic materials shows its potential applications in the treatment of industrial and agricultural wastes after carrying out careful investigations on how it can be applied for achieving efficient outcomes.

Author Contributions: Conceptualization, M.D.S. and N.N.; methodology, M.D.S.; validation, M.D.S., A.V. and N.N.; formal analysis, M.D.S., R.P. and A.V.; investigation, N.N.; resources, N.N.; data curation, M.D.S., N.N. and R.P.; writing—original draft preparation, N.N.; writing—review and editing, M.D.S., R.P. and A.V.; visualization, M.D.S.; supervision, M.D.S. and R.P.; project administration, M.D.S. and A.V. All authors have read and agreed to the published version of the manuscript.

Funding: This research received no external funding.

Institutional Review Board Statement: Not applicable.

Informed Consent Statement: Not applicable.

Data Availability Statement: All available data are reported in the paper.

Acknowledgments: The authors would like to thank Amity University Uttar Pradesh, India, for all support facilitating the research.

Conflicts of Interest: The authors declare no conflicts of interest.

References

1. Beaucamp, A.; Muddasar, M.; Amiinu, I.S.; Leite, M.M.; Culebras, M.; Latha, K.; Gutiérrez, M.C.; Rodriguez-Padron, D.; del Monte, F.; Kennedy, T.; et al. Lignin for energy applications—State of the art, life cycle, technoeconomic analysis and future trends. *Green Chem.* **2022**, *24*, 8193–8226. [CrossRef]
2. Achyuthan, K.E.; Achyuthan, A.M.; Adams, P.D.; Dirk, S.M.; Harper, J.C.; Simmons, B.A.; Singh, A.K. Supramolecular self-assembled chaos: Polyphenolic lignin's barrier to cost-effective lignocellulosic biofuels. *Molecules* **2010**, *15*, 8641–8688. [CrossRef] [PubMed]
3. Bilal, M.; Iqbal, H.M. Ligninolysis potential of ligninolytic enzymes: A green and sustainable approach to bio-transform lignocellulosic biomass into high-value entities. In *Alternative Energy Resources: The Way to a Sustainable Modern Society*; Springer: Cham, Switzerland, 2021; pp. 151–171. [CrossRef]
4. Brink, D.P.; Ravi, K.; Lidén, G.; Gorwa-Grauslund, M.F. Mapping the diversity of microbial lignin catabolism: Experiences from the eLignin database. *Appl. Microbiol. Biotechnol.* **2019**, *103*, 3979–4002. [CrossRef] [PubMed]
5. Gillet, S.; Aguedo, M.; Petitjean, L.; Morais, A.R.C.; da Costa Lopes, A.M.; Łukasik, R.M.; Anastas, P.T. Lignin transformations for high value applications: Towards targeted modifications using green chemistry. *Green Chem.* **2017**, *19*, 4200–4233. [CrossRef]
6. Mandal, D.D.; Singh, G.; Majumdar, S.; Chanda, P. Challenges in developing strategies for the valorization of lignin—A major pollutant of the paper mill industry. *Env. Sci. Pollut. Res.* **2023**, *30*, 11119–11140. [CrossRef]
7. Mattoo, A.J.; Nonzom, S. Endophytes in lignin valorization: A novel approach. *Front. Bioeng. Biotech.* **2022**, *10*, 895414. [CrossRef]
8. Challacombe, J.F.; Hesse, C.N.; Bramer, L.M.; McCue, L.A.; Lipton, M.; Purvine, S.; Nicora, C.; Gallegos-Graves, L.V.; Porras-Alfaro, A.; Kuske, C.R. Genomes and secretomes of Ascomycota fungi reveal diverse functions in plant biomass decomposition and pathogenesis. *BMC Genom.* **2019**, *20*, 976. [CrossRef]
9. Shary, S.; Ralph, S.A.; Hammel, K.E. New insights into the ligninolytic capability of a wood decay ascomycete. *Appl. Environ. Microbiol.* **2007**, *73*, 6691–6694. [CrossRef]
10. Dicko, M.; Ferrari, R.; Tangthirasunun, N.; Gautier, V.; Lalanne, C.; Lamari, F.; Silar, P. Lignin degradation and its use in signaling development by the coprophilous ascomycete *Podospora anserina*. *J. Fungi* **2020**, *6*, 278. [CrossRef]
11. Falade, A.O.; Nwodo, U.U.; Iweriebor, B.C.; Green, E.; Mabinya, L.V.; Okoh, A.I. Lignin peroxidase functionalities and prospective applications. *Microbiologyopen* **2017**, *6*, e00394. [CrossRef]
12. Weng, C.; Peng, X.; Han, Y. Depolymerization and conversion of lignin to value-added bioproducts by microbial and enzymatic catalysis. *Biotechnol. Biofuels* **2021**, *14*, 84. [CrossRef] [PubMed]
13. Tan, F.; Cheng, J.; Zhang, Y.; Jiang, X.; Liu, Y. Genomics analysis and degradation characteristics of lignin by *Streptomyces thermocarboxydus* strain DF3-3. *Biotechnol. Biofuels Biop.* **2022**, *15*, 78. [CrossRef] [PubMed]
14. Higuchi, Y.; Aoki, S.; Takenami, H.; Kamimura, N.; Takahashi, K.; Hishiyama, S.; Lancefield, C.S.; Ojo, O.S.; Katayama, Y.; Westwood, N.J.; et al. Bacterial catabolism of β-hydroxypropiovanillone and β-hydroxypropiosyringone produced in the reductive cleavage of arylglycerol-β-aryl ether in lignin. *Appl. Environ. Microbiol.* **2018**, *84*, e02670-17. [CrossRef] [PubMed]
15. Gutiérrez, A.; Rencoret, J.; Cadena, E.M.; Rico, A.; Barth, D.; del Río, J.C.; Martínez, Á.T. Demonstration of laccase-based removal of lignin from wood and non-wood plant feedstocks. *Bioresour. Technol.* **2012**, *119*, 114–122. [CrossRef] [PubMed]
16. Wackett, L.P. Microbial-based motor fuels: Science and technology. *Microb. Biotechnol.* **2008**, *1*, 211–225. [CrossRef]

7. Riyadi, F.A.; Tahir, A.A.; Yusof, N.; Sabri, N.S.A.; Noor, M.J.M.M.; Akhir, F.N.M.; Othman, N.A.; Zakaria, Z.; Hara, H. Enzymatic and genetic characterization of lignin depolymerization by *Streptomyces* sp. S6 isolated from a tropical environment. *Sci. Rep.* **2020**, *10*, 7813. [CrossRef]
8. Civzele, A.; Stipniece-Jekimova, A.A.; Mezule, L. Fungal ligninolytic enzymes and their application in biomass lignin pre-treatment. *J. Fungi* **2023**, *9*, 780. [CrossRef]
9. Liu, H.; Liu, Z.H.; Zhang, R.K.; Yuan, J.S.; Li, B.Z.; Yuan, Y.J. Bacterial conversion routes for lignin valorization. *Biotechnol. Adv.* **2022**, *60*, 108000. [CrossRef]
0. Tamura, K.; Stecher, G.; Kumar, S. MEGA11: Molecular evolutionary genetics analysis version 11. *Mol. Biol. Evol.* **2021**, *38*, 3022–3027. [CrossRef]
1. Liu, Y.; Zheng, J.; Xiao, J.; He, X.; Zhang, K.; Yuan, S.; Peng, Z.; Chen, Z.; Lin, X. Enhanced enzymatic hydrolysis and lignin extraction of wheat straw by triethylbenzyl ammonium chloride/lactic acid-based deep eutectic solvent pre-treatment. *ACS Omega* **2019**, *4*, 19829–19839. [CrossRef]
2. Schütz, G.; Haltrich, D.; Atanasova, L. Influence of spore morphology on spectrophotometric quantification of *Trichoderma* inocula. *BioTechniques* **2020**, *68*, 279–282. [CrossRef] [PubMed]
3. Kristensen, J.B.; Thygesen, L.G.; Felby, C.; Jørgensen, H.; Elder, T. Cell-wall structural changes in wheat straw pre-treated for bioethanol production. *Biotechnol. Biofuels* **2008**, *1*, 5. [CrossRef] [PubMed]
4. Abd El Monssef, R.A.; Hassan, E.A.; Ramadan, E.M. Production of laccase enzyme for their potential application to decolorize fungal pigments on aging paper and parchment. *Ann. Agric. Sci.* **2016**, *61*, 145–154. [CrossRef]
5. Barapatre, A.; Jha, H. Degradation of kraft lignin by the bacterium *Pseudomonas* sp. isolated from sediment of lignin rich pulp paper mill sludge. *BioResources* **2017**, *12*, 6054–6064. [CrossRef]
6. Raybarman, A.; Rahman, K.A.; Vincent, R.M.; Chatterjee, S.; Sen, U.; Mitra, A.K.; Choudhury, S.; Majumder, R. Isolation and Characterisation of Lignin-Degrading Fungus from Coir. *IOSR-JESTFT* **2014**, *8*, 2319. [CrossRef]
7. Zainith, S.; Purchase, D.; Saratale, G.D.; Ferreira, L.F.R.; Bilal, M.; Bharagava, R.N. Isolation and Characterization of Lignin-Degrading Bacterium *Bacillus aryabhattai* from Pulp and Paper Mill Wastewater and Evaluation of Its Lignin-Degrading Potential. *3 Biotech* **2019**, *9*, 92. [CrossRef]
8. Rosales-Castillo, J.A.; Oyama, K.; Vázquez-Garcidueñas, M.S.; Aguilar-Romero, R.; García-Oliva, F.; Vázquez-Marrufo, G. Fungal Community and Ligninolytic Enzyme Activities in *Quercus deserticola* Trel. Litter from Forest Fragments with Increasing Levels of Disturbance. *Forests* **2017**, *9*, 11. [CrossRef]
9. Gao, H.; Wang, Y.; Zhang, W.; Wang, W.; Mu, Z. Isolation, Identification, and Application in Lignin Degradation of an Ascomycete GHJ-4. *Afr. J. Biotechnol.* **2011**, *10*, 4166–4174.
0. Dashtban, M.; Schraft, H.; Syed, T.A.; Qin, W. Fungal Biodegradation and Enzymatic Modification of Lignin. *Int. J. Biochem. Mol. Bio.* **2010**, *1*, 36.
1. Li, T.; Cui, L.; Song, X.; Cui, X.; Wei, Y.; Tang, L.; Xu, Z. Wood Decay Fungi: An Analysis of Worldwide Research. *J. Soil. Sediment.* **2022**, *22*, 1688–1702. [CrossRef]
2. Liers, C.; Arnstadt, T.; Ullrich, R.; Hofrichter, M. Patterns of Lignin Degradation and Oxidative Enzyme Secretion by Different Wood- and Litter-Colonizing Basidiomycetes and Ascomycetes Grown on Beech-Wood. *FEMS Microbiol. Ecol.* **2011**, *78*, 91–102. [CrossRef] [PubMed]
3. Salam, M.D.; Nilza, N.; Varma, A. Draft Genome Sequence of *Aspergillus ochraceus* Strain DY1, a Lignin-Degrading Fungus Isolated from Wood Rot. *Microbiol. Resour. Announc.* **2023**, *12*, e01278-22. [CrossRef] [PubMed]
4. Esposito, E.; Canhos, V.P.; Durán, N. Screening of Lignin-Degrading Fungi for Removal of Color from Kraft Mill Wastewater with No Additional Extra Carbon Source. *Biotechnol. Lett.* **1991**, *13*, 571–576. [CrossRef]
5. Ma, J.; Li, Q.; Wu, Y.; Yue, H.; Zhang, Y.; Zhang, J.; Shi, M.; Wang, S.; Liu, G.Q. Elucidation of Ligninolysis Mechanism of a Newly Isolated White-Rot Basidiomycete *Trametes hirsuta* X-13. *Biotechnol. Biofuels* **2021**, *14*, 189. [CrossRef]
6. Ulmer, D.C.; Leisola, M.S.; Schmidt, B.H.; Fiechter, A. Rapid Degradation of Isolated Lignins by *Phanerochaete chrysosporium*. *Appl. Environ. Microbiol.* **1983**, *45*, 1795–1801. [CrossRef]
7. Su, Y.; Yu, X.; Sun, Y.; Wang, G.; Chen, H.; Chen, G. Evaluation of Screened Lignin-Degrading Fungi for the Biological Pre-Treatment of Corn Stover. *Sci. Rep.* **2018**, *8*, 5385. [CrossRef]
8. Chandra, R.; Bharagava, R.N. Bacterial Degradation of Synthetic and Kraft Lignin by Axenic and Mixed Culture and Their Metabolic Products. *J. Environ. Biol.* **2013**, *34*, 991.
9. Cui, T.; Yuan, B.; Guo, H.; Tian, H.; Wang, W.; Ma, Y.; Fei, Q. Enhanced Lignin Biodegradation by Consortium of White Rot Fungi: Microbial Synergistic Effects and Product Mapping. *Biotechnol. Biofuels* **2021**, *14*, 162. [CrossRef]
0. Xu, Z.; Peng, B.; Kitata, R.B.; Nicora, C.D.; Weitz, K.K.; Pu, Y.; Shi, T.; Cort, J.R.; Ragauskas, A.J.; Yang, B. Understanding of Bacterial Lignin Extracellular Degradation Mechanisms by Pseudomonas putida KT2440 via Secretomic Analysis. *Biotechnol. Biofuels* **2022**, *15*, 117. [CrossRef]

41. Zhu, D.; Zhang, P.; Xie, C.; Zhang, W.; Sun, J.; Qian, W.J.; Yang, B. Biodegradation of Alkaline Lignin by Bacillus ligniniphilus L1. *Biotechnol. Biofuels* **2017**, *10*, 44. [CrossRef]
42. Wendt, L.M.; Wahlen, B.D.; Walton, M.R.; Nguyen, J.A.; Lin, Y.; Brown, R.M.; Zhao, H. Exploring Filamentous Fungi Depolymerization of Corn Stover in the Context of Bioenergy Queuing Operations. *Food Energy Secur.* **2022**, *11*, e333. [CrossRef]

Journal of
Fungi

MDPI

Article

Comparative Evaluation of Mechanical and Physical Properties of Mycelium Composite Boards Made from *Lentinus sajor-caju* with Various Ratios of Corn Husk and Sawdust

Praween Jinanukul [1], Jaturong Kumla [2,3,4], Worawoot Aiduang [2,3], Wandee Thamjaree [5], Rawiwan Oranratmanee [1], Umpiga Shummadtayar [1], Yuttana Tongtuam [1], Saisamorn Lumyong [3,4,6], Nakarin Suwannarach [2,3,4,*] and Tanut Waroonkun [1,*]

1 Faculty of Architecture, Chiang Mai University, Chiang Mai 50200, Thailand; praween_ji@cmu.ac.th (P.J.); rawiwan.o@cmu.ac.th (R.O.); umpiga.sh@cmu.ac.th (U.S.); yuttana.t@cmu.ac.th (Y.T.)
2 Office of Research Administration, Chiang Mai University, Chiang Mai 50200, Thailand; jaturong_yai@hotmail.com (J.K.); worawoot.aiduang@cmu.ac.th (W.A.)
3 Department of Biology, Faculty of Science, Chiang Mai University, Chiang Mai 50200, Thailand; saisamorn.l@cmu.ac.th
4 Center of Excellence in Microbial Diversity and Sustainable Utilization, Chiang Mai University, Chiang Mai 50200, Thailand
5 Department of Physics and Materials Science, Faculty of Science, Chiang Mai University, Chiang Mai 50200, Thailand; wandee.th@cmu.ac.th
6 Academy of Science, The Royal Society of Thailand, Bangkok 10300, Thailand
* Correspondence: suwan.462@gmail.com (N.S.); tanut.w@cmu.ac.th (T.W.); Tel.: +66-865-127-518 (N.S.)

Abstract: Mycelium-based composites (MBCs) exhibit varied properties as alternative biodegradable materials that can be used in various industries such as construction, furniture, household goods, and packaging. However, these properties are primarily influenced by the type of substrate used. This study aims to investigate the properties of MBCs produced from *Lentinus sajor-caju* strain CMU-NK0427 using different ratios of sawdust to corn husk in the development of mycelium composite boards (MCBs) with thicknesses of 8, 16, and 24 mm. The results indicate that variations in the ratios of corn husk to sawdust and thickness affected the mechanical and physical properties of the obtained MCBs. Reducing the corn husk content in the substrate increased the modulus of elasticity, density, and thermal conductivity, while increasing the corn husk content increased the bending strength, shrinkage, water absorption, and volumetric swelling. Additionally, an increase in thickness with the same substrate ratio only indicated an increase in density and shrinkage. MCBs have sound absorption properties ranging from 61 to 94% at a frequency of 1000 Hz. According to the correlation results, a reduction in corn husk content in the substrate has a significant positive effect on the reduction in bending strength, shrinkage, and water absorption in MCBs. However, a decrease in corn husk content shows a strong negative correlation with the increase in the modulus of elasticity, density, and thermal conductivity. The thickness of MCBs with the same substrate ratio only shows a significant negative correlation with the modulus of elasticity and bending strength. Compared to commercial boards, the mechanical (bending strength) and physical (density, thermal conductivity, and sound absorption) properties of MCBs made from a 100% corn husk ratio are most similar to those of softboards and acoustic boards. The results of this study can provide valuable information for the production of MCBs and will serve as a guide to enhance strategies for further improving their properties for commercial manufacturing, as well as fulfilling the long-term goal of eco-friendly recycling of lignocellulosic substrates.

Keywords: biodegradable materials; lignocellulosic substrate; mycelium-based composites; mushroom mycelium; waste recycling

1. Introduction

A current issue with synthetic materials, including cement, concrete, and polymers, is their environmental impact, particularly their contribution to pollution problems [1–3]. These materials are incapable of deterioration and take hundreds of years to degrade, causing them to accumulate in landfills, oceans, and the natural environment [4,5]. Furthermore, the production of synthetic materials frequently requires complex equipment, high costs for raw materials, processing, and product development, as well as high energy consumption during production [6–8]. Consequently, a worldwide campaign has been established to create and use more sustainable materials and technologies [7,9,10]. In recent years, there has been a growing interest in lignocellulose-based composites due to their renewable and biodegradable materials, which often renders them more environmentally friendly compared to synthetic materials [11,12]. However, the widespread use of formaldehyde-based resins as adhesives for bonding lignocellulose-based composites limits the advancement of fully natural composite materials. Furthermore, formaldehyde emissions from these products are classified as carcinogenic and harmful to human health [13–15]. In response to these concerns, researchers have shifted their focus towards developing natural, ecologically sustainable alternatives to replace toxic chemicals used in lignocellulose-based materials [16,17].

One of the promising alternatives is mycelium, derived from fungi, which shows great potential as a green adhesive material, replacing synthetic adhesives in lignocellulose-based composite production and creating mycelium-based composites (MBCs) [18–24]. MBCs represent the biological processes during the vegetative growth phase of fungal mycelium, characterized by a complex network with the unique ability to digest and adhere to organic substrate surfaces under ambient conditions, acting as a natural self-assembling adhesive [25]. Beyond its adhesive properties, mycelium exhibits structural binding characteristics by generating interconnecting fibrous threads consisting of chitin- and beta-glucan-based structural oligosaccharides. These oligosaccharides contribute to forming a robust and cohesive network within the mycelial structure [26]. This transformative process of MBCs converts lignocellulose waste into valuable resources, offering an eco-friendly alternative material [6,27,28]. Moreover, the process of MBC production is characterized by low energy consumption, low processing costs, a low carbon footprint, biodegradability, and an attractive range of properties for construction materials, board products, packaging, and foam-like materials, contributing to an overall reduction in waste and pollution [21,23,24,29–31]. Based on the fungal mycelial network, the monomitic hyphal system consists solely of generative hyphae. While the dimitic hyphal system forms both generative and skeletal hyphae and the trimitic hyphal system comprises three types of hyphae (generative, skeletal, and binding hyphae) [32]. MBCs made from dimitic and trimitic hyphae have higher qualities than those formed from monomitic hyphae because the skeletal and binding hyphae in these types have thicker walls, higher densities, and greater hardness [22]. Various lignocellulosic materials, particularly from the agricultural sector, including cotton, straw, sawdust, woodchips, and rice husk, have been utilized as organic substrates in the production of MBCs [6,22–25,28]. Additionally, lignocellulosic materials have been selected and used in the production of MBCs based on their availability in each country. Remarkably, the characteristics of MBCs can be directly influenced by different fungal mycelial network and substrate types [8,14,21,24,27–29]. In our previous study, *Lentinus sajor-caju* (dimitic hyphal system) was found to have a great deal of potential for developing MBCs from sawdust and corn husk [33]. However, the different ratios of sawdust to corn husk in the production process are still necessary for the development of MBCs. To date, there are no reports on MBCs made from varying ratios of corn husks and sawdust. Therefore, the purpose of this study was to investigate the mechanical (modulus of elasticity and bending strength) and physical properties (density, shrinkage, water absorption, volumetric swelling, thermal conductivity, and sound absorption) of mycelium composite boards (MCBs) produced from *L. sajor-caju* using different ratios of sawdust to corn husk. The information obtained from this study can be valuable for developing MBCs as a type of bio-board with properties suitable for environmentally friendly architecture.

2. Materials and Methods

2.1. Fungal Strain

The pure culture of *L. sajor-caju* strain CMU-NK0427 was obtained from the Culture Collection of the Center of Excellence in Microbial Diversity and Sustainable Utilization, Faculty of Science, Chiang Mai University, Thailand, and used in this study. This fungus was cultured on potato dextrose agar (PDA, Conda, Madrid, Spain) and incubated at 30 °C.

2.2. Sources of Substrate and Preparation

Two different types of wood and agricultural residues, namely rubber tree sawdust and corn husk, were used as substrates for the present study. These residues were from a sawmill and agricultural regions situated in Chiang Mai Province, Thailand. Both substrates were dried at 60 °C in an oven for 72 h. Subsequently, each substrate was then ground in a woodchipper and sieved. Particles ranging in size from 5 to 20 mm were collected and used.

2.3. Preparation of Mycelial Inoculum

The mycelial inoculum was prepared using sorghum seeds as a nutrient source. Initially, sorghum seeds were boiled for 20 min and then cooled. Subsequently, 100 g of boiled grains was placed in a glass bottle with a cotton swab inserted. The bottle was autoclaved at 121 °C for 20 min and then gradually cooled at room temperature over 24 h. Following this process, mycelial plugs (1 × 1 cm) obtained from pure culture of *L. sajor-caju* on PDA were transferred to a 325 mL glass bottle (5 plugs per bottle). After that, the inoculated bottle was incubated for two weeks at 30 °C to allow the sorghum seeds to be completely covered with mycelium for use as an inoculum.

2.4. Preparation of MBCs

The preparation of MBCs in this study followed the methods of Aiduang et al. [33]. Five different ratios based on the dry weight of corn husk and sawdust were designed in this study, including 100% corn husk, 75% corn husk with 25% sawdust, 50% corn husk with 50% sawdust, 25% corn husk with 75% sawdust, and 100% sawdust. Each mixed substrate was added with 5% rice bran, 1% calcium carbonate, 2% calcium sulfate, and 0.2% sodium sulfate. The moisture content of each mixed substrate was adjusted at 60% relative humidity by adding water. Then, 500 g of each mixed substrate was carefully placed in a polypropylene bag with dimensions of 3.5 inches in width and 12.5 inches in length. The bag was securely sealed. These sealed bags were then inserted into polyvinyl chloride rings and covered with paper before being subjected to autoclaving at 121 °C for 60 min. After autoclaving, a cooling period of 24 h at room temperature was allowed. Five grams of mycelial inoculum was inoculated onto the top of the substrate of each bag. The inoculated bags were incubated at 30 °C and fungal mycelia were observed to completely cover the substrate after 35 days of incubation.

2.5. Preparation of MCB Molds

The molds used in MCB production were meticulously fabricated from acrylic sheets with dimensions of 410 mm × 570 mm, featuring a side margin of 20 mm, resulting in MCB specimens with sizes of 370 mm × 530 mm. The MCBs were then cut to standard sizes for testing material properties. It is important to note that the molds have varying thicknesses, mirroring the diversity observed in current board materials. For this study, three representative thicknesses of 8 mm, 16 mm, and 24 mm were intentionally selected to assess the influence of mycelium board thicknesses on its properties.

2.6. MCB Production

The substrate covered by fungal mycelia inside the bags were carefully transferred to designated molds corresponding to each thickness. A total of 800, 1600, and 2400 g were used for 8, 16, and 24 mm thick molds, respectively. Then, cold pressing was conducted using a press (Shop press ZX0901 E–1, New Taipei, Taiwan). The pressing process involved exerting a controlled force of 2 MPa for a precise duration of 10 min. Following cold pressing, the molds underwent incubation at 30 °C for a period of three weeks to allow further development of the samples. Then, specimens were removed and incubated for another 7 days. Subsequently, the samples were carefully transferred to an oven at 70 °C to dry completely for 72 h. The scheme for the MCB production process in this study is shown in Figure 1.

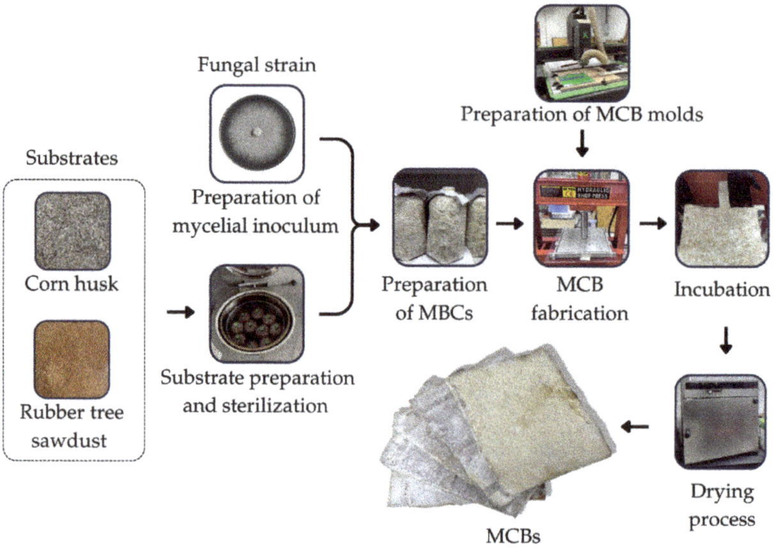

Figure 1. Scheme for the MCB production process in this study.

2.7. Specimen Preparation for Determination of Mechanical and Physical Properties

Each MBC, with varying substrate ratios and thicknesses (Figure 2A), was cut into specimens for testing mechanical and physical properties (Table 1). Three specimens were used for testing each property. Specimens for testing the modulus of elasticity (MOE) and bending strength (BS) were cut to a width of 50 mm and a length of 530 mm, as shown in Figure 2B. Specimens for testing density, water absorption, swelling thickness, and thermal conductivity were cut to a width and length of 50 mm, as shown in Figure 2C. Additionally, specimens for measuring the sound absorption coefficient were cut to a diameter of 50 mm, as shown in Figure 2D. Note that specimens for measuring thermal conductivity and sound absorption coefficient were cut from an MCB with a thickness of 24 mm due to limitations with the testing equipment. The material properties were examined at the Science and Technology Service Center and the Department of Physics and Materials Science, Faculty of Science, Chiang Mai University, Chiang Mai, Thailand.

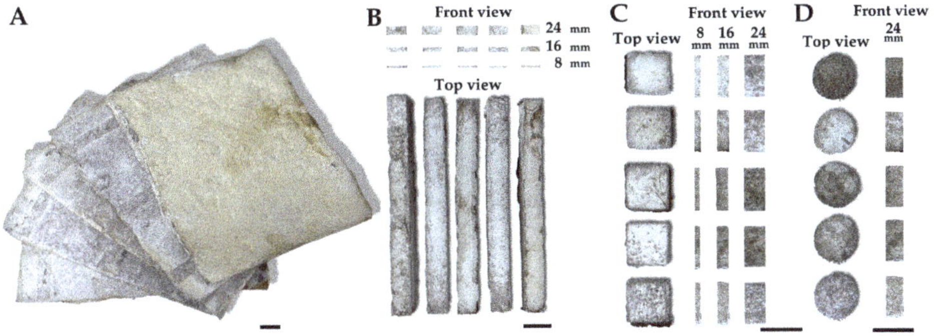

Figure 2. Example specimens of MBCs obtained from *L. sajor-caju* and corn husk and sawdust in this study. Mycelium composite boards (**A**). Specimens for modulus of elasticity and bending strength testing (**B**), specimens for water absorption measurement, thickness of swelling, shrinkage, and thermal conductivity measurement (**C**), and specimens for sound absorption coefficient measurement (**D**). Scale bar = 50 mm.

Table 1. Information of MCBs created in this study.

Thickness (mm)	Ratio of Substrate Based on Dry Weight (%)		Specimen Name
	Corn Husk	Sawdust	
8	100	0	8C100S0
	75	25	8C75S25
	50	50	8C50S50
	25	75	8C25S75
	0	100	8C0S100
16	100	0	16C100S0
	75	25	16C75S25
	50	50	16C50S50
	25	75	16C25S75
	0	100	16C0S100
24	100	0	24C100S0
	75	25	24C75S25
	50	50	24C50S50
	25	75	24C25S75
	0	100	24C0S100

2.8. Determination of Mechanical Properties

2.8.1. Modulus of Elasticity

The testing procedure was conducted utilizing a universal testing machine employing the centralized concentration loading method, as outlined in BS EN 310:1993 [34,35]. The modulus of elasticity (MOE) was determined using the formula $[l_1^3 (F_2 - F_1)]/4\,bt^3 (a_2 - a_1)$, where $(F_2 - F_1)$ represents the increased load applied along the linear segment of the load–deflection curve, measured in newtons (N). 'F_1' corresponds to approximately 10% of the maximum load, while 'F_2' corresponds to approximately 40% of the maximum load. The width of the specimen ('b') and its thickness ('t') are measured in millimeters (mm), and '$a_2 - a_1$' denotes the deflection of the specimen at mid-span. The MOE is expressed in units of MPa.

2.8.2. Bending Strength

The experiments were conducted using a universal testing machine based on the centralized concentration loading method, enabling the evaluation of bending strength (BS) [36,37]. The formula for BS is derived from $(3F_{max}l_1)/(2\ bt^2)$ according to BS EN 310:1993, where 'F_{max}' denotes the maximum load measured in newtons (N), 'l_1' refers to the distance between the centers of the supports measured in millimeters (mm), 'b' represents the width of the specimen in millimeters (mm), and 't' indicates the thickness of the specimen. The BS test yields resulted in units of MPa.

2.9. *Determination of Physical Properties*

2.9.1. Density

Density tests were conducted on the MBCs to determine the mean density associated with each specific board type. The testing procedure strictly adhered to the guidelines stipulated in the British code of standards, specifically BS EN 323 [36,38]. The equation of density $(kg/m^3) = M/V$, where W = the mass of the specimen (kg) and V = the volume of the specimen (m^3).

2.9.2. Shrinkage

The shrinkage of each specimen was assessed and computed using wet and dry volumes, following the procedure outlined by Elsacker et al. [20]. This reduction was quantified as a shrinkage percentage (%), calculated as follows: Shrinkage (%) = $[(V_1 - V_2)/V_1] \times 100$, where V_1 represents the wet volume of the specimen (m^3) and V_2 is the dry volume of the specimen (m^3).

2.9.3. Water Absorption

Water absorption tests followed the ASTM D1037:2002 standard [37,39]. The specimens were dried at 70 °C until achieving a stable mass. Following this, the initial mass of each specimen was precisely measured to ensure data accuracy. After measurement, the specimens were submerged in deionized water for a total duration of 84 h, with weight measurements taken at precise time intervals of 2, 4, 6, 12, 18, 24, 36, 48, 60, 72, and 84 h. Water absorption (%) was calculated using the formula $(W - D) \times 100)/D$, where W represents the wet mass (kg) and D represents the dry mass (kg) [40].

2.9.4. Volumetric Swelling

The determination of volumetric swelling of MCB specimens followed the protocols of the ASTM D1037:2002 standard [37]. Initial sizing (width × length × thickness) measurements (T_I) were accurately recorded using a vernier caliper. Subsequently, the samples underwent immersion in water for designated durations of 84 h, respectively. Post-immersion, the final sizing (T_f) of each sample was measured and documented. The percentage of size change by each sample was calculated using the equation for volumetric swelling: $(T_f - T_i) \times 100/T_i$, where T_f represents the final volume (m^3) and T_i denotes the initial volume (m^3) [41].

2.9.5. Thermal Conductivity

Thermal conductivity in this study was conducted following ISO 22007-2:2022 [41,42]. The measurement was conducted at room temperature employing the hot wire technique utilizing the Transient Hot Bridge apparatus (LINSEIS, THB-1). Two specimens were measured, with a sensor probe positioned between their surfaces in a sandwich setup configuration [43].

2.9.6. Sound Absorption Coefficient

The sound absorption coefficient in this study was quantified following the established protocols delineated in ISO 10534-2 [44], employing a Kundt's tube apparatus [45,46].

Measurements were conducted at discrete frequencies of 250, 500, and 1000 Hz, and outcomes were represented as a percentage.

2.10. Statistical Analysis

The data obtained from each experiment were analyzed through one-way analysis of variance (ANOVA) using the SPSS program, v22, for Windows. Subsequently, Duncan's multiple range test was utilized to ascertain significant differences ($p \leq 0.05$) among the mean values. The Pearson correlation coefficients (r) for the ratio of substrate and thickness with the properties of MCBs were analyzed using the SPSS program at a significance level of $p < 0.05$.

3. Results and Discussion

3.1. Mechanical Properties of MCBs

3.1.1. Modulus of Elasticity

The MOE values obtained in 8 mm, 16 mm, and 24 mm thick specimens ranged from 5.44 to 7.88 MPa, 0.75 to 2.36 MPa, and 0.17 to 0.77 MPa, respectively (Figure 3A). According to our results, the MOE values ranged from 0.17 to 7.88 MPa and fell within the previously documented ranges of MOE values observed in MBCs (0.14 to 97.00 MPa) [20,22,31,46]. The results showed that the different ratios of corn husks to sawdust affected the MOE value. An increase in the ratio of sawdust led to an increase in the MOE value for each thickness. These outcomes were consistent with previous studies showing that the type of lignocellulosic residues and the ratio mixture used in the bio-fabrication of MBCs affect their modulus of elasticity [20,22,31,46]. Results showed that there was no significant difference in the MOE values of 8C25S75 and 8C0S100 specimens. Furthermore, the MOE value decreased as MBC thickness increased (Figure 3A). The maximum MOE value for the same substrate ratio was found at a thickness of 8 mm, followed by 16 mm and 24 mm. This result was supported by previous studies that found the elastic modulus of the material increases with decreasing thickness [47,48].

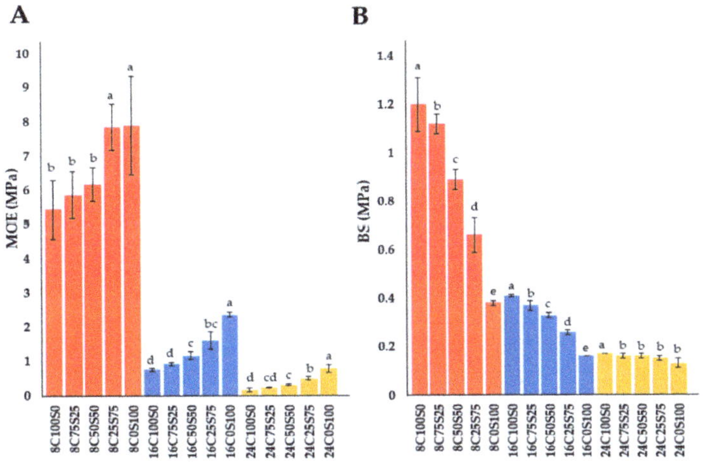

Figure 3. Modulus of elasticity [MOE; (**A**)] and bending strength [BS; (**B**)] of MCBs in this study. Red, blue, and yellow bars indicate MCBs with thicknesses of 8 mm, 16 mm, and 24 mm, respectively. The different letters within the same group of MCB thickness indicate a significant difference according to Duncan's multiple range test ($p \leq 0.05$).

Kuznetsova et al. [49] found that an increase in the thickness of mycelial films made from *Ganoderma lucidum* and *Pleurotus eryngii* was associated with decreased MOE values. Additionally, Appels et al. [22] explained that thinner MBCs have a higher distribution and density of mycelial networks per unit volume compared to thicker MBCs, which enhances the material's stiffness and resistance to deformation, contributing to a higher elastic modulus. According to the Pearson correlation ($p < 0.05$), the MOE value showed a significant strong negative correlation with a decreasing in the ratio of corn husk at the same thickness ($r = -0.937$ to -0.772, $p < 0.001$) (Table 2). This indicates that lower corn husk content is associated with a higher MOE value. The MOE value also showed a significant negative correlation with an increase in thickness at the same substrate ratio ($r = -0.929$ to -0.898, $p < 0.001$) (Table 3). Appels et al. [22] also found that the hot-pressing process increased the MOE value compared to the cold-pressing process.

Table 2. Pearson correlation coefficients between the properties and a decreasing in the ratio of corn husk of MCBs with the same thickness.

Properties	Pearson Correlation Coefficients (r/p-Value)		
	8 mm	16 mm	24 mm
MOE	−0.772 */<0.001	−0.937 */<0.001	−0.923 */<0.001
BS	0.977 */<0.001	0.967 */<0.001	0.764 */<0.001
DS	−0.635 */<0.001	−0.975 */<0.001	−0.975 */<0.001
SK	0.992 */<0.001	0.973 */<0.001	0.968 */0.007
WA	0.973 */0.005	0.963 */0.009	0.937 */0.019
VS	0.943/0.160	0.949/0.140	0.944/0.160
TC	ND	ND	−0.921 */0.026
SAC (250 Hz)	ND	ND	0.281/0.647
SAC (500 Hz)	ND	ND	0.228/0.712
SAC (1000 Hz)	ND	ND	0.740/0.153

"*" indicates a significant correlation at a significance level of $p < 0.05$. r = Pearson correlation coefficients, MOE = modulus of elasticity, BS = bending strength, DS = density, SK = shrinkage, WA = water absorption, VS = volume swelling, SAC = sound absorption coefficient, and "ND" = not determined.

Table 3. Pearson correlation coefficients between the properties and an increase in thickness of MCBs with different substrate ratios.

Properties	Pearson Correlation Coefficients (r/p-Value)				
	C100S0	C75S25	C50S50	C25S75	C0S100
MOE	−0.898 */<0.001	−0.908 */<0.001	−0.921 */<0.001	−0.922 */<0.001	−0.929 */<0.001
BS	−0.945 */<0.001	−0.951 */<0.001	−0.957 */<0.001	−0.937 */<0.001	−0.919 */<0.001
DS	0.635/0.066	0.555/0.120	0.929 */<0.001	0.898 */<0.001	0.949 */<0.001
SK	0.980/0.125	0.752/0.549	0.921/0.256	0.990/0.089	0.958/0.186
WA	−0.946/0.200	−0.946/0.210	−0.955/0.193	−0.896/0.293	−0.951/0.200
VS	−0.894/0.297	−0.917/0.262	−0.850/0.353	−0.093/0.940	−0.999/0.220

"*" indicates a significant correlation at a significance level of $p < 0.05$. r = Pearson correlation coefficients, MOE = modulus of elasticity, BS = bending strength, DS = density, SK = shrinkage, WA = water absorption, and VS = volume swelling.

3.1.2. Bending Strength

The BS values of specimens obtained in this study are shown in Figure 3B. The BS values were 0.38 to 1.26 MPa, 0.16 to 0.41 MPa, and 0.13 to 0.17 MPa for specimens with thicknesses of 8 mm, 16 mm, and 24 mm, respectively. When compared to previous studies, the BS values from this study (0.13 to 1.26 MPa) fall within the reported range of 0.07 to 4.40 MPa [6,22,50–52]. The results revealed that the different ratios of corn husks to sawdust affected the BS value, and an increase in the ratio of sawdust led to a decrease in the BS value for each thickness. The highest BS values of the obtained MCB in this study was discovered in the MCB made from 100% corn husk in each thickness. In contrast, the 100% sawdust MCBs had the lowest bending strength. Similarly, numerous previous studies have demonstrated that the flexural strength of MBCs was influenced by the type of substrate, type of mycelia network, and pressing method [22,53]. Generally, bending strength is influenced by the proportion of lignin and cellulose in the substrate. The proportion of lignin and cellulose in rubber tree sawdust ranges from 0.86 to 1.35, which is higher than in corn husks, which ranges from 0.22 to 0.33 [54–57]. Previous studies have found that substrates with a higher ratio of lignin and cellulose exhibit lower bending strength [58,59]. Similarly, this study found that MCBs made from substrates with increasing sawdust content showed a higher ratio of lignin to cellulose, which resulted in reduced bending strength. Furthermore, *L. sajor-caju* is a white-rot fungus that primarily breaks down lignin in its growing substrate [60]. This may enhance lignin degradation and lead to a higher cellulose content in the substrate, which is also associated with increased bending strength. In this study, the bending strength (BS) value showed a significant positive correlation with a decrease in the ratio of corn husks at the same thickness ($r = 0.764$ to 0.977, $p < 0.001$) (Table 2), indicating that lower corn husk content is associated with a lower BS value. A strong negative correlation between the BS value and an increase in thickness at the same substrate ratio ($r = -0.957$ to -0.919, $p < 0.001$) was observed, according to Pearson correlation analysis ($p < 0.05$) (Table 3).

3.2. Physical Properties of MCBs

3.2.1. Density

The density of the MCBs in this study is displayed in Figure 4A. The results exhibited that the obtained density value varied according to the ratio of substrates used. The highest density was found in the MCB made from 100% sawdust in each thickness. The findings indicated that the decrease in MCB density varied with the increase in the amount of corn husk added. Density values in this investigation ranged from 209.17 to 256.28 kg/m^3, and were in the range of 25 to 954 kg/m^3 from previous reports [15,22,57,61–65]. In addition, the density at the same substrate ratio increased with increasing thickness. The results are consistent with several previous studies, which reported that substrate type and ratio, substrate particles, volume fraction, and pressing process have a major impact on the density of MBCs [62,66,67].

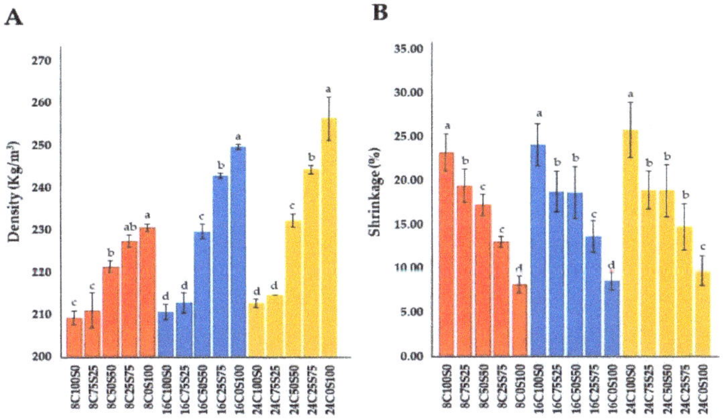

Figure 4. Density (**A**) and shrinkage (**B**) of MCBs in this study. Red, blue, and yellow bars indicate MCBs with thicknesses of 8 mm, 16 mm, and 24 mm, respectively. The different letters within the same group of MCB thickness indicate a significant difference according to Duncan's multiple range test ($p \leq 0.05$).

According to the Pearson correlation ($p < 0.05$), the density showed a significant strong negative correlation with a decrease in the ratio of corn husks at the same thickness ($r = -0.975$ to -0.635, $p < 0.001$). This indicates that a lower corn husk content has a higher density value. It also showed a significant positive correlation with an increase in thickness at substrate ratios of 50% corn husks and 50% sawdust, 25% corn husk and 75% sawdust, and 100% sawdust ($r = 0.898$ to 0.929, $p < 0.001$) (Table 2). However, the substrate ratios of

100% corn husk ($r = -0.635$, $p = 0.006$) and 75% corn husk and 25% sawdust ($r = -0.555$, $p = 0.120$) showed a non-significant correlation with an increase in thickness (Table 3).

3.2.2. Shrinkage

Shrinkage value is an important aspect of the physical properties of MCBs, primarily resulting from the dehydration process during drying and depending on the substrate type [20,53,62,64]. According to this study, different ratios of corn husks to sawdust affected the shrinkage value, as shown in Figure 4B. MCBs produced from 100% corn husk exhibited the highest shrinkage values at each thickness, while the lowest values were observed in MCBs made from 100% sawdust. An increase in the ratio of sawdust led to a decrease in shrinkage values for each thickness. Additionally, the obtained shrinkage values ranged between 8.15% and 25.78%. Prior to this study, the shrinkage values of MBCs were reported to be in the range of 2.78% to 17% [12,20,62,65–69]. Remarkably, the shrinkage values obtained from ratios of 100% corn husk (23.20% to 25.78%), 75% corn husk with 25% sawdust (18.89% to 19.39%), and 50% corn husk with 50% sawdust (17.19% to 18.86%) at 8 mm, 16 mm, and 24 mm thicknesses were higher than previously reported values. Therefore, it is crucial to select a suitable substrate in order to control the shrinkage value of MBCs. According to the Pearson correlation ($p < 0.05$), a decrease in the ratio of corn husks at the same thickness showed a significantly strong positive correlation with the shrinkage value ($r = 0.968$ to 0.992, $p < 0.001$ to 0.007) (Table 2). This indicates that lower corn husk content is associated with a lower shrinkage value. However, the shrinkage value did not show a significant positive correlation with thickness at the same substrate ratio ($p < 0.05$) (Table 3).

3.2.3. Water Absorption

The water absorption ability of the MCBs, obtained from different ratios of corn husks and sawdust and varying thicknesses, was assessed by immersing the MCB specimens in water for 84 h. The percentages of water absorption values over the period from 0 to 84 h are presented in Figure 5A–C. It was found that the water absorption ability of the MCBs made from 100% corn husks, in all thicknesses, increased over a 16 h period and slowly stabilized after 24 h. In contrast, the water absorption ability of the MCBs made from 100% sawdust increased over a period of 36 h and slowly stabilized after 48 h. Additionally, the water absorption ability of the mixture of corn husks and sawdust increased mostly over 24 h and slowly stabilized after 48 h. After 4 h, it was found that the water absorption for specimens with a 100% corn husk ratio showed the highest value, followed by specimens with a 75% corn husk and 25% sawdust ratio, a 50% corn husk and 50% sawdust ratio, and a 100% sawdust ratio. The water absorption for specimens with a 75% corn husk and 25% sawdust ratio and a 50% corn husk and 50% sawdust ratio was not significantly different over a period of 2 to 84 h. In addition, the increase in sawdust in MCBs decreased the water absorption. After 84 h, it was observed that MCBs produced from 100% corn husks displayed water absorption rates ranging between 204.01 and 256.11%, while MBCs produced from 100% sawdust exhibited water absorption rates ranging from 96.86 to 161.68%. The percentage of water absorbed by MCBs made of various sawdust and corn husk combinations ranged from 146.32 to 219.01%. The water absorption ability of MCBs in this study was within the range of 24.45 to 560%, as reported in previous studies when submerged in water for 24 to 192 h [6,20,22,23,30,33,66]. According to the results of several studies, MBCs are classified as hydrophilic materials, and the substrate type and fungal species influence their water absorption ability [20,22,23,30,33,66].

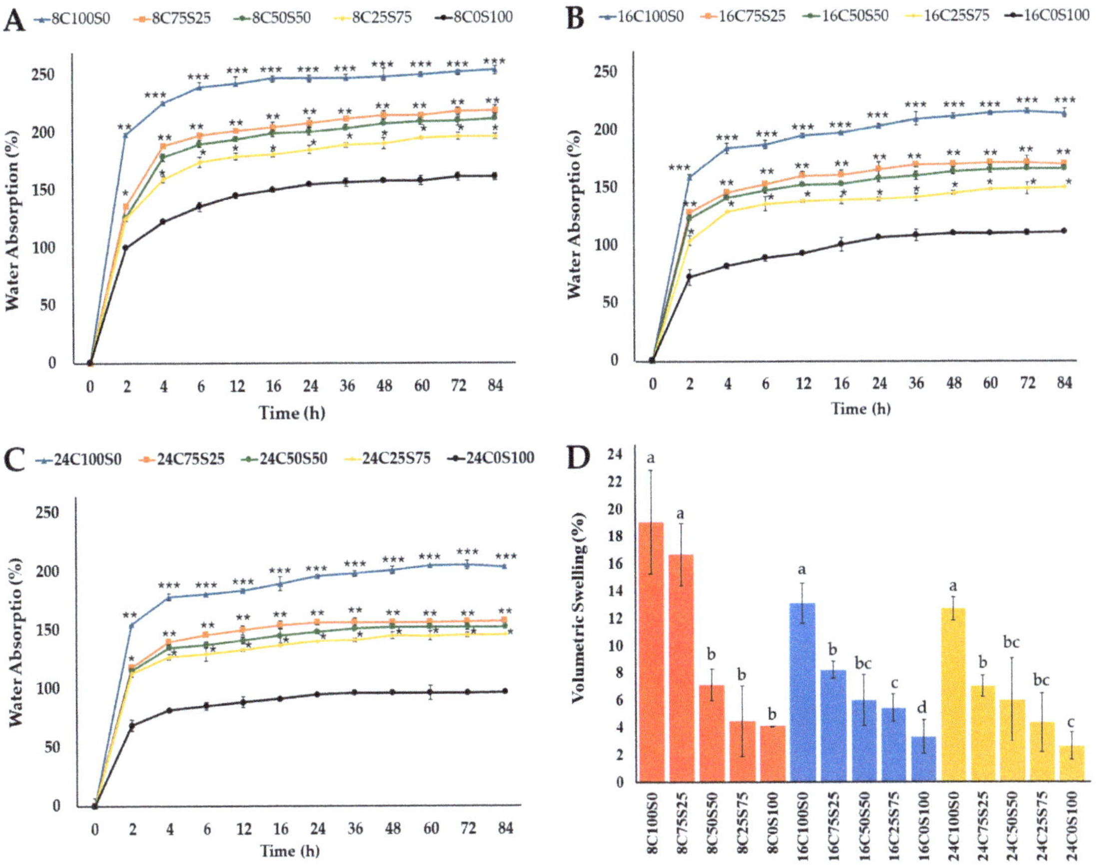

Figure 5. Water absorption behavior of MBC specimen's thickness at 8 mm (**A**), 16 mm (**B**), and 24 mm (**C**) and volumetric swelling (**D**) in this study. *, **, *** in subfigures (**A**–**C**) indicate significant differences according to Duncan's multiple range test ($p \leq 0.05$). Red, blue, and yellow bars in subfigures (**D**) indicate MCBs with thicknesses of 8 mm, 16 mm, and 24 mm, respectively. The different letters within the same group of MCB thickness in subfigures (**D**) indicate a significant difference according to Duncan's multiple range test ($p \leq 0.05$).

Based on the morphology of MBCs, water absorption capacity was found to correlate with the porous structure of the composites. Corn husks exhibit higher porosity compared to rubber tree sawdust [32]. Consequently, the higher ratio of corn husks used in the MCBs in this study led to increased water absorption. The water absorption capacity of MBCs is typically linked to the proportions of lignin and cellulose. Several previous studies have reported that lignin is a hydrophobic component, while cellulose is hydrophilic. Thus, a high proportion of lignin and cellulose in the substrate can decrease water absorption [32,33,70]. According to a comparison of the lignin and cellulose content of the two materials employed in this study, rubber tree sawdust has a larger lignin and cellulose proportion (0.86 to 1.35) than corn husks (0.22 to 0.33) [53–55,71]. As a result, MCBs with larger corn husk content have higher water absorption. Furthermore, the enhanced lignin degradation by *L. sajor-caju* results in a higher cellulose content in the substrate, which is associated with increased water absorption capacity. In this study, a decrease in the ratio of corn husks at the same thickness showed a significantly strong positive correlation with the water absorption ability ($r = 0.973$ to 0.937, $p = 0.005$ to 0.019) according to the Pearson correlation analysis ($p < 0.05$) (Table 2).

This indicates that a lower corn husk content is associated with lower water absorption values. However, water absorption did not show a significant negative correlation with an increase in thickness at the same substrate ratio ($r = -0.955$ to -0.896, $p = 0.193$ to 0.210) (Table 3).

3.2.4. Volumetric Swelling

After soaking in water for 84 h, the volumetric swelling of the MCBs in this study varied from 2.63 to 19.03% (Figure 5D) and fell within the previously reported ranges of volumetric swelling observed in MBCs (0.28 to 21%) [62–69]. The volumetric swelling rate of MCBs made from 100% corn husk was higher for all thicknesses compared to that of other MCBs. Subsequently, a decrease in corn husk ratio content in MCBs led to a reduction in volumetric swelling rate. It was found that MCBs with high water absorption also have high volumetric swelling. The outcome is in accordance with previous studies that found a direct correlation between volumetric swelling and the water absorption capability of the mycelium-based material [7,20]. In Table 2, the volumetric swelling rate showed a non-significant positive correlation ($r = 0.949$ to 0.943, $p = 0.140$ to 0.160) with a decrease in the ratio of corn husk at the same thickness. The volumetric swelling also exhibited a non-significant negative correlation with an increase in thickness at the same substrate ratio ($r = -0.999$ to -0.850, $p = 0.220$ to 0.940), as shown in Table 3.

3.2.5. Thermal Conductivity Values

Thermal conductivity is the capacity of a material to conduct or transfer heat. In this study, only the 24 mm thick specimens were used due to limitations of the testing equipment. According to our results, the thermal conductivity values ranged from 0.037 to 0.079 W/m·K (Table 4). These obtained values were within the previously documented range of thermal conductivity values observed in MBCs, which is 0.029 to 0.124 W/m·K [20,24,72–74]. The results showed that variations in the ratio of corn husk to sawdust led to different thermal conductivity values. An increase in sawdust ratio in MCBs led to an increase in the thermal conductivity value. The correlation between thermal conductivity value and a reduction in corn husk content showed a significant negative effect ($r = -0.921$, $p = 0.026$) (Table 2). In this study, it was found that the thermal conductivity values of MCBs were related to density, with higher thermal conductivity values associated with higher density values. The results are consistent with several previous studies that reported that the type of substrate used in mycelium-based material production influences thermal conductivity values and is related to density [53]. Based on our previous study [32], the morphological characteristics of MBCs produced from *L. sajor-caju* and lignocellulosic residues (including corn husks, sawdust, and rice straw) include larger pores and a dense mycelial network. Therefore, the possible mechanisms influencing the thermal conductivity of MCBs include porosity and air pockets within the material acting as thermal insulators, as well as the potential enhancement of heat transfer by a dense mycelial network, as reported in previous studies [75,76].

Table 4. Average thermal conductivity values of MCBs obtained in this study.

Specimen Name	Thermal Conductivity (W/m·K)
24C100S0	0.037
24C75S25	0.048
24C50S50	0.052
24C25S75	0.054
24C0S100	0.079

3.2.6. Sound Absorption Coefficient

Sound absorption is one of the most important factors in producing suitable materials with MBCs. MCBs typically feature a porous structure due to the growth of mycelium and the presence of fibrous materials [77]. The possible mechanisms of sound absorption by MCBs include trapping sound waves within their porous structure; converting sound energy into heat through friction and viscous effects of the fibrous materials; and reducing sound intensity through disruption and scattering of sound waves by the mycelial network [78–81]. In this study, only specimens with a thickness of 24 mm were employed due to testing equipment restrictions. The percentage of sound absorption of MCBs at frequencies of 250, 500, and 1000 Hz is shown in Table 5. It was found that the percentage of sound absorption of MCBs varied with different substrate ratios used. The highest percentage of sound absorption at frequencies of 250, 500, and 1000 Hz was found in MCBs made using a 75% corn husk and 25% sawdust ratio. It was found that the percentage of sound absorption of the mixed substrate showed higher values than individual substrates, consistent with previous findings [6,20]. Moreover, the sound absorption coefficient of the substrate is influenced by substrate size and the pressing method [82]. In addition, Barta et al. [83] found that MBCs made from *Ganoderma lucidum* and *Trametes versicolor* showed different sound absorption abilities, indicating that the fungal species affects this property. Pelletier et al. [26] found that MBCs made from varied substrates (including residues of flax shive, hemp, kenaf, rice straw, sorghum stalk, and switch glass) showed 70–75% sound absorption at 1000 Hz. Interestingly, MCB specimens in this study made from pure corn husk and a mixture of corn husk with sawdust showed 84–94% sound absorption at 1000 Hz. According to the Pearson correlation ($p < 0.05$), a decrease in the ratio of corn husk content in MCBs showed a non-significant positive correlation with sound absorption at frequencies of 250 ($r = 0.281$, $p = 0.647$), 500 ($r = 0.228$, $p = 0.712$), and 1000 Hz ($r = 0.740$, $p = 0.153$) (Table 2). Our results suggest that the selection of substrate combination and sound frequencies should be taken into consideration when developing suitable MCBs for sound absorption materials. However, reports on the sound absorption coefficient of MBCs are limited; therefore, further studies are required.

Table 5. Average sound absorption coefficient at each frequency for MCBs obtained in this study.

Specimen Name	Sound Absorption Coefficient (%)		
	250 Hz	500 Hz	1000 Hz
24C100S0	40	34	85
24C75S25	51	49	94
24C50S50	43	45	86
24C25S75	43	43	84
24C0S100	40	41	61

3.3. Comparison of Properties with Commercial Boards

The properties of MCBs obtained in this study were compared to previous reports of MBCs and other commercial boards, as shown in Table 6. Based on mechanical properties, the BS values of MCBs were within the range of softboards and acoustic boards. However, the MOE values were lower than those of other conventional boards. Based on physical properties, the density values of some MCBs, including 8C0S100, 16C25S75, 16C0S100, 24C50S50, 24C25S75, and 24C0S100 in this study, fall within the density range of softboards. However, the shrinkage, water absorption, and volumetric swelling of MCBs were higher than those of other conventional boards. It was found that the thermal conductivity properties of MCBs were within the range of softboards and insulated boards. In addition, MCBs made from 100% corn husk exhibited thermal conductivity within the range of acoustic boards. According to the sound absorption coefficient, MCBs were in the range of acoustic boards, insulated boards, and softboards, with higher values than medium-density fiberboards, gypsum boards, and fiber cement boards.

Table 6. Comparison of MBC properties in this study with commercial boards.

Properties		MBCs		Commercial Boards *					
		This Study	Previous Studies	Medium-Density Fiber Board	Softboard	Gypsum Board	Fiber Cement Board	Insulated Board	Acoustic Board
MOE (MPa)		0.17–7.88	0.14–97.00	2500–4000	80–150	1500–3000	6000–15,000	3000–7000	20–100
BS (MPa)		0.13–1.26	0.02–4.40	20–40	0.70–1.20	1.5–3.5	10–30	0.1–0.3	0.5–2.0
DS (kg/m^3)		209.17–256.28	25–954	640–800	230–400	600–800	1300–1700	15–200	60–400
SK (%)		8.15–25.78	2.78–17	0.01–0.30	0.2–0.5	0.05–0.10	0.02–0.05	0.015–0.030	0.02–0.10
WA (%)		96.86–254.11	24.45–560	5–15	30–70	30–50	10–25	1–4	5–30
VS (%)		2.63–19.03	0.28–21	10–25	15–40	0.02–0.10	0.1–2.0	1.0–2.5	0.1–1.0
TC (W/m·K)		0.037–0.079	0.029–0.124	0.10–0.18	0.035–0.060	0.16–0.25	0.20–0.40	0.022–0.040	0.03–0.05
SAC (%)	250 Hz	40–51	16–20	10–20	15–30	5–10	10–20	30–60	50–80
	500 Hz	34–49	50–51	8–15	30–50	5–8	8–15	65–90	80–100
	1000 Hz	61–94	70–75	6–12	50–70	4–7	6–12	85–100	90–100

MOE = modulus of elasticity, BS = bending strength, DS = density, SK = shrinkage, WA = water absorption, VS = volume swelling, SAC = sound absorption coefficient. * Seddeq [80], Hsu et al. [84], Cai et al. [85], Ayrilmis et al. [86–88], Suchsland et al. [89], Sonderegger et al. [90], Nemli et al. [91], BS EN 622-4:2009 [92], Papadopoulos and Hill [93], Ganev et al. [94], Berardi et al. [95], Cramer et al. [96], Bénichou et al. [97], Karni et al. [98], Wakili et al. [99], Camino et al. [100], Feng et al. [101], Warnock [102], Ardanuy et al. [103], Mohr et al. [104], Akers. [105], Biot et al. [106], Jelle [107], Papadopoulos [108], Schiavoni et al. [109], and Asdrubali et al. [110].

4. Conclusions

This study investigated various mechanical and physical properties of MCBs derived from *L. sajor-caju*. The results showed that mechanical and physical properties of MCBs varied in substrate ratio and thickness. The different ratios of corn husk and sawdust showed a significant correlation with the MOE, BS, density, shrinkage, water absorption, and thermal conductivity of MCBs. The thickness of MCBs showed a significant correlation with MOE, BS, and density (with corn husk at less than 50%). It was found that the MOE (0.17–7.88 MPa), BS (0.13–1.26 MPa), density (209.17–256.08 kg/m^3), water absorption (96.86–256.11%), volumetric swelling (2.63–19.03%), thermal conductivity (0.037–0.079 W/m·K), and sound absorption (61–94% at 1000 Hz) of MCBs were within the ranges reported in several previous studies. However, MCBs showed a higher shrinkage value (8.15–25.78%) compared to previous reports. Based on the mechanical (BS) and physical properties (density, thermal conductivity, and sound absorption), MCBs made from 100% corn husk are most similar to softboard and acoustic boards. Further improvement of the MOE, shrinkage, water absorption, and volumetric swelling of this MCB is still required. Improvement in the qualities of mycelium-based materials will be achieved through continued research and biotechnological developments, which will address the challenges of standardization, production costs, and commercial acceptability.

Author Contributions: Conceptualization, P.J., Y.T., S.L., T.W. and N.S.; methodology, P.J., Y.T., W.A., J.K., T.W. and N.S.; investigation, P.J., W.A., J.K., W.T., T.W. and N.S.; software, P.J. and W.A.; validation, P.J., Y.T., W.A., J.K., T.W. and N.S.; formal analysis, P.J., Y.T., J.K., W.T., T.W. and N.S.; data curation, P.J., Y.T., W.A., J.K., T.W. and N.S.; writing—original draft preparation—P.J., J.K. and N.S.; writing—review and editing—P.J., Y.T., W.A., J.K., S.L., W.T., T.W., U.S. and N.S.; supervision, Y.T., R.O., J.K., S.L., T.W., U.S. and N.S.; project administration, R.O., S.L., T.W. and N.S.; funding acquisition, R.O. and N.S. All authors have read and agreed to the published version of the manuscript.

Funding: This research was supported by the Fundamental Fund 2023 (FF66/056), Chiang Mai University. P.J. was supported by CMU Presidential Scholarship, Chiang Mai University (2566-075).

Institutional Review Board Statement: Not applicable.

Informed Consent Statement: Not applicable.

Data Availability Statement: The original contributions presented in the study are included in the article, further inquiries can be directed to the corresponding author.

Acknowledgments: P.J. acknowledges the CMU Presidential Scholarship for the Doctor of Philosophy Program in Architecture (International Program) at the Faculty of Architecture, Chiang Mai University. The authors are grateful to Wachirawit Bunma and Punyavee Lugsanagul for the substrate preparation.

Conflicts of Interest: The authors declare no conflicts of interest.

References

1. Hammond, G.; Jones, C.; Lowrie, F. *Embodied Carbon: The Inventory of Carbon and Energy (ICE) (Version 2.0)*; University of Bath with BSRIA: Bracknell, UK, 2011.
2. Symons, K. *Book Review: Embodied Carbon: The Inventory of Carbon and Energy (ICE)*; University of Bath with BSRIA: Bracknell, UK, 2011.
3. Xanthos, D.; Walker, T.R. International policies to reduce plastic marine pollution from single-use plastics (plastic bags and microbeads): A review. *Mar. Pollut. Bull.* **2017**, *118*, 17–26. [CrossRef] [PubMed]
4. Lambert, S.; Wagner, M. Characterisation of nanoplastics during the degradation of polystyrene. *Chemosphere* **2016**, *145*, 265–268. [CrossRef] [PubMed]
5. Suwanmanee, U.; Varabuntoonvit, V.; Chaiwutthinan, P.; Tajan, M.; Mungcharoen, T.; Leejarkpai, T. Life cycle assessment of single use thermoform boxes made from polystyrene (PS), polylactic acid (PLA), and PLA/starch: Cradle to consumer gate. *Int. J. Life Cycle Assess.* **2013**, *18*, 401–417. [CrossRef]
6. Rendón-Villalobos, R.; Ortíz-Sánchez, A.; Tovar-Sánchez, E.; Flores-Huicochea, E. *The Role of Biopolymers in Obtaining Environmentally Friendly Materials*; Ghaly, A.A., Ed.; Composites from Renewable and Sustainable Materials; InTech: Vladivostok, Russia, 2016; pp. 151–179.
7. Teeraphantuvat, T.; Jatuwong, K.; Jinanukul, P.; Thamjaree, W.; Lumyong, S.; Aiduang, W. Improving the physical and mechanical properties of mycelium-based green composites using paper waste. *Polymers* **2024**, *16*, e262. [CrossRef]
8. Garcia-Garcia, D.; Quiles-Carrillo, L.; Balart, R.; Torres-Giner, S.; Arrieta, M.P. Innovative solutions and challenges to increase the use of Poly (3-hydroxybutyrate) in food packaging and disposables. *Eur. Polym. J.* **2022**, *178*, e111505. [CrossRef]
9. Balaeș, T.; Radu, B.-M.; Tănase, C. Mycelium-composite materials—A promising alternative to plastics? *J. Fungi* **2023**, *9*, e210. [CrossRef] [PubMed]
10. Acharjee, S.A.; Bharali, P.; Gogoi, B.; Sorhie, V.; Walling, B.; Alemtoshi. PHA-based bioplastic: A potential alternative to address microplastic pollution. *Water Air Soil Pollut.* **2023**, *234*, e21. [CrossRef] [PubMed]
11. Khalil, H.P.S.A.; Bhat, A.H.; Yusra, A.F.I. Green composites from sustainable cellulose nanofibrils: A review. *Carbohydr. Polym.* **2012**, *87*, 963–979. [CrossRef]
12. Chen, Y.; Dang, B.; Jin, C.; Sun, Q. Processing lignocellulose-based composites into an ultrastrong structural material. *ACS Nano* **2018**, *13*, 371–376. [CrossRef]
13. Ferdosian, F.; Pan, Z.; Gao, G.; Zhao, B. Bio-based adhesives and evaluation for wood composites application. *Polymers* **2017**, *9*, e70. [CrossRef]
14. Hemmilä, V.; Adamopoulos, S.; Karlsson, O.; Kumar, A. Development of sustainable bio-adhesives for engineered wood panels—A review. *RSC Adv.* **2017**, *7*, 38604–38630. [CrossRef]
15. Sun, W.; Tajvidi, M.; Hunt, C.G.; McIntyre, G.; Gardner, D.J. Fully bio-based hybrid composites made of wood, fungal mycelium and cellulose nanofibrils. *Sci. Rep.* **2019**, *9*, e3766. [CrossRef]
16. Arikan, E.B.; Ozsoy, H.D. A review: Investigation of bioplastics. *J. Civil. Eng. Archit.* **2015**, *9*, 188–192.
17. Moshood, T.D.; Nawanir, G.; Mahmud, F.; Mohamad, F.; Ahmad, M.H.; AbdulGhani, A. Biodegradable plastic applications towards sustainability: Recent innovations in the green product. *Cleaner Eng. Technol.* **2022**, *6*, e100404. [CrossRef]
18. Girometta, C.; Picco, A.M.; Baiguera, R.M.; Dondi, D.; Babbini, S.; Cartabia, M.; Pellegrini, M.; Savio, E. Physico-mechanical and thermodynamic properties of mycelium-based biocomposites: A review. *Sustainability* **2019**, *11*, e281. [CrossRef]
19. Ali, G.; Paniz, F.; Fabricio, V.; John, P.; Benay, G. Mycelium-based bio-composites for architecture: Assessing the effects of cultivation factors on compressive strength. Architecture in the Age of the 4th Industrial Revolution. In Proceedings of the 37th eCAADe and 23rd SIGraDi Conference, University of Porto, Porto, Portugal, 11–13 September 2019; pp. 505–514.
20. Elsacker, E.; Vandelook, S.; Brabcart, J.; Peeters, E.; Laet, L.D. Mechanical, physical and chemical characterisation of mycelium-based composites with different types of lignocellulosic substrates. *PLoS ONE* **2019**, *14*, e0213954. [CrossRef]
21. Attias, N.; Danai, O.; Tarazi, E.; Pereman, I.; Grobman, Y.J. Implementing bio-design tools to develop mycelium-based products. *Des. J.* **2019**, *22*, 1647–1657. [CrossRef]
22. Appels, F.V.W.; Camere, S.; Montalti, M.; Karana, E.; Jansen, K.M.B.; Dijksterhuis, J.; Krijgsheld, P.; Wösten, H.A.B. Fabrication factors influencing mechanical, moisture- and water-related properties of mycelium-based composites. *Mater. Des.* **2019**, *161*, 64–71. [CrossRef]

23. Aiduang, W.; Jatuwong, K.; Jinanukul, P.; Suwannarach, N.; Kumla, J.; Thamjaree, W.; Teeraphantuvat, T.; Waroonkun, T.; Oranratmanee, R.; Lumyong, S. Sustainable innovation: Fabrication and characterization of mycelium-based green composites for modern interior materials using agro-industrial wastes and different species of fungi. *Polymers* **2024**, *16*, e550. [CrossRef]
24. Xing, Y.; Brewer, M.; El-Gharabawy, H.; Griffith, G.; Jones, P. Growing and testing mycelium bricks as building insulation materials. *IOP Conf. Ser. Earth Environ. Sci.* **2018**, *121*, e022032. [CrossRef]
25. Jiang, L.; Walczyk, D.; McIntyre, G.; Chan, W.K. Cost modeling and optimization of a manufacturing system for mycelium-based biocomposite parts. *J. Manuf. Syst.* **2016**, *41*, 8–20. [CrossRef]
26. Pelletier, M.G.; Holt, G.A.; Wanjura, J.D.; Bayer, E.; McIntyre, G. An evaluation study of mycelium based acoustic absorbers grown on agricultural by-product substrates. *Ind. Crops Prod.* **2013**, *51*, 480–485. [CrossRef]
27. Robertson, O.; Høgdal, F.; Mckay, L.; Lenau, T. Fungal future: A review of mycelium biocomposites as an ecological alternative insulation material. In *Proceedings of NordDesign*; Technical University of Denmark: Lyngby, Denmark, 2020; pp. 1–13.
28. Ongpeng, J.M.C.; Inciong, E.; Sendo, V.; Soliman, C.; Siggaoat, A. Using waste in producing bio-composite mycelium bricks. *Appl. Sci.* **2020**, *10*, e5303. [CrossRef]
29. Bagheriehnajjar, G.; Yousefpour, H.; Rahimnejad, M. Environmental impacts of mycelium-based bio-composite construction materials. *Int. J. Environ. Sci. Technol.* **2024**, *21*, 5437–5458. [CrossRef]
30. Alaneme, K.K.; Anaele, J.U.; Oke, T.M.; Kareem, S.A.; Adediran, M.; Ajibuwa, O.A.; Anabaranze, Y.O. Mycelium based composites: A review of their bio-fabrication procedures, material properties and potential for green building and construction applications. *Alex. Eng. J.* **2023**, *83*, 234–250. [CrossRef]
31. Yang, L.; Qin, Z. *Bioprospects of Macrofungi*; Qin, Z., Ed.; Experimental analysis of the mechanics of mycelium-based biocomposites; CRC Press: Boca Raton, FL, USA, 2023; pp. 205–232.
32. Aiduang, W.; Jatuwong, K.; Luangharn, T.; Jinanukul, P.; Thamjaree, W.; Teeraphantuvat, T.; Waroonkun, T.; Lumyong, S. A Review delving into the factors influencing mycelium-based green composites (MBCS) production and their properties for long-term sustainability targets. *Biomimetics* **2024**, *9*, 337. [CrossRef]
33. Aiduang, W.; Kumla, J.; Srinuanpan, S.; Thamjaree, W.; Lumyong, S.; Suwannarach, N. Mechanical, physical, and chemical properties of mycelium-based composites produced from various lignocellulosic residues and fungal species. *J. Fungi* **2022**, *8*, e1076. [CrossRef]
34. Kadir, Y.A.; Samsi, H.W.; Masseat, K.; Rashid, N.A.A.; Ali, S.Z.; Karim, S.R.A.; Karim, A.A. Physical and mechanical properties of plywood made from Meranti temak nipis (Shorea roxburghii). In *Proceedings of the Forest Products Seminar Lignocellulosic Materials Towards a Sustainable Future, FRIM Proceedings No. 25*; Forest Research Institute Malaysia: Kuala Lumpur, Malaysia, 2023; pp. 46–49.
35. *BS EN 310:1993*; Wood-Based Panels. Determination of Modulus of Elasticity in Bending and of Bending Strength. The British Standards Institution (BSI): London, UK, 1993.
36. Gezer, E.D.; Gümüşkaya, E.; Uçar, E.; Ustaömer, E. Mechanical properties of mycelium based MDF. *Sigma J. Eng. Nat. Sci.* **2020**, *11*, 135–140.
37. Espinoza-Herrera, R.; Cloutier, A. Physical and mechanical properties of gypsum particleboard reinforced with portland cement. *Eur. J. Wood Wood Prod.* **2011**, *69*, 247–254. [CrossRef]
38. *BS EN 323:1993*; Wood-Based Panels. Determination of Density. The British Standards Institution (BSI): London, UK, 1993.
39. *ASTM D1037:2002*; Standard Test Methods for Evaluating Properties of Wood-Base Fiber and Particle Panel Materials. ASTM: West Conshohocken, PA, USA, 2002.
40. Dotun, A.O.; Olalekan, O.S.; Olugbenga, A.L.; Emmanuel, M.A. Physical and mechanical properties evaluation of corncob and sawdust cement bonded ceiling boards. *Int. J. Eng. Res. Africa* **2019**, *42*, 65–75.
41. Faouel, J.; Mzali, F.; Jemni, A.; Nasrallah, A.B. Thermal conductivity and thermal diffusivity measurements of wood in the three anatomic directions using the transient hot-bridge method. *Spec. Top. Rev. Porous Media Int. J.* **2012**, *3*, 229–237. [CrossRef]
42. *ISO 22007-2:2022*; Plastics—Determination of Thermal Conductivity and Thermal Diffusivity—Part 2: Transient Plane Heat Source (Hot Disc) Method. ISO: Geneva, Switzerland, 2022.
43. Keawprak, N.; Arkom, P.; Lao-auyporn, P.; Kaewpreak, N.; Wannasut, P.; Watcharapasorn, A. Development of expanded perlite board for thermal insulation. *Chiang Mai J. Sci.* **2022**, *49*, 1644–1652. [CrossRef]
44. *ISO 10534-2:2023*; Acoustics—Determination of Acoustic Properties in Impedance Tubes—Part 2: Two-Microphone Technique for Normal Sound Absorption Coefficient and Normal Surface Impedance. ISO: Geneva, Switzerland, 2023.
45. Yang, W.D.; Li, Y. Sound absorption performance of natural fibers and their composites. *Sci. China Technol. Sci.* **2012**, *55*, 2278–2283. [CrossRef]
46. Siano, D.; Viscardi, M.; Chimenti, M.; Panzaa, M.A.; Molaro, A.; Lanzillo, S. Experimental analysis of innovative acoustic materials for engine noise control in vehicles. In Proceedings of the International Congress on Sound and Vibration. ICSV No. 26, Montreal, France, 7–11 July 2019; pp. 1–8.
47. Haneef, M.; Ceseracciu, L.; Canale, C.; Bayer, I.S.; Heredia-Guerrero, J.A.; Athanassiou, A. Advanced materials from fungal mycelium: Fabrication and tuning of physical properties. *Sci. Rep.* **2017**, *7*, e41292. [CrossRef]
48. Shi, J.; Peng, J.; Huang, Q.; Cai, L.; Shi, S.Q. Fabrication of densified wood via synergy of chemical pretreatment, hot-pressing and post mechanical fixation. *J. Wood Sci.* **2020**, *66*, e5. [CrossRef]

49. Kuznetsova, I.; Zaitsev, B.; Krasnopolskaya, L.; Teplykh, A.; Semyonov, A.; Avtonomova, A.; Ziangirova, M.; Smirnov, A.; Kolesov, V. Influence of humidity on the acoustic properties of mushroom mycelium films used as sensitive layers for acoustic humidity sensors. *Sensors* **2020**, *20*, e2711. [CrossRef]
50. Johnston, J.C.; Kuczmarski, M.A.; Olszko, G. Development of instrumentation for the measurement of the performance of acoustic absorbers. *Open J. Acoust.* **2015**, *5*, 172–192. [CrossRef]
51. Liu, R.; Li, X.; Long, L.; Sheng, Y.; Xu, J.; Wang, Y. Improvement of mechanical properties of mycelium/cotton stalk composites by water immersion. *Compos. Interfaces* **2020**, *27*, 953–966. [CrossRef]
52. Kuribayashi, T.; Lankinen, P.; Hietala, S.; Mikkonen, K.S. Dense and continuous networks of aerial hyphae improve flexibility and shape retention of mycelium composite in the wet state. *Compos. Part A Appl. Sci. Manuf.* **2022**, *152*, e106688. [CrossRef]
53. Jones, M.; Mautner, A.; Luenco, S.; Bismarck, A.; John, S. Engineered mycelium composite construction materials from fungal biorefineries: A critical review. *Mater. Des.* **2020**, *187*, e108397. [CrossRef]
54. Pointner, M.; Kuttner, P.; Obrlik, T.; Jäger, A.; Kahr, H. Composition of corncobs as a substrate for fermentation of biofuels. *Agron. Res.* **2014**, *12*, 391–396.
55. El-Tayeb, T.S.; Abdelhafez, A.A.; Ali, S.H.; Ramadan, E.M. Effect of acid hydrolysis and fungal biotreatment on agro-industrial wastes for obtainment of free sugars for bioethanol production. *Braz. J. Microbiol.* **2012**, *43*, 1523–1535. [CrossRef] [PubMed]
56. Song, S.T.; Saman, N.; Johari, K.; Mat, H.B. Removal of mercury (II) from aqueous solution by using rice residues. *J. Teknol.* **2013**, *63*, 67–73. [CrossRef]
57. Islam, M.R.; Tudryn, G.; Bucinell, R.; Schadler, L.; Picu, R.C. Morphology and mechanics of fungal mycelium. *Sci. Rep.* **2017**, *7*, e13070. [CrossRef]
58. Kokubo, A.; Kuraishi, S.; Sakurai, N. Culm strength of barley 1: Correlation among maximum bending stress, cell wall dimensions, and cellulose content. *Plant Physiol.* **1989**, *91*, 876–882. [CrossRef]
59. Ye, H.; Zhang, Y.; Yu, Z.; Mu, J. Effects of cellulose, hemicellulose, and lignin on the morphology and mechanical properties of metakaolin-based geopolymer. *Constr. Build. Mater.* **2018**, *173*, 10–16. [CrossRef]
60. Kirk, T.K.; Farrell, R.L. Enzymatic "combustion": The microbial degradation of lignin1, 2. *Ann. Rev. Micro.* **1987**, *41*, 465–505. [CrossRef]
61. Irbe, I.; Loris, G.D.; Filipova, I.; Andze, L.; Skute, M. Characterization of self-growing biomaterials made of fungal mycelium and various lignocellulose-containing ingredients. *Materials* **2022**, *15*, e7608. [CrossRef]
62. Mbabali, H.; Lubwama, M.; Yiga, V.A.; Were, E.; Kasedde, H. Development of rice husk and sawdust mycelium-based bio-composites: Optimization of mechanical, physical and thermal properties. *J. Inst. Eng. India Ser. D* **2023**, *105*, 97–117. [CrossRef]
63. Aiduang, W.; Suwannarach, N.; Kumla, J.; Thamjaree, W.; Lumyong, S. Valorization of agricultural waste to produce myco-composite materials from mushroom mycelia and their physical properties. *Agr. Nat. Resour.* **2020**, *56*, 1083–1090.
64. Butu, A.; Rodino, S.; Miu, B.; Butu, M. Mycelium-based materials for the ecodesign of bioeconomy. *Dig. J. Nanomater. Biostruct.* **2020**, *15*, 1129–1140. [CrossRef]
65. Ascensão, G.; Beersaerts, G.; Marchi, M.; Segata, M.; Faleschini, F.; Pontikes, Y. Shrinkage and mitigation strategies to improve the dimensional stability of CaO-FeO_x-Al_2O_3-SiO_2 inorganic polymers. *Materials* **2019**, *12*, e3679. [CrossRef] [PubMed]
66. Holt, G.A.; McIntyre, G.; Flagg, D.; Bayer, E.; Wanjura, J.D.; Pelletier, M.G. Fungal mycelium and cotton plant materials in the manufacture of biodegradable molded packaging material: Evaluation study of select blends of cotton byproducts. *J. Biobased Mater. Bioenergy* **2012**, *6*, 431–439. [CrossRef]
67. Houette, T.; Maurer, C.; Niewiarowski, R.; Gruber, P. Growth and mechanical characterization of mycelium-based composites towards future bioremediation and food production in the material manufacturing cycle. *Biomimetics* **2022**, *7*, e103. [CrossRef]
68. Rigobello, A.; Ayres, P. Compressive behaviour of anisotropic mycelium-based composites. *Sci. Rep.* **2022**, *12*, e6846. [CrossRef] [PubMed]
69. Angelova, G.; Brazkova, M.; Goranov, B. Effect of the lignocellulose substrate type on mycelium growth and biocomposite formation by *Ganoderma lucidum* GA3P. *Food Sci. Appl. Biotechnol.* **2022**, *5*, 211–218. [CrossRef]
70. Agustiany, E.A.; Ridho, M.R.; Rahmi, M.D.N.; Madyaratri, E.W.; Falah, F.; Lubis, M.A.R.; Solihat, N.N.; Syamani, F.A.; Karungamye, P.; Sohail, A.; et al. Recent developments in lignin modification and its application in lignin-based green composites: A review. *Polym. Comp.* **2022**, *8*, 4848–4865. [CrossRef]
71. Chuayplod, P.; Aht-Ong, D. A study of microcrystalline cellulose prepared from parawood (*Hevea brasiliensis*) sawdust waste using different acid types. *J. Met. Mater. Miner.* **2018**, *28*, 106–114.
72. Răut, I.; Călin, M.; Vuluga, Z.; Oancea, F.; Paceagiu, J.; Radu, N.; Doni, M.; Alexandrescu, E.; Purcar, V.; Gurban, A.-M.; et al. Fungal based biopolymer composites for construction materials. *Materials* **2021**, *14*, e2906. [CrossRef]
73. Dias, P.P.; Jayasinghe, L.B.; Waldmann, D. Investigation of mycelium-Miscanthus composites as building insulation material. *Results Mater.* **2021**, *10*, e100189. [CrossRef]
74. Vette, J.F.; Böttger, W.O.J. Properties of mycelium-composite biobased insulation materials with local organic waste streams as substrate and after use for mushroom cultivation. RILEM Proceedings Pro 131. In Proceedings of the 3rd International Conference on Bio-Based Building Materials, Belfast, UK, 26–28 June 2019; pp. 649–650.
75. Rao, S.E.; Ray, L.; Khan, T.; Ravi, G. Thermal conductivity, density and porosity of sedimentary and metamorphic rocks from the Lower and Higher Himalaya, Western Himalaya. *India. Geo. J. Inter.* **2022**, *231*, 459–473. [CrossRef]

76. Liu, H.; Zhao, X. Thermal Conductivity Analysis of High Porosity Structures with Open and Closed Pores. *Int. J. Heat Mass Transf.* **2022**, *183*, e122089. [CrossRef]
77. Olivero, E.; Gawronska, E.; Manimuda, P.; Jivani, D.; Chaggan, F.Z.; Corey, Z.; de Almeida, T.S.; Kaplan-Bie, J.; McIntyre, G.; Wodo, O.; et al. Gradient porous structures of mycelium: A quantitative structure–mechanical property analysis. *Sci. Rep.* **2023**, *13*, e19285. [CrossRef] [PubMed]
78. Cao, L.; Fu, Q.; Si, Y.; Ding, B.; Yu, J. Porous materials for sound absorption. *Compos. Commun.* **2018**, *10*, 25–35. [CrossRef]
79. Kalauni, K.; Pawar, S.J. A review on the taxonomy, factors associated with sound absorption and theoretical modeling of porous sound absorbing materials. *J. Porous Mater.* **2019**, *26*, 1795–1819. [CrossRef]
80. Seddeq, H.S. Factors influencing acoustic performance of sound absorptive materials. *Aust. J. Basic Appl. Sci.* **2009**, *3*, 4610–4617.
81. Walter, N.; Gürsoy, B. Simulating Acoustic Performance of Mycelium-Based Sound Absorption Panels. In Proceedings of the 41st Conference on Education and Research in Computer Aided Architectural Design in Europe, eCAADe 2023, Education and research in Computer Aided Architectural Design in Europe, Graz, Austria, 20–23 September 2023; pp. 277–286.
82. Castagnède, B.; Aknine, A.; Brouard, B.; Tarnow, V. Effects of compression on the sound absorption of fibrous materials. *Appl. Acoust.* **2000**, *61*, 173–182. [CrossRef]
83. Barta, D.-G.; Simion, I.; Tiuc, A.-E.; Vasile, O. Mycelium-based composites as a sustainable solution for waste management and circular economy. *Materials* **2024**, *17*, e404. [CrossRef]
84. Hsu, T.; Dessi-Olive, J. A design framework for absorption and diffusion panels with sustainable materials. In Proceedings of the 2021 Inter-Noise Conference, Washington, DC, USA, 1–4 August 2021.
85. Cai, J.; Han, J.; Ge, F.; Lin, Y.; Pan, J.; Ren, A. Development of impact-resistant mycelium-based composites (MBCs) with agricultural waste straws. *Constr. Build. Mater.* **2023**, *389*, e131730. [CrossRef]
86. Ayrilmis, N.; Buyuksari, U.; As, N. Bending strength and modulus of elasticity of wood-based panels at cold and moderate temperatures. *Cold Reg. Sci. Technol.* **2010**, *63*, 40–43. [CrossRef]
87. Ayrilmis, N.; Winandy, J.E. Effects of post heat-treatment on surface characteristics and adhesive bonding performance of medium density fiberboard. *Mater. Manuf. Process.* **2009**, *24*, 594–599. [CrossRef]
88. Ayrilmis, N.; Jarusombuti, S.; Fueangvivat, V.; Bauchongkol, P. Effects of thermal treatment of rubberwood fibres on physical and mechanical properties of medium density fibreboard. *J. Trop. For. Sci.* **2011**, *23*, 10–16.
89. Suchsland, O.; Woodson, G.E. *Fiberboard Manufacturing Practices in the United States*; United States Department of Agriculture, Forest Service: Washington, DC, USA, 1987.
90. Sonderegger, W.; Niemz, P. Thermal conductivity and water vapour transmission properties of wood-based materials. *Eur. J. Wood Prod.* **2009**, *67*, 313–321. [CrossRef]
91. Nemli, G.; Çolakoğlu, G. Effects of mimosa bark usage on some properties of particleboard. *Turk. J. Agric. For.* **2005**, *29*, 227–230.
92. *BS EN 622-4:2009*; Fibreboards, Specifications, Requirements for Softboards. British Standards Institution (BSI): London, UK, 2009.
93. Papadopoulos, A.N.; Hill, C.A.S. The sorption of water vapour by anhydride modified softwood. *Wood Sci. Technol.* **2003**, *37*, 221–231. [CrossRef]
94. Ganev, S.; Cloutier, A.; Beauregard, R.; Gendron, G. Linear expansion and thickness swell of MDF as a function of panel density and sorption state. *Wood Fiber Sci.* **2005**, *37*, 327–336.
95. Berardi, U.; Iannace, G. Acoustic characterization of natural fibers for sound absorption applications. *Build. Environ.* **2015**, *94*, 840–852. [CrossRef]
96. Cramer, S.M.; Friday, O.M.; White, R.H.; Sriprutkiat, G. Mechanical properties of gypsum board at elevated temperatures. In Proceedings of the Fire and Materials 2003: 8th International Conference, San Francisco, CA, USA, 27–28 January 2003; pp. 33–42.
97. Bénichou, N.; Sultan, M.A. Thermal properties of lightweight-framed construction components at elevated temperatures. *Fire Mater.* **2005**, *29*, 165–179. [CrossRef]
98. Karni, J.; Karni, E.Y. Gypsum in construction: Origin and properties. *Mater. Struct.* **1995**, *28*, 92–100. [CrossRef]
99. Wakili, K.G.; Tanner, C. U-value of a dried wall made of perforated porous clay bricks: Hot box measurement versus numerical analysis. *Energy Build.* **2003**, *35*, 675–680. [CrossRef]
100. Camino, G.; Maffezzoli, A.; Braglia, M.; de Lazzaro, D.; Zammarano, M. Effect of hydroxides and hydroxycarbonate structure on fire retardant effectiveness and mechanical properties in ethylene-vinyl acetate copolymer. *Polym. Degrad. Stab.* **2001**, *74*, 457–464. [CrossRef]
101. Feng, C.; Janssen, H.; Wu, C.; Feng, Y.; Meng, Q. Validating various measures to accelerate the static gravimetric sorption isotherm determination. *Build. Environ.* **2013**, *69*, 64–71. [CrossRef]
102. Warnock, A.C.C. Controlling sound transmission through concrete block walls. *J. Acoust. Soc. Am.* **1998**, *103*, 41–47.
103. Ardanuy, M.; Claramunt, J.; Filho, R.D.T. Cellulosic fiber reinforced cement-based composites: A review of recent research. *Constr. Build. Mater.* **2015**, *79*, 115–128. [CrossRef]
104. Mohr, B.J.; Nanko, H.; Kurtis, K.E. Durability of kraft pulp fiber–cement composites to wet/dry cycling. *Cem. Concr. Compos.* **2005**, *27*, 435–448. [CrossRef]
105. Akers, S.A.S. Cracking in fibre cement products. *Constr. Build. Mater.* **2010**, *4*, 202–207. [CrossRef]
106. Biot, M.A.; Tolstoy, I. Formulation of wave propagation in infinite media by normal coordinates with an application to diffraction. *J. Acoust. Soc. Am.* **1957**, *29*, 381–391. [CrossRef]

107. Jelle, B.P. Traditional, state-of-the-art and future thermal building insulation materials and solutions–Properties, requirements and possibilities. *Energy Build.* **2011**, *43*, 2549–2563. [CrossRef]
108. Papadopoulos, A.M. State of the art in thermal insulation materials and aims for future developments. *Energy Build.* **2005**, *37*, 77–86. [CrossRef]
109. Schiavoni, S.; Alessandro, F.D.; Bianchi, F.; Asdrubali, F. Insulation materials for the building sector: A review and comparative analysis. *Renew. Sustain. Energy Rev.* **2016**, *62*, 988–1011. [CrossRef]
110. Asdrubali, F.; Alessandro, F.D.; Schiavoni, S. A review of unconventional sustainable building insulation materials. *Sustain. Mater. Technol.* **2015**, *4*, 1–17. [CrossRef]

Journal of *Fungi*

MDPI

Article

Effects of Light on the Fruiting Body Color and Differentially Expressed Genes in *Flammulina velutipes*

Ji-Hoon Im [1,2], Che-Hwon Park [3], Ju-Hyeon Shin [3], Youn-Lee Oh [2], Minji Oh [2], Nam-Chon Paek [1,*] and Young-Jin Park [3,*]

[1] Department of Agriculture, Forestry and Bioresources, College of Agriculture and Life Sciences, Seoul National University, Seoul 08826, Republic of Korea; jihooni24@korea.kr
[2] Mushroom Research Division, National Institute of Horticultural and Herbal Science, Rural Development Administration, Eumseong-gun 27709, Republic of Korea; o5ne2@korea.kr (Y.-L.O.); minji1228@korea.kr (M.O.)
[3] Department of Medicinal Biosciences, Research Institute for Biomedicinal & Health Science, College of Biomedicinal and Health Science, Konkuk University, Chungju 27478, Republic of Korea; chehwon9798@kku.ac.kr (C.-H.P.); shin99@kku.ac.kr (J.-H.S.)
* Correspondence: ncpaek@snu.ac.kr (N.-C.P.); yjpark@kku.ac.kr (Y.-J.P.); Tel.: +82-43-840-3601 (Y.-J.P.)

Abstract: Light plays vital roles in fungal growth, development, reproduction, and pigmentation. In *Flammulina velutipes*, the color of the fruiting body exhibits distinct changes in response to light; however, the underlying molecular mechanisms remain unknown. Therefore, in this study, we aimed to analyze the *F. velutipes* transcriptome under red, green, and blue light-emitting diode (LED) lights to identify the key genes affecting the light response and fruiting body color in this fungus. Additionally, we conducted protein–protein interaction (PPI) network analysis of the previously reported fruiting body color-related gene, *Fvpal1*, to identify the hub genes. Phenotypic analysis revealed that fruiting bodies exposed to green and blue lights were darker than those untreated or exposed to red light, with the color intensifying more after 48 h of exposure to blue light compared to that after 24 h of exposure. Differentially expressed gene (DEG) analyses of all light treatments for 24 h revealed that the numbers of DEGs were 17, 74, and 257 under red, green, and blue lights, respectively. Subsequently, functional enrichment analysis was conducted of the DEGs identified under green and blue lights, which influenced the color of *F. velutipes*. In total, 103 of 168 downregulated DEGs under blue and green lights were included in the enrichment analysis. Among the DEGs enriched under both green and blue light treatments, four genes were related to monooxygenases, with three genes annotated as cytochrome P450s that are crucial for various metabolic processes in fungi. PPI network analysis of Fvpal1 revealed associations with 11 genes, among which the expression of one gene, pyridoxal-dependent decarboxylase, was upregulated in *F. velutipes* exposed to blue light. These findings contribute to our understanding of the molecular mechanisms involved in the fruiting body color changes in response to light and offer potential molecular markers for further exploration of light-mediated regulatory pathways.

Keywords: *Flammulina velutipes*; light; fruiting body color; transcriptome; cytochrome P450; pyridoxal-dependent decarboxylases

1. Introduction

The edible fungus *Flammulina velutipes* exhibits morphological diversity with taxonomic classification under division Basidiomycota, class Agaricomycetes, order Agaricales, family *Physalacriaceae*, and genus *Flammulina*. Although its commercial varieties typically exhibit white coloration, wild species with yellow-to-dark brown coloration are commonly found in nature. Moreover, commercial varieties have slender and elongated stipes and small pilei, whereas wild species exhibit various shapes and sizes for both stipes and pilei, with considerable morphological variation [1]. These mushrooms are commonly found on decaying old trees or remnants of various broadleaf trees from late autumn until the subsequent spring. This particular mushroom species exhibits a remarkable ability to withstand cold temperatures, allowing it to flourish even during the harsh winter months, and thus it is also known as "Winter mushroom." The resilience displayed by *F. velutipes* in surviving adverse environmental conditions further highlights its adaptive nature and ecological significance in its natural habitat.

Fungi, as dependent heterotrophs, have evolved morphologically to respond to various environmental factors, such as soil composition and climate; generate fruiting bodies of mushrooms; and effectively disperse spores at the most suitable locations and times. Environmental factors that influence the form and coloration of mushrooms include temperature, humidity, light, and CO_2 levels. Light, in particular, is a crucial determinant that influences various facets of fungal biology, including the growth and development of mycelia and fruiting bodies [2–4]. The mycelial growth of *Auricularia heimuer* was strong under a red light to blue light ratio of 1:1 [5], but the mycelia of most fungi thrive under low light or even completely dark conditions. In *Lentinula edodes*, the cultivation medium did not produce a brown film under 80 days of darkness, whereas it did form a brown film after 30 days of darkness followed by 50 days of light exposure [6]. Blue light significantly promoted the growth and development of the stipe, pileus, and gill of *Pleurotus ostreatus*, especially the pileus, while red light had a minor inhibitory effect on pileus growth [7]. In addition, the growth response of *Hypsizygus marmoreus* varied depending on the light treatment during cultivation, primordia formation, and growth stages [8]. It also enhances fruiting body uniformity, yield, and biological efficiency in *F. filiformis* and increases the primordia and fruiting body yields in *H. marmoreus* [8,9].

These photomorphogenesis phenomena, wherein light serves as a key environmental cue that regulates growth and development, affect mushroom pigmentation. Understanding the intricate interplay between light exposure and mushroom pigmentation is necessary to explore their ecology, physiology, and potential applications [10–12]. Light intensity, quality, duration, and directionality strongly influence the expression of genes involved in pigment biosynthesis pathways [7,13,14]. For example, ultraviolet (UV) light exposure stimulates the production of melanin-like pigments in some mushroom species as a protective response to UV radiation [15,16]. Spectral composition of light modulates the synthesis of specific pigments in mushrooms. Different wavelengths of light, ranging from blue to red, elicit differential responses for metabolite production and accumulation [14,17,18]. For example, in *P. ostreatus*, blue light leads to an accumulation of the aromatic amino acid shikimic acid due to the increased activity of 3-deoxy-D-arabinoheptulosonate 7-phosphate synthase (DAHPS), phosphofructokinase (PFK), and glucose-6-phosphate dehydrogenase (G6PD) [17]. In *L. edodes*, blue light induces changes in the expression of genes related to morphological development and pigment production in fruiting bodies [13]. Additionally, in *Terana caerulea*, exposure to blue light triggers the production of inkblue pentacyclic natural products, known as corticin pigments [14]. Therefore, influence of light on mushroom pigmentation extends beyond coloration, encompassing ecological interactions and adaptive strategies. In natural habitats, mushrooms exhibit phenotypic plasticity in response to varying light conditions, with coloration patterns tailored to optimize their survival and fitness [19]. Mushrooms inhabiting shaded environments exhibit lighter pigmentation to maximize light absorption, whereas those exposed to direct sunlight exhibit darker pigmentation for UV protection and heat dissipation. Moreover, effects of environmental factors on mushroom pigmentation extend beyond natural ecosystems to include various agricultural and industrial applications [11].

Advancements in genomic, transcriptomic, and proteomic technologies have facilitated elucidation of the molecular basis of light-induced pigmentation in mushrooms. High-throughput sequencing, gene expression profiling, and functional analyses have enabled the identification and characterization of key genes and regulatory elements governing pigment biosynthesis pathways [7,12–14,20,21]. For example, in *P. ostreatus*, blue light increased the activation of glycolysis and the pentose phosphate pathway, including enzymes encoding genes like 6-phosphogluconate dehydrogenase (6PGD) and phosphoenolpyruvate carboxykinase (PEPCK) [7]. In *Terana caerulea*, exposure to blue light resulted in significantly higher transcription of the gene encoding the gateway enzyme, polyporic acid synthetase CorA, which catalyzes the formation of the pigment core structure [14]. In *F. velutipes*, the gene *FvbHLH1* exhibited unique expression specifically in the yellow cap, suggesting its potential regulatory role in phenolic acid biosynthesis [22]. And isobavachalcone D, a yellow compound, and the control of riboflavin transportation by MCH5 play crucial roles in the development of the yellow cap in *P. citrinopileatus* [23]. Previous studies have explored mushroom responses to light and investigated color-related aspects; however, the molecular mechanisms underlying the relationship between light and mycelial color remain unclear. As the light response mechanisms of mushrooms vary among species [24], the functional significance of these mechanisms needs to be assessed in different conditions. Therefore, in this study, we performed transcriptome analysis of *F. velutipes* irradiated using red, green, and blue light-emitting diodes (LEDs) to identify the key genes associated with light response and fruiting body color. Additionally, we conducted protein–protein interaction (PPI) network analysis of a previously identified fruiting body color-related gene, *Fvpal1* [25], to identify the hub genes. The light-responsive and fruiting body color-related genes identified in this study provide new insights into the light response mechanisms of fungi and may be used as selective markers for color changes in *F. velutipes*.

2. Materials and Methods

2.1. Fungal Strains, Fruiting Body Cultivation, and Light Treatment

F. velutipes strain ASI 4232 was obtained from the Mushroom Division of the Rural Development Administration (Eumsung-gun, Republic of Korea) and cultured on potato dextrose agar (Difco, Seoul, Republic of Korea) at 25 °C for 20 d. Then, *F. velutipes* mycelia were inoculated into a fruiting medium (80% sawdust and 20% rice bran) and incubated at 18 °C under 65% humidity for 35 d. After incubation, the mycelia were scraped from the surface of the bottles, and primordia formed at 14 °C under 90–95% humidity were grown at 6–7 °C under 80–85% humidity for fruiting. They were exposed to red (639 nm), green (522 nm), and blue (470 nm) LEDs at a distance of 20 cm with the same light intensity of 500 lx for 24 and 48 h immediately before harvesting. Each treatment was applied to four bottles containing 350 g of fruiting bodies per bottle. Colorimetric analysis was conducted using a compact portable colorimeter, the Minolta CR-400 (Konica Minolta, Osaka, Japan), for all bottles to quantify the color of the samples.

2.2. Transcriptome Analysis and Differentially Expressed Gene (DEG) Identification

For transcriptome analysis, fruiting bodies harvested from three bottles per each of the four light treatments were ground in liquid nitrogen. Total RNA was isolated using the TRIzol reagent (Thermo Fisher Scientific Korea, Seoul, Republic of Korea) and purified using the RNeasy kit (Qiagen Korea, Seoul, Republic of Korea) with RNase-free DNase (Qiagen Korea), according to the manufacturers' instructions. Purified RNA (1 μg) was sequenced on the HiSeq 2000 platform (Illumina Korea, Seoul, Republic of Korea). Quality-trimmed short reads (Phred quality score: 30) were processed using the Trinity pipeline (version 2.15.0) [26]. Additionally, TransDecoder (version 5.5.0) [27] was used for gene modeling, CD-HIT (version 4.8.1) [28] for clustering and deduplication, and DESeq2 (in Trinity pipeline) [26] for the identification of DEGs ($|\log2FC| \geq 1$ and $p < 0.001$). All next-generation sequencing reads have been deposited into the National Center for Biotechnology Information Sequence Read Archive (SRA) (PRJNA1107629).

2.3. Functional Enrichment and PPI Network Analysis of DEGs

All coding sequences (CDSs) predicted via Trinity analysis were annotated using the protein families (Pfam) [29], InterPro [30], and Kyoto Encyclopedia of Genes and Genomes (KEGG) [31] databases. The criterion for statistical significance was set as $p < 0.001$ for all database searches. The predicted CDSs and DEGs were used for functional enrichment and PPI analyses using the Search Tool for the Retrieval of Interacting Genes/Proteins (STRING) database [32]. PPI hub genes were identified using the cytoHubba application in Cytoscape software (version 3.10.1) [33].

2.4. Quantatitive Polymerase Chain Reaction (PCR) of DEGs

Next, qPCR was performed to evaluate the expression levels of DEGs. cDNA was synthesized by genomic DNA elimination and reverse-transcription reaction using the QuantiTect-Reverse Transcription kit (Qiagen) according to the manufacturer's protocol. After genomic DNA elimination reaction with gDNA wipeout buffer at 42 °C for 2 min, the first-strand synthesis was conducted at 42 °C for 15 min, followed by incubation at 95 °C for 5 min to inactivate the reverse transcriptase. The synthesized cDNA (200 ng) and 1 μM of each primer (Table S1) were mixed with the QuantiTect SYBR green PCR kit (Qiagen) and amplified using the Rotor-Gene Q instrument (Qiagen). Relative quantification of gene expression was performed using the $^{\Delta\Delta}$Ct method.

3. Results

3.1. Effect of Light on the Fruitng Body Color

Here, three different LED lights (red, green, and blue) were used to irradiate the *F. velutipes* fruiting bodies immediately before harvest. As shown in Figure 1, colors of fruiting bodies exposed to green and blue lights were darker than those of the unexposed and red light-exposed fruiting bodies. Moreover, under blue light treatment, color of the fruiting body was darker after 48 h than after 24 h (Figure 1b). Notably, red light treatment did not affect the fruiting body color, which was similar to the color of the untreated fruiting bodies (Figure 1c,d). After determining the light treatment affected the fruiting body color, we performed transcriptome analysis to identify the specific genes involved in the response to light treatment.

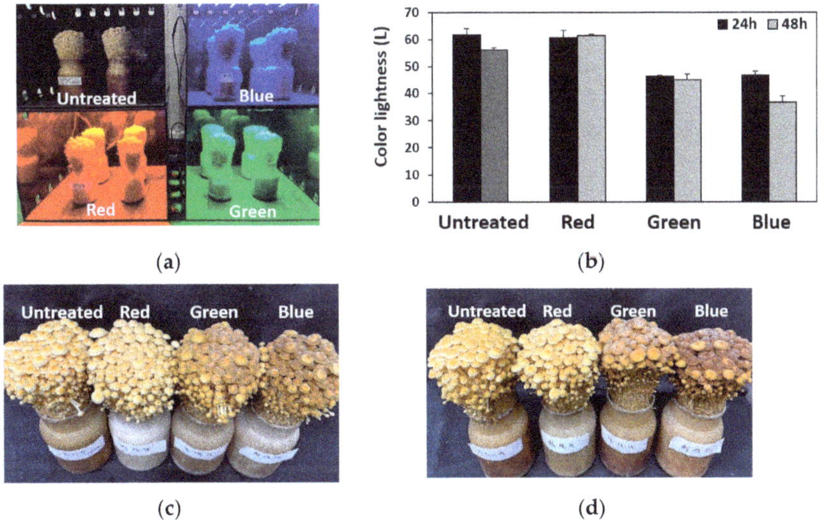

Figure 1. Light treatment of *Flammulina velutipes* fruiting bodies. (**a**) Light-emitting diode (LED) treatment. (**b**) Colors of the light-treated fruiting bodies (**c**) 24 h and (**d**) 48 h after light treatment.

3.2. DEG Identification

Next, DEGs were identified to assess the response of genes to light treatment as well as the resulting changes in the fruiting body color of *F. velutipes*. Figure 2 shows the numbers of DEGs after red, green, and blue light treatments compared to those in untreated *F. velutipes* using $|\log2FC| \geq 1$ and $p < 0.05$ as the threshold (Table S2). DESeq2 analysis revealed 17, 74, and 257 DEGs in the red-, green-, and blue-light-treated *F. velutipes*, respectively. Specifically, two DEGs (TRINITY_DN3502_c0_g1_i6 and TRINITY_DN363_c0_g1_i12) were detected in all light treatments, and their levels were upregulated in red-, green-, and blue-light-treated *F. velutipes* (Figure 2; Table S2). Of these, TRINITY_DN3502_c0_g1_i6 was annotated as a fungus-specific transcription factor (TF) domain (PF04082.22) via the Pfam database search. As shown in Figure 2b, the highest number of DEGs was observed in blue-light-treated *F. velutipes*. A relatively small number of DEGs was identified in red-light-treated *F. velutipes*.

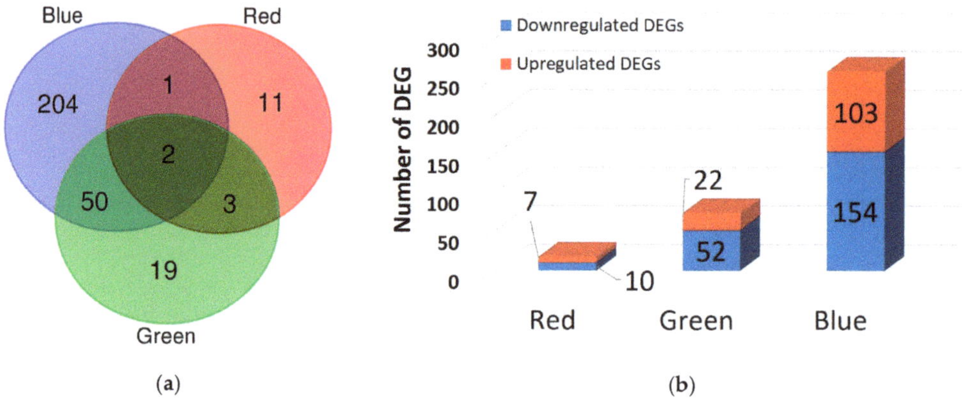

Figure 2. Venn diagram (**a**) and number (**b**) of differentially expressed genes (DEGs) identified after the red, green, and blue light treatments of *F. velutipes*.

3.3. Functional Enrichment of DEGs

In total, 290 genes were identified as DEGs in light-treated *F. velutipes*, and 273 DEGs were identified in samples treated with only blue and green lights, but not in those treated with red light. Figure 3a shows a heatmap of the 105 upregulated and 168 downregulated genes among the identified 273 DEGs. Then, DEG enrichment analysis was conducted using the STRING database. Among the 273 DEGs, only 168 downregulated DEGs in blue- and green-light-treated *F. velutipes* were enriched and classified using the Gene Ontology (GO) molecular function (MF), STRING cluster, and UniProt keywords (Figure 3b; Table S3). These downregulated DEGs were significantly enriched in three categories of the STRING database: monooxygenase activity (GO:0004497) in MF, alkaloid metabolic process in the STRING cluster, and monooxygenase in UniProt keywords. Enrichment analysis revealed that the downregulated DEGs were enriched in monooxygenase activity in all three categories (Table S3).

A total of 103 of the 168 downregulated DEGs in blue- and green-light-treated *F. velutipes* were included in the enrichment analysis (Table S4). Among these, 30, 31, and 79 genes were annotated using the KEGG, InterPro, and Pfam databases, respectively. In addition, 4 out of 16 monooxygenase-related genes were commonly identified as enriched DEGs in both blue- and green-light-treated *F. velutipes* (Tables 1 and S4). Among these four genes, three were annotated as cytochrome P450 (CYP) enzymes using the Pfam database.

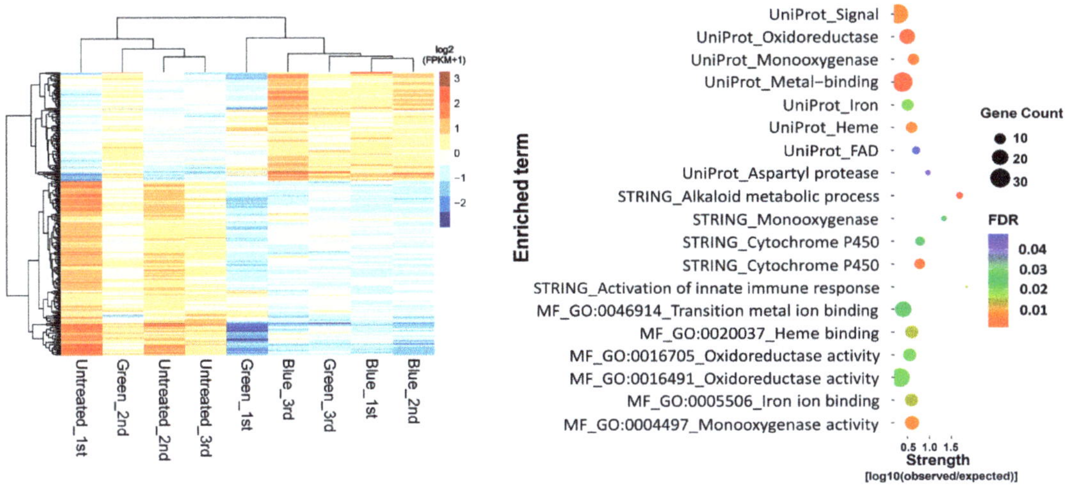

(**a**) (**b**)

Figure 3. Heat map (**a**) and significantly enriched terms (**b**) of 168 downregulated DEGs in blue- and green-light-treated *F. velutipes*.

Table 1. Significantly enriched differentially expressed genes (DEGs) in blue- and green-light-treated *Flammulina velutipes*.

Gene ID	Treated light	log2FC	FDR [1]	Pfam Database ($p < 0.001$)		Family
				ID	Description	
TRINITY_DN1447_c0_g1_i1	Blue	−2.0926	0.0000	PF00067.26	Cytochrome P450	CYP620
	Green	−1.5818	0.0002			
TRINITY_DN6930_c0_g1_i17	Blue	−1.3246	0.0000	PF00067.26	Cytochrome P450	CYP53
	Green	−1.2260	0.0015			
TRINITY_DN5360_c0_g1_i15	Blue	−1.5705	0.0000	PF00067.26	Cytochrome P450	CYP620
	Green	−1.4023	0.0212			
TRINITY_DN5015_c0_g3_i1	Blue	−2.2012	0.0001	-	-	
	Green	−2.2610	0.0158			

[1] False discovery rate (<0.05).

3.4. Expression Patterns and PPI Networks of the F. velutipes Fruiting Body Color-Related Genes

We previously identified a gene related to the fruiting body color of *F. velutipes* [25]. Comparative genomic analysis of different *F. velutipes* strains revealed 70 white strain-specific variations, including single-nucleotide polymorphisms and indels, with one mutation causing a deletion in the phenylalanine ammonia-lyase 1 (*Fvpal1*; EC 4.3.1.24) gene. This mutation is color-specific for white strains and plays an important role in determining the color of the fruiting body. Here, *Fvpal1* was identified as an upregulated DEG (log2FC 1.27) only in blue-light-treated *F. velutipes* (Table S2). Further analysis, including enrichment and PPI network analyses of *Fvpal1* (TRINI-TY_DN7330_c0_g3_i11), was performed using the STRING database and Cytoscape software. As shown in Figure 4a, 11 nodes and 44 edges were obtained with scores > 0.4. CytoHubba application of Cytoscape software was used to identify the hub genes in the PPI network (Figure 4b). All identified hub genes with closeness and degree of connectivity are listed in Tables 2 and S5.

Among the identified genes, only one gene (TRINITY_DN5724_c0_g1_i3) acted as an upregulated DEG in blue-light-treated *F. velutipes*. This gene was annotated to have a pyridoxal-dependent decarboxylase-conserved domain using the Pfam database.

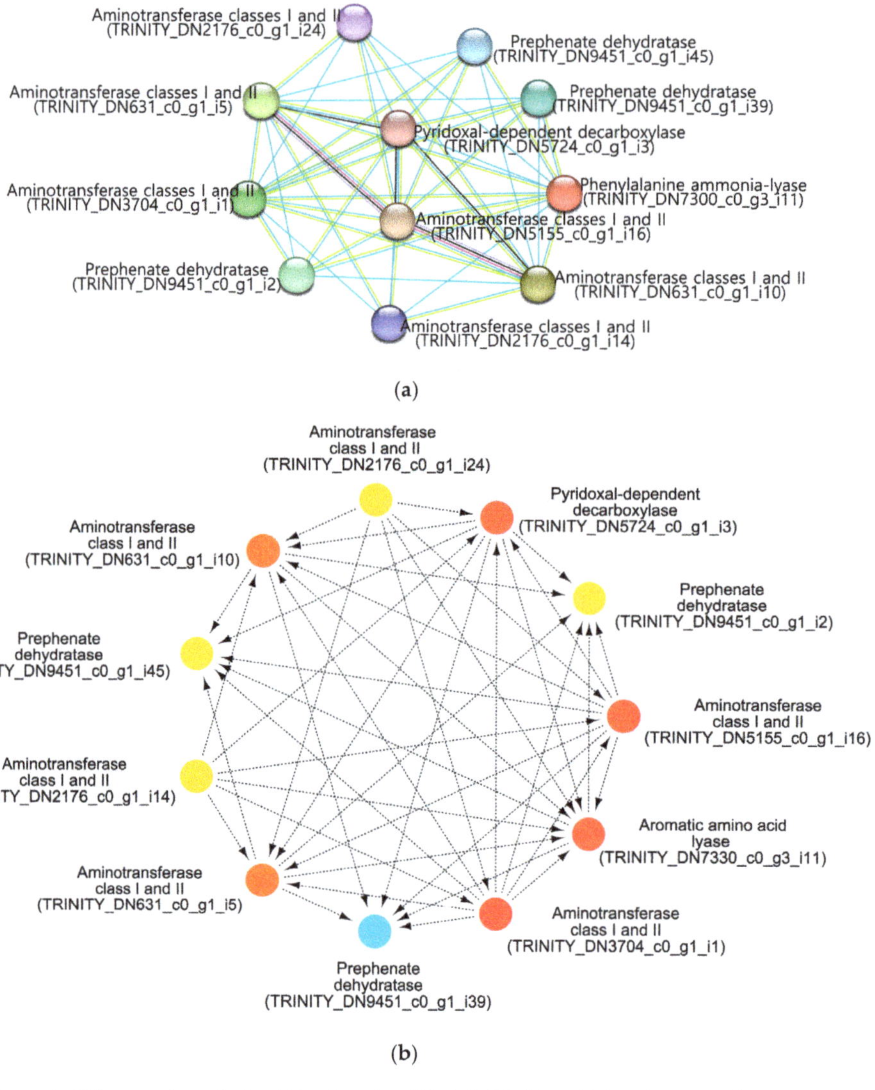

Figure 4. PPI network analysis of Fvpal1 using STRING DB (**a**) and identification of hub genes associated with *Fvpal1* by CytoHubba application; arrows indicate interactions (**b**).

Table 2. Hub genes in the *Fvpal1* protein–protein interaction (PPI) network identified using cytoHubba in Cytoscape software.

Gene ID	Closeness	Degree	Pfam Database		KEGG [1] Database	
			ID	Description	ID	Description
TRINITY_DN3704_c0_g1_i1	10	10	PF00155.25	Aminotransferase classes I and II	K00817	Histidinol-phosphate aminotransferase [EC:2.6.1.9]
TRINITY_DN5155_c0_g1_i16	10	10	PF00155.25		K14455	Aspartate aminotransferase, mitochondrial [EC:2.6.1.1]
TRINITY_DN5724_c0_g1_i3	10	10	PF00282.23	Pyridoxal-dependent decarboxylase conserved domain	-	-
TRINITY_DN631_c0_g1_i10	9.5	9	PF00155.25	Aminotransferase classes I and II	K14454	Aspartate aminotransferase, mitochondrial [EC:2.6.1.1]
TRINITY_DN631_c0_g1_i5	9.5	9	PF00155.25		K14454	
TRINITY_DN2176_c0_g1_i14	8	6	PF00155.25		K00838	Aromatic amino acid aminotransferase I/2-aminoadipate transaminase [EC:2.6.1.57 2.6.1.39 2.6.1.27 2.6.1.5]
TRINITY_DN2176_c0_g1_i24	8	6	PF00155.25		K00838	
TRINITY_DN9451_c0_g1_i2	8	6	PF00800.22	Prephenate dehydratase		-
TRINITY_DN9451_c0_g1_i39	8	6	PF00800.22			
TRINITY_DN9451_c0_g1_i45	8	6	PF00800.22			

[1] Kyoto Encyclopedia of Genes and Genomes.

3.5. qPCR Validation of DEGs

The DEGs identified from the fruiting bodies of *F. velutipes* exposed to light were analyzed using quantitative reverse transcription-PCR (qRT-PCR). One upregulated gene (TRINITY_DN3502_c0_g1_i6) under all light conditions (Figure 5a), four downregulated genes (TRINITY_DN1447_c0_g1_i1, TRINITY_DN6930_c0_g1_i7, TRINITY_DN5360_c0_g1_i15, and TRINITY_DN5015_c0_g3_i1) enriched only in blue- and green-light-treated *F. velutipes* via functional enrichment analysis (Figure 5b–e), and one upregulated gene (TRINITY_DN5724_c0_g1_i3) only under blue light (Figure 5f) via PPI network analysis of *Fvpal1* were used for validation of the transcriptome data. These genes showed similar expression patterns in qPCR and transcriptome analysis results (Figure 5). These results confirmed the reliability of transcriptome data and suggested the potential of the identified DEGs, responsive to light and associated with the color of fruiting bodies in *F. velutipes*, for further exploration of genes of interest.

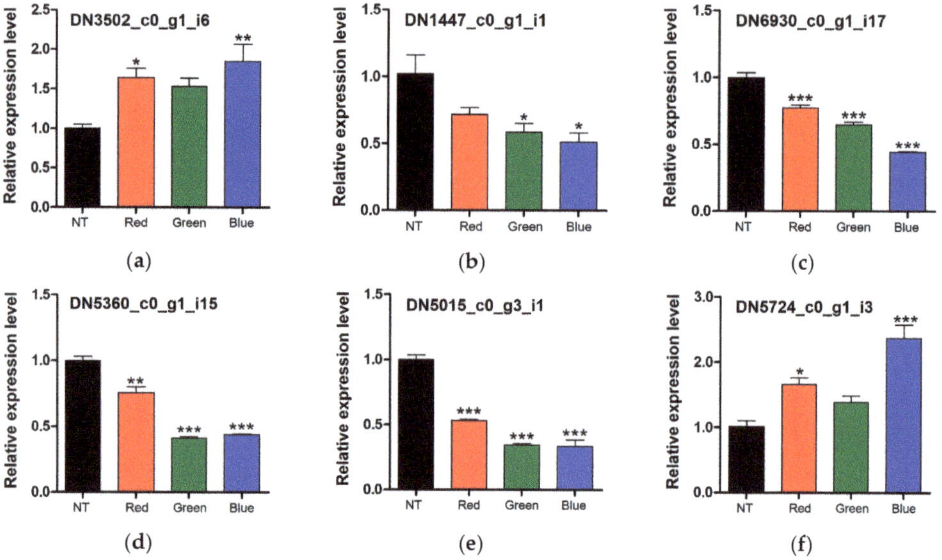

Figure 5. Relative expression levels of the upregulated genes under all light conditions (**a**), downregulated genes under blue and green lights (**b–e**), and upregulated genes under blue light (**f**). Data were analyzed using one-way analysis of variance (ANOVA) followed by Tukey's test. * $p < 0.05$, ** $p < 0.01$, and *** $p < 0.001$ vs. non-treated sample. NT, non-treated; Red, red LED treated; Green, green LED treated; Blue, blue LED treated.

4. Discussion

Fruiting body forming fungi exhibit a sophisticated ability to perceive various environmental cues, enabling them to meticulously assess the factors influencing the selection of ideal sites and timing of sexual reproduction. Additionally, these fungi possess a remarkable capacity to craft suitable fruiting bodies to ensure efficient dispersal of spores, thereby enhancing their reproductive success in the ecosystem. Light plays key roles in plant and fungal morphogenesis, primarily because of its significance in photosynthesis and morphogenesis. The effects of light on various biological systems have been extensively studied in ascomycetes [34–36]. Light perception is essential for the spatial recognition of mushroom-forming fungi, particularly in sexual reproduction. However, the correlation between light exposure and fruiting body formation in basidiomycetes remains unknown.

The necessity of light for the induction of fruiting bodies varies among basidiomycetes types. Some species can initiate fruiting body production in the absence of light, suggesting that light is not a critical factor for this process [37–39]. Nevertheless, light has the capability to trigger or enhance the production of fruiting bodies [40–43]. Wavelengths that effectively induce the formation of fruiting bodies include UV (280 nm) and blue (520 nm) lights [7,44,45].

Here, to examine the effects of different LED light sources on the morphological characteristics of *F. velutipes* fruiting bodies, *F. velutipes* was irradiated with blue, green, and red LED lights for 24 and 48 h after fruiting body formation. The color of the fruiting body darkened when exposed to both green and blue lights, but a more pronounced effect was noted with blue light. This suggests that different LED lights affect the fruiting body pigmentation. Subsequently, transcriptome analysis of fruiting bodies exposed to LED irradiation was conducted, focusing on the light-responsive genes. A total of 340 DEGs ($|\log2FC| \geq 1$ and $p < 0.05$) were identified, including 17 DEGs for red light, 74 DEGs for green light, and 257 DEGs for blue light. Two DEGs were consistently identified across all light treatments, with one of them annotated as a fungus-specific TF domain based on a

Pfam search. White collar-1 in *Neurospora crassa* acts as a fungus-specific TF domain that responds to light and serves as a blue light photoreceptor for circadian clock regulation [46]. The fungal-specific TF domain that responds to light stimuli is the GAL4-like Zn2C6 DNA-binding domain [47]. This domain is a characteristic feature of a significant TF class in fungi and plays a crucial role in regulating gene expression in response to light signals. Previous studies have shown that this domain, along with other fungus-specific domains, such as the middle homology domain, is involved in controlling various biological processes, including growth, development, and secondary metabolite production in fungi [47–49]. The interaction of these specific domains within TFs forms the basis of gene regulatory networks that orchestrate light-mediated responses in fungi, such as *N. crassa* and *Cordyceps militaris* [48,49]. Therefore, our results suggest that one of the two DEGs may act as a TF involved in the light response in *F. velutipes*.

As the color of *F. velutipes* fruiting bodies darkened when exposed to green and blue light, we performed an enrichment analysis for DEGs whose expression changed in response to the two light treatments. Among the 273 DEGs, 168 downregulated genes were enriched in the online STRING database. Among the 273 DEGs, only 168 downregulated DEGs in the blue- and green-light-treated samples were enriched in GO MF, STRING cluster, and UniProt keywords. The downregulated DEGs were enriched in monooxygenases across all three categories. Among the 168 downregulated DEGs, three were identified as CYPs and downregulated in *F. velutipes* treated with both blue and green light. CYP enzymes are important for various fungal metabolic processes, including sterol synthesis, steroid oxidation, and xenobiotic degradation. These enzymes exhibit diverse functions, including hydroxylation, dealkylation, and ring formation, and play a key role in detoxification and ecological survival [50–52]. It has been reported that the expression of CYP in blue-light-irradiated *L. edodes* increases with fruiting body development and shows consistently high expression at all growth stages [13]. The role of CYPs in pigment biosynthesis in plants is well known [53–55], but the relationship between CYPs and fruiting body color in basidiomycetes is still unclear. The biological clock controls many metabolic processes in various organisms, including microorganisms, plants, and animals. Circadian rhythms are important for physiological phenomena related to the growth and development of higher plants [56–58]. Additionally, circadian regulation by CYP monooxygenases is important for the synthesis of diverse secondary metabolites, including phenylpropanoids, carotenoids, glucosinolates, and brassinosteroids, in *Arabidopsis thaliana* [59]. Although further studies are required, these results suggest that these downregulated CYPs are probably regulated in response to light as circadian reporters but are unlikely to directly affect *F. velutipes* fruiting body color.

We previously identified a phenylalanine ammonia lyase (*Fvpal1*) gene with a mutation (ΔGCGCAC) specific to white *F. velutipes* strains [25]. Phenylalanine ammonia-lyase facilitates the deamination of L-phenylalanine into trans-cinnamic acid. It is frequently found in plants and fungi and plays an important role in various metabolic pathways and biological processes [60]. Interestingly, this gene was upregulated in *F. velutipes* treated with blue light (Table S2). In addition, as shown in Figure 1d, among the samples treated with light for 48 h, the color of the fruiting bodies treated with blue light was darker than that of the fruiting bodies treated with red or green light. Additionally, PPI network construction and hub gene identification revealed that this gene was mainly associated with aminotransferase genes (Table 2). Among them, one gene (TRINITY_DN5724_c0_g1_i3) possessing a pyridoxal-dependent decarboxylase conserved domain was upregulated in blue-light-treated *F. velutipes* (Table S2). Pyridoxal-dependent decarboxylases are crucial enzymes in metabolic pathways involved in the biosynthesis of amino acids, amino-acid-derived metabolites, and amino sugars of living organisms, utilizing pyridoxal 5′-phosphate (PLP) as a coenzyme for decarboxylation reactions [61]. Previously, the catalytic activity of this enzyme has been reported to be light-sensitive [62]. In addition, pyridoxal-dependent decarboxylase conserved domain is important for the function of enzymes, such as DOPA decarboxylase (DDC), tyrosine decarboxylase (TDC), glutamate decarboxylase (GAD), and histidine decarboxylase (HDC) [63]. DDC (EC4.1.1.26) and TDC (EC4.1.1.25) produce tyramine using the amino acid l-tyrosine as a substrate. GAD (EC4.1.1.15) catalyzes the decarboxylation of Glu to gamma-aminobutyric acid. HDC (EC 4.1.1.22) catalyzes the decarboxylation of histidine to produce histamine. Tyrosine, phenylalanine, and tryptophan are three AAAs involved in protein synthesis and are substrates for the synthesis of various secondary metabolites. AAAs are substrates for numerous anabolic pathways responsible for the synthesis of pigment compounds, plant hormones, and biological macromolecules [64]. Therefore, the pyridoxal-dependent decarboxylase conserved domain of the gene (TRINITY_DN5724_c0_g1_i3) sequence, which was identified as a hub gene by PPI network analysis and was also upregulated by blue light treatment, suggests that this gene may be involved in the metabolism of AAAs and the pigment biosynthetic pathway.

This study conducted a comprehensive bioinformatics analysis of DEGs involved in the light response of *F. velutipes*. We identified seven genes closely related to the light response and fruiting body color changes of *F. velutipes*. These genes are key candidate markers for further investigation of the molecular mechanisms underlying the light response and fruiting body color changes in this mushroom.

5. Conclusions

Here, exposure to different LED light sources, particularly blue and green lights, influenced the pigmentation of *F. velutipes* fruiting bodies. Using transcriptome analysis, we identified a set of DEGs involved in the light response of this mushroom. Among the identified 168 downregulated genes, three acted as CYP genes that are important for various metabolic processes in fungi. Circadian regulation by CYP monooxygenases affects the synthesis of secondary metabolites in various organisms. Here, light exposure led to the downregulation of CYPs as circadian reporters but did not directly affect the *F. velutipes* fruiting body color. Moreover, levels of some metabolic-pathway-related genes, such as *Fvpal1* and pyridoxal-dependent decarboxylase, were significantly upregulated under blue light. Upregulation of the expression of *Fvpal1*, an enzyme involved in phenylpropanoid metabolism, suggests its potential roles in pigment biosynthesis and fruiting body color changes in *F. velutipes* under blue light. Here, a pyridoxal-dependent decarboxylase gene associated with AAA metabolism and secondary metabolite biosynthesis was identified, further emphasizing the intricate relationship between light exposure and metabolic processes in mushroom development. These findings not only contribute to our understanding of fungal biology but also reveal new molecular markers for further exploration of light-mediated regulatory pathways in fruiting body forming fungi.

Supplementary Materials: The following supporting information can be downloaded at https://www.mdpi.com/article/10.3390/jof10060372/s1. Table S1: List of primers used in this study. Table S2: Differentially expressed genes (DEGs) in *Flammulina velutipes* treated with different light-emitting diodes (LEDs). Table S3: Enrichment of 168 downregulated DEGs in *Flammulina velutipes* treated with blue and green lights. Table S4: List of enriched genes in *Flammulina velutipes* treated with blue and green lights. Table S5: List of hub genes identified in this study.

Author Contributions: Conceptualization, J.-H.I. and Y.-J.P.; methodology, J.-H.I. and Y.-J.P.; software, Y.-J.P.; validation, J.-H.I., C.-H.P. and J.-H.S.; formal analysis, J.-H.I.; investigation, J.-H.I. and Y.-J.P.; resources, Y.-L.O. and M.O.; data curation, J.-H.I. and Y.-J.P.; writing—original draft preparation, J.-H.I. and Y.-J.P.; writing—review and editing, Y.-J.P.; visualization, Y.-J.P.; supervision, Y.-J.P. and N.-C.P.; project administration, J.-H.I.; funding acquisition, J.-H.I. and Y.-J.P. All authors have read and agreed to the published version of the manuscript.

Funding: This research was supported by the Rural Development Administration research project (project number: PJ01733103).

Institutional Review Board Statement: Not applicable.

Informed Consent Statement: Not applicable.

Data Availability Statement: Data are contained within the article and Supplementary Materials.

Conflicts of Interest: The authors declare no conflicts of interest.

References

1. Wang, P.M.; Liu, X.B.; Dai, Y.C.; Horak, E.; Steffen, K.; Yang, Z.L. Phylogeny and species delimitation of *Flammulina*: Taxonomic status of winter mushroom in East Asia and a new European species identified using an integrated approach. *Mycol. Prog.* **2018**, *17*, 1013–1030. [CrossRef]
2. Chaisaena, W. Light Effects on Fruiting Body Development of Wildtype in Comparison to Light-Insensitive Mutant Strains of the Basidiomycete *Coprinopsis cinerea*, Grazing of Mites (*Tyrophagus putrescentiae*) on the Strains and Production of Volatile Organic Compounds during Fruiting Body Development. Ph.D. Thesis, Georg-August-University, Göttingen, Germany, 2009.
3. Nmom, F.W.; Amadi, L.O.; Ngerebara, N.N. Influences of light regimes on reproduction, germination, pigmentation, pathogenesis and overall development of a variety of Filamentous Fungi—A Review. *Asian J. Biol.* **2021**, *11*, 25–34. [CrossRef]
4. Bayram, Ö.S.; Bayram, Ö. An anatomy of fungal eye: Fungal photoreceptors and signaling mechanisms. *J. Fungi* **2023**, *9*, 591. [CrossRef] [PubMed]
5. Hou, L.; Zhengpeng, L.I.; Lin, J.; Lin, M.A.; Huiping, L.I.; Shaoxuan, Q.U.; Jiang, J.; Zou, X.; Yang, H.; Changtian, L.I.; et al. Effects of different light quality of LED light source on growth rate, mycelium branch and biomass of straw mushroom mycelium. *Acta Agric. Zhejiangensis* **2021**, *33*, 1110–1116. [CrossRef]
6. Tang, L.H.; Jian, H.H.; Song, C.Y.; Bao, D.P.; Shang, X.D.; Wu, D.Q.; Tan, Q.; Zhang, X.H. Transcriptome analysis of candidate genes and signaling pathways associated with light-induced brown film formation in Lentinula edodes. *Appl. Microbiol. Biotechnol.* **2013**, *97*, 4977–4989. [CrossRef] [PubMed]
7. Wang, H.; Tong, X.; Tian, F.; Jia, C.; Li, C.; Li, Y. Transcriptomic profiling sheds light on the blue-light and red-light response of oyster mushroom (*Pleurotus ostreatus*). *AMB Express* **2020**, *10*, 10. [CrossRef] [PubMed]
8. Song, H.; Li, Y.; Huang, J.; Duan, J.; Zhang, Q.; Xie, B. Growth and development of *Hypsizygus marmoreus* and response expression of light receptor white collar genes under different light quality irradiation. *Acta Hortic. Sin.* **2020**, *47*, 467–476. [CrossRef]
9. Wang, H.; Zhao, S.; Han, Z.; Qi, Z.; Han, L.; Li, Y. Integrated transcriptome and metabolome analysis provides insights into blue light response of *Flammulina filiformis*. *AMB Express* **2024**, *14*, 21. [CrossRef]
10. Damaso, E.J.; Dulay, R.M.; Kalaw, S.; Reyes, R. Effects of color light emitting diode (LED) on the mycelial growth, fruiting body production, and antioxidant activity of *Lentinus tigrinus*. *CLSU Int. J. Sci. Technol.* **2018**, *3*, 9–16. [CrossRef]
11. Roshita, I.; Goh, S.Y. Effect of exposure to different colors light emitting diode on the yield and physical properties of grey and white oyster mushrooms. In Proceedings of the AIP Conference, Ho Chi Minh, Vietnam, 29–30 April 2018. [CrossRef]
12. Park, Y.J.; Jang, M.J. Blue light induced edible mushroom (*Lentinula edodes*) proteomic analysis. *J. Fungi* **2020**, *6*, 127. [CrossRef]
13. Kim, J.Y.; Kim, D.Y.; Park, Y.J.; Jang, M.J. Transcriptome analysis of the edible mushroom *Lentinula edodes* in response to blue light. *PLoS ONE* **2020**, *15*, e0230680. [CrossRef] [PubMed]
14. Lawrinowitz, S.; Wurlitzer, J.M.; Weiss, D.; Arndt, H.D.; Kothe, E.; Gressler, M.; Hoffmeister, D. Blue light-dependent pre-mRNA splicing controls pigment biosynthesis in the mushroom *Terana caerulea*. *Microbiol. Spectr.* **2022**, *10*, e01065-22. [CrossRef] [PubMed]
15. Eisenman, H.C.; Casadevall, A. Synthesis and assembly of fungal melanin. *Appl. Microbiol. Biotechnol.* **2012**, *93*, 931–940. [CrossRef] [PubMed]
16. Henson, J.M.; Butler, M.J.; Day, A.W. The Dark Side of the Mycelium: Melanins of Phytopathogenic Fungi. *Annu. Rev. Phytopathol.* **1999**, *37*, 447–471. [CrossRef] [PubMed]
17. Kojima, M.; Kimura, N.; Miura, R. Regulation of primary metabolic pathways in oyster mushroom mycelia induced by blue light stimulation: Accumulation of shikimic acid. *Sci. Rep.* **2015**, *5*, 8630. [CrossRef] [PubMed]
18. Tiniola, R. Light-emitting diode enhances the biomass yield and antioxidant activity of Philippine wild mushroom *Lentinus swartzii*. *Asian J. Agric. Biol.* **2021**, *2021*, 20210411682. [CrossRef]
19. Desjardin, D.E.; Oliveira, A.G.; Stevani, C.V. Fungi bioluminescence revisited. *Photochem. Photobiol. Sci.* **2008**, *7*, 170–182. [CrossRef] [PubMed]
20. Pawlik, A.; Mazur, A.; Wielbo, J.; Koper, P.; Żebracki, K.; Kubik-Komar, A.; Janusz, G. RNA sequencing reveals differential gene expression of *Cerrena Unicolor* in response to variable lighting conditions. *Int. J. Mol. Sci.* **2019**, *20*, 290. [CrossRef] [PubMed]
21. Yoo, S.I.; Lee, H.Y.; Markkandan, K.; Moon, S.; Ahn, Y.J.; Ji, S.; Ko, J.; Kim, S.J.; Ryu, H.; Hong, C.P. Comparative transcriptome analysis identified candidate genes involved in mycelium browning in *Lentinula edodes*. *BMC Genom.* **2019**, *20*, 121. [CrossRef]
22. Zeng, J.; Shi, D.; Chen, Y.; Bao, X.; Zong, Y. *FvbHLH1* Regulates the Accumulation of Phenolic Compounds in the Yellow Cap of *Flammulina velutipes*. *J. Fungi.* **2023**, *9*, 1063. [CrossRef]
23. Bao, X.; Ke, D.; Wang, W.; Ye, F.; Zeng, J.; Zong, Y. High fatty acid accumulation and coloration molecular mechanism of the elm mushroom (*Pleurotus citrinopileatus*). *Biosci. Biotechnol. Biochem.* **2024**, *88*, 437–444. [CrossRef] [PubMed]
24. Sakamoto, Y. Influences of environmental factors on fruiting body induction, development and maturation in mushroom forming fungi. *Fungal Biol. Rev.* **2018**, *32*, 236–248. [CrossRef]
25. Im, J.H.; Yu, H.W.; Park, C.H.; Kim, J.W.; Shin, J.H.; Jang, K.Y.; Park, Y.J. Phenylalanine ammonia-lyase: A key gene for color discrimination of edible mushroom *Flammulina velutipes*. *J. Fungi* **2023**, *9*, 339. [CrossRef] [PubMed]

26. Grabherr, M.G.; Haas, B.J.; Yassour, M.; Levin, J.Z.; Thompson, D.A.; Amit, I.; Adiconis, X.; Fan, L.; Raychowdhury, R.; Zeng, Q.; et al. Full-length transcriptome assembly from RNA-Seq data without a reference genome. *Nat. Biotechnol.* **2011**, *29*, 644–652. [CrossRef] [PubMed]
27. Hass, B.J. TransDecoder version 5.7.1. 2023. Available online: https://github.com/TransDecoder/TransDecoder (accessed on 2 December 2023).
28. Fu, L.; Niu, B.; Zhu, Z.; Wu, S.; Li, W. CD-HIT: Accelerated for clustering the next-generation sequencing data. *Bioinformatics* **2012**, *28*, 3150–3152. [CrossRef] [PubMed]
29. Mistry, J.; Chuguransky, S.; Williams, L.; Qureshi, M.; Salazar, G.A.; Sonnhammer, E.L.L.; Tosatto, S.C.E.; Paladin, L.; Raj, S.; Richardson, L.J.; et al. Pfam: The protein families database in 2021. *Nucleic Acids Res.* **2020**, *49*, D412–D419. [CrossRef] [PubMed]
30. Paysan-Lafosse, T.; Blum, M.; Chuguransky, S.; Grego, T.; Pinto, B.L.; Salazar, G.A.; Bileschi, M.L.; Bork, P.; Bridge, A.; Colwell, L.; et al. InterPro in 2022. *Nucleic Acids Res.* **2022**, *51*, D418–D427. [CrossRef] [PubMed]
31. Kanehisa, M.; Goto, S. KEGG: Kyoto Encyclopedia of Genes and Genomes. *Nucleic Acids Res.* **2000**, *28*, 27–30. [CrossRef] [PubMed]
32. Szklarczyk, D.; Gable, A.L.; Nastou, K.C.; Lyon, D.; Kirsch, R.; Pyysalo, S.; Doncheva, N.T.; Legeay, M.; Fang, T.; Bork, P.; et al. The STRING database in 2021: Customizable protein–protein networks, and functional characterization of user-uploaded gene/measurement sets. *Nucleic Acids Res.* **2020**, *49*, D605–D612. [CrossRef] [PubMed]
33. Shannon, P.; Markiel, A.; Ozier, O.; Baliga, N.; Wang, J.; Ramage, D.; Amin, N.; Schwikowski, B.; Ideker, T. Cytoscape: A software environment for integrated models of biomolecular interaction networks. *Genome Res.* **2003**, *13*, 2498–2504. [CrossRef] [PubMed]
34. Fuller, K.; Loros, J.; Dunlap, J. Fungal photobiology: Visible light as a signal for stress, space and time. *Curr. Genet.* **2014**, *61*, 275–288. [CrossRef] [PubMed]
35. Corrochano, L.M. Fungal photoreceptors: Sensory molecules for fungal development and behaviour. *Photochem. Photobiol. Sci.* **2007**, *6*, 725–736. [CrossRef] [PubMed]
36. Thind, T.S.; Schilder, A.C. Understanding photoreception in fungi and its role in fungal development with focus on phytopathogenic fungi. *Indian Phytopathol.* **2018**, *71*, 169–182. [CrossRef]
37. Sakamoto, Y.; Tamai, Y.; Yajima, T. Influence of light on the morphological changes that take place during the development of the *Flammulina velutipes* fruit body. *Mycoscience* **2004**, *45*, 333–339. [CrossRef]
38. Tsusué, Y.M. Experimental control of fruit-body formation in *Coprinus macrohizus*. *Dev. Growth Differ.* **1969**, *11*, 164–178. [CrossRef] [PubMed]
39. Kamada, T.; Sano, H.; Nakazawa, T.; Nakahori, K. Regulation of fruiting body photomorphogenesis in *Coprinopsis cinerea*. *Fungal Genet.* **2010**, *47*, 917–921. [CrossRef] [PubMed]
40. Chi, J.H.; Kim, J.H.; Won, S.Y.; Seo, G.S.; Ju, Y.C. Studies on favorable light condition for artificial cultivation of *Grifola frondosa*. *Kor. J. Mycol.* **2008**, *36*, 31–35. [CrossRef]
41. Berger, R.G.; Bordewick, S.; Krahe, N.K.; Ersoy, F. Mycelium vs. Fruiting Bodies of Edible Fungi—A Comparison of Metabolites. *Microorganisms* **2022**, *10*, 1379. [CrossRef] [PubMed]
42. Leatham, G.F.; Stahmann, M.A. Effect of light and aeration on fruiting of *Lentinula edodes*. *Trans. Br. Mycol. Soc.* **1987**, *88*, 9–20. [CrossRef]
43. Liu, C.; Kang, L.; Lin, M.; Bi, J.; Liu, Z.; Yuan, S. Molecular Mechanism by Which the GATA Transcription Factor CcNsdD2 Regulates the Developmental Fate of *Coprinopsis cinerea* under Dark or Light Conditions. *mBio* **2022**, *13*, e03626-21. [CrossRef] [PubMed]
44. Durand, R.; Furuya, M. Action Spectra for Stimulatory and Inhibitory Effects of UV and Blue Light on Fruit-Body Formation in *Coprinus congregatus*. *Plant Cell Physiol.* **1985**, *26*, 1175–1183. [CrossRef]
45. Kitamoto, Y.; Suzuki, A.; Furukawa, S. An Action Spectrum for Light-Induced Primordium Formation in a Basidiomycete, Favolus arcularius (Fr) Ames. *Plant Physiol.* **1972**, *49*, 338–340. [CrossRef] [PubMed]
46. He, Q.; Cheng, P.; Yang, Y.; Wang, L.; Gardner, K.H.; Liu, Y. White Collar-1, a DNA binding transcription factor and a light sensor. *Science* **2002**, *297*, 840–843. [CrossRef] [PubMed]
47. Mayer, C.; Vogt, A.; Uslu, T.; Scalzitti, N.; Chennen, K.; Poch, O.; Thompson, J.D. CeGAL: Redefining a Widespread Fungal-Specific Transcription Factor Family Using an In Silico Error-Tracking Approach. *J. Fungi* **2023**, *9*, 424. [CrossRef] [PubMed]
48. Ruger-Herreros, C.; Gil-Sánchez, M.M.; Sancar, G.; Brunner, M.; Corrochano, L.M. Alteration of light-dependent gene regulation by the absence of the RCO-1/RCM-1 repressor complex in the fungus *Neurospora crassa*. *PLoS ONE* **2014**, *9*, e95069. [CrossRef]
49. In-on, A.; Thananusak, R.; Ruengjitchatchawalya, M.; Vongsangnak, W.; Laomettachit, T. Construction of Light-Responsive Gene Regulatory Network for Growth. development and Secondary Metabolite Production in *Cordyceps militaris*. *Biology* **2022**, *11*, 71. [CrossRef] [PubMed]
50. Permana, D.; Kitaoka, T.; Ichinose, H. Conversion and synthesis of chemicals catalyzed by fungal cytochrome P450 monooxygenases: A review. *Biotechnol. Bioeng.* **2023**, *120*, 1725–1745. [CrossRef] [PubMed]
51. Syed, K.; Shale, K.; Pagadala, N.S.; Tuszynski, J. Systematic identification and evolutionary analysis of catalytically versatile cytochrome p450 monooxygenase families enriched in model basidiomycete fungi. *PLoS ONE* **2014**, *9*, e86683. [CrossRef]
52. Ichinose, H. Cytochrome P450 of wood-rotting basidiomycetes and biotechnological applications. *Biotechnol. Appl.* **2013**, *60*, 71–81. [CrossRef]

53. Naveed, A.; Li, H.; Liu, X. Cytochrome P450s: Blueprints for Potential Applications in Plants. *J. Plant Biochem. Physiol.* **2018**, *6*, 204. [CrossRef]
54. Tanaka, Y.; Brugliera, F. Metabolic Engineering of Flower Color Pathways Using Cytochromes P450. In *Fifty Years of Cytochrome P450 Research*, 1st ed.; Yamazaki, H., Ed.; Springer: Tokyo, Japan, 2014; pp. 207–229. [CrossRef]
55. Rasool, S.; Mohamed, R. Plant cytochrome P450s: Nomenclature and involvement in natural product biosynthesis. *Protoplasma* **2016**, *253*, 1197–1209. [CrossRef]
56. Harmer, S.L.; Hogenesch, J.B.; Straume, M.; Chang, H.-S.; Han, B.; Zhu, T.; Wang, X.; Kreps, J.A.; Kay, S.A. Orchestrated transcription of key pathways in Arabidopsis by the circadian clock. *Science* **2000**, *290*, 2110–2113. [CrossRef]
57. Covington, M.F.; Harmer, S.L. The circadian clock regulates auxin signaling and responses in Arabidopsis. *PLoS Biol.* **2007**, *5*, e222. [CrossRef]
58. Michael, T.P.; Mockler, T.C.; Breton, G.; McEntee, C.; Byer, A.; Trout, J.D.; Hazen, S.P.; Shen, R.; Priest, H.D.; Sullivan, C.M.; et al. Network discovery pipeline elucidates conserved time-of-day-specific cis-regulatory modules. *PLoS Genet.* **2008**, *4*, e14. [CrossRef]
59. Pan, Y.; Michael, T.P.; Hudson, M.E.; Kay, S.A.; Chory, J.; Schuler, M.A. Cytochrome P450 monooxygenases as reporters for circadian-regulated pathways. *Plant Physiol.* **2009**, *150*, 858–878. [CrossRef]
60. Kong, J.Q. Phenylalanine ammonia-lyase, a key component used for phenylpropanoids production by metabolic engineering. *RSC Adv.* **2015**, *5*, 62587–62603. [CrossRef]
61. Mozzarelli, A.; Bettati, S. Exploring the pyridoxal 5′-phosphate-dependent enzymes. *Chem. Rec.* **2006**, *6*, 275–287. [CrossRef]
62. Gerlach, T.; Nugroho, D.L.; Rother, D. The Effect of Visible Light on the Catalytic Activity of PLP-Dependent Enzymes. *ChemCatChem.* **2021**, *13*, 2398–2406. [CrossRef]
63. Liang, J.; Han, Q.; Tan, Y.; Ding, H.; Li, J. Current Advances on Structure-Function Relationships of Pyridoxal 5′-Phosphate-Dependent Enzymes. *Front. Mol. Biosci.* **2019**, *6*, 4. [CrossRef]
64. Parthasarathy, A.; Cross, P.J.; Dobson, R.C.J.; Adams, L.E.; Savka, M.A.; Hudson, A.O. A Three-Ring Circus: Metabolism of the Three Proteogenic Aromatic Amino Acids and Their Role in the Health of Plants and Animals. *Front. Mol. Biosci.* **2018**, *5*, 29. [CrossRef]

Journal of
Fungi

MDPI

Review

Engineering of Global Transcriptional Regulators (GTRs) in *Aspergillus* for Natural Product Discovery

Yujie Zhao [1,2], Qing Gong [1,2] and Huawei Zhang [1,2,*]

1 School of Pharmaceutical Sciences, Zhejiang University of Technology, Hangzhou 310014, China; 211122070041@zjut.edu.cn (Y.Z.); 221124070243@zjut.edu.cn (Q.G.)
2 State Key Laboratory of Green Chemical Synthesis and Conversion, Zhejiang University of Technology, Hangzhou 310014, China
* Correspondence: hwzhang@zjut.edu.cn; Tel.: +86-571-8832-0913

Abstract: The *Aspergillus* genus is an important group of filamentous fungi, and the various biological activities of its secondary metabolites (SMs) have great biosynthetic potential. Despite over 4200 SMs having been isolated from *Aspergillus* spp., their metabolic potential remains unexplored due to the presence of numerous silent biosynthetic gene clusters (BGCs) in their genomes. Fortunately, over the last two decades, the global transcriptional regulator (GTR) engineering strategy has emerged as a powerful tool for activating these cryptic BGCs in order to synthesize previously undiscovered SMs from *Aspergillus* spp. This review highlights recent advances in fungal GTR engineering techniques, the regulatory mechanisms of GTRs, and current challenges and future perspectives for their application in natural product discovery in the genus *Aspergillus*.

Keywords: global transcriptional regulators; secondary metabolites; regulatory mechanisms; CRISPR/Cas9; homologous recombination

1. Introduction

The genus *Aspergillus*, with 446 known species, stands as one of the most prevalent fungal genera in nature and demonstrates significant biosynthetic potential for producing secondary metabolites (SMs) [1]. According to the Dictionary of Natural Products database, over 4200 SMs produced by *Aspergillus* spp. have been deposited to date [2]. These natural products possess various chemical structures including terpenes, polyketides, peptides, and alkaloids [3–6]. However, the biosynthetic potential of *Aspergillus* spp. remains underexploited owing to the presence of a large number of silent biosynthetic gene clusters (BGCs) in their genomes. For instance, *A. nidulans* possesses over 50 BGCs, producing significantly fewer characterized SMs under standard cultivation conditions [7,8]. This pronounced disparity between BGC abundance and SM number strongly indicates the prevalence of transcriptionally silent gene clusters awaiting activation. Various strategies have been developed to mine these silent BGCs, such as one strain many compounds (OSMAC) strategies, microbial co-culture, promoter engineering, metabolic shunting, heterologous expression, and transcriptional regulation strategies [9–13].

Transcriptional regulators, including pathway-specific regulators and global transcriptional regulators (GTRs), can bind to promoters of target genes for gene activation and repression [14]. Gene expression regulation mediated by GTRs is an effective strategy for facilitating natural product discovery (Figure 1). Genes encoding GTRs are generally located outside secondary metabolic gene clusters, and the encoded GTRs directly or indirectly regulate the expression of most genes in *Aspergillus* spp. [15]. Typical GTRs in *Aspergillus* spp. include LaeA, VeA, PacC, CreA, AreA, and CBC (the CCAAT-binding complex). In addition, GTRs such as McrA, LaeB, HbxA/Hbx1, RsmA, RimO (SrrB), Sirtuin E (SirE), StuA and RlcA have been progressively identified and characterized. They regulate various physiological processes in *Aspergillus* spp. such as spore formation, mycelial growth primary metabolism and secondary metabolism, and they can be regulated in response to a variety of environmental signals such as light, pH, carbon source, and nitrogen source [16]. With the continuous innovation and advancement of genetic engineering technology, the clustered regularly interspaced short palindromic repeat/CRISPR-associated protein 9 (CRISPR/Cas9) system, homologous recombination (HR) system-based gene knockout strategies, and gene overexpression strategies using strong promoter or overexpression vectors have been widely used to regulate the expression of GTRs, providing more opportunities to obtain improved strains with secondary metabolic alterations [17,18]. This review systematically summarizes recent advances in various GTR engineering techniques of *Aspergillus* spp. with particular emphasis on elucidating the molecular mechanisms of GTRs and their regulatory effects on SM biosynthesis. It provides a methodological reference for discovering novel natural products and enhancing the production of valuable natural products from *Aspergillus* spp.

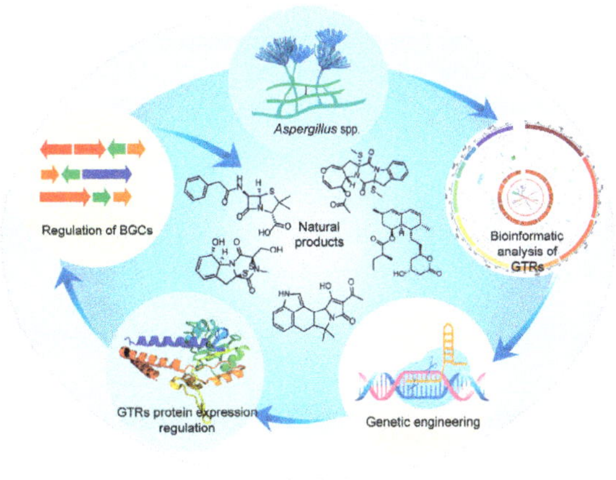

Figure 1. Global transcriptional regulator engineering strategy for secondary metabolite production in *Aspergillus* spp.

2. GTR Engineering Approaches

Regulating GTRs expression through methods such as gene knockout and overexpression is an effective strategy for studying their function and regulating downstream gene expression. This section describes key engineering methods for realizing this strategy.

2.1. Knockout Strategies for GTR

2.1.1. Knockout of GTR Based on HR

The most common method for the gene knockout of GTRs in *Aspergillus* spp. is using HR to replace the target gene in the strain. This technique uses double-joint polymerase chain reaction (PCR) or fusion PCR to fuse a drug-resistance marker and homologous sequence at both ends into an expression cassette, which is then transformed into the protoplast of the target strain and subsequently screened to construct a successful mutant strain [19–21] (Figure 2A).

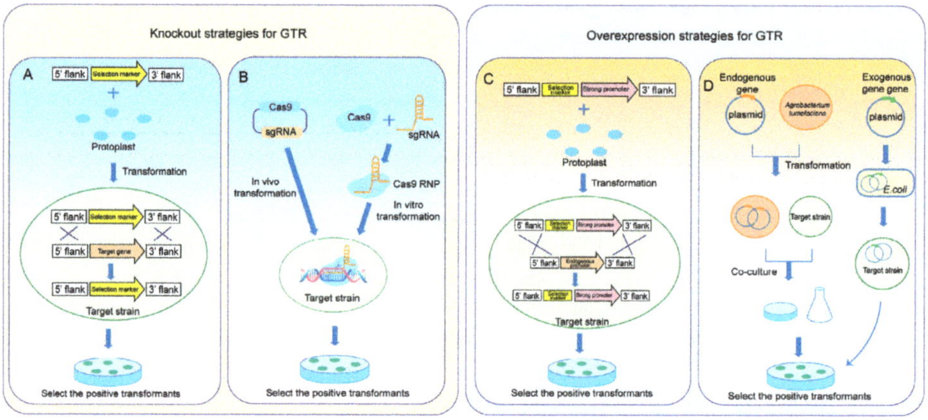

Figure 2. Knockout and overexpression strategies for GTRs. (**A**) Knockout of GTR based on homologous recombination. (**B**) Knockout of GTR based on the CRISPR/Cas9 system. (**C**) Overexpression of GTR using strong promoters. (**D**) Construction of overexpression vectors for GTR expression.

High efficiency in gene knockout and transformation is key to successfully constructing mutant strains. Non-homologous end joining (NHEJ) and HR are both mechanisms of DNA double-strand break repair [22]. While NHEJ often leads to the random integration of DNA fragments, which can result in transformation failure, disrupting *ku* genes effectively inhibits this error-prone repair mechanism [23,24]. Therefore, constructing *ku*-deficient strains has become a reliable strategy for increasing the success rate of targeted HR. This method was first applied in *A. nidulans* by Nayak et al. to improve the gene targeting efficiency [24]. In *A. niger*, the use of *akuA* (the homologous gene for *ku70*) knockout strains significantly increased the efficiency of HR, resulting in a gene knockout efficiency of over 80% [25]. In *A. fumigatus*, the deletion of *akuB*, the homologous gene for *ku80*, increased the homologous recombination frequency of *pksP* deletion to approximately 80% with 1.5 kb and 2.0 kb flanking regions compared to 5% in the wild-type strain. *akuA* deletion showed similar results [26,27].

The transformation efficiency can also be improved using split-marker technology, which was first developed in yeast [28]. In this method, the selectable marker gene is divided into two overlapping fragments, which replace the target gene and enable gene knockout [29]. In *A. nidulans*, using split-marker-targeting substrates increased the gene targeting efficiency by 2- to 3-fold compared to conventional continuous targeting substrates [30]. In particular, one way to improve the transformation efficiency is by changing the transformation method, such as using *Agrobacterium tumefaciens*-mediated transformation (ATMT), in addition to PEG-mediated transformation and electroporation [31]. ATMT was first applied to filamentous fungi by De Groot et al. A binary vector containing the marker and target genes was constructed and introduced into *Agrobacterium tumefaciens*, which was then co-cultured with the target strains, thereby integrating the target genes into their genome [32]. This transformation system is characterized by a high transformation rate, genetic stability, and a high single-copy integration rate.

2.1.2. Knockout of GTR Based on the CRISPR/Cas9 System

The CRISPR/Cas9 system is an important tool for genome modification. Since its first application as a gene-editing tool in filamentous fungi in 2015, it has been adapted for use in a wide range of filamentous fungal species [33]. CRISPR/Cas9 technology has also been widely used for targeted gene knockout in *Aspergillus* spp. The classic method involves constructing a plasmid containing the Cas9 enzyme and sgRNA to achieve targeted knockout through in vivo transformation [34] (Figure 2B). Recently, the in vitro CRISPR/Cas9 system has also been applied for knocking out GTRs in *Aspergillus* spp. For example, Yuan et al. knocked out *mcrA* in *A. wentii* using an in vitro CRISPR/Cas9 system [35]. In addition, the use of two crRNAs to achieve dual-target cleavage, along with functional protospacer-adjacent motif sequence screening, reduced off-target effects. The targeted knockout efficiency of *mcrA* in *A. wentii* reached 62.5% and was applicable for double-gene knockouts. This method was also successfully applied in *A. melleus* [36].

2.2. *Overexpression Strategies for GTR*

2.2.1. Overexpression of GTR Using Strong Promoters

Strong promoters efficiently recruit RNA polymerase to promote target gene transcription [37]. Therefore, replacing the GTR promoter with a strong promoter can enhance its expression. A classical strategy involves introducing a strong promoter via HR, which is similar to the approach used in HR-based gene knockout (Figure 2C). Promoter types include both natural and synthetic promoters; however, synthetic promoters have not yet been applied to the GTRs expression in *Aspergillus* spp. Commonly used natural promoter types include constitutive promoters and inducible promoters. The glyceraldehyde-3-phosphate dehydrogenase promoter *gpdA* from *A. nidulans* is a representative constitutive promoter, which drives gene stable expression that is relatively unaffected by external conditions [38]. The glucoamylase promoter *glaA*, identified from *A. niger*, is a representative inducible promoter that requires specific inducing factors to function [39]. Recently, the quantitative characterization of filamentous fungal promoters has opened up new possibilities for the discovery of strong promoters [40].

2.2.2. Construction of Overexpression Vectors for GTR Expression

GTRs overexpression can be achieved using overexpression vectors. This strategy involves inserting the target gene into a plasmid and transforming it into the host strain, thereby enhancing target gene expression [41,42] (Figure 2D). Using this technique, Wang et al. constructed the llm1 overexpression vector by cloning the putative methyltransferase gene *llm1* from *A. cristatus* into a plasmid containing the strong promoter *gpdA*. The mutant strain overexpressing *llm1* was subsequently obtained through ATMT, resulting in alterations to its SM profile [43]. This result indicates that this method is applicable for *Aspergillus* spp.

Given the conserved nature of GTRs in *Aspergillus* spp., in addition to overexpressing the endogenous GTR gene in the same strain, compatible heterologous GTRs can also exert global regulatory functions in target strains [44]. Khan et al. used this approach to express the *AnlaeA* gene from *A. nidulans* in *Aspergillus* sp. Z5. [45].

3. GTRs in *Aspergillus* spp.

The process of SM biosynthesis is regulated by a complex multilevel molecular network. The GTRs of *Aspergillus* spp. represented by LaeA, VeA, PacC, CreA, AreA, and CBC affect SMs gene expression through different mechanisms, such as the response to environmental signals and chromatin remodeling [15]. Other regulators such as McrA and LaeB have also been successfully identified and confirmed to be GTRs in *Aspergillus* spp. This section summarizes the mechanisms of the different GTRs and the SMs that they regulate.

3.1. LaeA

3.1.1. Regulatory Mechanisms of LaeA

LaeA, first identified in *A. nidulans*, functions as a nuclear protein methyltransferase characterized by a conserved S-adenosylmethionine binding site and is conserved across *Aspergillus* spp. [46–48]. Although a nuclear localization sequence (NLS) is absent, it is exclusively localized to the nucleus [46].

The regulation of secondary metabolism is the most extensively studied aspect of LaeA. In most cases, LaeA functions as a positive regulator of SMs, and the deletion of *laeA* inhibits the production of many SMs. Chromatin remodeling via histone methylation and velvet complex formation are currently proposed as the primary mechanisms by which LaeA regulates the expression of other genes [49–52] (Figure 3). It was found that the loss of *laeA* in *A. nidulans* led to a significant increase in histone 3 lysine 9 (H3K9) trimethylation and heterochromatin protein-1 levels, which promoted heterochromatin formation and consequently inhibited sterigmatocystin (ST, **1**) production (Figure S1). Chromatin immunoprecipitation (ChIP) further confirmed that LaeA plays a key role in reversing heterochromatin formation [49]. A study on *A. luchuensis* mut. *Kawachii* with *laeA* knockout demonstrated that LaeA regulates histone 3 lysine 4 (H3K4) and H3K9 methylation to control the expression of the citric acid transporter protein CexA [50]. The loss of laeA in *A. fumigatus* also confirmed this regulatory mechanism, showing the repression of heterochromatin formation at the fungal spore-forming gene *brlA*'s promoter and an increased level of H3K9 trimethylation, suggesting that LaeA regulates *brlA* transcription through chromatin modification [51]. The velvet complex is a protein complex formed in the nucleus by LaeA, VelB, and VeA through which LaeA regulates the expression of other genes [52]. Furthermore, the LaeA regulation of secondary metabolism genes is location-specific, primarily regulating clustered genes in the subtelomeric regions of chromosomes [53,54].

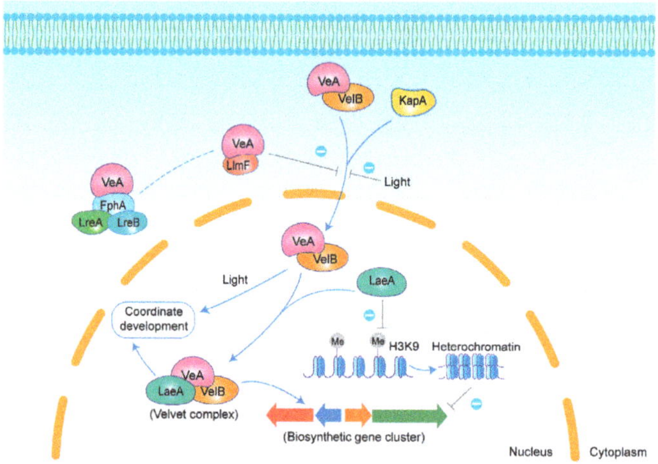

Figure 3. Mechanism of secondary metabolism regulation by LaeA and VeA in *Aspergillus* spp. LlmF, FphA, and light all inhibit the nuclear import of VeA; however, it remains unclear whether LlmF and FphA regulate VeA's subcellular localization independently or synergistically. In darkness, the VeA-VelB complex is transported into the nucleus and interacts with LaeA to form the velvet complex, which regulates secondary metabolism and development. Additionally, LaeA can regulate secondary metabolism by inhibiting histone methylation.

3.1.2. LaeA Regulation-Derived SMs

LaeA plays an essential role in secondary metabolism regulation in *Aspergillus* spp. with examples of it regulating SMs in *Aspergillus* spp. shown in Table 1. Figure S1 provides the structures of all SMs that are regulated.

Table 1. Examples of secondary metabolites regulated by LaeA in *Aspergillus* spp.

***Aspergillus* Strain**	**Deletion or Overexpression**	**Upregulated SMs**	**Downregulated SMs**	**Ref.**
A. nidulans FGSC 26	Deletion	-	Sterigmatocystin (ST, **1**), monacolin J (MONJ, **2**), and penicillin G (**3**)	[46]
	Overexpression	Penicillin G (**3**) and MONJ (**2**)	-	
A. fumigatus AF293	Deletion	-	Gliotoxin (**4**)	
A. terreus ATCC 20542	Deletion	-	Lovastatin (**5**)	
	Overexpression	Lovastatin (**5**)	-	
A. fumigatus AF293	Deletion	-	ST (**1**)	[55]
A. flavus CA14	Deletion	-	Aflatoxin B1 (**6**) and aflatoxin B2 (**7**)	[56]
A. fumigatus CEA17	Deletion	-	Endocrocin (**8**)	[57]
A. oryzae RIB40	Deletion	-	Kojic acid (**9**)	[58]
A. niger ATCC9029	Deletion	BMS-192548 (**13**) and aspernigrin A (**14**)	Asperrubrol (**10**), atromentin (**11**) and JBIR-86 (**12**)	[59]
A. carbonarius UdL-TA 3.83	Deletion	-	Ochratoxin A (OTA, **15**)	[60]
A. ochraceus fc-1	Deletion	-	OTA (**15**)	[61]
A. carbonarius Ac ITEM 5010	Deletion	-	OTA (**15**)	[62]
A. pseudoterreus ATCC 32359	Deletion	-	Itaconic acid (**16**)	[63]
A. flavipes (507)	Deletion	Flavipamide A and B (**17**,**18**), *N*-benzoylphenylalaniny-*N*-benzoylphenyl-alaninate (**19**), 4′-OMe-asperphenamate (**20**), cyclic Pro-Gly-Val-Gly-Try (/8-OH, 3-prenyl)-Gly-Trp (**21**)	-	[64]
Aspergillus sp. Z5	Overexpression	Diorcinol (**22**) and quinolactacin A (**23**)	-	[45]
A. fumisynnematus F746	Overexpression	Cyclopiazonic acid (**24**)	-	[65]
A. versicolor 0312	Overexpression	Versicolor A (**25**), acetylaranotin (**26**), acetylapoaranotin (**27**), ergosterol (**28**) and diisobutyl phthalate (**29**)	-	[66]
A. niger FGSC A1279	Overexpression	Flaviolin (**30**), orlandin (**31**), and kotanin (**32**)	-	[67]
A. terreus RA2905	Overexpression	Dihydroisoflavipucines 1 and 2 (**33**, **34**)	-	[68]
A. niger L14	Overexpression	Aspochracin (**35**), JBIR-15 (**36**), sclerotiotide C (**37**), kojic acid (**9**), and penicillic acid (**38**)	-	[69]

The deletion of *laeA* significantly suppresses the production of ST (**1**), monacolin J (MONJ, **2**), and the β-lactam antibiotic penicillin G (**3**) in *A. nidulans* FGSC 26; gliotoxin (**4**) in *A. fumigatus* AF293; and the anti-hypercholesterolemic agent lovastatin (**5**) in *A. terreus* ATCC 20542. Conversely, the overexpression of *laeA* significantly enhances penicillin G (**3**) and MONJ (**2**) production in *A. nidulans* FGSC 26 with MONJ (**2**) levels increasing 4-fold. In *A. terreus* ATCC 20542, lovastatin (**5**) production increases 4- to 7-fold, whereas no significant change in ST (**1**) production is observed in *A. fumigatus* AF293 [46]. Similarly, the knockout of *laeA* in *A. fumigatus* AF293 reduces ST (**1**) production to 20% of the wild-type level [55].

The production of aflatoxin B1 and B2 (**6,7**) is suppressed in the laeA deletion mutant of *A. flavus* CA14 by the downregulation of early biosynthetic genes involved in aflatoxin biosynthesis, including *aflR*, *nor1*, and *aflJ* [56]. In the *A. fumigatus* CEA17 *laeA* deletion mutant, key genes (*encA–D*) involved in endocrocin (**8**) biosynthesis are no longer expressed, resulting in the complete loss of endocrocin production [57]. In *A. oryzae* RIB40, the knockout of laeA resulted in a complete loss of kojic acid (**9**) production, while the *laeA* complementation strain restored compound **9** production [58]. The loss of *laeA* in *A. niger* ATCC9029 affected the yields of SMs, resulting in decreased yields of asperrubrol (**10**), atromentin (**11**), and JBIR-86 (**12**) as well as increased yields of BMS-192548 (**13**) and aspernigrin A (**14**) [59]. Biosynthesis of the carcinogenic mycotoxin ochratoxin A (OTA, **15**) is significantly regulated by LaeA, as demonstrated by multiple studies. The deletion of *laeA* in *A. carbonarius* UdL-TA 3.83 led to a decrease in the expression of the key non-ribosomal peptide synthetase (NRPS) involved in OTA (**15**) biosynthesis, leading to a 97% decrease in OTA production relative to the wild-type strain in the dark and a 68.5% decrease in production in the light [60]. The same was found in *A. ochraceus* fc-1 and *A. carbonarius* Ac ITEM 5010 [61,62].

LaeA is a positive regulator of itaconic acid (**16**). In *A. pseudoterreus* ATCC 32359, construction of the *laeA* knockout strain led to a 94% reduction in itaconic acid (**16**) production on glucose/xylose medium compared to that in the wild-type strain. In contrast, *laeA* overexpression increased itaconic acid production by 13% [63]. Simultaneously, LaeA seems to have negative role in the regulation of SM production. For instance, the knockout of *laeA* in *A. flavipes* (507) upregulated the expression of two NRPS-like BGCs and the new piperazine derivatives flavipamide A and B (**17,18**) as well as three known non-ribosomal peptides: N-benzoylphenylalaniny-N-benzoylphenyl-alaninate (**19**), 4′-OMe-asperphenamate (**20**), and cyclic Pro-Gly-Val-Gly-Try (/8-OH, 3-prenyl)-Gly-Trp (**21**) [64].

The overexpression of *laeA* from *A. nidulans* in *Aspergillus* sp. Z5 significantly increased the yields of diorcinol (**22**) and quinolactacin A (**23**). Diorcinol (**22**) exhibited cytotoxic activity against the HCT116 human colon cancer cell line with a half-maximal inhibitory concentration (IC_{50}) of 10 μg [45]. LaeA expression enhancement in *A. fumisynnematus* F746 under the control of the strong promoter *alcA* led to the production of a novel metabolite, cyclopiazonic acid (CPA, **24**), which had not previously been identified in *A. fumisynnematus* F746 [65].

Chemical investigation of the *A. versicolor* 0312 *laeA* overexpression mutant led to the identification of a new compound, versicolor A (**25**), as well as four known compounds acetylaranotin (**26**), acety-lapoaranotin (**27**), ergosterol (**28**), and diisobutyl phthalate (**29**); among these, **25** showed cytotoxic activity against the MOLT-4 cell line (IC_{50} = 29.6 μM), while **26** showed good cytotoxic activity against different cell lines such as MOLT-4, CaCo-2, and MCF-7 with IC_{50} values ranging from 7.8 to 19.9 μM [66]. The overexpression of *laeA* in *A. niger* FGSC A1279 resulted in the increased production of three identified compounds including flaviolin (**30**), orlandin (**31**), and kotanin (**32**) [67].

An enhanced expression of *laeA* in *A. terreus* RA2905 activated a silent gene cluster, leading to the production of two small-molecule compounds, dihydroisoflavipucines 1 and 2 (**33, 34**), which exhibited notable anti-*Vibrio* activity with MIC values ranging from 16 to 64 μg/mL [68]. In the marine fungus *A. niger* L14, the overexpression of *laeA* led to the discovery of five compounds not produced by the wild-type strain, including aspochracin (**35**), JBIR-15 (**36**), sclerotiotide C (**37**), kojic acid (**9**), and penicillic acid (**38**). Among them, JBIR-15 (**36**) exhibited antifungal activity against *Candida albicans* (MIC = 32 μg/mL), while kojic acid (**9**) showed significant DPPH (2,2-diphenyl-1-picrylhydrazyl) radical scavenging activity (IC_{50} = 5 μg/mL) [69]. The *laeA*-like gene *llm1* was identified in *A. cristatus* CM1303, and the construction of the *llm1* overexpression mutant revealed that llm1 not only positively regulates sexual development but also reduces oxidative stress tolerance to hydrogen peroxide while simultaneously enhancing the production of terpenoids and flavonoids. Unfortunately, the newly generated SMs were not isolated [43].

3.2. *VeA*

3.2.1. Regulatory Mechanisms of VeA

Originally identified in *A. nidulans*, VeA is a member of the velvet family protein that contains a conserved velvet domain and functions as a polyphosphorylated protein whose expression is influenced by its own phosphorylation status [70,71]. Early studies on VeA identified a functional bipartite NLS within its amino acid sequence that is conserved in *Aspergillus* spp. [72].

Compared to LaeA, which is exclusively localized in the nucleus, the subcellular localization of VeA is light-dependent. VeA is predominantly localized in the nucleus under dark conditions, whereas it is mainly found in the cytoplasm in the presence of light [73] (Figure 3). Red and blue light also affect the action of VeA. The red light-sensitive phytochrome FphA modulates VeA activity. VeA interacts with FphA, which then binds to the complex formed by the blue light-sensitive elements LreA and LreB. This cascade ultimately enables VeA to positively regulate sexual development [71,74]. In addition, the LaeA-like methyltransferase LlmF directly interacts with VeA to form a transient complex, thereby impeding the nuclear translocation of VeA and mediating its negative regulation of secondary metabolism in *A. nidulans* [75]. In 2008, the heterotrimeric velvet complex VelB/VeA/LaeA was identified, in which VeA functions as a bridge between VelB and LaeA. In the dark, the NLS is specifically recognized by KapA, which is a homologue of the general nuclear transporter importin α. KapA transports VeA and VelB into the nucleus, where they form a complex with LaeA to jointly regulate sexual development and secondary metabolism in *Aspergillus* spp. Under light conditions, VeA and VelB remain in the cytoplasm and promote asexual development [52].

3.2.2. VeA Regulation-Derived SMs

ST (**1**) and penicillin G (**3**) production in *A. nidulans* FGSC4 is regulated by VeA. The pathway-specific transcription factor AlfR, which controls the mycotoxin ST (**1**), is positively regulated by VeA. The knockout of *veA* leads to reduced transcript levels of *aflR*, thereby further inhibiting ST (**1**) production [76]. Penicillin production is influenced by the delta-(*L*-alpha-aminoadipyl)-*L*-cysteinyl-*D*-valine synthetase gene *acvA*, which positively regulates penicillin biosynthesis. In *A. nidulans* AXB4A2, both the overexpression and deletion of veA inhibit *acvA* expression, thereby reducing penicillin production. These findings suggest that veA negatively regulates penicillin biosynthesis [77]. However, another study demonstrated that VeA positively regulates penicillin production in *A. oryzae* RIB40. This indicates the complexity of the regulatory role of VeA [78].

The knockout of *veA* in *A. flavus* ATCC MYA384 resulted in no further expression of the genes key to aflatoxin biosynthesis, thereby eliminating aflatoxin production. Furthermore, the production of the mycotoxins aflatrem (**39**) and CPA (**24**) was also drastically reduced with CPA (**24**) yields decreasing by 48% in the light and 66% in the dark [79]. To determine the effect of veA on *A. flavus* NRRL 3357, a microarray analysis of *veA* knockout and overexpression mutants revealed that VeA positively regulates genes within the aflatoxin BGC [80]. The deletion of *veA* in *A. fumigatus* CEA10 resulted in a decrease in the expression levels of *gliZ* and *gliP*, genes involved in gliotoxin (**4**) biosynthesis, as well as a 5-fold decrease in gliotoxin (**4**) production compared to the wild-type strain, suggesting that VeA in *A. fumigatus* CEA10 positively regulates gliotoxin (**4**) biosynthesis [81] (Table 2).

Table 2. Examples of secondary metabolites regulated by VeA in *Aspergillus* spp.

***Aspergillus* Strain**	**Deletion or Overexpression**	**Upregulated SMs**	**Downregulated SMs**	**Ref.**
A. nidulans FGSC4	Deletion	-	ST (**1**) and penicillin G (**3**)	[76]
A. nidulans AXB4A2	Deletion Overexpression	- -	Penicillin G (**3**) Penicillin G (**3**)	[77]
A. oryzae RIB40	Deletion	-	Penicillin G (**3**)	[78]
A. flavus ATCC MYA384	Deletion	-	Aflatoxin, aflatrem (**39**) and CPA (**24**)	[79]
A. fumigatus CEA10	Deletion	-	Gliotoxin (**4**)	[81]
A. carbonarius UdL-TA 3.83	Deletion	-	OTA (**15**)	[60]
A. niger CICC 41702	Deletion	-	OTA (**15**)	[82]
A. fumigatus CEA10	Deletion Overexpression	- -	Fumagillin (**40**), fumitremorgin G (**41**), fumigaclavine C (**42**), and glionitrin A (**43**) fumagillin (**40**), fumitremorgin G (**41**), fumigaclavine C (**42**), and glionitrin A (**43**)	[83]
A. pachycristatus NRRL 11440	Deletion	-	Echinocandin B (ECB,**44**) and ST (**1**)	[84]
A. nidulans RDIT9.32	Deletion	F9775A (**45**), F9775B (**46**), and orsellinic acid (**47**)	-	[85]

The knockout of *veA* in *A. carbonarius* UdL-TA 3.83 led to the downregulation of NRPS, a key enzyme in the biosynthesis of OTA (**15**), resulting in a 90% reduction in OTA (**15**) production compared to the wild-type strain under light conditions. This may be due to VeA's inability to enter the nucleus and form a complex with LaeA in the presence of light [60]. VeA has been shown to positively regulate conidia production, oxidative stress tolerance, and OTA (**15**) biosynthesis in *A. niger* CICC 41702. The *veA* disruption strain exhibited a significant reduction in OTA production. Quantitative reverse transcription polymerase chain reaction (qRT-PCR) analysis confirmed that VeA influences OTA production by modulating the expression of another key polyketide synthase (PKS) gene within OTA biosynthesis [82].

The analysis of SMs in *A. fumigatus* CEA10 *veA* knockout and overexpression mutants revealed that fumagillin (**40**), fumitremorgin G (**41**), fumigaclavine C (**42**), and glionitrin A (**43**) were among the most significantly altered. In the *veA* knockout mutant, the production of these SMs decreased to approximately 20% of the wild-type level, while in the veA overexpression strain, these SMs were produced at nearly undetectable levels [83]. *A. pachycristatus* NRRL 11440 produces the antifungal drug echinocandin B (ECB, **44**), and the deletion of *Apc.veA* reduced ECB yield by 93% and almost eliminated the by-product

ST (**1**). A qRT-PCR analysis revealed a significant reduction in the expression of nine key biosynthetic genes involved in ECB (**44**) and four key biosynthetic genes associated with ST (**1**), demonstrating that VeA positively regulates the production of both ECB and ST [84]. Although VeA generally functions as a positive regulator of SMs, the deletion of *veA* in *A. nidulans* RDIT9.32 activated the production of F9775A (**45**), F9775B (**46**), and orsellinic acid (**47**), suggesting that VeA acts as a negative regulator of these SMs. Further experiments revealed that the gene *gncE*, which encodes GcnE (H3K9 acetyltransferase of the Spt-Ada-Gcn5 acetyltransferase/Ada acetyltransferase complex), was upregulated, indicating that this regulatory effect involves histone H3 acetylation [85].

3.3. *PacC*

3.3.1. Regulatory Mechanisms of PacC

pH is an essential factor influencing microbial secondary metabolism. The Pal/Rim signaling pathway can regulate gene expression by sensing changes in the extracellular pH. PacC is a key transcriptional regulator of the Pal/Rim signaling pathway with regulatory functions in *Aspergillus* spp. growth, metabolism, and virulence [86]. *pacC* expression is more active at alkaline pH and also activates the transcription of other alkaline-expressed genes [87]. PacC contains three Cys2His2 zinc finger domains. Zinc finger 2 and zinc finger 3 directly bind the consensus sequence 5′-GCCARG-3′, while zinc finger 1 indirectly participates by enhancing the binding ability of zinc finger 3. Additionally, zinc finger 3 contains a nuclear localization signal (NLS) that regulates the subcellular localization of PacC. PacC activates the expression of target genes by directly binding to the consensus sequence [88,89]. Ambient pH alters the active form and localization of PacC, which, in turn, affects its ability to transcriptionally regulate other genes [90]. PacC's C-terminal region has been shown to negatively regulate its activity, rendering the full-length PacC protein (72 kDa) inactive [91] (Figure 4).

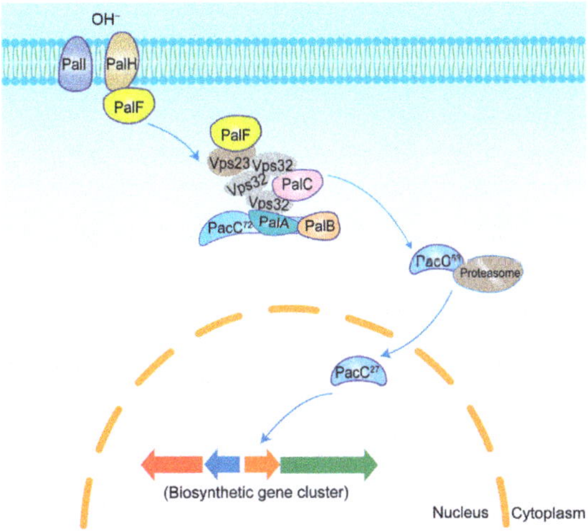

Figure 4. Mechanism of secondary metabolism regulation by PacC in *Aspergillus* spp. PalI and PalH sense the OH^- signal and promote PalF ubiquitination. Ubiquitinated PalF then interacts with the ESCRT complex and, with the assistance of PalC, recruits PalA. Subsequently, PalB is incorporated into the complex and cleaves $PacC^{72}$ to produce $PacC^{53}$. $PacC^{53}$ is then further processed by proteases into the activated form $PacC^{27}$, which in turn regulates fungal secondary metabolism.

PalI, PalH and PalF are three integral components of the pH-sensing signaling complex localized to the *A. nidulans* plasma membrane [92]. Under alkaline conditions, PalI and PalH at the membrane sense the extracellular OH^- signal and subsequently promote PalF ubiquitination. Ubiquitinated PalF then facilitates the recruitment of the ESCRT-I (endosomal sorting complex required for transport I) component Vps23, which, in turn, recruits additional ESCRT required for transport [93]. The ESCRT-III core subunit Vps32 interacts with PalC (72 Ka), leading to the subsequent recruitment of PalA. The complex formed by PalA and full-length PacC associates with the calpain-like protease PalB, resulting in the proteolytic cleavage of the C-terminal domain of PacC and generation of $PacC^{53}$ (53 kDa), which is further processed by proteasome-mediated partial hydrolysis into the active form $PacC^{27}$ (27 kDa), which contains a DNA-binding domain and functions as a transcriptional regulator, activating alkaline-expressed genes while repressing acid-expressed genes [94–96]. The active form of PacC also affects its subcellular localization. When in its full-length form, it is mainly localized in the cytoplasm, whereas the C-terminally truncated active form, $PacC^{27}$, is predominantly localized in the nucleus [97].

3.3.2. PacC Regulation-Derived SMs

PacC is involved in secondary metabolic processes in *Aspergillus* spp. In *A. nidulans*, PacC can directly bind to the sequence between *acvA* and *ipnA*, thereby modulating *ipnA* expression and ultimately governing penicillin biosynthesis [98]. A series of PacC gain-of-function strains were obtained in *A. nidulans* by mimicking alkaline mutations, and both their penicillin yield and *ipnA* gene expression levels were significantly higher than those of the wild-type strain [99]. O-methylsterigmatocystin (**48**) in *A. parasiticus* RHN1 is regulated by PacC, and it is reduced in strains with deletion of the PacC binding site in the *aflR* promoter [100]. When *AopacC* was knocked out of *A. ochraceus* fc-1, both acidic and alkaline conditions significantly reduced OTA (**15**) production and the expression of the related biosynthetic gene *Aopks* compared to that in neutral conditions [101]. PacC also modulates OTA (**15**) biosynthesis in *A. carbonarius* NRRL 368, as proven by the *pacC* deletion strain completely ceasing OTA production at pH 7.0 [102] (Table 3).

Table 3. Examples of secondary metabolites regulated by PacC, CreA, AreA, and CBC in *Aspergillus* spp.

GTRs	*Aspergillus* Strain	Deletion or Overexpression	Upregulated SMs	Downregulated SMs	Ref.
PacC	*A. nidulans*	Overexpression	Penicillin G (**3**)	-	[99]
	A. parasiticus RHN1	Deletion of the PacC binding site	-	O-methylsterigmatocystin (**48**)	[100]
	A. ochraceus fc-1	Deletion	-	OTA (**15**)	[101]
	A. carbonarius NRRL 368	Deletion	-	OTA (**15**)	[102]
CreA	*A. flavus* CA14PTS	Overexpression Deletion	Aflatoxin B1 (**6**) -	- Aflatoxin B1 (**6**)	[103]
	A. ochraceus fc-1	Deletion	-	OTA (**15**)	[104]
AreA	*A. terreus* SBUG844	Deletion	-	Terrein (**49**)	[105]
	A. flavus SRRC1709	Overexpression Deletion	Aflatoxin B1 (**6**) Aflatoxin B1 (**6**)	-	[106]
CBC	*A. nidulans* MH8193	Deletion of *hapC*	-	Penicillin (**3**)	[107]
	A. fumigatus A1160P$^+$	Deletion of *hapE*	Ergosta-5,7,24(28)-trien-3β-ol (**50**) and episterol (**51**)	-	[108]

3.4. CreA

3.4.1. Regulatory Mechanisms of CreA

CreA is a major regulator of carbon source utilization in *Aspergillus* spp. and has been shown to be a repressor protein containing two C2H2-like zinc finger domains, which can regulate the expression of other genes by binding to the consensus sequence 5′-SYGGRG-3′ [109,110]. CreA regulates gene expression both directly and indirectly. In *A. nidulans*, CreA can indirectly regulate the expression of other xylanase genes by directly binding to the consensus sequence in *xlnR*'s promoter region [111]. In addition, changes in carbon source availability indirectly influence the CreA-mediated regulation of other genes. Carbon catabolite repression (CCR) is a mechanism by which microorganisms are able to preferentially utilize favored carbon sources (such as D-glucose, D-fructose, or D-xylose) during microbial growth, thereby inhibiting the utilization of other carbon sources. CreA was shown to be a transcriptional repressor involved in this mechanism. Alteration of the carbon source modulates the transcript level of *creA* [112]. In the presence of favored carbon sources, the *creA* expression levels increase, leading to the negative regulation of other genes. Conversely, under repressing carbon source conditions, creA expression is suppressed, resulting in de-repression [113] (Figure 5).

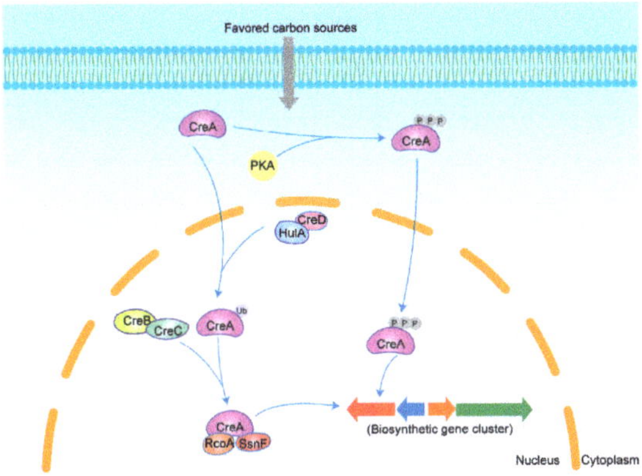

Figure 5. Mechanism of secondary metabolism regulation by CreA in *Aspergillus* spp. CreA is ubiquitinated by the CreD–HulA complex and subsequently deubiquitinated by the CreB–CreC complex. The deubiquitinated CreA then associates with RcoA and SsnF to regulate secondary metabolism. In addition, favored carbon sources can further govern secondary metabolism by promoting CreA phosphorylation.

The nuclear localization of CreA is independent of its activity but is affected by different carbon sources. CreA localizes to the nucleus in the presence of favored carbon sources and to the cytoplasm when favored carbon sources are absent [114]. CreA activation is influenced by both ubiquitination and phosphorylation. In the presence of favored carbon sources, the complex formed by the interaction between the ubiquitin ligase HulA and the arrestin-like protein CreD mediates CreA ubiquitination [115]. Subsequently, the CreB–CreC deubiquitination complex, composed of the deubiquitinating enzyme CreB and the scaffold protein CreC, mediates CreA deubiquitination, and the deubiquitinated CreA then associates with the corepressors SsnF and RcoA, resulting in the activation of CreA's repressive function [116–118]. In addition, favored carbon sources induce CreA phosphorylation, mediated by protein kinase A (PKA), which also contributes to CreA activation [119,120]. Furthermore, autogenous regulation is also an important mode of *creA* expression, and CreA negatively regulates its own expression [121].

3.4.2. CreA Regulation-Derived SMs

Aflatoxin B1 (**6**) biosynthesis in *A. flavus* CA14PTS is regulated by CreA. In the *creA* overexpressing mutant, aflatoxin B1 (**6**) production was reported to increase from 0.096 μg/g to 0.105 μg/g when grown in complete medium. In contrast, the deletion of *creA* resulted in the almost complete loss of aflatoxin B1 (**6**) production [103]. In *A. ochraceus* fc-1, the knockout of *creA* inhibits conidiation and OTA (**15**) production with different carbon sources [104]. Analysis of the CreA regulatory network revealed that more than 35% of the genes encoding transcription factors in the *A. nidulans* genome are under CreA control. This highlights the global regulatory role of CreA [122] (Table 3).

3.5. AreA

3.5.1. Regulatory Mechanisms of AreA

AreA is a GATA-type positive transcriptional regulator that contains a C-terminal zinc finger domain and a DNA-binding domain recognizing the consensus sequence. Early studies identified AreA as the major regulatory protein in nitrogen metabolite repression (NMR). It not only indirectly regulates the expression of other genes by influencing nitrogen source utilization but also directly regulates gene expression by binding to the consensus sequence [123–125]. Furthermore, AreA had been shown to regulate other genes through histone H3 acetylation-mediated chromatin remodeling [126,127] (Figure 6). Additional regulatory factors were shown to interact with AreA to achieve indirect regulation. For example, the co-activator TamA interacts with AreA to promote gene expression [128]. Another GATA factor in the NMR pathway, AreB, has been shown to positively regulate AreA expression [129]. Furthermore, the NMR pathway repressor NmrA binds to AreA to inhibit its activity, and bZIP transcription factor MeaB indirectly regulates AreA activity by activating NmrA expression [130,131].

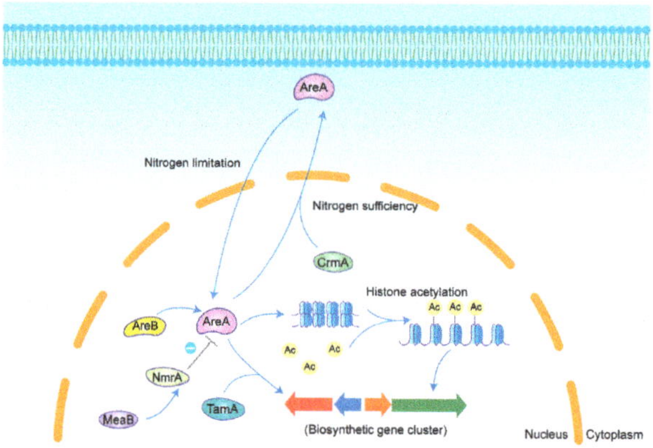

Figure 6. Mechanism of secondary metabolism regulation by AreA in *Aspergillus* spp. Under nitrogen limitation, AreA undergoes nuclear import and subsequently activates silent BGCs by directly binding to secondary metabolite genes or by facilitating chromatin remodeling through promoted histone acetylation.

The nuclear localization of AreA is affected by the nitrogen source. In *A. nidulans*, there are five classical NLSs in the AreA sequence as well as an atypical bipartite NLS within the DNA-binding domain. These six NLSs act together to regulate the nuclear localization of AreA [132]. In the NMR mechanism, nitrogen starvation causes AreA to be primarily localized in the nucleus, where it activates genes dependent on AreA. Upon detection of the nitrogen signal, CrmA facilitates the nuclear export of AreA, leading to its rapid accumulation in the cytoplasm. This suggests that AreA acts as a positive regulator in NMR [133].

3.5.2. AreA Regulation-Derived SMs

Most studies on AreA have focused on its role in NMR, while other studies have suggested that AreA also plays a regulatory role in *Aspergillus* spp. secondary metabolism. The deletion of *areA* in *A. terreus* SBUG844 results in a reduced expression of *terR*, a specific activator of terrein (**49**), which in turn decreases terrein (**49**) biosynthesis. The loss of *areA* in *A. terreus* SBUG844 also inhibits growth on media where methionine is the only nitrogen source [105]. The nitrogen source affects aflatoxin B1 (**6**) production in *A. flavus* SRRC1709. AreA, as a major regulator of nitrogen metabolism, can further affect aflatoxin B1 metabolism by influencing nitrogen metabolism, and it was found that the presence of glutamine, the optimal nitrogen source for aflatoxin B1 synthesis, with either the overexpression or deletion of *area*, resulted in increased aflatoxin B1 production [106] (Table 3).

3.6. CBC

3.6.1. Regulatory Mechanisms of CBC

CBC is a transcriptional regulator comprising three core subunits: HapB, HapC, and HapE. CBC binds to the CCAAT box in the promoter regions of target genes, thereby regulating their transcription [134]. The bZIP-type transcriptional regulator HapX is activated under iron-restricted conditions and subsequently interacts with CBC to form a complex that co-regulates genes involved in iron homeostasis in *Aspergillus* spp. [135]. HapC and HapE, which do not possess NLS, depend on their interaction with the NLS-bearing HapB to assemble the CBC. CBC is subsequently imported into the nucleus and performs its regulatory roles [136,137] (Figure 7). It has been demonstrated that the knockout of *hapB* resulted in a rate of 46.3% for differentially expressed genes in *A. fumigatus*, further indicating the global regulatory role of CBC [138].

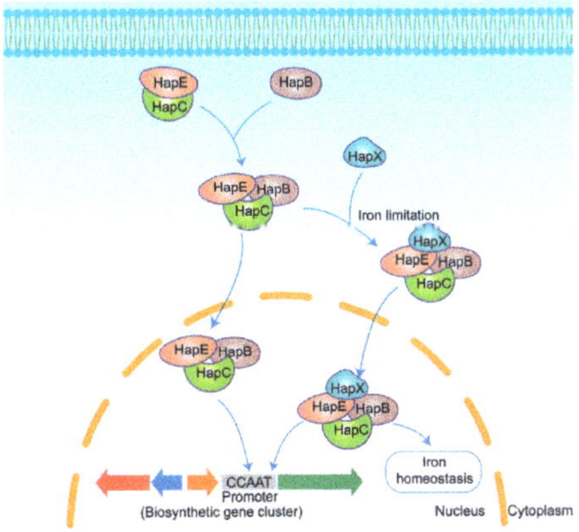

Figure 7. Mechanism of secondary metabolism regulation by CBC in *Aspergillus* spp. The HapE–HapC complex associates with HapB to form CBC, which translocates into the nucleus and regulates the expression of biosynthetic genes by binding to the CCAAT box within promoters. Furthermore, it can form a complex with HapX to regulate secondary metabolism and iron homeostasis.

3.6.2. CBC Regulation-Derived SMs

In *A. nidulans* MH8193, penicillin biosynthesis is regulated by CBC. The deletion of *hapC* significantly reduces *ipnA* and *aat* (the gene encoding isopenicillin N acyltransferase) expression and also decreases penicillin production by approximately 30% [107]. In *A. fumigatus* A1160P, *hapC* deletion mutants showed a 2.4-fold increase in ergosterol levels, and *hapE* loss-of-function mutations increased ergosta-5,7,24(28)-trien-3β-ol (**50**) and episterol (**51**) production [108] (Table 3). These results revealed that CBC negatively regulates sterol biosynthesis and positively regulates penicillin biosynthesis.

3.7. Other GTRs

McrA, a GTR containing the $Zn(II)_2Cys_6$ structural domain, is a negative regulator of fungal SMs identified through a genetic screening approach in *A. nidulans*. In the *A. nidulans* FGSC A4 *mcrA* knockout mutant, the production of nine SMs was found to be upregulated, and two new compounds (**52,53**) were identified [139]. However, its regulatory mechanisms have not been precisely studied. The knockout of *mcrA* in *A. nidulans* FGSCA442 upregulated the production of 15 SMs and led to the biosynthesis of a new cichorine intermediate (**54**) [140]. The deletion of *mcrA* in *A. wentii* IMI 49129 upregulates the production of 15 SMs and led to the identification of nine novel chemicals (**55–63**), while knockout of the polyketide synthase gene resulted in the production of aspergillus acids B (**64**) and E (**65**) [35]. In the *mcrA* deletion mutant of *A. melleus* IMV 01140, the neoaspergillic acid (**66**) and neohydroxyaspergillic acid (**67**) yields were increased by 1.7- and 1.6-fold, respectively [36]. In the *A. oryzae* NSAR1 *mcrA* knockout mutant, although no obvious production of new SM was observed, the production of kojic acid (**9**) increased from 1.23 g/L to 2.52 g/L [141]. The overexpression of *mcrA* in *A. nidulans* FGSC4 completely inhibited ST (**1**) biosynthesis, highlighting the role of McrA as a negative regulator [142] (Table 4).

Table 4. Examples of secondary metabolites regulated by other GTRs in *Aspergillus* spp.

GTRs	*Aspergillus* Strain	Deletion or Overexpression	Upregulated SMs	Downregulated SMs	Ref.
McrA	*A. Nidulans* FGSC A4	Deletion	1,3-Dihydro-6-hydroxy-4-methoxy-5-methyl-1-oxo-2H-isoindole-2-pentanoic acid (**52**), 4-[hydroxy(4-hydroxy-2-methoxy-3,6-dimethylphenyl)methoxy]-2-methoxy-3,6-dimethylbenzaldehyde (**53**)	-	[139]
	A. nidulans FGSCA442	Deletion	Cichorine intermediate (**54**)	-	[140]
	A. wentii IMI 49129	Deletion	Emodin (**55**), physcion (**56**), sulochrin (**57**), 14-O-demethylsulochrin (**58**), physcion bianthrone (**59**), (trans)-emodin bianthrone (**60**), (cis)-emodin bianthrone (**61**), (trans)-emodin physcion bianthrone (**62**), (cis)-emodin physcion bianthrone (**63**), aspergillus acid B (**64**), and aspergillus acid E (**65**)	-	[35]
	A. melleus IMV 01140	Deletion	Neoaspergillic acid (**66**), and neohydroxyaspergillic acid (**67**)	-	[36]
McrA	*A. oryzae* NSAR1	Deletion	Kojic acid (**9**)	-	[141]
	A. nidulans FGSC4	Overexpression	-	ST (**1**)	[142]
LaeB	*A. Nidulans* BTP69	Deletion	-	ST (**1**)	[143]
	A. flvus NRRL3357	Deletion	-	Aflatoxin	
LaeB	*A. nidulans* RJMP1.49	Deletion	Gibellulin C (**68**), gibellulin D (**69**), 3-methoxyporriolide (**70**), 7-methoxyporriolide (**71**), porriolide (**72**), cichorine (**73**), asperthecin (**74**), microperfuranone (**75**)	ST (**1**)	[144]
HbxA/Hbx1	*A. flavus* CA14	Deletion of *hbx1*	-	CPA (**24**), aflatrem (**39**) and aflatoxin	[145]
	A. nidulans FGSC4	Deletion of *hbxA*	ST (**1**)	-	[146]

Table 4. *Cont.*

GTRs	*Aspergillus* Strain	Deletion or Overexpression	Upregulated SMs	Downregulated SMs	Ref.
RsmA	*A. nidulans* RJMP1.49	Overexpression	ST (**1**)	-	[147]
	A. fumigatus AF293	Overexpression	Gliotoxin (**4**) and cyclo(L-Phe-L-Ser) (**76**)	-	[148]
	A. flavus CA14	Overexpression	Aflatoxin B1 (**6**)	-	[149]
StuA	*A. terreus* RA2905	Deletion	Terrein(**49**)	Butyrolactones	[150]
SirE	*A. nidulans* A26	Deletion	-	ST (**1**)	[151]
	A. flavus CA14 PTS	Deletion	Aflatoxin	-	[152]
RimO	*A. nidulans* FGSC4	Overexpression	ST (**1**) and penicillin G (**3**)	-	[153]
		Deletion	-	ST (**1**)	

A new GTR, LaeB, was identified through a forward genetic screening approach in *A. nidulans*. LaeB has a significant effect on SMs, and the production of ST (**1**) in *A. nidulans* BTP69 and aflatoxin in *A. flavus* NRRL3357 was abolished as a result of *laeB* knockout [143]. The deletion of *laeB* activated silent BGCs in *A. nidulans* RJMP1.49 and led to the production of four new substances (**68–71**) and four known metabolites (**72–75**). This demonstrates the potential of LaeB to unlock the production of cryptic natural products in *Aspergillus* spp. [144].

HbxA/Hbx1 is a class of GTRs with homeobox protein domains. In *A. flavus* CA14, CPA (**24**), aflatrem (**39**), and aflatoxin cannot be detected after the deletion of *hbx1* [145]. In contrast, the *A. nidulans* FGSC4 *hbxA* knockout mutant increased ST (**1**) production, suggesting that HbxA acts as a negative regulator of ST biosynthesis [146]. The mutagenic screening of *laeA* deletion *A. nidulans* RJMP1.49 strain led to the identification of the bZIP-type GTR RsmA [154]. RsmA positively regulates ST (**1**) production in *A. nidulans* RJMP1.49 [147]. The overexpression of *rsmA* in *A. fumigatus* AF293 increased gliotoxin (**4**) and cyclo(L-Phe-L-Ser) (**76**) production [148]. AflrsmA, a homolog of RsmA from *A. flavus* CA14, was shown to regulate the production of the secondary metabolite aflatoxin B1 (**6**) through the oxidative stress pathway [149]. The APSES family transcription factor StuA was originally recognized as a key protein involved in filamentous fungal development [155,156]. Recent studies on StuA in *A. terreus* RA2905 have demonstrated that the deletion of *stuA* significantly increases the yield of the anticancer drug terrein (**49**) by 21-fold compared to that from the wild-type strain while reducing the content of the by-product butyrolactones [150].

The NAD+-dependent histone deacetylase SirE has been identified as a GTR associated with epigenetic regulation. SirE positively regulates ST (**1**) biosynthesis in *A. nidulans* A26 and negatively regulates aflatoxin production in *A. flavus* CA14 PTS [151,152]. The novel chromatin-associated GTR, RimO, was discovered through an RNA sequencing (RNA-seq) analysis of *A. nidulans* FGSC4. RimO has been shown to positively regulate the starvation-induced production of SMs. RimO deletion led to ST (**1**) being nearly undetectable, while the RimO overexpression mutant not only promoted the biosynthesis of ST but also enhanced the production of penicillin G (**3**) [153]. The recently identified RlcA is a regulator of light sensing and chromatin remodeling, and a transcriptome analysis revealed that RlcA regulates more than one third of the genes in *A. nidulans* [157]. Although there is currently no specific research on the secondary metabolism regulation by RlcA, it is speculated that RlcA may indirectly influence the expression of secondary metabolism genes through chromatin remodeling [158].

4. Conclusions and Perspectives

The regulatory network of GTRs is complex, and regulating GTR expression can affect SMs production in *Aspergillus* spp. In this review, we summarize engineering approaches for the knockout and overexpression of GTRs in *Aspergillus* spp. LaeA, VeA, PacC, CreA, AreA, and CBC are typical GTRs in *Aspergillus* spp. We have elucidated in detail the possible mechanisms by which they exert their regulatory functions based on previous studies and have described the SMs regulated by them. In addition, further exploration of the regulatory network of GTRs has led to the identification of several novel GTRs. These new GTRs were identified either through genetic screening or from previously known transcriptional regulators that were revealed to function as potential GTRs. These GTRs have been used to discover novel compounds, including flavipamide A and B (**17**,**18**), versicolor A (**25**), 3-methoxyporriolide (**70**), and 7-methoxyporriolide (**71**). Additionally, the production of some natural product drugs, such as penicillin G (**3**) and lovastatin (**5**), has also been increased.

Multi-omics technologies have been applied for GTR identifications and investigations into their regulatory mechanisms, and a deeper understanding of the GTRs regulatory networks will promote the natural products discovery [157,159]. Furthermore, with the continuous development of genetic engineering technologies, it is important to improve gene editing precision and to apply emerging technologies more broadly to the regulation of GTR expression in *Aspergillus* spp. Efficient genetic targeting methods can accelerate gene overexpression or knockout and enable the rapid verification of gene functions. Currently, some recently identified GTRs, such as RlcA, have not been applied for the activation of silent BGCs, despite their predicted potential to regulate secondary metabolism. In the future, more attention could be directed toward exploring the regulatory potential of these factors in secondary metabolism.

Supplementary Materials: The following supporting information can be downloaded at https://www.mdpi.com/article/10.3390/jof11060449/s1, Figure S1: Structures of secondary metabolite 1-76 regulated by global transcriptional regulators in *Aspergillus* spp.

Author Contributions: Conceptualization, H.Z. and Y.Z.; investigation, Y.Z. and Q.G.; visualization, Y.Z. and Q.G.; writing—original draft, Y.Z.; writing—review and editing, H.Z.; funding acquisition, H.Z. All authors have read and agreed to the published version of the manuscript.

Funding: This work was financially supported by the National Key Research and Development Program of China (2022YFC2804203).

Institutional Review Board Statement: Not applicable.

Informed Consent Statement: Not applicable.

Data Availability Statement: The data presented in this study are available in the Supplementary Materials.

Conflicts of Interest: The authors declare no conflicts of interest.

Abbreviations

The following abbreviations are used in this manuscript:

ATMT	*Agrobacterium tumefaciens*-mediated transformation
BGCs	biosynthetic gene clusters
Cas9	CRISPR-Associated Protein 9
CBC	the CCAAT-binding complex
CCR	carbon catabolite repression
ChIP	chromatin immunoprecipitation
CPA	cyclopiazonic acid

CRISPR	clustered Regularly Interspaced Short Palindromic Repeats
DPPH	2,2-diphenyl-1-picrylhydrazyl
ECB	echinocandin B
ESCRT	endosomal sorting complex required for transport
GTRs	global transcriptional regulators
HR	homologous recombination
H3K4	histone 3 lysine 4
H3K9	histone 3 lysine 9
IC_{50}	half maximal inhibitory concentration
MIC	minimum Inhibitory Concentration
MONJ	monacolin J
NHEJ	non-homologous end joining
NLS	nuclear localization sequence
NMR	nitrogen metabolite repression
NRPS	non-ribosomal peptide synthetase
OSMAC	one strain many compounds
OTA	ochratoxin A
PCR	polymerase chain reaction
PKA	protein kinase A
PKS	polyketide synthase
qRT-PCR	quantitative reverse transcription polymerase chain reaction
RNA-seq	RNA sequencing
SirE	sirtuin E
SMs	secondary metabolites
ST	sterigmatocystin

References

1. Houbraken, J.; Kocsubé, S.; Visagie, C.M.; Yilmaz, N.; Wang, X.C.; Meijer, M.; Kraak, B.; Hubka, V.; Bensch, K.; Samson, R.A.; et al. Classification of *Aspergillus, Penicillium, Talaromyces* and related genera (Eurotiales): An overview of families, genera, subgenera, sections, series and species. *Stud. Mycol.* **2020**, *95*, 5–169. [CrossRef] [PubMed]
2. The Dictionary of Natural Products. Available online: https://dnp.chemnetbase.com (accessed on 27 March 2025).
3. Bai, X.; Sheng, Y.; Tang, Z.; Pan, J.; Wang, S.; Tang, B.; Zhou, T.; Shi, L.; Zhang, H. Polyketides as secondary metabolites from the genus *Aspergillus*. *J. Fungi* **2023**, *9*, 261. [CrossRef] [PubMed]
4. Xu, K.; Yuan, X.L.; Li, C.; Li, X.D. Recent discovery of heterocyclic alkaloids from marine-derived *Aspergillus* species. *Mar. Drugs* **2020**, *18*, 54. [CrossRef]
5. Yu, R.; Liu, J.; Wang, Y.; Wang, H.; Zhang, H. *Aspergillus niger* as a secondary metabolite factory. *Front. Chem.* **2021**, *9*, 701022. [CrossRef]
6. Bladt, T.T.; Frisvad, J.C.; Knudsen, P.B.; Larsen, T.O. Anticancer and antifungal compounds from *Aspergillus*, penicillium and other filamentous fungi. *Molecules* **2013**, *18*, 11338–11376. [CrossRef]
7. Andersen, M.R.; Nielsen, J.B.; Klitgaard, A.; Petersen, L.M.; Zachariasen, M.; Hansen, T.J.; Blicher, L.H.; Gotfredsen, C.H.; Larsen, T.O.; Nielsen, K.F.; et al. Accurate prediction of secondary metabolite gene clusters in filamentous fungi. *Proc. Natl. Acad. Sci. USA* **2013**, *110*, E99–E107. [CrossRef]
8. Bergmann, S.; Schümann, J.; Scherlach, K.; Lange, C.; Brakhage, A.A.; Hertweck, C. Genomics-driven discovery of PKS-NRPS hybrid metabolites from *Aspergillus nidulans*. *Nat. Chem. Biol.* **2007**, *3*, 213–217. [CrossRef]
9. Xu, S.; Li, M.; Hu, Z.; Shao, Y.; Ying, J.; Zhang, H. The potential use of fungal co-culture strategy for discovery of new secondary metabolites. *Microorganisms* **2023**, *11*, 464. [CrossRef]
10. Zhao, F.; Sun, C.; Liu, Z.; Cabrera, A.; Escobar, M.; Huang, S.; Yuan, Q.; Nie, Q.; Luo, K.L.; Lin, A.; et al. Multiplex base-editing enables combinatorial epigenetic regulation for genome mining of fungal natural products. *J. Am. Chem. Soc.* **2023**, *145*, 413–421. [CrossRef]
11. Wei, Q.; Bai, J.; Yan, D.; Bao, X.; Li, W.; Liu, B.; Zhang, D.; Qi, X.; Yu, D.; Hu, Y. Genome mining combined metabolic shunting and OSMAC strategy of an endophytic fungus leads to the production of diverse natural products. *Acta Pharm. Sin. B* **2021**, *11*, 572–587. [CrossRef]
12. Hautbergue, T.; Jamin, E.L.; Debrauwer, L.; Puel, O.; Oswald, I.P. From genomics to metabolomics, moving toward an integrated strategy for the discovery of fungal secondary metabolites. *Nat. Prod. Rep.* **2018**, *35*, 147–173. [CrossRef] [PubMed]

13. Lyu, H.N.; Liu, H.W.; Keller, N.P.; Yin, W.B. Harnessing diverse transcriptional regulators for natural product discovery in fungi. *Nat. Prod. Rep.* **2020**, *37*, 6–16. [CrossRef] [PubMed]
14. Brakhage, A.A. Regulation of fungal secondary metabolism. *Nat. Rev. Microbiol.* **2013**, *11*, 21–32. [CrossRef]
15. Yu, W.; Pei, R.; Zhou, J.; Zeng, B.; Tu, Y.; He, B. Molecular regulation of fungal secondary metabolism. *World J. Microbiol. Biotechnol.* **2023**, *39*, 204. [CrossRef]
16. Macheleidt, J.; Mattern, D.J.; Fischer, J.; Netzker, T.; Weber, J.; Schroeckh, V.; Valiante, V.; Brakhage, A.A. Regulation and role of fungal secondary metabolites. *Annu. Rev. Genet.* **2016**, *50*, 371–392. [CrossRef]
17. Wang, S.; Chen, H.; Tang, X.; Zhang, H.; Chen, W.; Chen, Y.Q. Molecular tools for gene manipulation in filamentous fungi. *Appl. Microbiol. Biotechnol.* **2017**, *101*, 8063–8075. [CrossRef]
18. Mei, Y.Z.; Zhu, Y.L.; Huang, P.W.; Yang, Q.; Dai, C.C. Strategies for gene disruption and expression in filamentous fungi. *Appl. Microbiol. Biotechnol.* **2019**, *103*, 6041–6059. [CrossRef]
19. Yu, J.H.; Hamari, Z.; Han, K.H.; Seo, J.A.; Reyes-Domínguez, Y.; Scazzocchio, C. Double-joint PCR: A PCR-based molecular tool for gene manipulations in filamentous fungi. *Fungal Genet. Biol.* **2004**, *41*, 973–981. [CrossRef]
20. Szewczyk, E.; Nayak, T.; Oakley, C.E.; Edgerton, H.; Xiong, Y.; Taheri-Talesh, N.; Osmani, S.A.; Oakley, B.R. Fusion PCR and gene targeting in *Aspergillus nidulans*. *Nat. Protoc.* **2006**, *1*, 3111–3120. [CrossRef]
21. Son, Y.E.; Park, H.S. Genetic Manipulation and transformation methods for *Aspergillus* spp. *Mycobiology* **2021**, *49*, 95–104. [CrossRef]
22. Shrivastav, M.; De Haro, L.P.; Nickoloff, J.A. Regulation of DNA double-strand break repair pathway choice. *Cell Res.* **2008**, *18*, 134–147. [CrossRef] [PubMed]
23. Carvalho, N.D.S.P.; Arentshorst, M.; Jin Kwon, M.; Meyer, V.; Ram, A.F.J. Expanding the *ku70* toolbox for filamentous fungi: Establishment of complementation vectors and recipient strains for advanced gene analyses. *Appl. Microbiol. Biotechnol.* **2010**, *87*, 1463–1473. [CrossRef]
24. Nayak, T.; Szewczyk, E.; Oakley, C.E.; Osmani, A.; Ukil, L.; Murray, S.L.; Hynes, M.J.; Osmani, S.A.; Oakley, B.R. A versatile and efficient gene-targeting system for *Aspergillus nidulans*. *Genetics* **2006**, *172*, 1557–1566. [CrossRef]
25. Meyer, V.; Arentshorst, M.; El-Ghezal, A.; Drews, A.C.; Kooistra, R.; van den Hondel, C.A.M.J.J.; Ram, A.F.J. Highly efficient gene targeting in the *Aspergillus niger kusA* mutant. *J. Biotechnol.* **2007**, *128*, 770–775. [CrossRef]
26. Da Silva Ferreira, M.E.; Kress, M.R.V.Z.; Savoldi, M.; Goldman, M.H.S.; Härtl, A.; Heinekamp, T.; Brakhage, A.A.; Goldman, G.H. The *akuB(KU80)* mutant deficient for nonhomologous end joining is a powerful tool for analyzing pathogenicity in *Aspergillus fumigatus*. *Eukaryot. Cell* **2006**, *5*, 207–211. [CrossRef]
27. Krappmann, S.; Sasse, C.; Braus, G.H. Gene targeting in *Aspergillus fumigatus* by homologous recombination is facilitated in a nonhomologous end- joining-deficient genetic background. *Eukaryot. Cell* **2006**, *5*, 212–215. [CrossRef]
28. Fairhead, C.; Llorente, B.; Denis, F.; Soler, M.; Dujon, B. New vectors for combinatorial deletions in yeast chromosomes and for gap-repair cloning using 'split-marker' recombination. *Yeast* **1996**, *12*, 1439–1457. [CrossRef]
29. Goswami, R.S. Targeted gene replacement in fungi using a split-marker approach. In *Plant Fungal Pathogens: Methods and Protocols*; Bolton, M.D., Thomma, B.P.H.J., Eds.; Humana Press: Totowa, NJ, USA, 2012; pp. 255–269.
30. Nielsen, M.L.; Albertsen, L.; Lettier, G.; Nielsen, J.B.; Mortensen, U.H. Efficient PCR-based gene targeting with a recyclable marker for *Aspergillus nidulans*. *Fungal Genet. Biol.* **2006**, *43*, 54–64. [CrossRef]
31. Li, D.; Tang, Y.; Lin, J.; Cai, W. Methods for genetic transformation of filamentous fungi. *Microb. Cell Fact.* **2017**, *16*, 168. [CrossRef]
32. De Groot, M.J.A.; Bundock, P.; Hooykaas, P.J.J.; Beijersbergen, A.G.M. *Agrobacterium tumefaciens*-mediated transformation of filamentous fungi. *Nat. Biotechnol.* **1998**, *16*, 839–842. [CrossRef]
33. Nødvig, C.S.; Nielsen, J.B.; Kogle, M.E.; Mortensen, U.H. A CRISPR-Cas9 system for genetic engineering of filamentous fungi. *PLoS ONE* **2015**, *10*, e0133085. [CrossRef] [PubMed]
34. Wassano, N.S.; da Silva, G.B.; Reis, A.H.; Gerhardt, J.A.; Antoniel, E.P.; Akiyama, D.; Rezende, C.P.; Neves, L.X.; Vasconcelos, E.J.R.; de Figueiredo, F.L.; et al. Sirtuin E deacetylase is required for full virulence of *Aspergillus fumigatus*. *Commun. Biol.* **2024**, *7*, 704. [CrossRef] [PubMed]
35. Yuan, B.; Keller, N.P.; Oakley, B.R.; Stajich, J.E.; Wang, C.C.C. Manipulation of the global regulator mcrA upregulates secondary metabolite production in *Aspergillus wentii* using CRISPR-Cas9 with in vitro assembled ribonucleoproteins. *ACS Chem. Biol.* **2022**, *17*, 2828–2835. [CrossRef] [PubMed]
36. Yuan, B.; Grau, M.F.; Murata, R.M.; Torok, T.; Venkateswaran, K.; Stajich, J.E.; Wang, C.C.C. Identification of the neoaspergillic acid biosynthesis gene cluster by establishing an in vitro CRISPR-Ribonucleoprotein genetic system in *Aspergillus melleus*. *ACS Omega* **2023**, *8*, 16713–16721. [CrossRef]
37. Juven-Gershon, T.; Hsu, J.-Y.; Theisen, J.W.; Kadonaga, J.T. The RNA polymerase II core promoter—The gateway to transcription. *Curr. Opin. Cell Biol.* **2008**, *20*, 253–259. [CrossRef]

38. Punt, P.J.; Dingemanse, M.A.; Kuyvenhoven, A.; Soede, R.D.M.; Pouwels, P.H.; van den Hondel, C.A.M.J.J. Functional elements in the promoter region of the *Aspergillus nidulans* gpdA gene encoding glyceraldehyde-3-phosphate dehydrogenase. *Gene* **1990**, *93*, 101–109. [CrossRef]
39. Fowler, T.; Berka, R.M.; Ward, M. Regulation of the glaA gene of *Aspergillus niger*. *Curr. Genet.* **1990**, *18*, 537–545. [CrossRef]
40. Wei, P.L.; Fan, J.; Yu, J.; Ma, Z.; Guo, X.; Keller, N.P.; Li, E.; Lou, C.; Yin, W.B. Quantitative characterization of filamentous fungal promoters on a single-cell resolution to discover cryptic natural products. *Sci. China Life Sci.* **2023**, *66*, 848–860. [CrossRef]
41. Nishikawa, R.; Yoshida, M.; Noda, T.; Okuhara, T.; Taguchi, G.; Inatomi, S.; Shimosaka, M. pFungiway: A series of plasmid vectors used for gene manipulation in fungi. *Ann. Microbiol.* **2016**, *66*, 825–832. [CrossRef]
42. Roux, I.; Woodcraft, C.; Sbaraini, N.; Pepper, A.; Wong, E.; Bracegirdle, J.; Chooi, Y.H. Next-generation AMA1-based plasmids for enhanced heterologous expression in filamentous fungi. *Microb. Biotechnol.* **2024**, *17*, e70010. [CrossRef]
43. Wang, Y.; Chen, Y.; Zhang, J.; Zhang, C. Overexpression of *llm1* affects the synthesis of secondary metabolites of *Aspergillus cristatus*. *Microorganisms* **2022**, *10*, 1707. [CrossRef] [PubMed]
44. Wang, T.; Liang, C.; Xing, W.; Wu, W.; Hou, Y.; Zhang, L.; Xiao, S.; Xu, H.; An, Y.; Zheng, M.; et al. Transcriptional factor engineering in microbes for industrial biotechnology. *J. Chem. Technol. Biotechnol.* **2020**, *95*, 3071–3078. [CrossRef]
45. Khan, I.; Xie, W.L.; Yu, Y.C.; Sheng, H.; Xu, Y.; Wang, J.Q.; Debnath, S.C.; Xu, J.Z.; Zheng, D.Q.; Ding, W.J.; et al. Heteroexpression of *Aspergillus nidulans laeA* in marine-derived fungi triggers upregulation of secondary metabolite biosynthetic genes. *Mar. Drugs* **2020**, *18*, 652. [CrossRef]
46. Bok, J.W.; Keller, N.P. LaeA, a regulator of secondary metabolism in *Aspergillus* spp. *Eukaryot. Cell* **2004**, *3*, 527–535.
47. Jin, W.B.; Balajee, S.A.; Marr, K.A.; Andes, D.; Nielsen, K.F.; Frisvad, J.C.; Keller, N.P. LaeA, a regulator of morphogenetic fungal virulence factors. *Eukaryot. Cell* **2005**, *4*, 1574–1582.
48. Keller, N.; Bok, J.; Chung, D.; Perrin, R.M.; Keats Shwab, E. LaeA, a global regulator of *Aspergillus* toxins. *Med. Mycol.* **2006**, *44*, 83–85. [CrossRef]
49. Reyes-Dominguez, Y.; Bok, J.W.; Berger, H.; Shwab, E.K.; Basheer, A.; Gallmetzer, A.; Scazzocchio, C.; Keller, N.; Strauss, J. Heterochromatic marks are associated with the repression of secondary metabolism clusters in *Aspergillus nidulans*. *Mol. Microbiol.* **2010**, *76*, 1376–1386. [CrossRef]
50. Kadooka, C.; Nakamur, E.; Mori, K.; Okutsu, K.; Yoshizaki, Y.; Takamine, K.; Goto, M.; Tamaki, H.; Futagami, T. LaeA controls citric acid production through regulation of the citrate exporter-encoding *cexA* gene in *Aspergillus luchuensis* mut. kawachii. *Appl. Environ. Microbiol.* **2020**, *86*, e01950-19. [CrossRef]
51. Lind, A.L.; Lim, F.Y.; Soukup, A.A.; Keller, N.P.; Rokas, A. An LaeA- and BrlA-dependent cellular network governs tissue-specific secondary metabolism in the human pathogen *Aspergillus fumigatus*. *mSphere* **2018**, *3*, e00050-18. [CrossRef]
52. Bayram, Ö.; Krappmann, S.; Ni, M.; Bok, J.W.; Helmstaedt, K.; Valerius, O.; Braus-Stromeyer, S.; Kwon, N.-J.; Keller, N.P.; Yu, J.H.; et al. VelB/VeA/LaeA complex coordinates light signal with fungal development and secondary metabolism. *Science* **2008**, *320*, 1504–1506. [CrossRef]
53. Bok, J.W.; Noordermeer, D.; Kale, S.P.; Keller, N.P. Secondary metabolic gene cluster silencing in *Aspergillus nidulans*. *Mol. Microbiol.* **2006**, *61*, 1636–1645. [CrossRef] [PubMed]
54. Bok, J.W.; Jr., R.A.C.; Wortman, J.R.; Kim, H.S.; Nierman, W.C.; Keller, N.P. Transcriptional regulation of chemical diversity in *Aspergillus fumigatus* by LaeA. *PLoS Pathog.* **2007**, *3*, e50.
55. Sugui, J.A.; Pardo, J.; Chang, Y.C.; Müllbacher, A.; Zarember, K.A.; Galvez, E.M.; Brinster, L.; Zerfas, P.; Gallin, J.I.; Simon, M.M.; et al. Role of laeA in the regulation of alb1, gliP, conidial morphology, and virulence in *Aspergillus fumigatus*. *Eukaryot. Cell* **2007**, *6*, 1552–1561. [CrossRef] [PubMed]
56. Chang, P.K.; Scharfenstein, L.L.; Ehrlich, K.C.; Wei, Q.; Bhatnagar, D.; Ingber, B.F. Effects of laeA deletion on *Aspergillus flavus* conidial development and hydrophobicity may contribute to loss of aflatoxin production. *Fungal Biol.* **2012**, *116*, 298–307. [CrossRef]
57. Lim, F.Y.; Hou, Y.; Chen, Y.; Oh, J.H.; Lee, I.; Bugni, T.S.; Keller, N.P. Genome-based cluster deletion reveals an endocrocin biosynthetic pathway in *Aspergillus fumigatus*. *Appl. Environ. Microbiol.* **2012**, *78*, 4117–4125. [CrossRef]
58. Oda, K.; Kobayashi, A.; Ohashi, S.; Sano, M. *Aspergillus oryzae* laeA regulates kojic acid synthesis genes. *Biosci. Biotechnol. Biochem.* **2011**, *75*, 1832–1834. [CrossRef]
59. Niu, J.; Arentshorst, M.; Nair, P.D.S.; Dai, Z.; Baker, S.E.; Frisvad, J.C.; Nielsen, K.F.; Punt, P.J.; Ram, A.F.J. Identification of a classical mutant in the industrial host *Aspergillus niger* by systems genetics: LaeA is required for citric acid production and regulates the formation of some secondary metabolites. *G3* **2016**, *6*, 193–204. [CrossRef]
60. Crespo-Sempere, A.; Marín, S.; Sanchis, V.; Ramos, A.J. VeA and LaeA transcriptional factors regulate ochratoxin a biosynthesis in *Aspergillus carbonarius*. *Int. J. Food Microbiol.* **2013**, *166*, 479–486. [CrossRef]
61. Wang, G.; Zhang, H.; Wang, Y.; Liu, F.; Li, E.; Ma, J.; Yang, B.; Zhang, C.; Li, L.; Liu, Y. Requirement of LaeA, VeA, and VelB on asexual development, ochratoxin A biosynthesis, and fungal virulence in *Aspergillus ochraceus*. *Front. Microbiol.* **2019**, *10*, 2759. [CrossRef]

62. Iliadi, M.K.; Varveri, M.; Kapetanakou, A.E.; Skandamis, P.N.; Tsitsigiannis, D.I. The global secondary metabolite regulator AcLaeA modulates *Aspergillus carbonarius* virulence, ochratoxin biosynthesis, and the mode of action of biopesticides and essential oils. *Toxins* **2025**, *17*, 2. [CrossRef]
63. Pomraning, K.R.; Dai, Z.; Munoz, N.; Kim, Y.M.; Gao, Y.; Deng, S.; Lemmon, T.; Swita, M.S.; Zucker, J.D.; Kim, J.; et al. Itaconic acid production is regulated by LaeA in *Aspergillus pseudoterreus*. *Metab. Eng. Commun.* **2022**, *15*, e00203. [CrossRef] [PubMed]
64. Liu, Y.; Li, P.; Qi, C.; Zha, Z.; Meng, J.; Liu, C.; Han, J.; Zhou, Q.; Luo, Z.; Wang, J.; et al. Cryptic piperazine derivatives activated by knocking out the global regulator LaeA in *Aspergillus flavipes*. *Bioorg. Med. Chem.* **2024**, *103*, 117685. [CrossRef] [PubMed]
65. Hong, E.J.; Kim, N.K.; Lee, D.; Kim, W.G.; Lee, I. Overexpression of the laeA gene leads to increased production of cyclopiazonic acid in *Aspergillus fumisynnematus*. *Fungal Biol.* **2015**, *119*, 973–983. [CrossRef] [PubMed]
66. He, T.; Wang, Y.; Du, L.; Li, F.; Hu, Q.; Cheng, G.; Wang, W. Overexpression of global regulator LaeA induced secondary metabolite production in *Aspergillus versicolor* 0312. *Rec. Nat. Prod.* **2020**, *14*, 387–394. [CrossRef]
67. Wang, B.; Li, X.; Tabudravu, J.; Wang, S.; Deng, H.; Pan, L. The chemical profile of activated secondary metabolites by overexpressing LaeA in *Aspergillus niger*. *Microbiol. Res.* **2021**, *248*, 126735. [CrossRef]
68. Bingxin, Z.; Wen, S.; Qin, X.Y.; Wang, Z.H.; Yao, G.S. Activating biosynthesis by overexpressing LaeA and anti-Vibrio activity of dihydroisoflavipucines of marine-derived *Aspergillus terreus*. *Mycosystema* **2023**, *42*, 562–569.
69. Zhou, W.; Li, M.; Wang, W.; Bai, X.; Zhang, H. Overexpression of the global transcriptional regulator LaeA leads to production of cyclic lipopeptides in marine-derived *Aspergillus niger* L14. *Chem. Biodivers.* **2025**, *22*, e202402704. [CrossRef]
70. Käfer, E. Origins of translocations in *Aspergillus nidulans*. *Genetics* **1965**, *52*, 217–232. [CrossRef]
71. Rauscher, S.; Pacher, S.; Hedtke, M.; Kniemeyer, O.; Fischer, R. A phosphorylation code of the *Aspergillus nidulans* global regulator Velvet A (VeA) determines specific functions. *Mol. Microbiol.* **2016**, *99*, 909–924. [CrossRef]
72. Stinnett, S.M.; Espeso, E.A.; Cobeño, L.; Araújo-Bazán, L.; Calvo, A.M. Aspergillus nidulans VeA subcellular localization is dependent on the importin α carrier and on light. *Mol. Microbiol.* **2007**, *63*, 242–255. [CrossRef]
73. Kim, H.S.; Han, K.Y.; Kim, K.J.; Han, D.M.; Jahng, K.Y.; Chae, K.S. The veA gene activates sexual development in Aspergillus nidulans. *Fungal Genet. Biol.* **2002**, *37*, 72–80. [CrossRef] [PubMed]
74. Purschwitz, J.; Müller, S.; Kastner, C.; Schöser, M.; Haas, H.; Espeso, E.A.; Atoui, A.; Calvo, A.M.; Fischer, R. Functional and physical interaction of blue- and red-light sensors in *Aspergillus nidulans*. *Curr. Biol.* **2008**, *18*, 255–259. [CrossRef] [PubMed]
75. Palmer, J.M.; Theisen, J.M.; Duran, R.M.; Grayburn, W.S.; Calvo, A.M.; Keller, N.P. Secondary metabolism and development is mediated by LlmF control of VeA subcellular localization in *Aspergillus nidulans*. *PLoS Genet.* **2013**, *9*, e1003193. [CrossRef]
76. Kato, N.; Brooks, W.; Calvo, A.M. The expression of sterigmatocystin and penicillin genes in *Aspergillus nidulans* is controlled by *veA*, a gene required for sexual development. *Eukaryot. Cell* **2003**, *2*, 1178–1186. [CrossRef]
77. Spröte, P.; Brakhage, A.A. The light-dependent regulator velvet A of *Aspergillus nidulans* acts as a repressor of the penicillin biosynthesis. *Arch. Microbiol.* **2007**, *188*, 69–79. [CrossRef]
78. Marui, J.; Ohashi-Kunihiro, S.; Ando, T.; Nishimura, M.; Koike, H.; Machida, M. Penicillin biosynthesis in *Aspergillus oryzae* and its overproduction by genetic engineering. *J. Biosci. Bioeng.* **2010**, *110*, 8–11. [CrossRef]
79. Duran, R.M.; Cary, J.W.; Calvo, A.M. Production of cyclopiazonic acid, aflatrem, and aflatoxin by *Aspergillus flavus* is regulated by veA, a gene necessary for sclerotial formation. *Appl. Microbiol. Biotechnol.* **2007**, *73*, 1158–1168. [CrossRef]
80. Cary, J.W.; OBrian, G.R.; Nielsen, D.M.; Nierman, W.; Harris-Coward, P.; Yu, J.; Bhatnagar, D.; Cleveland, T.E.; Payne, G.A.; Calvo, A.M. Elucidation of veA-dependent genes associated with aflatoxin and sclerotial production in *Aspergillus flavus* by functional genomics. *Appl. Microbiol. Biotechnol.* **2007**, *76*, 1107–1118. [CrossRef]
81. Dhingra, S.; Andes, D.; Calvo, A.M. VeA regulates conidiation, gliotoxin production, and protease activity in the opportunistic human pathogen *Aspergillus fumigatus*. *Eukaryot. Cell* **2012**, *11*, 1531–1543. [CrossRef]
82. Zhang, J.; Chen, H.; Sumarah, M.W.; Gao, Q.; Wang, D.; Zhang, Y. *veA* gene acts as a positive regulator of conidia production, ochratoxin A biosynthesis, and oxidative stress tolerance in *Aspergillus niger*. *J. Agric. Food Chem.* **2018**, *66*, 13199–13208. [CrossRef]
83. Dhingra, S.; Lind, A.L.; Lin, H.C.; Tang, Y.; Rokas, A.; Calvo, A.M. The fumagillin gene cluster, an example of hundreds of genes under *veA* control in *Aspergillus fumigatus*. *PLoS ONE* **2013**, *8*, e77147. [CrossRef] [PubMed]
84. Lan, N.; Yue, Q.; An, Z.; Bills, G.F. Apc.LaeA and Apc.VeA of the velvet complex govern secondary metabolism and morphological development in the echinocandin-producing fungus *Aspergillus pachycristatus*. *J. Ind. Microbiol. Biotechnol.* **2020**, *47*, 155–168. [CrossRef] [PubMed]
85. Bok, J.W.; Soukup, A.A.; Chadwick, E.; Chiang, Y.-M.; Wang, C.C.C.; Keller, N.P. VeA and MvlA repression of the cryptic orsellinic acid gene cluster in *Aspergillus nidulans* involves histone 3 acetylation. *Mol. Microbiol.* **2013**, *89*, 963–974. [CrossRef]
86. Selvig, K.; Alspaugh, J.A. pH Response Pathways in Fungi: Adapting to host-derived and environmental signals. *Mycobiology* **2011**, *39*, 249–256. [CrossRef]
87. Tilburn, J.; Sarkar, S.; Widdick, D.A.; Espeso, E.A.; Orejas, M.; Mungroo, J.; Peñalva, M.A.; Arst, H.N. The *Aspergillus* PacC zinc finger transcription factor mediates regulation of both acid- and alkaline-expressed genes by ambient pH. *EMBO J.* **1995**, *14*, 779–790. [CrossRef]

8. Espeso, E.A.; Tilburn, J.; Sánchez-Pulido, L.; Brown, C.V.; Valencia, A.; Arst, H.N.; Peñalva, M.A. Specific DNA recognition by the *Aspergillus nidulans* three zinc finger transcription factor PacC. *J. Mol. Biol.* **1997**, *274*, 466–480. [CrossRef]
9. Fernández-Martínez, J.; Brown, C.V.; Díez, E.; Tilburn, J.; Arst, H.N.; Peñalva, M.Á.; Espeso, E.A. Overlap of nuclear localisation signal and specific DNA-binding residues within the zinc finger domain of PacC. *J. Mol. Biol.* **2003**, *334*, 667–684. [CrossRef]
0. Ke, R.; Haynes, K.; Stark, J. Modelling the activation of alkaline pH response transcription factor PacC in *Aspergillus nidulans*: Involvement of a negative feedback loop. *J. Theor. Biol.* **2013**, *326*, 11–20. [CrossRef]
1. Díez, E.; Álvaro, J.; Espeso, E.A.; Rainbow, L.; Suárez, T.; Tilburn, J.; Arst, H.N.; Peñalva, M.Á. Activation of the Aspergillus PacC zinc finger transcription factor requires two proteolytic steps. *EMBO J.* **2002**, *21*, 1350–1359. [CrossRef]
2. Galindo, A.; Calcagno-Pizarelli, A.M.; Arst, H.N., Jr.; Peñalva, M.Á. An ordered pathway for the assembly of fungal ESCRT-containing ambient pH signalling complexes at the plasma membrane. *J. Cell Sci.* **2012**, *125*, 1784–1795. [CrossRef]
3. Hervás-Aguilar, A.; Galindo, A.; Peñalva, M.A. Receptor-independent Ambient pH Signaling by Ubiquitin Attachment to Fungal Arrestin-like PalF. *J. Biol. Chem.* **2010**, *285*, 18095–18102. [CrossRef] [PubMed]
4. Rodríguez-Galán, O.; Galindo, A.; Hervás-Aguilar, A.; Arst, H.N.; Peñalva, M.A. Physiological Involvement in pH Signaling of Vps24-mediated Recruitment of Aspergillus PalB Cysteine Protease to ESCRT-III. *J. Biol. Chem.* **2009**, *284*, 4404–4412. [CrossRef] [PubMed]
5. Orejas, M.; Espeso, E.A.; Tilburn, J.; Sarkar, S.; Arst, H.N.; Peñalva, M.A. Activation of the *Aspergillus* PacC transcription factor in response to alkaline ambient pH requires proteolysis of the carboxy-terminal moiety. *Genes Dev.* **1995**, *9*, 1622–1632. [CrossRef] [PubMed]
6. Hervás-Aguilar, A.; RodrÁguez, J.M.; Tilburn, J.; Arst, H.N.; Peñalva, M.A. Evidence for the direct involvement of the proteasome in the proteolytic processing of the *Aspergillus nidulans* zinc finger transcription factor PacC. *J. Biol. Chem.* **2007**, *282*, 34735–34747. [CrossRef]
7. Arst, H.N.; Peñalva, M.A. pH regulation in *Aspergillus* and parallels with higher eukaryotic regulatory systems. *Trends Genet.* **2003**, *19*, 224–231. [CrossRef]
8. Brakhage, A.A. Molecular regulation of penicillin biosynthesis in *Aspergillus (Emericella) nidulans*. *FEMS Microbiol. Lett.* **1997**, *148*, 1–10. [CrossRef]
9. Espeso, E.A.; Tilburn, J.; Arst, H.N.; Peñalva, M.A. pH regulation is a major determinant in expression of a fungal penicillin biosynthetic gene. *EMBO J.* **1993**, *12*, 3947–3956. [CrossRef]
00. Cary, J.W.; Ehrlich, K.C.; Wright, M.; Chang, P.K.; Bhatnagar, D. Generation of *aflR* disruption mutants of *Aspergillus parasiticus*. *Appl. Microbiol. Biotechnol.* **2000**, *53*, 680–684. [CrossRef]
01. Wang, Y.; Liu, F.; Wang, L.; Wang, Q.; Selvaraj, J.N.; Zhao, Y.; Wang, Y.; Xing, F.; Liu, Y. pH-signaling transcription factor AopacC regulates ochratoxin A biosynthesis in *Aspergillus ochraceus*. *J. Agric. Food Chem.* **2018**, *66*, 4394–4401. [CrossRef]
02. Barda, O.; Maor, U.; Sadhasivam, S.; Bi, Y.; Zakin, V.; Prusky, D.; Sionov, E. The pH-responsive transcription factor PacC governs pathogenicity and ochratoxin A biosynthesis in *Aspergillus carbonarius*. *Front. Microbiol.* **2020**, *11*, 210. [CrossRef]
03. Fasoyin, O.E.; Wang, B.; Qiu, M.; Han, X.; Chung, K.R.; Wang, S. Carbon catabolite repression gene creA regulates morphology, aflatoxin biosynthesis and virulence in *Aspergillus flavus*. *Fungal Genet. Biol.* **2018**, *115*, 41–51. [CrossRef] [PubMed]
04. Wang, G.; Wang, Y.; Yang, B.; Zhang, C.; Zhang, H.; Xing, F.; Liu, Y. Carbon catabolite repression gene *AoCreA* regulates morphological development and ochratoxin A biosynthesis responding to carbon sources in *Aspergillus ochraceus*. *Toxins* **2020**, *12*, 697. [CrossRef] [PubMed]
05. Gressler, M.; Meyer, F.; Heine, D.; Hortschansky, P.; Hertweck, C.; Brock, M. Phytotoxin production in *Aspergillus terreus* is regulated by independent environmental signals. *eLife* **2015**, *4*, e07861. [CrossRef] [PubMed]
06. Fasoyin, O.E.; Yang, K.; Qiu, M.; Wang, B.; Wang, S.; Wang, S. Regulation of morphology, aflatoxin production, and virulence of *Aspergillus flavus* by the major nitrogen regulatory gene *areA*. *Toxins* **2019**, *11*, 718. [CrossRef]
07. Litzka, O.; Papagiannopolous, P.; Davis, M.A.; Hynes, M.J.; Brakhage, A.A. The penicillin regulator PENR1 of *Aspergillus nidulans* is a HAP-like transcriptional complex. *Eur. J. Biochem.* **1998**, *251*, 758–767. [CrossRef]
08. Gsaller, F.; Hortschansky, P.; Furukawa, T.; Carr, P.D.; Rash, B.; Capilla, J.; Müller, C.; Bracher, F.; Bowyer, P.; Haas, H.; et al. Sterol biosynthesis and azole tolerance is governed by the opposing actions of SrbA and the CCAAT binding complex. *PLoS Pathog.* **2016**, *12*, e1005775.
09. Dowzer, C.E.A.; Kelly, J.M. Analysis of the *creA* gene, a regulator of carbon catabolite repression in *Aspergillus nidulans*. *Mol. Cell. Biol.* **1991**, *11*, 5701–5709.
10. Kulmburg, P.; Mathieu, M.; Dowzer, C.; Kelly, J.; Felenbok, B. Specific binding sites in the *alcR* and *alcA* promoters of the ethanol regulon for the CreA repressor mediating carbon cataboiite repression in *Aspergillus nidulans*. *Mol. Microbiol.* **1993**, *7*, 847–857. [CrossRef]
11. Tamayo, E.N.; Villanueva, A.; Hasper, A.A.; de Graaff, L.H.; Ramón, D.; Orejas, M. CreA mediates repression of the regulatory gene *xlnR* which controls the production of xylanolytic enzymes in *Aspergillus nidulans*. *Fungal Genet. Biol.* **2008**, *45*, 984–993. [CrossRef]

112. Strauss, J.; Horvath, H.K.; Abdallah, B.M.; Kindermann, J.; Mach, R.L.; Kubicek, C.P. The function of CreA, the carbon catabolite repressor of *Aspergillus nidulans*, is regulated at the transcriptional and post-transcriptional level. *Mol. Microbiol.* **1999**, *32*, 169–178. [CrossRef]
113. Prathumpai, W.; McIntyre, M.; Nielsen, J. The effect of CreA in glucose and xylose catabolism in *Aspergillus nidulans*. *Appl. Microbiol. Biotechnol.* **2004**, *63*, 748–753. [CrossRef] [PubMed]
114. De Assis, L.J.; Ries, L.N.A.; Savoldi, M.; dos Reis, T.F.; Brown, N.A.; Goldman, G.H. *Aspergillus nidulans* protein kinase A plays an important role in cellulase production. *Biotechnol. Biofuels* **2015**, *8*, 213. [CrossRef] [PubMed]
115. Boase, N.A.; Kelly, J.M. A role for creD, a carbon catabolite repression gene from *Aspergillus nidulans*, in ubiquitination. *Mol. Microbiol.* **2004**, *53*, 929–940. [CrossRef] [PubMed]
116. Alam, M.A.; Kelly, J.M. Proteins interacting with CreA and CreB in the carbon catabolite repression network in *Aspergillus nidulans*. *Curr. Genet.* **2017**, *63*, 669–683. [CrossRef]
117. Ma, X.; Jiang, Y.; Ma, L.; Luo, S.; Du, H.; Li, X.; Xing, F. Corepressors SsnF and RcoA regulate development and aflatoxin B1 biosynthesis in *Aspergillus flavus* NRRL 3357. *Toxins* **2022**, *14*, 174. [CrossRef]
118. Adnan, M.; Zheng, W.; Islam, W.; Arif, M.; Abubakar, Y.S.; Wang, Z.; Lu, G. Carbon catabolite repression in filamentous fungi. *Int. J. Mol. Sci.* **2018**, *19*, 48. [CrossRef]
119. Ribeiro, L.F.C.; Chelius, C.; Boppidi, K.R.; Naik, N.S.; Hossain, S.; Ramsey, J.J.J.; Kumar, J.; Ribeiro, L.F.; Ostermeier, M.; Tran, B.; et al. Comprehensive analysis of *Aspergillus nidulans* PKA phosphorylome identifies a novel mode of CreA regulation. *mBio* **2019**, *10*, e02825-18. [CrossRef]
120. De Assis, L.J.; Silva, L.P.; Bayram, O.; Dowling, P.; Kniemeyer, O.; Krüger, T.; Brakhage, A.A.; Chen, Y.; Dong, L.; Tan, K.; et al. Carbon catabolite repression in filamentous fungi is regulated by phosphorylation of the transcription factor CreA. *mBio* **2021**, *12*, e03146-20. [CrossRef]
121. Qin, L.; Guo, S.; Li, A.; Fan, L.; Tan, K.; Wong, K.H. An effective strategy for identifying autogenous regulation of transcription factors in filamentous fungi. *Microbiol. Spectr.* **2023**, *11*, e02347-23. [CrossRef]
122. Chen, Y.; Dong, L.; Alam, M.A.; Pardeshi, L.; Miao, Z.; Wang, F.; Tan, K.; Hynes, M.J.; Kelly, J.M.; Wong, K.H. Carbon catabolite repression governs diverse physiological processes and development in *Aspergillus nidulans*. *mBio* **2022**, *13*, e03734-21. [CrossRef]
123. Caddick, M.X.; Peters, D.; Platt, A. Nitrogen regulation in fungi. *Antonie Van Leeuwenhoek* **1994**, *65*, 169–177. [CrossRef] [PubMed]
124. Wilson, R.A.; Arst, H.N. Mutational analysis of AreA, a transcriptional activator mediating nitrogen metabolite repression in *Aspergillus nidulans* and a member of the "Streetwise" GATA family of transcription factors. *Microbiol. Mol. Biol. Rev.* **1998**, *62*, 586–596. [CrossRef] [PubMed]
125. Chant, A.; Provatopoulou, X.; Manfield, I.W.; Kneale, G.G. Structural and functional characterisation of the DNA binding domain of the *Aspergillus nidulans* gene regulatory protein AreA. *Biochim. Biophys. Acta* **2003**, *1648*, 84–89. [CrossRef]
126. Muro-Pastor, M.I.; Gonzalez, R.; Strauss, J.; Narendja, F.; Scazzocchio, C. The GATA factor AreA is essential for chromatin remodelling in a eukaryotic bidirectional promoter. *EMBO J.* **1999**, *18*, 1584–1597. [CrossRef]
127. Berger, H.; Basheer, A.; Böck, S.; Reyes-Dominguez, Y.; Dalik, T.; Altmann, F.; Strauss, J. Dissecting individual steps of nitrogen transcription factor cooperation in the *Aspergillus nidulans* nitrate cluster. *Mol. Microbiol.* **2008**, *69*, 1385–1398. [CrossRef]
128. Small, J.A.; Todd, R.B.; Zanker, M.C.; Delimitrou, S.; Hynes, M.J.; Davis, M.A. Functional analysis of TamA, a coactivator of nitrogen-regulated gene expression in *Aspergillus nidulans*. *Mol. Genet. Genom.* **2001**, *265*, 636–646. [CrossRef]
129. Chudzicka-Ormaniec, P.; Macios, M.; Koper, M.; Weedall, G.D.; Caddick, M.X.; Weglenski, P.; Dzikowska, A. The role of the GATA transcription factor AreB in regulation of nitrogen and carbon metabolism in *Aspergillus nidulans*. *FEMS Microbiol. Lett.* **2019**, *366*, fnz066. [CrossRef]
130. Kotaka, M.; Johnson, C.; Lamb, H.K.; Hawkins, A.R.; Ren, J.; Stammers, D.K. Structural analysis of the recognition of the negative regulator NmrA and DNA by the zinc finger from the GATA-type transcription factor AreA. *J. Mol. Biol.* **2008**, *381*, 373–382. [CrossRef]
131. Wong, K.H.; Hynes, M.J.; Todd, R.B.; Davis, M.A. Transcriptional control of nmrA by the bZIP transcription factor MeaB reveals a new level of nitrogen regulation in *Aspergillus nidulans*. *Mol. Microbiol.* **2007**, *66*, 534–551. [CrossRef]
132. Hunter, C.C.; Siebert, K.S.; Downes, D.J.; Wong, K.H.; Kreutzberger, S.D.; Fraser, J.A.; Clarke, D.F.; Hynes, M.J.; Davis, M.A.; Todd, R.B. Multiple nuclear localization signals mediate nuclear localization of the GATA transcription factor AreA. *Eukaryot. Cell* **2014**, *13*, 527–538. [CrossRef]
133. Todd, R.B.; Fraser, J.A.; Wong, K.H.; Davis, M.A.; Hynes, M.J. Nuclear accumulation of the GATA factor AreA in response to complete nitrogen starvation by regulation of nuclear export. *Eukaryot. Cell* **2005**, *4*, 1646–1653. [CrossRef] [PubMed]
134. Hortschansky, P.; Haas, H.; Huber, E.M.; Groll, M.; Brakhage, A.A. The CCAAT-binding complex (CBC) in *Aspergillus* species. *Biochim. Biophys. Acta Gene Regul. Mech.* **2017**, *1860*, 560–570. [CrossRef] [PubMed]
135. Furukawa, T.; Scheven, M.T.; Misslinger, M.; Zhao, C.; Hoefgen, S.; Gsaller, F.; Lau, J.; Jöchl, C.; Donaldson, I.; Valiante, V.; et al. The fungal CCAAT-binding complex and HapX display highly variable but evolutionary conserved synergetic promoter-specific DNA recognition. *Nucleic Acids Res.* **2020**, *48*, 3567–3590. [CrossRef]

136. Steidl, S.; Tüncher, A.; Goda, H.; Guder, C.; Papadopoulou, N.; Kobayashi, T.; Tsukagoshi, N.; Kato, M.; Brakhage, A.A. A single subunit of a heterotrimeric CCAAT-binding complex carries a nuclear localization signal: Piggy back transport of the pre-assembled complex to the nucleus. *J. Mol. Biol.* **2004**, *342*, 515–524. [CrossRef]
137. Goda, H.; Nagase, T.; Tanoue, S.; Sugiyama, J.; Steidl, S.; Tüncher, A.; Kobayashi, T.; Tsukagoshi, N.; Brakhage, A.A.; Kato, M. Nuclear translocation of the heterotrimeric CCAAT binding factor of *Aspergillus* oryzae is dependent on two redundant localising signals in a single subunit. *Arch. Microbiol.* **2005**, *184*, 93–100. [CrossRef]
138. Ren, Y.; Zhang, C.; Chen, Z.; Lu, L. The heterotrimeric transcription factor CCAAT-binding complex and Ca^{2+}-CrzA signaling reversely regulate the transition between fungal hyphal growth and asexual reproduction. *mBio* **2021**, *12*, e0300721. [CrossRef]
139. Oakley, C.E.; Ahuja, M.; Sun, W.W.; Entwistle, R.; Akashi, T.; Yaegashi, J.; Guo, C.J.; Cerqueira, G.C.; Russo Wortman, J.; Wang, C.C.C.; et al. Discovery of McrA, a master regulator of *Aspergillus* secondary metabolism. *Mol. Microbiol.* **2017**, *103*, 347–365. [CrossRef]
140. Grau, M.F.; Entwistle, R.; Oakley, C.E.; Wang, C.C.C.; Oakley, B.R. Overexpression of an LaeA-like methyltransferase upregulates secondary metabolite production in *Aspergillus nidulans*. *ACS Chem. Biol.* **2019**, *14*, 1643–1651. [CrossRef]
141. Dao, T.T.; de Mattos-Shipley, K.M.J.; Prosser, I.M.; Williams, K.; Zacharova, M.K.; Lazarus, C.M.; Willis, C.L.; Bailey, A.M. Cleaning the cellular factory-deletion of McrA in *Aspergillus oryzae* NSAR1 and the generation of a novel kojic acid deficient strain for cleaner heterologous production of secondary metabolites. *Front. Fungal Biol.* **2021**, *2*, 632542. [CrossRef]
142. Lee, M.K.; Son, Y.E.; Park, H.S.; Alshannaq, A.; Han, K.H.; Yu, J.H. Velvet activated McrA plays a key role in cellular and metabolic development in *Aspergillus nidulans*. *Sci. Rep.* **2020**, *10*, 15075. [CrossRef]
143. Pfannenstiel, B.T.; Zhao, X.; Wortman, J.; Wiemann, P.; Throckmorton, K.; Spraker, J.E.; Soukup, A.A.; Luo, X.; Lindner, D.L.; Lim, F.Y.; et al. Revitalization of a forward genetic screen identifies three new regulators of fungal secondary metabolism in the genus *Aspergillus*. *mBio* **2017**, *8*, e01246-17. [CrossRef] [PubMed]
144. Lin, H.; Lyu, H.; Zhou, S.; Yu, J.; Keller, N.P.; Chen, L.; Yin, W.B. Deletion of a global regulator LaeB leads to the discovery of novel polyketides in *Aspergillus nidulans*. *Org. Biomol. Chem.* **2018**, *16*, 4973–4976. [CrossRef] [PubMed]
145. Cary, J.W.; Harris-Coward, P.; Scharfenstein, L.; Mack, B.M.; Chang, P.K.; Wei, Q.; Lebar, M.; Carter-Wientjes, C.; Majumdar, R.; Mitra, C.; et al. The *Aspergillus flavus* homeobox gene, *hbx1*, is required for development and aflatoxin production. *Toxins* **2017**, *9*, 315. [CrossRef] [PubMed]
146. Son, S.H.; Son, Y.E.; Cho, H.J.; Chen, W.; Lee, M.K.; Kim, L.H.; Han, D.M.; Park, H.S. Homeobox proteins are essential for fungal differentiation and secondary metabolism in *Aspergillus nidulans*. *Sci. Rep.* **2020**, *10*, 6094. [CrossRef]
147. Yin, W.B.; Reinke, A.W.; Szilágyi, M.; Emri, T.; Chiang, Y.M.; Keating, A.E.; Pócsi, I.; Wang, C.C.C.; Keller, N.P. bZIP transcription factors affecting secondary metabolism, sexual development and stress responses in *Aspergillus nidulans*. *Microbiology* **2013**, *159*, 77–88. [CrossRef]
148. Sekonyela, R.; Palmer, J.M.; Bok, J.W.; Jain, S.; Berthier, E.; Forseth, R.; Schroeder, F.; Keller, N.P. RsmA regulates *Aspergillus fumigatus* gliotoxin cluster metabolites including cyclo(L-Phe-L-Ser), a potential new diagnostic marker for invasive aspergillosis. *PLoS ONE* **2013**, *8*, e62591. [CrossRef]
149. Wang, X.; Zha, W.; Liang, L.; Fasoyin, O.E.; Wu, L.; Wang, S. The bZIP transcription factor AflRsmA regulates aflatoxin B1 biosynthesis, oxidative stress response and sclerotium formation in *Aspergillus flavus*. *Toxins* **2020**, *12*, 271. [CrossRef]
150. Yao, G.; Bai, X.; Zhang, B.; Wang, L.; Chen, S.; Wang, Z. Enhanced production of terrein in marine-derived Aspergillus terreus by refactoring both global and pathway-specific transcription factors. *Microb. Cell Fact.* **2022**, *21*, 136. [CrossRef]
151. Itoh, E.; Odakura, R.; Oinuma, K.I.; Shimizu, M.; Masuo, S.; Takaya, N. Sirtuin E is a fungal global transcriptional regulator that determines the transition from the primary growth to the stationary phase. *J. Biol. Chem.* **2017**, *292*, 11043–11054. [CrossRef]
152. Wen, M.; Lan, H.; Sun, R.; Chen, X.; Zhang, X.; Zhu, Z.; Tan, C.; Yuan, J.; Wang, S. Histone deacetylase SirE regulates development, DNA damage response and aflatoxin production in *Aspergillus flavus*. *Environ. Microbiol.* **2022**, *24*, 5596–5610. [CrossRef]
153. Zehetbauer, F.; Seidl, A.; Berger, H.; Sulyok, M.; Kastner, F.; Strauss, J. RimO (SrrB) is required for carbon starvation signaling and production of secondary metabolites in *Aspergillus nidulans*. *Fungal Genet. Biol.* **2022**, *162*, 103726. [CrossRef] [PubMed]
154. Pandit, S.S.; Zheng, J.; Yin, Y.; Lorber, S.; Puel, O.; Dhingra, S.; Espeso, E.A.; Calvo, A.M. Homeobox transcription factor HbxA influences expression of over one thousand genes in the model fungus *Aspergillus nidulans*. *PLoS ONE* **2023**, *18*, e0286271. [CrossRef] [PubMed]
155. Sheppard, D.C.; Doedt, T.; Chiang, L.Y.; Kim, H.S.; Chen, D.; Nierman, W.C.; Filler, S.G. The *Aspergillus fumigatus* StuA protein governs the up-regulation of a discrete transcriptional program during the acquisition of developmental competence. *Mol. Biol. Cell* **2005**, *16*, 5866–5879. [CrossRef] [PubMed]
156. Miller, K.Y.; Wu, J.; Miller, B.L. StuA is required for cell pattern formation in *Aspergillus*. *Genes Dev.* **1992**, *6*, 1770–1782. [CrossRef]
157. Yu, Z.; Hübner, J.; Herrero, S.; Gourain, V.; Fischer, R. On the role of the global regulator RlcA in red-light sensing in *Aspergillus nidulans*. *Fungal Biol.* **2020**, *124*, 447–457. [CrossRef]

158. Collemare, J.; Seidl, M.F. Chromatin-dependent regulation of secondary metabolite biosynthesis in fungi: Is the picture complete? *FEMS Microbiol. Rev.* **2019**, *43*, 591–607. [CrossRef]
159. Moon, H.; Lee, M.K.; Bok, I.; Bok, J.W.; Keller, N.P.; Yu, J.H. Unraveling the gene regulatory networks of the global regulators VeA and LaeA in *Aspergillus nidulans*. *Microbiol. Spectr.* **2023**, *11*, e00166-23. [CrossRef]

Journal of
Fungi

MDPI

Review

A Review of Genomic, Transcriptomic, and Proteomic Applications in Edible Fungi Biology: Current Status and Future Directions

Muyun Xie †, Jing Wang †, Feixiang Wang, Jinfeng Wang, Yunjin Yan, Kun Feng * and Baixiong Chen *

School of Bioengineering, Zunyi Medical University, Jinwan Road No. 368, Zhuhai 519090, China; xiemujun@zmuzh.edu.cn (M.X.)
* Correspondence: fengk@zmu.edu.cn (K.F.); baixiong@zmu.edu.cn (B.C.)
† These authors contributed equally to this work.

Abstract: Edible fungi, a group of globally significant macrofungi, are highly valued for their unique flavors and substantial nutritional and medicinal properties. Understanding the molecular mechanisms governing their growth, development, gene function, biosynthesis of valuable compounds, and environmental adaptation is crucial for enhancing yield and quality, providing essential scientific support for industrial progress. Genomics, transcriptomics, and proteomics, as cornerstone life science technologies, offer powerful, integrated approaches to decipher genetic codes, reveal gene expression patterns, and elucidate complex metabolic networks in edible fungi. These advancements are transitioning research from traditional cultivation methods towards deeper molecular biology exploration. This review synthesizes key progress in applying genomics, transcriptomics, and proteomics to edible fungi, with a particular focus on metabolism-related research and the fundamentals of metabolic network construction. It discusses how these technologies, independently and in preliminary integration, uncover critical steps and regulatory mechanisms within endogenous metabolic pathways. While acknowledging the importance of metabolomics and epigenomics as cutting-edge areas, this review focuses on the "classical triad" of genomics, transcriptomics, and proteomics due to their technological maturity, data accessibility, and established application base in elucidating core metabolic mechanisms in edible fungi. The goal is to deepen the understanding of edible fungi metabolic mechanisms, providing a vital theoretical basis and practical insights for optimizing cultivation, enabling genetic improvement, harnessing bioactive substances, and promoting industrial upgrading, thereby boosting the overall efficiency and competitiveness of the edible fungi industry.

Keywords: genomics; transcriptomics; proteomics; edible fungi; metabolic networks; biosynthesis; molecular markers; stress resistance

1. Introduction

Edible fungi are generally defined as macroscopic edible fungi possessing distinct fruiting bodies or sclerotia and rich in nutrients and bioactive compounds. Globally, approximately 2300 species of edible fungi have been identified [1], primarily belonging to the phyla *Ascomycota* and *Basidiomycota*, with the latter constituting the majority [2]. Certain *Ascomycetes*, such as truffles (*Tuber* spp.) and *Cordyceps* species, are rich in active components like amino acids [3], nucleosides [4,5], polysaccharides [6], and sterols [7] and are recognized for various biological activities, including antioxidant [8], antitumor [9],

https://doi.org/10.3390/jof11060422

antiviral, and antimicrobial effects [10]. The more commonly consumed *Basidiomycetes*, including shiitake (*Lentinula edodes*), button mushroom (*Agaricus bisporus*), and wood ear (*Auricularia* spp.), are generally abundant in carbohydrates, minerals, proteins, and diverse amino acids [11,12], exhibiting significant physiological activities, such as antioxidant [13], antitumor [14], hypoglycemic [15], and immune-enhancing properties. Consequently, edible fungi play an indispensable role in the human diet and health maintenance, fueling a thriving cultivation industry and making basic biological research on these organisms a prominent scientific focus.

Against this backdrop, 'Omics' research has emerged, employing high-throughput platforms for the systematic acquisition and analysis of multi-dimensional biological data—including genomes, transcriptomes, proteomes, and metabolomes—from specific biological samples under defined conditions. The goal is to comprehend complex gene expression regulatory networks and metabolic systems holistically [16]. Initially achieving significant success in areas like disease etiology exploration [17], drug discovery [18], disease classification, and precision medicine/diagnostics [19], these high-throughput technologies have been increasingly integrated into edible fungi research. Studies now encompass a variety of important species, including *Agaricus bisporus*, *Auricularia* spp., *Flammulina velutipes*, *Ganoderma lucidum*, *Lentinula edodes*, *Pleurotus eryngii*, *Pleurotus ostreatus*, *Volvariella volvacea*, *Cordyceps militaris*, and *Morchella* spp. [20]. The application of multi-omics technologies is profoundly shifting edible fungi research from traditional reliance on morphological observation and cultivation experience towards fundamental investigations based on molecular evidence. This enables researchers to obtain complete DNA sequences, predict and identify key genes, and further investigate specific expression patterns, molecular genetic information, and functions using modern molecular biology tools like gene cloning, transgenesis, and gene editing [20]. These advancements are transitioning research from traditional cultivation methods towards deeper molecular biology exploration. While emerging technologies such as metabolomics (studying the global profile of small-molecule metabolites) and epigenomics, alongside comprehensive multi-omics integration, represent the cutting edge of systems biology research [16,20], the 'classical triad' of genomics, transcriptomics, and proteomics remains the cornerstone for elucidating life processes in edible fungi, particularly their core metabolic mechanisms (Figure 1). This is due to their technological maturity, relative data accessibility, and established application base. Genomics provides the genetic blueprint, transcriptomics reveals spatio-temporal gene expression dynamics [21], and proteomics directly reflects the status of functional effectors (proteins) and actual metabolic activity [22,23]. Information from these three levels is interconnected and hierarchical, providing indispensable insights into crucial questions: how edible fungi synthesize various nutrients, produce bioactive secondary metabolites, and respond to environmental changes through adaptive metabolic regulation. Such research, combined with genetic engineering, brings precision breeding for optimizing strain characteristics within reach. These advances not only deepen our understanding of gene function, molecular mechanisms, and regulatory networks at the micro-level but also enhance comprehension of macroscopic phenomena like phenotypic variation, growth, development, and pathological processes, ultimately serving molecularly informed breeding and scientific cultivation practices [24].

Genomics, transcriptomics, and proteomics each possess unique strengths and play pivotal roles in revealing gene function, regulating growth and development, managing stress responses, analyzing protein interaction networks, and elucidating biosynthetic pathways of active substances in edible fungi [25–28]. This review aims to summarize the current applications of genomics, transcriptomics, and proteomics in the basic research of common edible fungi (Figure 1). We will focus particularly on how these technologies, applied both individually and in initial combinations, are utilized to construct and analyze metabolic networks. We explore their contributions to dissecting key metabolic pathways, enzyme functions, and regulatory mechanisms. Through this review, we hope to provide valuable references and guidance for research focused on the genetic improvement of edible fungi, enhancement of bioactive compound yields, optimization of cultivation practices, and improvement of stress resistance, with the ultimate goal of facilitating quality enhancement and upgrading the entire edible fungi industry.

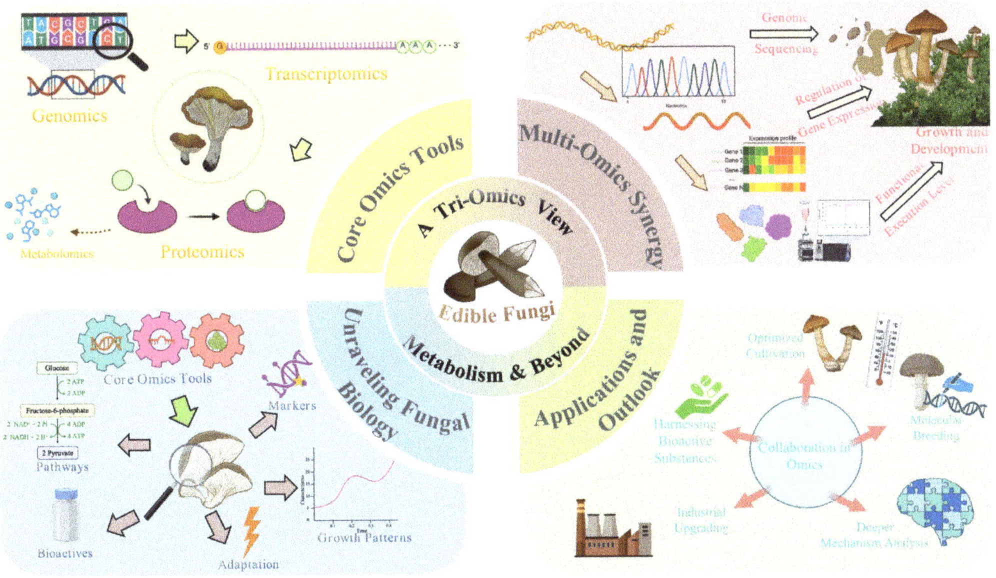

Figure 1. A framework illustrating genomic, transcriptomic, and proteomic applications in edible fungi: unraveling biology and metabolism, and guiding industrial advancement (created with BioGDP.com).

2. Genomics of Edible Fungi: Decoding the Genetic Basis of Metabolic Potential

The genome encompasses the entire genetic information of an organism, serving as the fundamental blueprint for its life activities. For edible fungi, resolving their genome sequences not only allows for the identification of species-specific characteristic sequences to support accurate taxonomic studies [29,30] but, more importantly, enables the mining of genes playing crucial roles in growth, development, environmental adaptation, and complex metabolic networks [31]. The composition and expression of these genes directly determine the biological characteristics of edible fungi, including growth rate, morphogenesis, stress tolerance, and the highly valued capacity for secondary metabolite biosynthesis [32]. Therefore, in-depth genomic research on edible fungi constitutes the logical starting point for understanding their metabolic potential and regulatory mechanisms [20]. With the advancement of gene editing technologies, targeted modification of key functional genes based on genomic information has become a significant pathway for the artificial selection of superior strains, enhancing their nutritional value and environmental adaptability [33] (Table 1).

Table 1. Applications of genomics in metabolic research of edible fungi.

Mushroom Species	Genome Research Type	Key Genes	Involved Metabolic Pathways	Significance	Ref.
Agrocybe aegerita	genome sequencing and hybrid assembly, gene annotation	genes for non-specific peroxidases, dye-decolorizing peroxidases, and Carbohydrate-active enzymes (CAZymes)	lignocellulose degradation	study of fruiting body development, biodegradation potential	[34]
Cordyceps guangdongensis	genome sequencing and annotation	CAZymes genes	biosynthesis, transport, and catabolism of secondary metabolites	provides basis for studying genetic and molecular mechanisms of fruiting body development	[35]
Cordyceps militaris	construction of genome-scale metabolic network (GSMM)	genes for biosynthesis of cordycepin and hydrolytic enzymes	biosynthesis of cordycepin, central carbon and amino acid metabolism	aids efficient production of high-value bioactive compounds with pharmaceutical potential	[36]
Cordyceps militaris	genome sequencing and annotation	genes for non-ribosomal peptide synthetases, type 1 polyketide synthases, and polyketide synthase-non-ribosomal peptide synthetase	biosynthesis of emercellamide and equisetin	further research and characterization of secondary metabolites produced by *C. militaris*	[37]
Cordyceps militaris	genome sequencing and annotation	genes for terpenoid cyclase, terpenoid synthase, fatty-acid synthase, and geranylgeranyl diphosphate synthase	biosynthesis of cordycepin	facilitates molecular studies on biology, fungicity, pathogenicity, medicinal compound production	[38]
Cordyceps militaris	genome bisulfite sequencing	genes for pyruvate metabolism, glycerophospholipid metabolism, DNA replication, and biosynthesis of N-glycan	pyruvate metabolism, glycerophospholipid metabolism, DNA replication, and N-glycan biosynthesis	helps understand potential mechanisms of strain degeneration in *C. militaris*	[39]
Dictyophora indusiate	secondary metabolite gene cluster analysis	genes for CYP450 family, and biosynthesis of terpene	biosynthesis of terpene	potential for medicinal compound production	[40]
Flammulina velutipes var. *lupinicola*	genome sequencing	lignocellulosic enzyme	lignocellulose degradation	understanding lignocellulose decomposition mechanisms	[41]
Ganderma lucidum	genome sequencing and annotation	genes for CAZymes and ligninolytic enzymes	biosynthesis of triterpene, wood degradation	aids bioengineering studies for active ingredient and bioenergy production	[42]
Ganoderma leucocontextum	genome sequencing and annotation, genome composition analysis	genes for CAZymes and biosynthesis of ergosterol	biosynthesis of sesquiterpenoid and triterpenoid	preliminary understanding of the biosynthesis of active secondary metabolites	[43]
Ganoderma Lucidum	genome resequencing and gene variation detection, annotation	genes for mycelial growth rate and synthesis of triterpene	synthesis of triterpene	reveals genetic mechanisms of *G. lucidum* mycelial growth and triterpene synthesis	[44]
Gomphus purpuraceus	genome sequencing and comparative genomics	genes for CAZymes and α-amylase family	starch/sucrose metabolism	studying biosynthesis of active compounds, understanding survival mechanisms and saprophytic ability	[30]
Hericium erinaceus	genome sequencing and annotation, gene identification	CAZymes genes	lignocellulose degradation	characterizes CAZymes and TFs in the genome, expands biological and genetic studies	[45]
Lentinula edodes	genome sequencing	genes for kinases and heat shock proteins	stress response	studying domestication process of Chinese *L. edodes* using population genomics	[46]
Lentinula edodes	genome sequence analysis	multicopper oxidase genes	expression of multicopper oxidases in mycelia, growing fruiting bodies, and fruiting bodies after harvest	suggests shared expression patterns and biological functions for laccases within the same group	[47]
Lyophyllum decastes	genome sequencing and annotation, CAZyme ID, mating type loci char.	laccase, quinone reductase, and CAZymes genes	lignocellulose degradation	understanding lignin degradation	[48]
Morchella importuna	genome sequencing, assembly and annotation, comparative analysis	CAZymes genes	starch degradation	better understanding of *Morchella* biology and evolution	[49]
Morchella sextelata	genome sequencing and annotation, comparative genomics	genes for ER repair protein 1 family, secondary metabolite, and CAZymes	posttranslational modification, protein turnover, amino acid transport and metabolism	discovery of bioactive compounds	[50]
Ophiocordyceps sinensis (Medicinal fungi)	genome sequencing and annotation	genes for type I polyketide synthases (PKSs), non-ribosomal peptide synthetases (NRPSs), and terpene synthase	biosynthesis of cordycepin, carbohydrate, amino acid, and energy metabolism	facilitates discovery and identification of novel secondary metabolite gene clusters, medicinal potential	[51]

Table 1. *Cont.*

Mushroom Species	Genome Research Type	Key Genes	Involved Metabolic Pathways	Significance	Ref.
Oudemansiella raphanipes	genome sequencing	CAZymes genes	biosynthesis of secondary metabolites, wood degradation	wood degradation and bioactive compound synthesis	[52]
Pleurotus giganteus	genome sequencing and assembly, comparative genomics	CAZymes genes	lignocellulose degradation, high-temperature adaptation	strong lignocellulolytic ability, development of molecular markers for ID and breeding	[53]
Sparassis latifolia	genome sequencing and comparison, SMGC analysis	genes encoding enzymes for carbohydrate and glycoconjugate metabolism, key SM biosynthesis enzymes	indole, terpene, and type I polyketide pathways	elucidates genetic basis of reported medicinal properties	[54]
Stropharia rugosoannulata	genome sequencing and comparative genomics	CAZymes genes	lignocellulose degradation	understanding lignocellulolytic ability	[55]
Tricholoma matsutake	comparative genomics and symbiotic adaptation	iron homeostasis genes	tryptophan metabolism	adaptation to symbiotic lifestyle and nutrient availability	[56]
Volvariella volvacea	genome sequencing and assembly, DEG identification	cellulase, hemicellulase, and pectinase genes	low temperature response, degradation of cellulose, hemicellulose, and pectin	provides evolutionary model for saprophytic nutrition and specific molecular mechanisms of cold sensitivity	[57]
Volvariella volvacea	genome sequencing and phylogenetic analysis	Flavin adenine dinucleotide (FAD)-binding proteins genes	energy transfer and utilization	understanding functional differences between homokaryons and heterokaryons	[58]
Flammulina velutipes	genome sequencing and key gene bioinformatics analysis	L-lysine biosynthesis genes	α-aminoadipate pathway	improving varieties for higher lysine content via genetic engineering	[59]

2.1. Advances in Genome Sequencing Technology and Information Mining

The maturation and declining cost of High-Throughput Sequencing (HTS) technologies have profoundly propelled genomic research in edible fungi. Whole Genome Sequencing (WGS) has laid a solid foundation for deciphering the complete genetic codes of numerous edible fungi species [20]. In recent years, the genomes of various important edible fungi have been sequenced, including *Flammulina velutipes* [60], *Ganoderma lucidum* [61], *Sparassis latifolia* [54], *Morchella* spp. [50], and *Pleurotus giganteus* [53]. Furthermore, the mitochondrial genome sequences of species such as *G. lucidum* [62], *Pleurotus eryngii* [63], *Pleurotus ostreatus* [64], *Volvariella volvacea* [65], *Lentinula edodes* [29], *Russula griseocarnosa* [66], *Coprinellus micaceus* [67], and *Grifola frondosa* [68] have also been determined. Obtaining high-quality, high-completeness genome sequences is a prerequisite for subsequent precise gene function annotation, comparative genomics analysis, and metabolic pathway reconstruction. For instance, by continuously optimizing sequencing strategies (e.g., combining second- and third-generation sequencing) and assembly algorithms, researchers have obtained more accurate and contiguous genome sequence maps for *F. velutipes* [69,70] and *V. volvacea* [71]. A high-quality WGS project for *Lyophyllum decastes*, an edible and medicinal fungus widely cultivated in China, not only predicted approximately 14,499 gene models but also specifically identified a substantial repository of Carbohydrate-Active enZyme (CAZyme) genes, particularly within the laccase gene family [48]. Similarly, sequencing the genome of *Dictyophora indusiata* revealed a vast gene repertoire (about 19,909 genes), including 369 cytochrome P450 (CYP450) family genes and as many as 64 secondary metabolism gene clusters associated with terpenoid synthesis, directly implying significant potential for producing medicinally valuable compounds [40]. The genome sequencing of *Oudemansiella raphanipes* identified about 21,308 protein-coding genes, with 56 predicted to participate in the biosynthesis of important secondary metabolites (e.g., terpenes, polyketides, non-ribosomal peptides, siderophores). Its genome also harbors abundant CAZyme genes, indicating strong lignocellulose degradation capabilities and potential for producing bioactive substances [52]. These examples clearly demonstrate that identifying gene families involved in substance degradation (e.g., Carbohydrate-active enzymes, CAZymes) and synthesis (e.g., secondary metabolite gene clusters) is a common focus in current edible fungi genomics research, highlighting the importance of these genes for understanding fungal ecological functions and tapping into their application potential.

Building upon reference genome sequences, Whole Genome Resequencing (WGR) has emerged as a critical tool for uncovering intra-specific genetic variation, understanding population genetic structure, and mining trait-associated genes [72]. By comparing differences (e.g., SNPs, InDels, SVs) between individuals or populations and the reference genome, WGR provides an effective means to dissect the genetic basis of variations in metabolic traits. For example, resequencing studies on albino mutants of *C. militaris* successfully localized several mutated genes associated with aberrant pigment metabolism (e.g., the CmPKS gene) and found that specific mutations affected light-induced melanin accumulation, offering new insights into the molecular mechanism of albinism in this species [73]. Another study utilized WGR to analyze heat-tolerant mutants of *L. edodes* generated via *Agrobacterium tumefaciens*-mediated transformation (ATMT). It was discovered that T-DNA insertion sites were sometimes located in non-coding regulatory regions, suggesting that variations in non-coding regions can also significantly impact important traits (like heat tolerance, potentially related to the thermostability of stress metabolism-associated enzymes) and underscoring the importance of searching for key regulatory elements within these regions [74]. Due to a longstanding lack of effective genetic and molecular tools for *Hericium erinaceus*, its strain improvement has progressed relatively slowly. By resequencing an F1 single-spore isolation population, researchers successfully constructed a high-resolution genetic map for *H. erinaceus*. This laid a solid foundation for subsequent localization and cloning of Quantitative Trait Loci (QTLs) associated with the synthesis of important medicinal compounds (e.g., hericenones, erinacines), strongly supporting molecular marker-assisted breeding in this species [75]. Furthermore, a population genomics resequencing study involving wild and commercially cultivated *L. edodes* from China and other parts of the world not only revealed the primary genetic origins of cultivated Chinese shiitake but also discovered unique gene pools within wild populations from Southwest and Northeast China. This provides valuable information for the conservation of *L. edodes* germplasm resources and the exploration of desirable genes (e.g., those related to flavor compound synthesis or environmental adaptation metabolism) [76]. Resequencing applications have also extended into genetic engineering. For instance, by combining different transformation and screening techniques, alongside obtaining more precise genome information for *F. velutipes* [49,50] and *V. volvacea* [71], researchers successfully tested and mapped insertion sites for multiple exogenous genes in *F. velutipes* transformants [77]. This is significant for understanding insertion site preferences, optimizing transformation systems, and ensuring transgenic safety. Collectively, these research outcomes underscore a paradigm shift: genome (re)sequencing transcends mere data acquisition. Its true analytical power lies in the deep mining of embedded information concerning genetic diversity, functional genes, regulatory elements, and evolutionary history. The broader implication is a more nuanced understanding of how genomic landscapes shape metabolic capacities and adaptive traits, paving the way for knowledge-based strain improvement and resource exploitation.

As edible fungi research continues to deepen and sequencing technologies undergo continuous innovation, genome and resequencing analyses will persistently generate vast amounts of data. Future research will increasingly focus on refining genome sequence maps. This includes integrating technologies like optical mapping and High-throughput Chromosome Conformation Capture (Hi-C) to construct chromosome-level physical maps, anchoring these to genetic maps and ultimately achieving the integration of genome sequence maps, physical maps, and high-density linkage maps ('three maps into one'). This will dramatically improve the precision of gene localization. Concurrently, utilizing high-quality reference genomes to thoroughly analyze developmentally abnormal mutants or compare the genomic variation patterns of specific strains under different environmental pressures (e.g., temperature, humidity, disease) will aid in decoding the formation mechanisms of complex traits. This will provide more precise guidance for the targeted improvement of desired traits and related breeding efforts. Furthermore, a persistent challenge impacting the utility of genomic data is the completeness of functional annotation. Even with advancements, the genomes of many edible fungi contain a significant number of genes with unknown functions, often labeled as "hypothetical proteins". This directly limits the downstream functional interpretation derived from genomic blueprints and complicates the integration with transcriptomic and proteomic data, especially for non-model edible fungi species. Ongoing efforts in genome sequencing and annotation projects are continually working to address this gap.

2.2. *Functional Annotation of Edible Fungi Genomes*

Following genome sequencing, the subsequent critical step is functional annotation: identifying various functional elements within the genome (such as protein-coding genes, non-coding RNA genes, regulatory sequences, etc.) and predicting their biological functions. Functional genomics, built upon the information provided by structural genomics, is the primary research field dedicated to interpreting genetic information and elucidating gene functions at a holistic level in the post-genomic era [78,79]. A core objective of functional annotation in edible fungi research is the identification of genes encoding key enzymes, transporters, and regulatory factors within metabolic networks, laying the groundwork for subsequent metabolic pathway analysis and functional validation. Currently, functional annotation primarily relies on bioinformatics approaches, predicting potential gene functions by comparing predicted sequences against major public databases (e.g., NCBI NR, Swiss-Prot, GO, KEGG, Pfam). However, limitations in functional annotation persist, especially for non-model edible fungi, where the genomes may contain numerous genes of unknown function or "hypothetical proteins", directly constraining the functional interpretation and integration of subsequent transcriptomic and proteomic data. Ongoing sequencing and annotation projects, such as those available through databases like MycoCosm (JGI) and FungiDB (VEuPathDB), are crucial for improving this aspect.

Throughout their life cycle, edible fungi synthesize and degrade various substances, a process reliant on a vast array of functionally diverse enzymes. Enzymes, acting as highly efficient biological catalysts under mild conditions with high substrate specificity, are deeply involved in nearly all life processes, including metabolism, nutrient acquisition, energy conversion, and signal transduction. Beyond the core metabolic enzymes essential for sustaining life, many edible fungi, particularly wood-rot species, synthesize numerous enzymes associated with the degradation of major plant cell wall components (lignocellulose). Enzymes such as cellulases, hemicellulases, and lignin-degrading enzymes not only assist fungi in obtaining carbon and energy from their environment but also possess significant industrial potential for applications in bioenergy, environmental protection, food processing, textiles, and paper manufacturing [80]. Consequently, the identification and annotation of various enzyme gene families within edible fungi genomes are paramount in functional genomics research.

2.2.1. Identification of Carbohydrate-Active enZymes (CAZymes)

The Carbohydrate-Active enZYmes (CAZymes) database is the authoritative resource for annotating enzymes involved in carbohydrate metabolism. CAZymes are a broad class of enzymes involved in the synthesis, modification, and breakdown of carbohydrates. In virtually all sequenced edible fungi genomes, extensive and diverse repertoires of CAZyme genes have been discovered. These genes encode enzymes primarily belonging to Glycoside Hydrolases (GHs), GlycosylTransferases (GTs), Polysaccharide Lyases (PLs), Carbohydrate Esterases (CEs), and enzymes with Auxiliary Activities (AAs, such as lignin-degrading oxidases) [45,81]. For instance, analysis of the *Hericium erinaceus* genome identified 341 candidate CAZyme genes, comprehensively covering the major enzyme families required for degrading cellulose, hemicellulose, and lignin [45]. Genomes of various other edible fungi, including *Flammulina velutipes* [41], *Lentinula edodes* [47], *Volvariella volvacea* [81], *Termitomyces albuminosus* [82], as well as the previously mentioned *Lyophyllum decastes* [48], *Oudemansiella raphanipes* [52], and *Stropharia rugosoannulata* [55], have been successfully annotated to contain numerous enzyme genes involved in lignocellulose or hemicellulose degradation, such as cellulases, xylanases, mannanases, laccases, and manganese peroxidases [34,41,47,81,82]. The composition and abundance of these enzymes directly dictate the fungus' ability to decompose and utilize substrates (especially plant fiber-rich materials), forming the basis of carbon and energy acquisition for primary metabolism. Concurrently, genome annotation also reveals members of GT families, which are involved not only in synthesizing oligosaccharides and glycoconjugates but also in the biosynthesis of the fungus' own cell wall polysaccharides (e.g., β-glucans, chitin) [83], crucial for maintaining cell structure and morphology.

2.2.2. Identification of Secondary Metabolism-Related Enzymes

Edible fungi are renowned for producing structurally diverse secondary metabolites with various bioactivities, often representing the source of their medicinal value. Genome annotation is the first step toward exploring this biosynthetic potential and identifying key enzyme-encoding genes. Major enzyme families involved in secondary metabolism include Cytochrome P450 monooxygenases (CYP450s, involved in diverse modification reactions), terpene synthases/cyclases (synthesizing terpene skeletons), Polyketide Synthases (PKSs, synthesizing polyketides), Non-Ribosomal Peptide Synthetases (NRPSs, synthesizing peptides), and various transferases (e.g., methyl-, acetyl-, glycosyltransferases, involved in structural modifications). For example, the significant expansion of the P450 gene family and the identification of multiple putative terpene synthase genes in the *Ganoderma lucidum* genome are thought to be closely related to its ability to synthesize an exceptionally rich array of ganoderic acids (a class of triterpenoids) [38,72]. Similarly, numerous P450 genes and terpene synthesis-related gene clusters were found in the *Dictyophora indusiata* genome [40]. Annotation of the *O. raphanipes* genome also predicted genes involved in the biosynthesis of terpenes, polyketides, NRPSs, and siderophores (a specific type of secondary metabolite) [52]. Phylogenetic analysis of the *Cordyceps militaris* genome identified not only 61 protease families but also genes involved in terpene cyclases, terpene synthases, fatty acid synthases, and geranylgeranyl diphosphate synthase, all potentially related to its synthesis of adenosine, cordycepin, and possibly terpenoids [38]. Key enzyme genes associated with L-lysine synthesis, involving potential diaminopimelate (DAP) and aminoadipate (AA) pathways, were annotated in the *F. velutipes* genome [59]. Functional annotation of *Agaricus blazei* indicated relatively active glycolysis, gluconeogenesis, and glyoxylate metabolism during the mycelial stage, whereas the fruiting body stage showed higher activity related to stress response, ribosome biogenesis, arginine synthesis, and steroid biosynthesis potentially linked to β-1,3-glucan synthesis [83]. Genome analysis of *Sparassis latifolia* also revealed, besides numerous carbohydrate metabolism enzyme genes, the presence of genes participating in the synthesis pathways of indole, terpene, and type I polyketide secondary metabolites [54]. Mining and annotating enzyme genes in edible fungi genomes, particularly CAZymes and secondary metabolism enzymes, not only facilitate a comprehensive understanding of their fundamental decomposition, synthesis, and metabolic capabilities at the micro-level but also provide guidance for subsequent in-depth investigations into gene regulatory networks, growth characteristics, and biosynthetic pathways of key active compounds. This information establishes a solid theoretical foundation for screening critical genes, performing functional validations, applying gene editing technologies, and ultimately serving the molecular breeding and efficient cultivation of edible fungi.

2.2.3. Identification of Metabolic Regulatory Genes

In addition to enzyme genes directly participating in metabolic reactions, genome annotation can identify numerous potential regulatory genes. These genes finely tune the direction and rate of metabolic flux by controlling the timing and intensity of structural gene expression. Key regulatory genes include various Transcription Factors (TFs) [84], such as members of the widely distributed Zn(II)2Cys6 zinc cluster protein family in fungi [85], as well as crucial components of signal transduction pathways, like G protein-coupled receptors (GPCRs) [86], G protein subunits [87], Mitogen-Activated Protein Kinase (MAPK) pathway components [88], and calcium signaling pathway-associated proteins [89]. Identifying these regulatory genes is crucial for understanding the overall control logic of metabolic networks.

2.2.4. Functional Annotation of the Mitochondrial Genome

The mitochondrion, the cell's "powerhouse", possesses its own relatively small genome that encodes essential proteins for energy metabolism (aerobic respiration). Functional annotation of the mitochondrial genome (mitogenome) primarily focuses on genes encoding subunits of the respiratory chain complexes (Complexes I to IV, e.g., NADH dehydrogenase subunits, cytochrome b, cytochrome c oxidase subunits) and ATP synthase (Complex V) subunits [41,43,48,81]. Furthermore, the mitogenome encodes rRNAs and some tRNAs required for mitochondrial ribosomes [63,68]. Resolving the coding content of the mitogenome helps elucidate its roles in cellular energy supply, maintaining mitochondrial replication, and potential involvement in processes like fungal drug resistance and cellular senescence [82,83]. For instance, sequencing and annotation of mitogenomes from species like *Pleurotus cornucopiae* [90], *P. eryngii* [63], *Grifola frondosa* [68], and *Sparassis* spp. [91] have identified these conserved protein-coding, rRNA, and tRNA genes. Although the amount of genetic information encoded by the mitogenome is far less than the nuclear genome, it still exerts significant regulatory influence on the growth, development, energy metabolism, and environmental adaptation of edible fungi. Exploring its functional genes contributes to a more comprehensive understanding of edible fungi biology.

In summary, genome functional annotation provides a "blueprint" or "parts list" of the metabolic capabilities of edible fungi. It systematically reveals their potential enzymatic capacities, secondary metabolite synthesis potential, and possible regulatory mechanisms. This serves as an indispensable foundation for subsequent transcriptomic, proteomic, and metabolomic studies, as well as for metabolic pathway reconstruction and experimental functional validation.

2.3. Influence of Edible Fungi Genomic Features and Comparative Genomics on Metabolic Network Construction

The structural characteristics of edible fungi genomes—such as gene arrangement, gene family size, and the presence of gene fusions or gene clusters—exert a profound influence on the constitution and evolutionary trajectory of their metabolic networks, ultimately shaping their unique metabolic product profiles. Comparative genomics, through the analysis of genomes from different species or distinct strains of the same species, reveals both conserved and specific elements arising during genome evolution. This approach elucidates how the expansion and contraction of gene families have molded the metabolic adaptability of species and identifies genetic variations linked to specific metabolic phenotypes.

2.3.1. Gene Fusion and Metabolic Pathway Evolution

Gene fusion is a significant evolutionary event that can generate proteins with novel functions or optimize existing metabolic pathways. During investigations into the biosynthesis of *Ganoderma lucidum* triterpenes (ganoderic acids), researchers identified a unique gene formed by the fusion of the gene encoding lanosterol synthase (catalyzing a key step in triterpene synthesis) with a gene encoding a membrane protein [42]. This fusion event, likely occurring during the evolution of the *Ganoderma* genus, is hypothesized to have had a significant impact on the efficiency or diversity of triterpene synthesis, representing a distinct mechanism of metabolic pathway evolution.

2.3.2. Gene Clusters (SMGCs) and Coordinated Regulation of Secondary Metabolite Synthesis

In fungi, including edible species, genes involved in the biosynthesis of specific secondary metabolites—comprising synthetic enzymes, regulatory genes, and transporter genes—are often physically clustered together on the genome, forming biosynthetic gene clusters (BGCs), also referred to as Secondary Metabolite Gene Clusters (SMGCs). This physical linkage is thought to facilitate the co-expression and coordinated regulation of these functionally related genes, thereby enabling the efficient and precise synthesis of specific metabolites. For example, the genome of *Cordyceps kyushuensis* contains the BGCs for cordycepin and pentostatin. These compounds function synergistically (pentostatin inhibits adenosine deaminase, thereby protecting cordycepin from degradation), and the clustering of their respective BGCs ensures their synchronized production [92]. The aforementioned identification of up to 64 SMGCs in the *Dictyophora indusiata* genome [40] also underscores its substantial potential for synthesizing a diverse array of secondary metabolites. Consequently, the systematic mining and annotation of SMGCs within edible fungi genomes has become a crucial strategy for predicting potential secondary metabolite repertoires and discovering novel bioactive substances.

2.3.3. Gene Family Expansion/Contraction and Metabolic Diversity/Adaptability

The expansion (increase in member number) or contraction (decrease in member number) of gene families is a common phenomenon in genome evolution and a primary driver of species adaptation to different environments and the development of distinct metabolic capabilities. Expansion of specific metabolic enzyme gene families often correlates with an enhanced capacity of the species to synthesize particular types of compounds. A quintessential example is the notable expansion of the cytochrome P450 (CYP450) gene family in the *G. lucidum* genome, widely considered a key factor contributing to its ability to produce an exceptionally diverse range of triterpenoid compounds (ganoderic acids) [61]. Comparative genomic analysis between *Lyophyllum decastes* and its close relative *Hypsizygus marmoreus* revealed that subsequent to their divergence, *L. decastes* experienced the expansion of 426 gene families and the contraction of 834. These alterations in gene content are highly likely responsible for the physiological differences observed between these species, potentially including variations in metabolic pathways [48]. Intriguingly, comparative genomics of the symbiotic edible fungus *Tricholoma matsutake* and its congeneric relatives revealed that, compared to saprophytic fungi, the *T. matsutake* genome has lost parts of the tryptophan metabolism pathway but gained more genes related to iron homeostasis. This is interpreted as a genetic signature of its adaptation to a symbiotic lifestyle (with pine trees) and nutrient-limited conditions (especially nitrogen sources), demonstrating that adaptive genome evolution directly reflects the unique ecological niche and life strategy of the fungus, accompanied by distinct metabolic network remodeling [56]. Similarly, variations in gene family composition among different edible fungi species account for differences in their metabolic profiles. For instance, comparative genomic analysis of *Ganoderma leucocontextum* and *G. lucidum* (both within the same genus) identified significant differences in the gene complements involved in the biosynthetic pathways of sesquiterpenoids, triterpenoids, and polyketides, supporting the observation that they possess distinct active compound profiles [43]. Collectively, these studies illustrate that the evolutionary dynamics of gene families provide a fundamental basis for the metabolic diversity and environmental adaptability of edible fungi.

2.3.4. Impact of Genetic and Epigenetic Variation on Metabolic Network Regulation

Beyond gene presence/absence and copy number variations, subtle genomic variations, such as single nucleotide polymorphisms (SNPs), small insertions/deletions (InDels), and structural variations (SVs), can directly affect the activity or stability of metabolic enzymes or the function of regulatory genes, consequently leading to changes in metabolic phenotypes. The gene mutations identified in the *C. militaris* albino mutant via resequencing [73] serve as an example of this principle. Furthermore, epigenetic modifications, including DNA methylation and histone modifications, represent another critical layer of gene expression regulation that can influence metabolic network activity. Research has indicated that the genomic DNA methylation level in a serially passaged mutant strain of *C. militaris* was significantly higher compared to the wild type. Differentially expressed genes associated with differentially methylated regions (DMRs) in this mutant were enriched in pathways such as pyruvate metabolism, glycerophospholipid metabolism, DNA replication, and N-glycan biosynthesis [39]. This finding suggests that alterations in the epigenetic landscape may constitute an important regulatory mechanism governing metabolism in edible fungi, warranting further in-depth investigation.

In conclusion, the genomes of edible fungi not only furnish the complete set of genetic elements (genes) required for constructing metabolic networks but also possess inherent structural features (e.g., gene fusions, gene clusters); undergo evolutionary dynamics (e.g., gene family expansion/contraction); and harbor internal variations (genetic and epigenetic) that profoundly shape the topology, evolutionary path, and regulatory modes of these networks. These genomic-level attributes collectively dictate the unique metabolic phenotypes, the rich array of bioactive compounds, and the environmental adaptability characteristic of edible fungi. Therefore, comprehensive genomics and comparative genomics research constitutes the essential foundation and prerequisite for systematically understanding the metabolic networks within edible fungi.

3. Transcriptomic Analysis: Revealing the Dynamic Expression Profile of Metabolic Activities

While the genome provides a static genetic blueprint, the transcriptome captures a dynamic snapshot of gene expression under specific temporal, spatial, or conditional contexts. The transcriptome encompasses the complete set of transcripts (primarily messenger RNA (mRNA), but also various non-coding RNAs like rRNA, tRNA, snRNA, snoRNA, miRNA, lncRNA, etc.) present in a particular cell, tissue, or organism at a specific state. Serving as a crucial bridge connecting static genomic information with dynamic cellular functional states, transcriptomics primarily investigates RNA-level gene expression patterns and their regulatory mechanisms [21]. By quantifying transcripts using high-throughput sequencing, transcriptomics provides vital information for understanding how the metabolic networks of edible fungi respond to internal developmental cues (e.g., different growth stages) and external environmental stimuli (e.g., temperature, light, nutrition, stress). As research progresses, the focus has shifted from merely describing static genome composition towards exploring how genes are dynamically regulated in response to diverse internal and external factors, which is particularly critical for understanding often tightly controlled metabolic processes [20] (Table 2).

Table 2. Transcriptomic studies revealing gene expression in edible fungi.

Mushroom Species	Samples Collected Condition	Key DEGs	Key Pathways	Significance	Ref.
Agaricus bisporus	browning susceptible vs. resistant	polyphenol oxidase (PPO) genes	browning-related pathways	reveals browning mechanisms	[93]
Agaricus bisporus	six strains at different developmental stages	aminodeoxychorismate synthase, transcriptional enhancer factor genes, among others	fatty acid metabolism, biosynthesis of steroid and folate	reveals important DEGs in fruiting body development from different perspectives	[94]
Agaricus bisporus	caps at different post-harvest storage times	PPO genes	browning-related pathways	understanding PPO regulation of browning mechanism in *A. bisporus*	[95]
Agaricus bisporus	four developmental stages of fruiting body	DNA replication, base repair, RNA transport, ribosome etc.	amino acid, carbohydrate, nucleotide, lipid, and energy metabolism	reveals genes related to fruiting body growth and development	[96]
Agaricus blazei	different developmental stages of fruiting body	carbohydrate metabolism, fatty acid degradation, amino acid metabolism genes, among others	DNA replication, base excision repair, mismatch repair, RNA transport, and ribosome	understanding gene expression at different fruiting body developmental stages	[97]
Agaricus blazei	mycelia vs. primordia	response to stress genes	energy production	understanding pathways for polysaccharide and benzaldehyde biosynthesis, and fruiting body formation genes	[83]
Auricularia auricula-judae	three morphologically distinct fruiting bodies	peroxidase-like genes	starch and sucrose metabolism, MAPK signaling (yeast), biosynthesis of amino acid, secondary metabolite, and antibiotic	contributes sequence data to public databases, establishes relationships between major varieties	[98]
Auricularia fibrillifera	drought stress, rehydration	Tyrosinase and homogenesate 1,2-dioxygenase genes, among others	resisting oxidative stress, osmotic adjustment	provides new potential targets for breeding and cultivation of drought-tolerant fungi	[88]
Auricularia polytricha	mycelia vs. mature fruiting bodies	tyrosinase genes	metabolism of amino acid, carbohydrate, energy, lipid, nucleotide	reveals candidate genes related to fruiting body formation	[99]
Cordyceps cicadae	three developmental stages	5′-nucleotidase and adenosine deaminase genes	cordycepin biosynthesis	enhancing understanding of biosynthesis of cordycepin and other characteristic secondary metabolites	[100]
Cordyceps militari	mycelia vs. fruiting bodies	genes for virulence and sexual development	energy metabolism, signaling pathways	suggests lncRNA expression regulates fungal virulence and sexual development by affecting gene expression	[101]
Cordyceps militaris	dark vs. light	*Cmtns* gene	biosynthesis of carotenoids	reveals carotenoid biosynthesis pathway and improving yield	[102]
Cordyceps militaris	dark vs. specific light	glyoxalase system genes	response to light/dark stimulation, biosynthesis of carotenoid and cordycepin	suggests transcriptional co-regulation plays a metabolic control role in light adaptation	[103]
Cordyceps militaris	six generations of artificial culture	genes for toxin biosynthesis, DNA methylation and chromatin remodeling, and energy metabolism	strain degeneration	reveals strain degeneration mechanism	[104]
Cordyceps militaris	albino mutant vs. normal strain	secondary metabolite backbone genes	response to light	understanding pigment biosynthesis pathway	[105]
Cordyceps militaris	submerged vs. liquid surface culture	adenylosuccinate synthetase and phosphoribosylaminoimidazole-succinocarboxamide (SAICAR) synthase genes	purine nucleotide metabolism	reveals cordycepin biosynthesis pathway and improving cordycepin production	[106]
Cordyceps militaris	different culture media	cell membrane, catalytic activity	carotenoid biosynthesis	understanding carotenoid biosynthesis pathway in *C. militaris*	[107]
Cordyceps militaris	culture with vs. without L-alanine addition	Zn2Cys6 type TFs	biological of energy production and amino acid transformation	supports improving cordycepin yield and strain breeding	[108]
Cordyceps militaris	different carbon sources	cordycepin biosynthesis genes	transcriptional regulation of central carbon metabolism	studying overall metabolic response of *C. militaris* for cordycepin production	[109]
Cordyceps militaris	xylose vs. other carbon sources	pentose and glucuronate interconversions	cordycepin biosynthesis	reveals response mechanism to xylose utilization	[110]

Table 2. *Cont.*

Mushroom Species	Samples Collected Condition	Key DEGs	Key Pathways	Significance	Ref.
Cordyceps militaris	wild-type vs. wc-1 deficient strain	Cmwc-1(a homolog of the blue-light receptor gene white collar-1 (wc-1) in Neurospora crassa)	steroid biosynthesis	investigating role of blue light receptor gene (wc-1) in *C. militaris* fruiting and secondary metabolism	[32]
Flammulina filiformis	mycelia vs. primordia	genes for sugar transporter and unsaturated fatty acid synthesis	glycolysis, phospholipid and sphingolipid metabolism	reveals energy source transition during primordium formation	[111]
Flammulina filiformis	four developmental stages	heat stress-induced hsp70, hsp90, fes1 genes	heat stress response	studying gene function and improving mushroom heat tolerance	[112]
Flammulina velutipes	monokaryotic vs. dikaryotic mycelia	genes for transcription factors, protein kinases, and WD40 repeat-like proteins	synthesis of fatty acid, amino acid, and most saccharide	screens DEGs before and after mating in *F. velutipes*	[113]
Flammulina velutipes	mycelia vs. primordia	ribosome and DNA replication, among others	glycolysis, pentose phosphate pathway	reveals DEGs between F. velutipes dikaryotic mycelia, and primordia	[114]
Ganoderma lucidum	three continuous developmental stages	genes for carbohydrate metabolism, triterpenoid and ergosterol biosynthesis	ganoderic acids and ergosterol biosynthesis	reveals genes potentially involved in meiotic transcriptional control, metabolic pathways for energy supply, and biosynthesis of ganoderic acids and ergosterol	[115]
Ganoderma lucidum	addition of L-phenylalanine	genes for L-phenylalanine metabolism and cell wall mannoprotein	fungal polysaccharide production	provides effective strategy for high-yield, low-cost fungal polysaccharide production	[116]
Ganoderma lucidum	addition of PDE inhibitor or AC activator	genes for squalene synthase and lanosterol synthase	cAMP-induced apoptosis, ganoderic acids (GAs) biosynthesis	reveals cAMP signaling induces fungal apoptosis, and secondary metabolite production	[117]
Ganoderma lucidum	mycelia vs. fruiting bodies	FOLymes and CAZymes genes	terpene backbone biosynthesis pathway	provides comprehensive gene expression information	[118]
Grifola frondosa	mycelia	polysaccharide synthesis-related genes	polysaccharide synthesis	provides basis for studying polysaccharide metabolic pathways and related functional genes	[119]
Lentinula edodes	dikaryotic mycelia vs. mature fruiting bodies	peptidases and phosphotransferases genes	oxidative stress, starvation stress response	elucidating molecular mechanisms of mature fruiting body development and beneficial properties	[120]
Lentinula edodes	blue light vs. dark	morphogenesis and photoreception genes	blue light signaling pathway	identification of light-responsive genes	[121]
Lentinula edodes	monokaryotic mycelia with different laccase activity	cytochrome P450, glycoside hydrolase, and UDPG dehydrogenase genes	lignin degradation and carbohydrate metabolism	deeper understanding of physiological metabolism in high-laccase-yielding *L. edodes* strains	[122]
Lentinula edodes	dark vs. blue light	CAZymes genes	pentose and glucuronate conversion, starch and sucrose metabolism	aids functional studies of genes involved in developmental control	[123]
Lentinula edodes	different vegetative mycelial growth phenotypes (light exposure)	kinases, tyrosinase, glucanase, chitinase, and laccase genes	melanogenesis, cell wall degradation, signaling	analysis of expression patterns of light-induced browning-related genes	[124]
Lyophyllum decastes	five developmental stages	extracellular enzymes and TFs genes	signaling pathways	understanding fruiting body development	[125]
Morchella importuna	three mycelial growth stages	transketolase (tktA) and glucose-6-phosphate dehydrogenase (G6PDH) genes, among others	carbohydrate metabolism	elucidating mechanisms of mycelial growth	[126]
Morchella importuna	mycelia vs. young fruiting bodies	CAZymes, mitochondrial proteins, oxidoreductases, and HSPs genes	carbohydrate catabolism and energy metabolism	exploring fruiting body formation mechanisms in *M. importuna*	[127]
Morchella importuna	three mycelial growth stages	CAZymes genes	metabolism of carbohydrates, polysaccharides, hydrolases, caprolactam, β-galactosidase, and disaccharides	reveals carbohydrate catabolism mainly occurs during vegetative mycelial stage	[128]
Morchella importuna	mature fruiting bodies	genes for respiration, carbohydrate metabolism, tissue softening, and oxidative browning	molecular mechanism of post-harvest quality changes	provides theoretical basis for post-harvest quality change mechanisms and preservation technology	[129]
Chinese cordyceps (Medicinal fungi)	mycelia, sclerotia, and primordia	pheromone receptor and amino acid sensing genes,	DNA synthesis, cell division, MAPK pathway	reveals potential mechanisms of fruiting body initiation	[130]
Ophiocordyceps sinensis (Medicinal fungi)	mycelia vs. fruiting bodies	adenosine metabolism enzyme genes	cordycepin biosynthesis	provides data supporting elucidation of cordycepin biosynthesis mechanisms	[131]
Ophiocordyceps sinensis (Medicinal fungi)	fruiting bodies	pheromone receptors, G protein γ-subunit, G protein α/β subunits, and cordycepin synthesis enzymes genes	signaling, cordycepin biosynthesis	reveals genes associated with fruiting body development, and cordycepin biosynthesis	[87]

Table 2. *Cont.*

Mushroom Species	Samples Collected Condition	Key DEGs	Key Pathways	Significance	Ref.
Pleurotus eryngii	different developmental stages	genes encoding enzymes involved in carbon and amino acid metabolism	carbon and amino acid metabolism	reveals gene expression changes during fruiting body growth and development	[132]
Pleurotus eryngii	transformed vs. wild-type strains	genes for mycelial growth and enzyme activity	MAPK signaling and inositol phosphate metabolism, among others	validates the functional role of GNAI in *P. eryngii* growth and development	[133]
Pleurotus eryngii	dark vs. blue light	CAZymes genes	carbon metabolism, glycolysis and biosynthesis of amino acids	understanding primordia response to blue light at developmental stage	[134]
Pleurotus eryngii subsp. *Tuoliensis*	cold stimulus, mycelia	genes for cell wall and membrane stability, Ca^{2+} signaling, MAPK pathway, and soluble sugar and protein biosynthesis	mycelia response to cold stimulation	understanding molecular mechanisms related to cold stimulus response	[135]
Pleurotus ostreatus	H_2O_2 regulation	genes in the respiratory chain	ATP synthesis	provides basis for breeding dark P. ostreatus strains and understanding pigmentation mechanism	[136]
Pleurotus ostreatus	heat stress	MYB gene family	mechanisms of fruiting body development and stress response	promotes further functional analysis of MYB genes	[137]
Pleurotus pulmonarius	mycelia vs. fruiting bodies	*PpFBD1* (a gene that is highly expressed during the early stages of primordium formation)	synthesis of fungal cell walls	studying DEGs and their functions at different growth stages	[138]
Pleurotus tuoliensis	three developmental stages	morphogenesis genes	primary carbohydrate metabolism, cold stimulus and blue light response	indicates vegetative-to-reproductive transition as most active and critical period for gene expression changes	[139]
Pleurotus tuoliensis	immature vs. mature mycelia	genes encoding nucleoside diphosphate kinase, GH family proteins, and extracellular polygalacturonase	nucleotide synthesis and energy metabolism	understanding molecular mechanisms of mycelial maturation	[140]
Pleurotus tuoliensis	monokaryotic vs. dikaryotic strains	phenylalanine ammonium lyase and aryl-alcohol oxidase genes	catalytic activity and metabolic process	exploring fruiting body development genes, providing efficient markers for MAS	[141]
Sparassis Primordia	dark vs. light	genes associated with metabolism of vitamin B6 and selenocompound metabolism, among others	cysteine synthesis, vitamin B6 metabolism, glycine metabolism	establishes DE map for light-induced primordium formation	[142]
Sparassis Primordia	dark vs. light	glutathione S-transferase and HSP 9/12 genes, among others	secondary metabolite synthesis, starch and sucrose metabolism, glutathione metabolism	analysis of molecular mechanisms of light response	[143]
Stropharia rugosoannulata	cold stress	carbohydrate enzyme genes	metabolism of carbohydrate, lipid, and xenobiotic	reveals cold resistance mechanisms	[144]
Tremella fuciformis	mycelia	L-iditol 2-dehydrogenase and butanol dehydrogenase genes	polysaccharide metabolic pathways	provides data support for studying polysaccharide and other product biosynthesis pathways and mechanisms	[145]
Volvariella volvacea	different commercial strains	heat shock proteins genes	stress response	identifying candidate genes involved in rapid growth requirements	[146]
Wolfiporia cocos	mycelia vs. sclerotia	diphosphomevalonate decarboxylase and farnesyl diphosphate synthase genes, among others	biosynthesis of triterpenoids	reveals genes related to triterpenoid biosynthesis	[147]

3.1. Principles, Technologies, and Basic Applications of Transcriptome Sequencing in Edible Fungi Research

Transcriptome Sequencing (RNA-Seq) is the current mainstream technology for studying gene expression. Its fundamental principle involves deep sequencing of total RNA or specific RNA fractions (e.g., enriched mRNA after rRNA depletion) extracted from cells or tissues using high-throughput sequencing platforms. The resulting short reads are then aligned to a reference genome or assembled de novo to obtain expression abundance information for each gene (or transcript), typically represented by metrics like FPKM (Fragments Per Kilobase of transcript per Million mapped reads), RPKM (Reads Per Kilobase of transcript per Million mapped reads), or TPM (Transcripts Per Million). In edible fungi metabolic research, the core applications of RNA-Seq include, first, identifying Differentially Expressed Genes (DEGs) by comparing transcriptomic data across different conditions (e.g., developmental stages, tissues, environmental treatments, strains); second, performing functional enrichment analyses (e.g., Gene Ontology (GO) and Kyoto Encyclopedia of Genes and Genomes (KEGG) pathway enrichment) on these DEGs using bioinformatics tools to infer which biological processes or metabolic pathways exhibit significantly altered activity under the compared conditions; third, discovering novel transcripts beyond known coding genes, including alternative splicing isoforms and non-coding RNAs potentially playing roles in metabolic regulation, such as long non-coding RNAs (lncRNAs) [148] and small non-coding RNAs (sncRNAs, including miRNAs) [149]. Although the specific functional mechanisms of non-coding RNAs in edible fungi metabolism require further investigation, existing studies suggest their potential importance. For instance, several long intergenic non-coding RNAs (lincRNAs) potentially involved in gene expression regulation were identified in the *Ganoderma lucidum* transcriptome [148], and sncRNAs have been found in *Pleurotus ostreatus* [149].

Beyond directly studying gene expression and regulation, transcriptomic data offer other significant applications, particularly for species lacking high-quality genomic information or when developing molecular markers is necessary. Edible fungi are highly valued for their nutritional content and unique flavors, but their extensive species diversity and complex genetic backgrounds necessitate accurate species identification and germplasm evaluation as prerequisites for industry development and basic research. Traditional morphology-based identification is often inefficient, susceptible to environmental influence, and inadequate for complex taxonomic delimitations. While genome sequencing provides the most comprehensive genetic information, it remains relatively costly and time-consuming. In this context, utilizing transcriptomic data for developing molecular markers or species identification presents an efficient and economical alternative. Researchers can mine Simple Sequence Repeats (SSRs), also known as microsatellites, from transcriptomic data. Due to their wide genomic distribution, high polymorphism, and codominant inheritance, SSRs are ideal markers for genetic diversity analysis, genetic map construction, and variety identification. By sequencing transcriptomes from mycelia and fruiting bodies of *Auricularia heimuer* and *A. polytricha*, Zhou et al. [150] not only analyzed transcriptional differences but also identified abundant SSR loci in both species' transcriptomes. Although differing slightly in quantity and repeat motif types, their overall composition was similar, providing a basis for marker development in these related species. Similarly, potential SSR loci were successfully identified from transcriptomic data in *A. polytricha* [99] and *Morchella* spp. Researchers developed 15 pairs of genetically polymorphic SSR primers based on *Morchella* transcriptome sequences, validating the utility of transcriptomic data for molecular marker development [151].

Furthermore, RNA-Seq data can be employed for species identification, notably, using Internal Transcribed Spacer (ITS) sequences of ribosomal RNA genes. The ITS region (ITS1 and ITS2), being relatively conserved within species but highly variable between species and having moderate length (typically 350 bp and 400 bp, respectively), is widely recognized as the core DNA barcoding marker for fungi, effectively distinguishing closely related species. Southwest China, including Yunnan Province, harbors exceptionally rich wild edible fungi resources. However, cultivation difficulties and volatile market prices lead to instances of fraud and adulteration, sometimes including toxic mushrooms, posing food safety risks. To rapidly and accurately identify mushrooms in the market, studies have utilized RNA-Seq to directly obtain and analyze ITS sequences from collected samples. Through phylogenetic analysis of the sequencing results, one study not only discovered potential new edible fungi resources but also successfully identified toxic mushroom species mixed within commercial samples, thus demonstrating the feasibility of RNA-Seq as a rapid, preliminary method for identifying edible fungi species in circulation, including differentiating toxic from non-toxic ones [152]. Additionally, transcriptomics combined with amplicon sequencing (e.g., targeting bacterial 16S rRNA and fungal ITS sequences) has been used to investigate the microbial community structure in edible fungi cultivation environments and its impact on fungal growth. For example, studies on *Morchella* cultivation soil revealed that continuous cropping significantly reduces microbial diversity, particularly affecting groups potentially interacting (mutualistically or antagonistically) with *Morchella*. A one-year fallow period was shown to help restore the soil microbial community structure [153].

Therefore, RNA-Seq technology functions as more than a tool for tracking gene expression; its outputs provide a dynamic molecular lens into fungal life. The collective findings demonstrate its analytical strength in not only elucidating metabolic regulation but also in generating foundational data for molecular marker development, species identification, and ecological interaction studies. The broader implication is the multifaceted utility of transcriptomics in providing critical support for strategic goals such as germplasm conservation, food safety assurance, and the sustainable optimization of cultivation systems for edible fungi.

3.2. Changes in Gene Expression Profiles Reveal Dynamic Metabolic Regulation

Metabolic activity in edible fungi is not static but undergoes precise dynamic adjustments according to their life cycle progression and changes in the external environment. By capturing variations in gene expression levels under different conditions, transcriptomics paints a molecular picture of this dynamic regulation, serving as a primary tool for understanding how edible fungi adaptively allocate resources and execute specific physiological functions.

3.2.1. Transcriptional Dynamics and Metabolic Adaptation Across Life Cycles and Tissue Specificity

The life cycle of edible fungi typically includes a vegetative growth phase (mycelia) and a reproductive growth phase encompassing primordium formation, fruiting body development, maturation, and spore production. Different developmental stages, and even distinct tissue parts within the same stage (e.g., pileus/cap, stipe/stalk, lamellae/gills, volva), often undertake different physiological functions, leading to distinct metabolic requirements and priorities. Transcriptomic studies have been widely applied to various edible fungi, including *Agaricus bisporus* [96], *Auricularia polytricha* [99], *Flammulina velutipes* [114,154], *Ganoderma lucidum* [115], *Lentinula edodes* [120], *Pleurotus eryngii* [132], *P. tuoliensis* [139,140], *P. ostreatus* [155], *Agaricus subrufescens* (often considered synonymous or related to *A. blazei*) [97], *Tricholoma matsutake* [156], *P. geesteranus* [138], *Wolfiporia cocos* [147], *Agrocybe aegerita* [157], and *Grifola frondose* [119,157], revealing dramatic yet ordered changes in the expression profiles of metabolism-related genes throughout their life cycles.

The transition from vegetative to reproductive growth represents a critical juncture, often accompanied by fundamental shifts in energy metabolism patterns and drastic morphological restructuring. For instance, studies on *F. velutipes* revealed that during cold-induced primordium formation from mycelia, the expression of genes related to nutrient uptake from the substrate (e.g., sugar transporter genes) tends to decrease, while genes associated with energy production (e.g., aerobic respiration genes), cell wall remodeling (e.g., chitin synthase and hydrolase genes), and stress responses become significantly more active [111]. This indicates that primordium formation is an energy-intensive process requiring mobilization of internal reserves and cellular reorganization. Similarly, comparative transcriptome analysis of *A. blazei* mycelia and primordia showed that the mycelial stage emphasizes gene expression related to carbohydrate metabolism (substrate decomposition for nutrition), transmembrane transport, and cell wall synthesis (nutrient acquisition and growth). In contrast, the primordial stage shifts focus towards stress responses, nucleic acid and protein synthesis (providing materials for rapid cell division and differentiation), and steroid biosynthesis (potentially affecting membrane fluidity or acting as signaling molecules), all associated with morphogenesis and stress tolerance [83]. Comparative transcriptomics across five developmental stages of *Lyophyllum decastes*, from vegetative mycelium to mature fruiting body, identified numerous DEGs encoding extracellular enzymes (nutrient acquisition/environmental modification), transcription factors (developmental regulation), and signaling components (responding to internal/external cues). This confirmed the complex regulatory network governing fruiting body development, and enzyme activity assays validated the correlation between differential expression and functional activity for some extracellular enzymes [125]. These transcriptomic insights are further substantiated by proteomic studies on *L. decastes*, which, for instance, confirmed the upregulation of specific laccases and quinone reductases at the protein level when the fungus is grown on lignin-rich substrates, directly linking genomic potential and transcriptional activity to functional protein execution in lignocellulose degradation (Section 4.2.1). Combined transcriptomic and metabolomic analysis of *Morchella importuna* across three stages, from vegetative growth to sclerotium formation (a dormant/storage structure and primordium precursor), also revealed key genes and metabolic pathways crucial for mycelial morphogenesis and transition, particularly highlighting dynamic changes in carbohydrate metabolism pathways such as starch/sucrose metabolism, pentose/glucuronate interconversions, and fructose/mannose metabolism [126]. The understanding of *M. importuna*'s developmental metabolism is further enriched by proteomic analyses (Table 3), which compared vegetative mycelia to sexual reproductive stages, identifying differential protein abundance in carbohydrate and amino acid metabolic pathways and highlighting the roles of specific proteins like lectins. Such integration provides a more robust picture of the molecular shifts occurring during its life cycle.

Table 3. Proteomic analysis of metabolic pathways and protein networks in edible fungi.

Mushroom Species	Proteomic Research	Key Proteins	Key Pathways	Significance	Ref.
Auricularia auricula-judae	effect of freezing on melanin accumulation mechanism	glycolysis/ gluconeogenesis proteins	tyrosine metabolism, ribosome and arginine biosynthesis	provides information on the mechanism of freezing treatment effects on color quality	[158]
Hypsizygus marmoreus	protein changes from "scratching" to primordia	oxidoreductases, peptidases, hydrolases	catabolic and carbohydrate -related processes	understanding developmental changes preceding primordia formation	[159]
Hypsizygus marmoreus	protein expression during mycelial growth	proteins involved in carbohydrate metabolism, catabolic processes, oxidoreductase activity	carbohydrate metabolism, catabolic processes, oxidoreductase activity	elucidating protein changes during development	[160]
Hypsizygus marmoreus	heat stress response	SOD and peroxidase, trehalose synthase, heat shock proteins (HSPs)	expression of catalase	understanding molecular mechanisms of heat stress response	[161]
Lentinula edodes	substrate effect on protein expression	proteins involved in carbohydrate and oxidoreductase activity pathways	carbohydrate and oxidoreductase activity pathways	cultivation effects on nutritional quality	[162]
Lyophyllum decastes	lignin degradation	laccase, quinone reductase	lignocellulose degradation	confirms lignin degradation mechanism	[48]
Morchella importuna	proteome analysis of vegetative vs. sexual stages	lectin proteins	carbohydrate and amino acid metabolic pathways	understanding nutrient metabolism and fruiting body development	[163]

Table 3. *Cont.*

Mushroom Species	Proteomic Research	Key Proteins	Key Pathways	Significance	Ref.
Ophiocordyceps sinensis (Medicinal fungi)	Proteome analysis of three key dev. stages	ROS-related proteins, proteins involved in carbon transport and metabolism, response to oxidative stress, antioxidant activity and translation	carbon transport and mechanism, response to oxidative stress, antioxidative activity	understanding biological traits of fruiting body development and high-altitude adaptation	[164]
Schizophyllum commun	trehalose biosynthesis pathway	trehalose synthase, trehalose phosphatase, trehalose phosphorylase	growth and development of mycelium, trehalose biosynthesis	potential for industrial trehalose production	[165]
Stropharia rugosoannulata	tissue-specific protein expression	proteins for carbon metabolism, energy production, stress response	fatty acid synthesis, mRNA splicing	understanding tissue-specific metabolism	[144]
Tremella fuciformis	mechanism and proteomics of conidia–mycelia transition	proteins involved in biosynthesis, DNA replication, and DNA damage repair, among others	MAPK signaling pathway	provides basis for studying dimorphism formation mechanisms	[166]
Volvariella volvacea	post-harvest autolysis mechanism	proteins involved in fatty acid metabolism and RNA transport	RNA transport, biosynthesis of fatty acid and amino acid	provides reference for further studies on senescence mechanisms	[167]

Even within the fruiting body development stage, significant tissue- or maturation-specific differences exist in gene expression and metabolic activity. For example, comparing transcriptomes from different parts (pileus, indusium, mycelial cord, stipe, volva) of *Dictyophora indusiata* revealed the highest overall gene expression levels in the pileus. DEGs associated with fruiting body differentiation (e.g., genes involved in tryptophan metabolism; tryptophan is a precursor for auxin/IAA, potentially involved in morphogenesis) were significantly enriched in the pileus, suggesting its central role in regulating formation and highlighting specialized metabolic functions among tissues [40]. Research on *Stropharia rugosoannulata* found that pathways for biosynthesis (protein, nucleic acid), glucose metabolism (energy/carbon skeletons), and amino acid metabolism remain active throughout fruiting body growth and development [168]. This study also examined post-transcriptional regulation, finding that alternative splicing events (especially intron retention and alternative 3′ splice sites) play significant roles during fruiting body development and maturation. Studies on the medicinally important *G. lucidum* indicated metabolic shifts during development: genes related to triterpene (ganoderic acid) and ergosterol (key membrane component) biosynthesis were highly expressed in young fruiting bodies. During sporulation, expression of genes involved in carbohydrate metabolism (providing energy/structure for spores, e.g., trehalose synthesis) significantly increased. In the later stages of spore release, genes encoding glycoside hydrolases (e.g., chitinases, glucanases) were upregulated, possibly related to the final maturation and modification of the spore wall [115].

Furthermore, different reproductive strategies (sexual vs. asexual) correspond to distinct gene expression profiles. Research on *C. militaris* showed that compared to its sexual phase (producing fruiting bodies and ascospores), the asexual phase (producing conidia) exhibited significantly higher upregulation of genes related to cellular metabolism, energy metabolism, and stress responses, possibly reflecting a strategy prioritizing rapid proliferation and dispersal [169]. In studies concerning fruiting body development and biosynthesis of the key medicinal compound cordycepin in *Ophiocordyceps sinensis* (a medicinal fungus, not typically edible, but studied similarly), RNA-Seq identified transcripts of genes commonly associated with fungal fruiting body development (e.g., pheromone receptors, G protein subunits, signaling kinases). It also pinpointed genes for key enzymes in the cordycepin pathway, such as adenosine kinase (ADK), 5′-nucleotidase (5′-NT), adenylate kinase (ADEK), and ribonucleotide reductase (RNR) subunits, enabling the preliminary construction of a cordycepin biosynthetic model based on this information [87].

In summary, by capturing gene expression dynamics across different stages, tissues, and reproductive modes throughout the life cycle, transcriptomic analysis provides a detailed map of shifting metabolic priorities, profoundly revealing the intricate control exerted by intrinsic developmental programs over metabolic networks.

3.2.2. Environmental Regulation of Transcription and Metabolic Remodeling

The growth, development, and metabolic activities of edible fungi are governed not only by internal genetic programs but are also constantly influenced by external environmental factors such as temperature, humidity, light, pH, nutrient status, gas composition, and biotic stress. When environmental conditions change, particularly under unfavorable stress conditions, edible fungi initiate stress response mechanisms, remodeling their metabolic networks through regulated gene expression to adapt and survive. Transcriptomics serves as the core tool for investigating the gene expression regulation underlying this adaptive metabolic remodeling.

Temperature is a critical environmental factor influencing growth and yield. Studies on *F. velutipes* revealed significant differences in transcriptional responses to heat stress among varieties (e.g., white vs. yellow), potentially correlating with variations in heat tolerance [112]. Conversely, under cold stress or cold induction conditions, transcriptomes undergo drastic changes. For instance, studies on *V. volvacea* under cold stress [170] and *F. velutipes* during cold-induced primordium formation [111] both identified significant alterations in genes related to stress response (e.g., heat shock proteins), signal transduction, energy metabolism (adjusting respiration/fermentation), and carbohydrate metabolism (accumulating protective sugars like trehalose). Transcriptome analysis of *S. rugosoannulata* under cold stress demonstrated a general downregulation of carbohydrate utilization enzyme genes and upregulation of antioxidant enzyme (SOD, CAT) and heat shock protein (Hsp) genes, regardless of strain sensitivity. This suggests a relatively conserved molecular mechanism for cold tolerance, although the intensity and efficiency of these responses may differ among strains with varying sensitivity [144]. While these transcriptomic changes suggest a conserved cold response mechanism, proteomic analyses from the same study on *S. rugosoannulata* (Table 3) also revealed tissue-specific protein expression patterns, for instance, with stipe proteins being enriched for carbon metabolism and stress response, suggesting that transcript-level stress responses manifest in distinct proteomic profiles within different parts of the organism. Similarly, multi-omics analysis of cold-treated *F. velutipes* mycelia revealed significant enrichment of pathways related to carbohydrate metabolism, unveiling its molecular strategy for coping with cold stress [171]. Following heat stress treatment of *P. ostreatus*, researchers specifically examined the expression changes within its large Zn(II)2Cys6 zinc cluster protein-encoding gene family (often transcription factors). They found that members from different clusters exhibited distinct expression patterns (up- or downregulation), implying diverse roles within the transcriptional regulatory network responding to heat stress [85].

Light is another crucial environmental factor influencing growth and development, especially fruiting body morphogenesis and pigmentation. Research on *L. edodes* demonstrated that blue light induces changes in numerous genes. Comparing transcriptomes under blue light versus dark conditions identified 762 DEGs. Several genes validated as upregulated under blue light included those involved in morphogenesis (encoding HSPs, fibronectin domain proteins, carbohydrate esterases) and light sensing (encoding photoreceptor-related proteins with FAD/NAD-binding domains) [121]. This indicates that light signals can modulate metabolic activity and developmental processes in shiitake by activating specific gene expression. Another study simulating intense sunlight exposure found significantly enhanced expression of genes involved in DNA repair, heat shock response, and potentially photoprotective secondary metabolic pathways in *L. edodes* samples exposed for 30 and 60 min, revealing specific molecular defense mechanisms against light radiation stress [172].

Heavy metal contamination represents a potential stressor in agricultural environments. To understand fungal responses, the transcriptome of *Hypomyces chrysospermus*, a fungus known to potentially contaminate edible fungi cultivation settings, was analyzed under cadmium stress. DEGs were primarily enriched in pathways related to protein synthesis (translation); amino acid metabolism; transport and catabolism; carbohydrate metabolism; and protein folding, sorting, and degradation. Notably, upregulation was observed for genes encoding various transporters (potentially for cadmium efflux or sequestration), antioxidant enzymes (counteracting cadmium-induced oxidative stress), and detoxification enzymes (e.g., glutathione S-transferases), elucidating complex molecular mechanisms for heavy metal tolerance and detoxification [173]. This research has implications for evaluating the safety of edible fungi grown in contaminated areas and exploring the potential of fungi in bioremediation.

Biotic stress, such as infection by pathogenic fungi, is a prevalent issue in edible fungi production. Understanding fungal responses to pathogens is essential for developing resistant varieties and control strategies. For instance, investigating the interaction and pathogenesis of *Paecilomyces penicillatus* on *M. importuna*, researchers analyzed RNA-Seq data from transcripts isolated at the interaction zone. They observed dynamic changes in *M. importuna* gene expression across different infection stages. Some potentially resistance-related genes, encoding glycosyltransferase family 31 proteins and ferric chelate reductases, exhibited initial upregulation followed by downregulation. Conversely, genes encoding mannitol-1-phosphate 5-dehydrogenase (mannitol is a key osmolyte/antioxidant in some fungi) and the glucose-repressible protein Grg1 showed an opposite trend. Intriguingly, pathways related to fatty acid metabolism and biosynthesis (including unsaturated fatty acids) were significantly upregulated early in the infection, suggesting a potential role for fatty acid metabolism in fungal interactions (e.g., defense or signaling) [174]. While this study focused on transcriptomic and genomic changes, complementary proteomic analyses of *M. importuna* under such biotic stress would further elucidate the functional protein effectors involved in defense and interaction, bridging the gap between gene expression and actual protein-mediated responses. For instance, proteomic studies on *M. sextelata* have provided insights into pathogen defense (Section 5.1.3). This study provides molecular insights for enhancing white mold resistance and optimizing cultivation management in *M. importuna*.

Other abiotic stresses, such as drought, and post-harvest physiological changes, like browning, also induce significant transcriptomic alterations. Transcriptome analysis of *Auricularia cornea* (potentially a variety or relative of *A. heimuer*) during drought stress and subsequent rehydration revealed distinct adaptive strategies. Drought primarily induced tyrosine metabolism (possibly related to stress-associated phenolics), caffeine metabolism, ribosome function adjustments, phagosome formation (autophagy/recycling), proline and arginine metabolism (proline as osmoprotectant), and activation of peroxisome and MAPK signaling pathways. During rehydration, pathways activated included diterpenoid biosynthesis (potentially related to growth recovery), butyrate metabolism (GABA production, potentially stress recovery signaling), C5-branched dibasic acid metabolism, and aflatoxin biosynthesis (possibly related to interaction or stress) [88]. Addressing post-harvest browning in *A. bisporus*, comparative transcriptomics combined with metabolomics on varieties differing in browning sensitivity identified differential expression of key enzyme genes (e.g., polyphenol oxidase, PPO) and correlated these changes with the accumulation levels of relevant metabolites (e.g., organic acids, trehalose), which collectively influence browning resistance [93].

In summary, by comprehensively analyzing gene expression profile changes in response to various environmental challenges (biotic and abiotic), transcriptomic studies clearly illustrate how edible fungi flexibly remodel their metabolic networks through transcriptional regulation. This adaptation allows them to cope with changing environments, ensure survival, or execute specific physiological processes. This knowledge is invaluable for understanding ecological adaptability, optimizing cultivation conditions, enhancing stress tolerance, and improving post-harvest quality.

3.2.3. Differential Gene Expression Between Strains/Varieties and Comparison of Metabolic Capabilities

Significant variations often exist between different strains or cultivated varieties of the same edible fungi species concerning growth rate, yield, fruiting body morphology, color, flavor, nutrient content, stress resistance (e.g., heat, cold, disease tolerance), and enzymatic properties. Underlying these phenotypic differences frequently lie variations at the gene expression level. Comparative transcriptomics, by directly comparing the gene expression profiles of strains with distinct genetic backgrounds or phenotypic traits, effectively links transcriptional-level differences to specific metabolic capabilities or biological characteristics. This approach helps unveil the molecular basis of these variations, facilitates the screening for superior strains, and aids in identifying functional genes controlling key traits. Such research has been undertaken in numerous edible fungi, comparing, for example, mycelia with different nuclear statuses (monokaryotic vs. dikaryotic) or varying mating type combinations [113,175–178], as well as contrasting commercial varieties or mutants exhibiting distinct features. Comparative analyses typically focus on the following:

Growth Rate and Nutrient Utilization Efficiency: Mycelial growth rate significantly impacts the production cycle length. By comparing commercial strains of *Volvariella volvacea* with differing growth rates, Liu et al. [146] found that faster-growing strains generally exhibited higher expression levels of genes related to carbohydrate utilization (e.g., encoding glycoside hydrolases, sugar transporters). This indicates that efficient nutrient uptake and decomposition capabilities are among the key factors supporting rapid growth.

Fruiting Body Quality, Flavor, and Appearance: Fruiting body quality is central to the commercial value. Transcriptome analysis of different *A. bisporus* varieties (e.g., browning-susceptible vs. resistant) across developmental stages helps identify important DEGs associated with quality formation (e.g., differential PPO expression controlling browning [93]), morphogenesis, and maturation processes [94]. The ability of *L. edodes* mycelium to undergo normal browning (transition from white to brown, necessary for subsequent fruiting) is a critical production issue. Tang et al. [179] compared transcriptomes of browning and non-browning shiitake strains, identifying expression differences in several transcription factors that potentially regulate pigment synthesis or related metabolic pathways, thus influencing the browning process. Addressing cap color diversity in *P. ostreatus*, Hou et al. [136] linked color variation to transcriptomic changes, proposing that H_2O_2 might act as a signal, influencing pigment accumulation by differentially regulating genes associated with the oxidative phosphorylation pathway. The texture and commercial value of *A. heimuer* relate to fruiting body wrinkling. Zhao et al. [98] compared gene expression profiles of varieties with different wrinkling characteristics, discovering that non-wrinkled varieties exhibited unique patterns in certain gene functions (potentially involving cell wall synthesis, turgor regulation) and overall expression levels, providing clues to the molecular basis of these morphological differences.

Key Enzyme Activity Differences and Regulation: Certain enzymes produced by edible fungi, such as laccases, possess significant industrial value. To identify high-laccase-yielding strains or understand their regulation, Lei et al. [122] performed comparative transcriptome analysis on *L. edodes* monokaryotic mycelia with differing laccase activities. Results revealed significant expression differences in genes related to lignin degradation (a primary laccase function), cytochrome P450-mediated metabolism (potentially substrate conversion or secondary metabolism), and carbohydrate metabolism, offering potential targets for screening high-yielding strains or enhancing laccase production by modulating related pathways.

Gene Function Validation and Regulatory Network Analysis: Comparative transcriptomics also serves as a vital tool for validating specific gene functions. By constructing overexpression (OE) or gene silencing (e.g., RNAi) strains for a target gene and comparing their transcriptomes with wild-type or control strains, the downstream regulatory network of the gene can be elucidated. For instance, Cao et al. [133] investigated the function of a G protein α-subunit gene (PeGNAI) in *P. eryngii*. Comparing transcriptomes of PeGNAI OE and RNAi strains demonstrated that altered expression levels significantly impacted genes in pathways related to amylase and laccase activities, thereby confirming the role of PeGNAI in regulating growth, development, and associated enzyme activities in king oyster mushroom.

Exploring Superior Germplasm Resources and Assisting Breeding: Comparing transcriptomes of wild germplasm versus cultivated varieties, or strains from different geographic origins or with varying stress tolerance levels, can identify specific genes or regulatory modules associated with desirable traits (e.g., high nutritional value, strong stress resistance, unique flavor). This provides candidate targets for molecular marker-assisted breeding (MAS) or direct genetic improvement.

In conclusion, employing transcriptomic techniques to compare gene expression differences between edible fungi strains or varieties constitutes a powerful research strategy. It not only aids in understanding the molecular origins of biological trait diversity, correlating specific metabolic capabilities or phenotypes with gene expression patterns, but also effectively facilitates the screening of strains possessing superior characteristics and the discovery of core genes regulating key pathways. This provides a robust theoretical basis and technical support for the genetic improvement and germplasm innovation of edible fungi.

4. Proteomic Research: Directly Addressing the Functional Execution Level of Metabolism

While the genome provides the genetic blueprint and the transcriptome reveals gene expression dynamics, proteomics offers direct insights into the proteome—the functional executors of cellular processes. Understanding the proteome is crucial, as it directly reflects metabolic activities and provides a critical layer for validating and complementing genomic and transcriptomic findings, particularly in the complex metabolic networks of edible fungi. The proteome refers to the entire complement of proteins expressed by a cell, tissue, or organism at a specific time point and under specific conditions [20,22]. Proteomics is the discipline dedicated to the systematic study of protein composition, abundance, post-translational modifications (PTMs), subcellular localization, and protein–protein interactions (PPIs) [22]. As proteins are the direct effectors of biological function, proteomic research offers information closer to the actual functional state of an organism compared to genomics and transcriptomics. Particularly for understanding complex metabolic activities, it directly detects the enzymes performing metabolic reactions and the proteins regulating these processes [23]. Consequently, proteomics is regarded as an essential complement and validation for genomic and transcriptomic studies, crucial for the comprehensive elucidation of metabolic networks and their regulatory mechanisms in edible fungi (Table 3).

4.1. Overview of Proteomic Research Methods

Currently, the most commonly employed core technologies in edible fungi proteomics are based on mass spectrometry (MS). Depending on the sample separation and quantification strategies, these can be broadly categorized into two main types:

Gel-based proteomics: Two-dimensional gel electrophoresis (2-DE) is the classic method within this category. The method 2-DE separates complex protein mixtures on a gel based on two properties, isoelectric point (pI) and molecular weight (MW), generating a protein spot map. By comparing protein spot patterns between different conditions (e.g., treatments, developmental stages), differentially expressed proteins can be initially screened. Subsequently, protein spots of interest are excised from the gel, enzymatically digested (typically with trypsin), and the resulting peptide mixtures are identified using MS techniques (e.g., MALDI-TOF MS or LC-ESI-MS/MS) [26]. The advantages of 2-DE include its ability to visualize changes in protein abundance intuitively and detect intact proteins with their major PTMs. However, its throughput is relatively low, detection sensitivity for low-abundance proteins and those with extreme pI/MW values is limited, the procedure is relatively laborious, and quantitative accuracy can be constrained.

Gel-free/Shotgun proteomics (LC-MS/MS based): This approach involves enzymatically digesting the protein sample into a peptide mixture first, followed by separation using high-performance liquid chromatography (HPLC, typically reversed-phase LC, RPLC). The separated peptides are directly introduced into the mass spectrometer for analysis and identification. Based on the quantification method, LC-MS/MS techniques can be further subdivided into label-based and label-free quantification (LFQ) methods. The former (label-based quantification) includes techniques such as iTRAQ (Isobaric Tags for Relative and Absolute Quantitation) and TMT (Tandem Mass Tags). These techniques utilize isotopically labeled chemical reagents to tag peptides from different samples (allowing for multiplexing, e.g., simultaneous comparison of up to 8 or 16 samples with iTRAQ or TMT, respectively). At the MS1 level (first mass spectrometry scan), tagged versions of the same peptide originating from different samples are isobaric (exhibit identical mass-to-charge ratios, m/z) and thus indistinguishable. However, upon fragmentation during tandem mass spectrometry (MS/MS), the tags yield reporter ions with distinct m/z values. Measuring the intensity ratios of these reporter ions enables the relative quantification of proteins across the different samples [26]. Label-based quantification offers high throughput and generally good quantitative accuracy. The latter (label-free quantification) does not require chemical labeling of peptides. Instead, relative quantification is achieved by comparing peptide signal intensities (e.g., peak areas or ion currents) or spectral counts detected during LC-MS/MS analysis across different samples. Label-free methods offer a simpler workflow, are not limited by multiplexing sample numbers, are comparatively less expensive, and can potentially cover a wider protein dynamic range [180]. However, achieving high quantitative accuracy and reproducibility places greater demands on experimental consistency and sophisticated data analysis workflows.

In edible fungi research, Differential Proteomics is the most widely applied strategy. Whether using 2-DE or LC-MS/MS, the core objective is to compare proteome expression profiles under different physiological states (e.g., developmental stages, tissues/organs), environmental conditions (e.g., stress treatments, cultivation substrates), or genetic backgrounds (e.g., different varieties, mutants). This allows for the screening of key Differentially Expressed Proteins (DEPs), particularly enzymes and regulatory proteins associated with metabolic pathways [26], providing direct functional-level evidence for the molecular mechanisms underlying specific biological processes. Recently, the scope of proteomics in edible fungi research has strategically expanded towards identifying medically relevant proteins, signifying a shift to explore bioactive potentials beyond nutrition [23]. A critical analytical challenge, however, remains: the often observed discrepancies between transcript and protein abundances, and the still limited coverage of the full proteome, especially dynamic PTMs, complicate direct gene-to-function inferences [181]. The broader implication is that, while proteomics provides invaluable functional data, its analytical power is maximized when integrated with other omics to navigate these complexities.

4.2. Protein Expression Abundance and Functional Characterization in Metabolic Analysis

By directly measuring the types and abundance changes in intracellular proteins, proteomics provides the most direct evidence for understanding which functional modules are activated and which metabolic pathways are operative during specific physiological processes. It can validate predictions made at the transcript level, reveal regulation occurring at the translational and post-translational levels, and elucidate the dynamic changes in proteins as they execute specific functions.

4.2.1. Protein-Level Analysis and Functional Confirmation of Important Enzymes

Enzymes are the core executors of metabolic networks. Proteomic techniques have been extensively used to directly detect changes in the abundance of enzyme proteins within key metabolic pathways, thereby confirming their actual roles in specific processes.

Pathway-Specific Enzyme Confirmation: When studying specific metabolic pathways, proteomics can directly confirm the involvement and activity changes in key enzymes. For instance, investigating the response of *Lyophyllum decastes* to lignin-rich substrates involved combined genomic and proteomic analysis. Genomics predicted numerous CAZyme genes, particularly laccases. Subsequent proteomic analysis (identifying 1071 proteins, including CAZymes and secreted enzymes) revealed that the abundance of several specific laccases and quinone reductases was indeed significantly upregulated in the lignin-rich medium [48]. This result directly confirmed, at the protein level, the central role of these enzymes in lignin degradation, strongly complementing and validating the genomic predictions. Similarly, proteomic analysis of *Schizophyllum commune*, aimed at exploring its biotechnological potential, successfully identified key enzymes involved in the trehalose biosynthesis pathway, such as trehalose synthase, trehalose phosphatase, and trehalose phosphorylase [165]. As trehalose is a high-value carbohydrate with broad applications, identifying these enzymes via proteomics opens possibilities for industrial enzymatic synthesis using S. commune or its enzyme systems.

Post-Harvest Processes: Edible fungi are susceptible to autolysis, browning, and decay during post-harvest storage, processes closely linked to internal enzyme activities. Proteomics offers powerful tools to uncover the enzymatic mechanisms involved. Quantitative proteomics analysis of post-harvest autolysis in *Coprinus comatus* found that hydrolytic enzymes involved in cell wall metabolism (e.g., chitinases, β-glucanases) significantly increased at the protein level during early storage (cap opening), regulated by ribosomal activity. This led to cell wall polysaccharide degradation, structural disruption, and ultimately autolysis. The abundance of these enzymes then rapidly declined later in the process. Concurrently, accumulation of reactive oxygen species (ROS) from respiratory metabolism throughout storage caused oxidative stress, activating MAPK signaling cascades in later stages, potentially inducing apoptosis and accelerating autolysis [182]. Using label-free proteomics to study *Flammulina velutipes* under commercial modified atmosphere packaging (MAP; low O_2, high CO_2) revealed significant enrichment of the tricarboxylic acid (TCA) cycle pathway mid-storage. The protein abundance of key enzymes like citrate synthase (CS) increased, indicating attempts to maintain energy supply against storage stress. (Note: AAS mentioned in the original text might be misidentified or mislabeled, potentially referring to other amino acid metabolism or TCA cycle enzymes). Towards the end of storage, antibiotic biosynthesis became prominent, alongside changes in carbon and amino acid metabolism, such as a significant decrease in the glycolytic enzyme triosephosphate isomerase (TPI) protein abundance [180]. These complex changes in the enzyme protein abundance collectively reflect the sophisticated energy metabolism regulation employed by F. velutipes to sustain viability during storage.

Environmental Stress Responses: Environmental stresses also trigger dramatic responses at the enzyme protein level. Proteomic analysis of frozen *Auricularia heimuer* mycelia directly quantified increased levels of antioxidant enzymes like superoxide dismutase (SOD) and peroxidase (POD), indicating cellular activation of defense mechanisms against freezing-induced oxidative stress. The study also found that freezing inhibited the activity of tyrosinase, a key enzyme in melanin synthesis, while enhancing laccase activity [158], revealing complex direct effects of cold stress on the melanin pathway. High CO_2 concentrations can be encountered during cultivation or storage of some species like *Pleurotus ostreatus*. Comparative proteomics of *P. ostreatus* fruiting bodies grown under high CO_2 revealed reduced metabolic activity related to kinases (involved in signaling and metabolic regulation) and a decreased abundance of several elongation factors involved in protein synthesis. Additionally, expression of proteins related to cell wall synthesis and sexual differentiation processes was significantly affected [183]. These results unveil, at the protein execution level, the specific regulatory mechanisms by which high CO_2 impacts *P. ostreatus* growth and metabolism.

In summary, proteomics provides "seeing is believing" evidence by directly measuring and comparing enzyme protein abundance. This validates transcript-level predictions and can uncover the impact of post-transcriptional regulation on enzyme levels, thereby more accurately elucidating the actual roles and importance of specific enzymes in the physiological activities and stress responses of edible fungi.

4.2.2. Global Proteome Responses and Metabolic Correlations Under Different Physiological States

Beyond focusing on specific enzymes, the power of proteomics lies in its ability to provide a global perspective. By comparing the overall proteome composition and abundance changes in edible fungi across different developmental stages, tissues, environmental conditions, or morphologies, it can depict the complete metabolic landscape and functional priorities under specific physiological states.

Developmental Dynamics: Fungal development is a continuous, complex process involving significant morphological changes and physiological transitions, inevitably reflected in proteome dynamics. iTRAQ-based quantitative proteomic analysis of *Volvariella volvacea* at different time points (0, 24, 60, 96 h) in post-harvest storage at 15 °C showed a significant decline in the abundance of proteins associated with stress response and cellular homeostasis (e.g., stress-induced PAB1, membrane integrity protein RPG1) as storage progressed. Proteins involved in gene expression regulation (RNA transcription-related) and the protein synthesis machinery (all differentially expressed ribosomal proteins) were also generally downregulated. Conversely, some proteins related to fatty acid metabolism showed an upward trend, potentially further affecting membrane function [167]. These global proteomic trends collectively depict a senescence process in stored *V. volvacea*, characterized by diminishing stress resistance, declining cellular structural stability, and attenuated overall metabolic activity (especially anabolism). Using label-free LC-MS/MS, research on the commercially valuable *Hypsizygus marmoreus* (white beech mushroom) compared protein expression profiles across developmental stages (pre-primordium to primordium emergence), identifying various CAZymes, transcription factors, and heat shock proteins with altered abundance [160]. Further studies examined three consecutive stages from mycelial physical treatment (inducing primordia) to primordium appearance, finding continuous evolution in differential protein expression involving various functional proteins like importins, dehydrogenases, and HSPs [159]. This highlights the capability of proteomics to map dynamic protein changes during development, potentially reflecting shifting metabolic demands, such as increased carbohydrate metabolism enzymes needed during rapid growth phases. Building upon the transcriptomic insights into *M. importuna* development (Section 3.2.1), proteomic comparisons between its vegetative mycelia and sexual reproductive stages further elucidated the molecular underpinnings. These proteomic studies revealed significant enrichment in protein abundance related to ribosome biogenesis in the reproductive stages [163], suggesting that the observed transcriptomic shifts translate into increased protein synthesis machinery to support rapid fruiting body development. iTRAQ comparative proteomics across different growth stages of *Ophiocordyceps sinensis* identified several differentially abundant proteins (DAPs) shared across stages, involved in pathways like carbon transport/metabolism and antioxidant activity, providing protein-level clues to its unique growth characteristics and medicinal compound synthesis [164]. Proteomic study of *Pleurotus giganteus* reported for the first time an upregulation of proteins associated with carbon metabolism and energy generation pathways in mature fruiting bodies, consistent with findings in other mushrooms, further confirming the dominant role of carbohydrate catabolism in supporting rapid growth and substance accumulation in fruiting bodies [184]. Beyond developmental stages, proteomics also dissects functional differences between distinct morphologies of the same species. For example, *Tremella fuciformis* exists in yeast-like and filamentous forms. Comparative proteomics (combining global and targeted approaches) of its dikaryotic yeast-like basidiospores (FBMds) and dikaryotic mycelia (DM) indicated that although the overall basal synthetic and metabolic activity might be lower in DM, the abundance and phosphorylation levels of certain signaling proteins (e.g., MAPK pathway components) were upregulated [166]. This suggests distinct signaling strategies might be employed in different morphologies to adapt to specific lifestyles or environments, revealing significant differences in synthesis, metabolic activity, and signaling responses between the two forms.

Environmental Adaptation and Cultivation Effects: Environmental adaptability is key to the survival and distribution of edible fungi. Proteomics has been used to investigate responses to different geographical environments or cultivation conditions. A proteomic study (validated with transcriptomic data) on fruiting bodies of the prized white truffle (*Tuber magnatum Pico*) from different Italian regions (North, Center, South) found significant geographic variation in the abundance of key enzymes in the sulfur metabolism pathway, such as cystathionine γ-lyase [185]. This strongly suggests that white truffles might adapt to habitats with varying soil conditions (e.g., sulfur content, redox state) by dynamically regulating their sulfur metabolism network (sulfur compounds contribute significantly to truffle aroma and may also be involved in stress response). Combining genomic potential (reflected by transcriptomics) with actual protein expression levels (proteomics) allows for a more comprehensive understanding of mushroom metabolic profiles and environmental responses [55]. The cultivation substrate significantly impacts yield and quality. Integrated proteomics and metabolomics studies on *Lentinula edodes* grown on different alternative substrates (e.g., supplemented with wolfberry or Sea buckthorn) showed that these substrates significantly increased the protein and fatty acid content in fruiting bodies. Proteomic analysis revealed concordant upregulation of proteins involved in carbohydrate metabolism and oxidoreductase activity (potentially related to energy supply and stress response) in mushrooms grown on alternative substrates [162,186]. This clearly demonstrates proteomics is an effective tool to understand how cultivation conditions (like substrate composition) influence nutritional quality and the metabolic state by analyzing protein expression changes linked to key metabolic pathways. Furthermore, proteomic analyses of fruiting bodies from two other important medicinal fungi, *Sparassis crispa* (cauliflower mushroom) and *Hericium erinaceum* (lion's mane), revealed a wealth of functionally diverse proteins, including 14-3-3 proteins involved in signal transduction and septins involved in cell division, providing deeper insight into their proteome complexity [187]. These proteomic explorations, especially when contextualized with genomic potential (e.g., Reference for *Sparassis* [54]) and transcriptomic responses (as seen in various examples in Section 3), highlight how edible fungi modulate their protein landscapes for specific functions, from environmental adaptation (e.g., *Tuber magnatum Pico*) to reflecting cultivation impacts on nutritional quality (e.g., *L. edodes*). Future focused proteomic studies, particularly on underrepresented species or delving into PTMs for common species like *L. edodes* and *P. ostreatus*, are crucial for building a more comprehensive, integrated understanding of fungal biology.

In conclusion, proteomics, by analyzing global protein changes across various conditions, offers a comprehensive and systematic understanding of the activity levels, regulatory priorities, and functional adaptations of metabolic networks in edible fungi. This provides crucial protein-level information for deepening our understanding of their biological characteristics, optimizing cultivation and management practices, and developing strategies for resource utilization.

5. Integrative Omics Analysis: Towards a Systems Understanding of Edible Fungi Metabolic Networks

Genomics provides the static repository of genetic information, transcriptomics reveals the dynamic changes in gene expression, and proteomics directly reflects the functional state of effector molecules. However, the relationships between these layers are not simply linear. Significant discrepancies can exist between the transcriptional level of a gene and the abundance of its encoded protein (due to factors like post-transcriptional regulation, translational control, and protein degradation), and protein abundance does not fully equate to its activity (requiring consideration of post-translational modifications, allosteric regulation, cofactor availability, etc.). Therefore, data from a single omics layer can only provide insights into specific facets of complex biological systems. To more systematically and comprehensively dissect the intricate metabolic networks of edible fungi and their sophisticated regulatory mechanisms, the integrated analysis (Integrative Omics) of genomic, transcriptomic, and proteomic data—sometimes also including metabolomic data (metabolomics, studying the global profile of small molecule metabolites)—has become an inevitable trend and forefront direction in current life science research [188]. This integrated approach helps trace the flow of information from genetic potential (genome) through transcriptional response (transcriptome) to functional execution (proteome), ultimately manifesting as metabolic output (metabolome). Such integration yields deeper and more systematic biological insights than single-omic studies can offer [20,40,189] (Figure 1, Table 4).

Table 4. Integrative omics studies providing insights into the metabolism of edible fungi.

Mushroom Species	Integrated Omics Technologies	Research Question/Focus	Key Insights Obtained	Ref.
Agaricus blazei	genomics, transcriptomics	gene function, DEG between mycelia and primordia stages	understanding pathways for polysaccharide and benzaldehyde biosynthesis, fruiting body formation related genes	[83]
Agrocybe aegerita	transcriptomics, proteomics	mycelia vs. fruiting bodies	provides clues for nutritional, pharmaceutical, and industrial applications	[157]
Auricularia cornea	genomics, transcriptomics, metabolomics	high-quality genome assembly and multi-omics analysis of pigment synthesis pathway	revealed genetic blueprint, evolutionary history, and pigment synthesis mechanism	[190]
Cordyceps cicadae	genomics, transcriptomics, metabolomics	asexual fruiting body formation characteristics and secondary metabolism	understanding fungal genetics and promoting medicinal research	[191]
Cordyceps kyushuensis	transcriptomics, proteomics	single gene cluster for cordycepin and pentostatin biosynthesis	for improving cordycepin yield and finding more functional proteins	[92]
Cordyceps militaris	genomics, transcriptomics	genome, transcriptome, and methylome of a new strain	provides basis for fungal molecular biology	[192]
Cordyceps militaris	transcriptomics, proteomics	gene expression differences between mycelia and fruiting bodies	facilitates systematic molecular studies of developmental stages	[193]
Dictyophora indusiate	genomics, transcriptomics, metabolomics	fruiting body differentiation	importance of tryptophan metabolism	[40]
Flammulina filiformis	genomics, comp. transcriptomics	expression of biosynthetic gene clusters (BGCs) in wild/cultivated strains and dev. stages	elucidating regulation of secondary metabolites	[194]
F. filiformis	genomics, transcriptomics	role of cytochrome c peroxidase gene (*FfCcP*) in dev. and stress response	upregulation of *FfCcP* might facilitate stipe elongation	[195]
F. velutiper, G. Lucidum, H. erinaceus, H. marmoreus, L. edodes, P. eryngii, P. geesteranus, V. volvacea	comp. metabolomics, proteomics	ameliorative effect of melatonin (MT) on Cd-induced oxidative stress	confirmed widespread presence of MT in edible fungi and its important role in physiological regulation	[196]
Flammulina velutipes	transcriptomics, metabolomics	regulation of ergosterol biosynthesis	understanding ergosterol biosynthesis regulation during fruiting and metabolite–gene relationships	[197]
Ganoderma lucidum	genomics, transcriptomics	RNA editing	elucidating role of transcriptional plasticity in dev., environmental adaptation, and SM pathway regulation	[198]
Ganoderma lucidum	genomics, transcriptomics	lignocellulolytic enzymes	understanding changes in lignocellulose degradation ability during *G. lucidum* growth	[199]
Ganoderma lucidum	genomics, transcriptomics	key genes (Cyt P450s, transporters, regulators), triterpenoid biosynthesis	potential model system for studying SM pathways and regulation in medicinal fungi	[61]
Lentinula edodes	proteomics, metabolomics	substrate effect on metabolism and nutrition	linking substrate to protein/metabolite changes	[162]
Morchella importuna	genomics, transcriptomics	response mechanisms at different stages post-*Paecilomyces* infection	fungus–fungus interactions involving edible fungi	[174]
Morchella sextelata	transcriptomics, proteomics	heat shock response, development, pathogen infection	heat shock adaptation network; role of sugar metabolism; defense mechanisms against pathogens	[200]
P. eryngii var. eryngii, Pleurotus tuoliensis	genomics, transcriptomics	DNA methylation and gene expression during three major developmental transitions	epigenetic and transcriptional regulatory mechanisms supporting gene expression during development	[201]
Phlebopus portentosus	genomics, transcriptomics	lignocellulose degradation system and fruiting body development	understanding key pathways and hub genes involved in development	[202]
Pleurotus giganteus	transcriptomics, proteomics, nutritional analysis	substrate and mRNA transport within fruiting body	complex interactions between gene expression, protein synthesis, and nutrient allocation	[184]
Pleurotus ostreatus	genomics, transcriptomics, proteomics	secreted proteins and environmental interaction	secreted protein transcription depends more on strain genotype than environment	[178]
Pleurotus ostreatus	genomics, proteomics, metabolomics	secretome of strains on different substrates	identifying enzymes overproduced in lignocellulose cultures	[203]
Volvariella volvacea	genomics, transcriptomics, proteomics	low-temperature autolysis	cold stress might trigger VvAgo1-mediated RNAi to promote low-temperature autolysis	[170]
Volvariella volvacea	genomics, transcriptomics	composition and expression of CAZyme-encoding genes	identified CAZymes, showed differential expression between heterokaryon and homokaryon	[81]

5.1. *Examples of Integrated Analysis of Key Metabolic Pathways and Biological Processes*

Multi-omics integration has demonstrated significant power in various aspects of edible fungi research, particularly in elucidating key metabolic pathways, understanding complex biological processes (e.g., development, stress response), and linking genotypes to phenotypes (Figure 1).

5.1.1. Integrative Analysis of Bioactive Substance Biosynthetic Pathways

Bioactive compounds produced by edible fungi, such as polysaccharides, terpenoids, sterols, and nucleosides, are sources of their significant nutritional and medicinal value. Integrated analysis can more clearly delineate the biosynthetic pathways of these compounds and their regulatory networks. For example, regarding the synthesis of widely studied fungal polysaccharides (mainly β-glucans), integrating genomic annotation identifying β-glucan synthase gene family members [61,83] with transcriptomic data revealing upregulated expression patterns of these genes during specific developmental stages (e.g., *Ganoderma lucidum* sporulation [115], *Agaricus blazei* primordia [83]) or under specific induction conditions (e.g., L-phenylalanine inducing extracellular polysaccharide production in *G. lucidum* [116]), and further confirming corresponding changes in synthase protein abundance via proteomics, would collectively construct a complete regulatory map of the polysaccharide synthesis pathway from gene to functional expression. Key targets in polysaccharide biosynthesis include β-glucan synthases and chitin synthases and their regulators. Similarly, for the synthesis of terpenoids and sterols like ganoderic acids (triterpenes) and ergosterol, combining genomic identification of synthases (e.g., terpene synthases, squalene epoxidase, CYP450s) [40,61,194] with transcriptomic studies of their expression patterns in specific stages (e.g., high expression in young *G. lucidum* fruiting bodies [115]) or tissues allows for the construction of more complete and reliable biosynthetic pathway models, predicting key rate-limiting steps or regulatory nodes. In terpenoid pathways, important enzymes such as HMG-CoA reductase, squalene synthase, lanosterol synthase, and various cytochrome P450 enzymes are crucial targets. Research on *C. militaris* went further; Vongsangnak et al. [36], based on its genome sequence, comparative genomics, and the literature information (potentially implicitly using transcriptomic or proteomic data), reconstructed six core metabolic pathway networks (carbohydrate, energy, amino acid, nucleotide, lipid, cofactor). They specifically focused on the adenine metabolic pathway and cordycepin biosynthesis, relevant to its medicinal value, performing cross-species comparisons to infer the evolutionary origins of the pathway and potential reasons for high yield (e.g., deletion of ribonucleotide reductase inhibitor sequences). Nucleoside compound biosynthesis, like cordycepin, involves purine de novo synthesis and modification enzymes (e.g., adenosine kinase, SAM-cycle-related enzymes) and potentially the pentose phosphate pathway. Ergothioneine biosynthesis, relevant for fungi like *P. ostreatus*, relies on specific enzymes like EgtB and EgtD. Some fungi also produce phenolic and polyphenolic compounds with antioxidant properties via the shikimate pathway and phenylalanine ammonia-lyase (PAL). Studies on *Hericium erinaceus* also emphasize using multi-omics approaches (starting from genomics) to understand the biosynthesis of complex secondary metabolites (e.g., hericenones, erinacines) and achieve their rapid identification and quantification [204]. Beyond these, other pathways holding significant promise for metabolic engineering based on multi-omics insights include the biosynthesis of fungal polysaccharides, particularly β-glucans, which involves enzymes like β-glucan synthases and chitin synthases and their regulators, and are prized for immunomodulatory activities [61,83,115,116] [Section 5.1.1]. Additionally, phenolic and polyphenolic compound biosynthesis pathways (via the shikimate pathway, PAL) are targets for enhancing antioxidant properties. The ergothioneine biosynthesis pathway, involving enzymes like EgtB and EgtD, is also a key target in mushrooms like *P. ostreatus* and *Agaricus bisporus* due to ergothioneine's potent antioxidant capabilities. Integrated omics approaches are pivotal for identifying rate-limiting steps and regulatory control points within these complex pathways to guide effective engineering strategies.

5.1.2. Integrated Regulation of Fruiting Body Development and Differentiation

Fruiting body formation is a critical event in the life cycle of edible fungi and the primary source of their edible and medicinal value. This process is controlled by complex gene regulatory networks. Multi-omics integration helps reveal the underlying molecular mechanisms. The study on *Dictyophora indusiata* [40] serves as a compelling illustration of how multi-omics integration (genomics, transcriptomics, and metabolomics) can unravel complex biological processes like fruiting body differentiation, moving beyond descriptive findings to mechanistic insights. Genomic analysis first established the genetic potential, revealing numerous terpene synthesis gene clusters potentially linked to its unique characteristics. Transcriptomics then pinpointed the pileus as a hub of gene activity, with significant upregulation of tryptophan metabolism pathway genes. Crucially, metabolomics provided functional validation by demonstrating the differential accumulation of tryptophan-derived metabolites, such as auxin (IAA), across tissues. This integrated approach did not just correlate disparate datasets but forged a cohesive molecular narrative: from the genetic blueprint (gene clusters), through dynamic transcriptional regulation (tissue-specific pathway activation), to the functional metabolic output (auxin accumulation). Such integrated evidence strongly supports a model where spatially regulated tryptophan metabolism plays a pivotal role in controlling differentiation processes within the fruiting body, a significantly deeper insight than any single-omics approach could have provided alone. Similarly, research on *Pleurotus giganteus* integrated transcriptomic, proteomic, and nutritional composition data, revealing bidirectional transport of substrates (nutrients) and mRNA molecules between the pileus and stipe during fruiting body development [184]. This highlights complex interactions and coordination mechanisms in gene expression, protein synthesis, and nutrient allocation between different parts of the fruiting body, offering new perspectives on the physiological processes controlling growth and morphogenesis. Combined transcriptomic and proteomic analysis of *Morchella sextelata* from the young to harvested stages, using GO and KEGG enrichment, identified significant differences in the expression of genes and proteins involved in metabolic pathways (especially sugar metabolism) between the two stages, depicting the metabolic shifts during fruiting body maturation [205].

5.1.3. Systematic Analysis of Stress Response Mechanisms

Edible fungi frequently encounter various stresses in natural or cultivated environments, such as temperature fluctuations or pathogen attacks. Integrated analysis facilitates a more comprehensive understanding of the complex regulatory networks underlying their stress responses. Integrated transcriptomic and proteomic analysis of *M. sextelata* strains with different heat tolerances under heat shock conditions not only identified conserved heat shock response patterns across all strains but also revealed heterogeneous response strategies among strains with varying thermotolerance. Based on these data, the researchers proposed the first molecular regulatory network model for heat shock adaptation in *M. sextelata*, providing a new framework for understanding thermotolerance mechanisms in large ascomycetes [200]. Also focusing on *M. sextelata*, another study integrated transcriptomic and proteomic data to analyze its response mechanism to infection by a pathogenic fungus (*Pseudodiploöspora longispora*). Results showed that pathways related to oxidoreductase activity, peroxisome function, lipid transport and metabolism, cell wall biogenesis, and cell membrane composition were significantly enriched or altered at the transcriptional and/or protein levels post-infection [189]. This clearly demonstrates the power of integrated omics in uncovering complex interactions between edible fungi and pathogens and elucidating the molecular mechanisms of their defense responses. Regarding post-harvest physiological changes, such as deterioration of *Pleurotus tuoliensis* during storage, Wu et al. [188] successfully employed integrated transcriptomic and proteomic analysis. They linked the upregulation of CAZyme gene expression (e.g., laccase, chitinase, glucanase) at the transcript level to increased protein abundance of these enzymes and observed fruiting body softening. Concurrently, they associated the upregulation of respiration-related enzymes (e.g., CS, ICDH, mitochondrial respiratory complex subunits) at both transcript and protein levels with increased ROS production and the response of the antioxidant system (e.g., Mn-SOD gene upregulation, Hsp60 protein upregulation possibly interacting with it), thus systematically dissecting the molecular events involved in post-harvest deterioration and their interrelationships. It is also important to acknowledge a common limitation in many current stress response studies: conventional 'bulk' omics analyses often average out molecular responses from different parts of the fungus (mycelia, cap, stipe) or even heterogeneous cell populations within the same tissue. This can mask the spatial specificity of metabolic regulation during stress. Future integration efforts could benefit significantly from spatial omics or single-cell omics technologies, although their application in edible fungi is currently limited, to achieve a more granular understanding of these responses.

5.1.4. Association Between Optimized Cultivation Conditions and Quality Attributes

Integrated analysis can also be used to understand how cultivation conditions, such as the substrate, influence the metabolism and quality of edible fungi. Integrated proteomics and metabolomics studies on *L. edodes* cultivated on different alternative substrates (supplemented with wolfberry or Sea buckthorn) successfully linked observed proteomic changes (e.g., upregulation of carbohydrate and oxidoreductase-related proteins) directly to alterations in the metabolite profile (e.g., increased protein and fatty acid content). This provided deeper insights into how substrate composition affects the nutritional quality of shiitake by modulating specific metabolic pathways [162]. This type of integrated analysis holds significant practical value for guiding breeding efforts (selecting strains that utilize specific substrates more efficiently) and optimizing cultivation processes (designing substrate formulas that promote the accumulation of target nutrients or bioactive compounds).

5.2. Challenges and Methods of Integrative Analysis

Despite the promising outlook for multi-omics integration, its practical application faces numerous challenges (Figure 1). First is persistent data heterogeneity and scale differences: different omics data types (gene sequences, RNA abundance, protein abundance, metabolite concentrations) possess distinct characteristics, distributions, dynamic ranges, and scales, making effective data standardization, weighted integration, and direct comparison problematic. Recent progress underscores that proteomic data, especially those involving post-translational modifications (PTMs), still exhibit a significant gap in dynamic range and coverage compared to transcriptomic data, making the linear correlation analysis of "gene-transcript-protein-function" particularly challenging [181]. Second is the complexity of biological time lags and feedback regulation: significant delays exist from gene transcription to protein translation, and further to protein function impacting metabolite levels. Moreover, complex intracellular feedback loops (e.g., protein degradation regulation, metabolite feedback inhibition of upstream gene expression) can mislead integration analyses based on simple correlations, complicating the establishment of causal relationships between different omics datasets. Third are limitations in functional annotation: the genomes of many edible fungi remain incompletely annotated, with numerous genes of unknown function ("hypothetical proteins"), which directly limits the functional interpretation and integration of transcriptomic and proteomic data [Section 2.2]. Fourth is spatial heterogeneity: differences in mycelia, various parts of the fruiting body, and even cellular heterogeneity within the same tissue are often averaged out in conventional "bulk" omics analyses, failing to accurately reflect the spatial specificity of metabolic regulation, an issue that emerging single-cell omics technologies aim to address [Section 6]. Fifth is the applicability of computational and statistical models: many existing multi-omics integration algorithms were initially developed for model organisms, and their direct application to edible fungi might not fully account for unique fungal biological characteristics, such as heterokaryosis or complex secondary metabolic networks. Finally, there are issues of data completeness and accuracy, particularly at the proteomic and metabolomic levels, where current technologies often fall short of comprehensive coverage and precise quantification for all molecules. To address these challenges, researchers have developed a range of bioinformatics methods and tools. Common strategies include the following:

Pathway-based integration: Utilizing public pathway databases like KEGG and MetaCyc, different omics data are mapped onto known metabolic or signaling pathways. Beyond KEGG and MetaCyc, valuable resources for fungal researchers include FungiDB (VEuPathDB), which consolidates genomic, transcriptomic, and some proteomic data, supporting comparative analyses (https://fungidb.org/fungidb/ accessed on 14 April 2025). MycoCosm by JGI is a crucial repository for fungal genome sequences and annotations (https://mycocosm.jgi.doe.gov/mycocosm/home accessed on 14 April 2025). The STRING database remains useful for predicting and visualizing protein–protein interaction networks (https://string-db.org/ accessed on 14 April 2025). Additionally, specialized databases like the Chinese Ganoderma Lucidum Database (https://nmdc.cn/resource/db/linzhiitem accessed on 14 April 2025) can provide targeted information for specific genera. Analysis then focuses on overall pathway activity changes or identifies pathway nodes (genes/proteins/metabolites) exhibiting significant alterations across multiple omics levels.

Network-based integration: This involves constructing gene regulatory networks, protein–protein interaction (PPI) networks, or metabolic networks, and then integrating various omics data into these models. Analysis focuses on changes in network topology, identifying key regulatory hubs or functional modules. Network visualization tools like Cytoscape (latest stable version from https://cytoscape.org, accessed 14 April 2025) play a vital role here. More advanced network-based methods might integrate Weighted Gene Co-expression Network Analysis (WGCNA) to identify cross-omics functional modules. For PPI, the STRING database is a useful resource.

Multivariate statistical methods: Techniques such as Partial Least Squares Discriminant Analysis (PLS-DA), Canonical Correlation Analysis (CCA), and Multi-block analysis can handle multiple high-dimensional datasets, seeking correlation patterns or common trends among different omics data. Multi-omics data integration platforms (e.g., the mixOmics R package (https://mixomics.org/ on 14 April 2025), functionalities within the MetaboAnalyst website (https://metaboanalyst.ca/ on 14 April 2025)) provide tools for these analyses. Recent developments also include deep learning-based multi-omics integration models for identifying non-linear relationships and predicting phenotypes [206].

Mathematical modeling: This particularly utilizes Genome-Scale Metabolic Models (GSMMs). Based on genomic annotation and stoichiometric principles, GSMMs are mathematical representations of all known metabolic reactions in an organism, including their gene–protein-reaction (GPR) associations. By integrating transcriptomic or proteomic data to constrain reaction rates (fluxes) within the model, GSMMs can predict cellular metabolic phenotypes (e.g., growth rates, byproduct secretion rates, metabolic flux distributions) under specific conditions. Representing one of the most advanced forms of integration, GSMMs enable prediction from genotype to phenotype, but their application in edible fungi research is still nascent, though recent progress includes the development and application of GSMMs for species like *C. militaris* using transcriptomic and exometabolomic data for flux balance analysis [207,208]. Other useful databases include FungiDB (VEuPathDB) for broad fungal omics data, MycoCosm (JGI) for fungal genomes, and potentially species-specific databases like the Chinese Ganoderma Lucidum Database. General integration platforms like OmicsIntegrator or functionalities within Galaxy and Tbtools-II may also assist researchers [36].

Addressing these multifaceted challenges necessitates the development of bespoke integration strategies and bioinformatics tools attuned to the unique biology of edible fungi, such as their complex life cycles and nuclear states. The broader implication of such advancements is transformative: successfully integrating multi-omics data (genomics, transcriptomics, proteomics, and often metabolomics) will shift the research paradigm from component-level analysis to a systems-level understanding of metabolic network complexity and regulation. Ultimately, this analytical leap is foundational for enabling the rational design, predictive modeling, and targeted optimization of fungal metabolic processes, offering insights far exceeding the sum of individual omic parts and paving the way for novel biotechnological applications.

6. Summary

Edible fungi, a treasured biological resource, offer diverse flavors and significant nutritional and medicinal benefits owing to their rich bioactive compounds like nucleosides, polysaccharides, terpenoids, and sterols, positioning them prominently in health foods and biopharmaceuticals. Growing consumer demand for enhanced quality, yield, and specific bioactives, alongside challenges such as stringent cultivation needs for superior varieties, high production costs, germplasm scarcity, and post-harvest spoilage, necessitates a profound understanding of their complex metabolic processes.

The "classical triad" of omics—genomics, transcriptomics, and proteomics—has been pivotal in advancing this understanding (Figure 2). Genomics furnishes the genetic blueprint, revealing metabolic gene repertoires and evolutionary signatures [20,49–51,66,81]. Transcriptomics demonstrates dynamic gene expression in response to developmental and environmental cues, uncovering regulatory logic [40,93,121,125,126,144,168,171]. Proteomics directly observes the execution of metabolic functions, validating predictions and revealing the impact of translational and post-translational regulation on metabolic activity [23,48,55,159,162,165,186,187]. These technologies, applied independently or through integrated approaches, have significantly advanced our comprehension of core metabolic pathways (e.g., carbohydrate, energy, amino acid metabolism) and the biosynthetic pathways of important secondary metabolites (e.g., polysaccharides, terpenoids, cordycepin), along with their intricate regulatory mechanisms. This collective knowledge forms the bedrock for future innovations in the edible fungi industry (Figure 1).

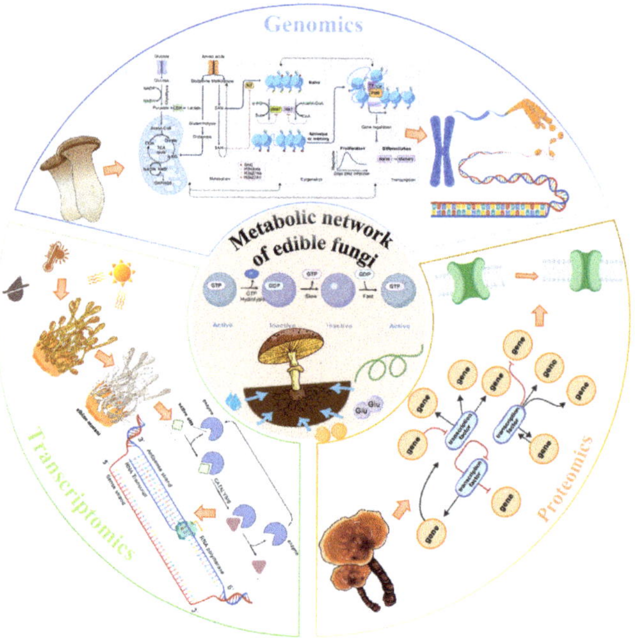

Figure 2. Application of omics technologies in understanding the metabolic network of edible fungi. This diagram illustrates how genomics provides the blueprint, transcriptomics reveals expression dynamics, and proteomics identifies functional protein effectors, all contributing to the elucidation and potential manipulation of metabolic pathways and cellular processes in edible fungi (created with BioGDP.com).

7. Future Directions, Challenges, and Translational Potential

The burgeoning knowledge derived from omics studies is progressively being translated into practical strategies and technologies aimed at advancing the edible fungi industry. However, significant challenges persist alongside vast potential for future research and application.

7.1. Translating Omics Discoveries into Applications

Guiding Precision Breeding and Genetic Improvement: Omics technologies efficiently identify genes or molecular markers [209] associated with key economic traits, such as high yield, superior quality [93], and stress resistance [111,148,154]. This information, when combined with gene editing tools like CRISPR-Cas9, enables targeted modification of specific genes for directed improvement of strain characteristics [209], facilitating the breeding of new varieties with enhanced yield, quality, and resilience. Marker-Assisted Selection (MAS) and Genomic Selection (GS) are already aiding the development of new edible mushroom lines exhibiting, for example, improved post-harvest quality [210]. The potential of using CRISPR to enhance cordycepin or ergothioneine production in *C. militaris* [33,211] and employing RNAi for gene expression regulation are prime examples of successful proof-of-concept studies, drawing inspiration from omics-assisted improvements in other crops [212]. Research institutes frequently report lab-scale successes in engineering strains with precisely edited genes for traits like increased metabolite production (e.g., 30–50% increase in cordycepin) or enhanced stress tolerance, though broad commercialization requires navigating regulatory hurdles, scale-up challenges, and market acceptance.

Optimizing Cultivation Processes and Active Substance Production: A deeper understanding of how environmental factors (e.g., light [121], temperature [117,118], humidity, CO_2 concentration [183], substrate composition [162]) regulate growth and metabolism informs the design of more efficient and tailored cultivation protocols. Precise environmental control can be used to induce specific metabolic pathways, thereby enhancing the content of target products or improving overall quality attributes. For medicinal fungi, integrated omics approaches are crucial for elucidating the complex biosynthetic pathways of key active compounds, such as hericenones/erinacines in *Hericium* spp. [152], ganoderic acids in *Ganoderma* spp. [42,61], and cordycepin in *Cordyceps* spp. [78,157]. This knowledge provides the theoretical basis and molecular targets for rationally designing and constructing engineered strains for overproduction via metabolic engineering strategies [19,172]. Optimizing edible fungi as 'cell factories' for sustainable production of nutraceuticals or pharmaceutical intermediates is an area of active development, with producers likely adopting precise cultivation parameters based on omics data to improve yield and quality.

Enhancing Overall Industry Efficiency and Competitiveness: Fundamentally, deepening molecular research acts as a driving force for industrial upgrading [20]. Omics-driven technological innovations can not only improve the yield and quality of traditional edible mushroom products but also facilitate the development of novel, high-value products, including functional foods and drug lead compounds. A thorough exploration of the metabolic potential of diverse edible fungi, including the utilization of agricultural or industrial waste biomass as cultivation substrates [213], can open up new application domains. This promotes industrial diversification, enhances economic benefits, and bolsters the core competitiveness of the edible fungi sector [20]. Furthermore, research on mitochondrial genomics continues to contribute valuable insights for improving cultivation biotechnology [214].

7.2. Persistent Challenges and Untapped Potential

Expanding Scope and Depth: Current research predominantly focuses on a limited number of commercially significant species, leaving the vast metabolic potential of numerous wild edible and medicinal fungi largely untapped. Species like *Tuber* spp. (truffles), with their complex symbiotic lifestyles, and *Morchella* spp. (morels), despite increasing genomic and transcriptomic data, require more extensive proteomic investigation, particularly concerning cultivation-relevant traits like sclerotium formation or fruiting body development, and comparative proteomics across ecotypes. Similarly, *Sparassis* spp. (cauliflower mushrooms), valued for their medicinal properties, would benefit from deeper proteomics focusing on bioactive compound synthesis (e.g., β-glucan) [54] and stress responses [187]. Many regionally important, traditionally consumed wild mushrooms (e.g., from *Boletaceae*, *Russulaceae*) remain molecularly uncharacterized, representing a reservoir of untapped metabolic diversity. Even for well-studied commercial species like *L. edodes* or *P. ostreatus*, there is considerable scope for advanced proteomic studies, such as investigating the role of Post-Translational Modifications (PTMs) in metabolic regulation or performing comparative proteomics of diverse strains with distinct traits. Furthermore, a more profound understanding of the fine-tuned regulatory mechanisms (e.g., transcription factor networks, PTMs, non-coding RNA functions) governing metabolic networks in known species is essential. Finer spatio-temporal resolution in omics studies, potentially through single-cell omics, is also needed, though its current application in edible fungi is limited due to challenges in cell wall disruption and obtaining sufficient high-quality material from distinct fungal cell types within complex tissues.

Deepening Multi-Omics Integration and Predictive Modeling: While multi-omics integration is advancing, effectively integrating heterogeneous data, handling differences in data dimensionality and scale, accounting for biological time lags, and improving data completeness and accuracy remain core challenges (as detailed in Section 5.2). Current integration efforts often culminate in correlation analyses rather than the construction of truly predictive systems models, such as dynamic metabolic models or comprehensive gene regulatory networks. The application of Genome-Scale Metabolic Models (GSMMs) in edible fungi is still relatively nascent but is showing promise for species like *C. militaris* when combined with transcriptomic and exometabolomic data [207]. Developing integration strategies, algorithms, and bioinformatics tools specifically tailored for the unique characteristics of fungi (e.g., heterokaryosis, complex secondary metabolism) is imperative.

Addressing Bottlenecks in Metabolic Engineering: For effective metabolic engineering to enhance bioactive compound production, several bottlenecks need to be addressed: (i) incomplete understanding of complex regulatory networks (transcription factors, signaling, PTMs), making engineering outcomes unpredictable; (ii) ensuring adequate precursor supply and balancing metabolic flux, which requires redirecting central carbon metabolism without overburdening cell growth (GSMMs can aid here, but model construction and validation are demanding); (iii) considering subcellular compartmentalization of biosynthetic pathways and the targeting and transport of enzymes and intermediates; (iv) the lack of efficient and versatile genetic tools, including multiplex gene editing systems (e.g., CRISPRa/i), for many non-model edible fungi; (v) limited knowledge of transport and secretion mechanisms for extracellularly targeted products; and (vi) ensuring the genetic stability and scalability of engineered strains under industrial conditions.

Improving Translation from Basic Research to Industrial Application: Effectively converting fundamental omics knowledge into practical breeding strategies, optimized cultivation techniques, and novel product development represents a critical "last mile" challenge. This necessitates closer collaboration between research institutions and industry. While omics provides foundational knowledge and proof-of-concept for improvements (e.g., browning resistance in *L. edodes* via PPO gene insights or cordycepin enhancement in *C. militaris*), widespread commercial products explicitly and solely derived from very recent omics/CRISPR translation are still emerging. Patent activity and pilot-scale successes suggest active development, but the path to market is often long, involving scale-up, regulatory approvals, and market acceptance. This translational research is key to realizing the full industrial potential of edible fungi.

7.3. Concluding Outlook

Looking ahead, with ongoing advances in sequencing technologies (including long-read and single-molecule sequencing), the maturation of single-cell omics, enhanced bioinformatics capabilities, and the increasingly sophisticated application of gene editing tools like CRISPR-Cas9/Cas12a, research on edible fungi metabolism based on genomics, transcriptomics, proteomics, and their integration (as depicted conceptually in Figure 2) will enter a new era characterized by greater precision, systematic approaches, and improved efficiency. A profound understanding of how edible fungi construct and regulate their complex metabolic networks will not only satisfy fundamental scientific inquiry but also provide the crucial scientific foundation for the rational development and utilization of these valuable biological resources. This, in turn, will promote sustainable development of the edible fungi industry and contribute significantly to human health and well-being.

Author Contributions: Conceptualization, B.C. and M.X.; Methodology, M.X. and J.W. (Jin Wang); Investigation, M.X., J.W. (Jin Wang), F.W., J.W. (Jinfeng Wang) and Y.Y.; Writing—Original Draft Preparation, M.X. and J.W. (Jinfeng Wang); Writing—Review & Editing, K.F., B.C., M.X.; Visualization, J.W. (Jinfeng Wang) and F.W.; Supervision, K.F. and B.C.; Project Administration, K.F.; Funding Acquisition, K.F. and B.C. All authors have read and agreed to the published version of the manuscript.

Funding: This work was supported by the National Natural Science Foundation of China (No. 32360787); the Postdoctoral Training Fund of Zunyi Medical University, China (No. F-ZH-020); and Zunyi Medical University Zhuhai Campus Key Construction Discipline (No. ZHGY2024-1).

Acknowledgments: During the preparation of this draft, the authors used Gemini 2.5 Pro in order to checking grammar and linguistic touches. After using this tool, the authors reviewed and edited the content as needed and take full responsibility for the content of the publication.

Conflicts of Interest: The authors report there are no competing interests to declare.

References

1. Boa, E. *Wild Edible Fungi: A Global Overview of Their Use and Importance to People*; Non-Wood Forest Products: Rome, Italy, 2007.
2. Dai, Y.-C.; Yang, Z.-L.; Cui, B.-K.; Yu, C.-J.; Zhou, L.-W. Species Diversity and Utilization of Medicinal Mushrooms and Fungi in China (Review). *Int. J. Med. Mushrooms* **2009**, *11*, 287–302. [CrossRef]
3. Hu, X.-R.; Xiong, H.-K.; Xue, W.T. Research progress on the composition and functional activity of truffles. *Sci. Technol. Food Ind.* **2017**, *38*, 341–345+352.
4. Ma, W.W.; Pan, S.-H. Research Advancement of *Cordyceps militaris*. *Asia-Pac. Tradit. Med.* **2008**, *4*, 148–150.
5. Ai, Z.; Qian, Z.-M.; Li, W.J.; Yang, F.Q.; Chen, S.L.; Li, E.-W. Recent advances in the analysis of nucleosides in Chinese cordyceps. *Mycosystema* **2016**, *35*, 388–403.
6. Yan, W.-J.; Lee, T.-H.; Tang, F.-Y.; Zheng, B.-S.; Jiang, Z.-D. Extraction and Contents Determination of Polysaccharide in *Cordyceps guangdongensis*. *J. South China Agric. Univ.* **2009**, *30*, 4.
7. Qian, Z.-M.; Li, W.-Q.; Sun, M.-T.; Liu, X.-Z.; Li, E.-W.; Li, W.-J. Analysis of chemical compounds in Chinese cordyceps. *Mycosystema* **2016**, *35*, 476–490.
8. Guo, T.; Hou, C.-L.; Wei, L.; Sun, J.; Fan, L. Antioxidant activities of extracts and sub-fractions from Tuber indicum. *Mycosystema* **2010**, *29*, 569–575.
9. Wang, L.-M.; Du, X.-F.; Li, X.-J.; Li, S.-Y. Study on inhibition of proliferation and induction of apoptosis in BGC-823 cell by cordycep militaris. *Chin. J. Microecol.* **2009**, *21*, 118–121.
10. Wu, Z.-W.; Wang, Y.-B.; Zhao, X.-F.; Dou, Y.-P. Antibacterial Activity of Fermentation Broth by *Cordyceps sinensis* and *Cordyceps militaris*. *J. Microbiol.* **2008**, *28*, 47–50.
11. Shi, Q.-Y.; Shao, W.-P. Determination of nutritive components of eight edible fungi. *J. Gansu Agric. Univ.* **2003**, *38*, 336–339+345.
12. Yang, G. Nutritional Component of Einght Species of Fungi of Basidiomycetes. *J. Food Sci. Biotechnol.* **2000**, *19*, 173–176.
13. Xiao, X.-N.; Yuan, Y.; Liao, X.; Wang, L.-Y.; Shi, F.; Ming, J. Research advance on antioxidant active composition and antioxidant mechanisms of edible fungi. *Food Mach.* **2016**, *32*, 207–211.
14. Chen, K.-X.; Wang, W.-L.; Liu, J.; Zhang, F.-C.; Zheng, X.-F. Advance in Antitumor Mechanism of Bioactive Compounds in Edible Mushrooms. *Biotechnol. Bull.* **2015**, *31*, 35–42.
15. Niu, J.; Wang, A.; Shi, C.; Zhuo, L.-J.; Xie, Y.-Z.; Li, W.-Z.; Hu, H.-P. Advance Research on Hypoglycemic Effect of Edible Fungi in China. *Edible Fungi China* **2020**, *39*, 1–7+15.
16. Arshadi, N.; Nouri, H.; Moghimi, H. Increasing the production of the bioactive compounds in medicinal mushrooms: An omics perspective. *Microb. Cell Factories* **2023**, *22*, 11. [CrossRef]
17. Pang, Y.-J.; Lyu, J.; Yu, C.-Q.; Sun, D.-J.Y.; Li, L.-M. A multi-omics approach to investigate the etiology of non-communicable diseases:recent advance and applications. *Chin. J. Epidemiol.* **2021**, *42*, 1–9.
18. Cheng, S.; Zeng, Y.-L.; Li, H.; Wang, Y.-J. Multiomics Studies Facilitate Stroke Drug Development. *Chin. J. Stroke* **2022**, *17*, 213–215.
19. Dong, L.-Q.; Gao, Q. Multi-omics molecular subgrouping of hepatocellular carcinoma and its application in precision diagnosis and treatment. *J. Clin. Hepatol.* **2022**, *38*, 510–514.
20. Cao, L.; Zhang, Q.; Miao, R.; Lin, J.; Feng, R.; Ni, Y.; Li, W.; Yang, D.; Zhao, X. Application of omics technology in the research on edible fungi. *Curr. Res. Food Sci.* **2023**, *6*, 100430. [CrossRef]
21. Velculescu, V.E.; Zhang, L.; Zhou, W.; Vogelstein, J.; Basrai, M.A.; Bassett, D.E., Jr.; Hieter, P.; Vogelstein, B.; Kinzler, K.W. Characterization of the Yeast Transcriptome. *Cell* **1997**, *88*, 243–251. [CrossRef]
22. Wang, Y.-C.; Dang, Y.; Li, X.-Y.; Wang, X.-L. Proteomics and the development of Proteomics Techniques. *Lett. Biotechnol.* **2010**, *21*, 139–144.
23. Al-Obaidi, J.R. Proteomics of edible mushrooms: A mini-review. *Electrophoresis* **2016**, *37*, 1257–1263. [CrossRef] [PubMed]
24. Yuan, X.-H.; Fu, Y.-P.; Xiao, S.-J.; Li, C.-T.; Wang, D.; Li, Y. Research progress on mushroom phenotyping. *Mycosystema* **2021**, *40*, 721–742.
25. Agrawal, G.K.; Sarkar, A.; Righetti, P.G.; Pedreschi, R.; Carpentier, S.; Wang, T.; Barkla, B.J.; Kohli, A.; Ndimba, B.K.; Bykova, N.V.; et al. A decade of plant proteomics and mass spectrometry: Translation of technical advancements to food security and safety issues. *Mass. Spectrom. Rev.* **2013**, *32*, 335–365. [CrossRef] [PubMed]

26. Cha, L.; Zhao, R.-H.; Yu, C.-X.; Zhao, Y.; Chen, H.; Zhang, J.-J.; Chen, M.-J. Proteomics and Its Advance in Edible Fungi. *Mol. Plant Breed.* **2017**, *15*, 2656–2661.
27. Yu, Y.; Zhang, Y.-L. Research progress of omics analysis techniques in edible fungi. *China Food Saf. Mag.* **2021**, *13*, 176–177.
28. Cao, Z.-J.; Hu, B.; Xu, Z.-X.; Zheng, S.-Y.; Wang, C.-X. Research Progress on Transcriptome of Edible Fungi Under Adversity Stress. *North. Hortic.* **2022**, *6*, 131. [CrossRef]
29. Yang, R.; Li, Y.; Song, X.; Tang, L.; Li, C.; Tan, Q.; Bao, D. The complete mitochondrial genome of the widely cultivated edible fungus *Lentinula edodes*. *Mitochondrial DNA Part B* **2017**, *2*, 13–14. [CrossRef]
30. Geng, Y.; Zhang, S.; Yang, N.; Qin, L. Whole-Genome Sequencing and Comparative Genomics Analysis of the Wild Edible Mushroom (*Gomphus purpuraceus*) Provide Insights into Its Potential Food Application and Artificial Domestication. *Genes* **2022**, *13*, 1628. [CrossRef]
31. Yin, L.-W.; Chi, Y.-J. Cloning and Bioinformatics Analysis of Mn P1 cDNA Gene from Hericium erinaceum. *Sci. Silvae Sin.* **2015**, *51*, 68–77.
32. Yang, T.; Guo, M.; Yang, H.; Guo, S.; Dong, C. The blue-light receptor CmWC-1 mediates fruit body development and secondary metabolism in *Cordyceps militaris*. *Appl. Microbiol. Biotechnol.* **2016**, *100*, 743–755. [CrossRef]
33. Katrolia, P.; Liu, X.; Zhao, Y.; Kopparapu, N.K.; Zheng, X. Gene cloning, expression and homology modeling of first fibrinolytic enzyme from mushroom (*Cordyceps militaris*). *Int. J. Biol. Macromol.* **2020**, *146*, 897–906. [CrossRef] [PubMed]
34. Gupta, D.K.; Rühl, M.; Mishra, B.; Kleofas, V.; Hofrichter, M.; Herzog, R.; Pecyna, M.J.; Sharma, R.; Kellner, H.; Hennicke, F.; et al. The genome sequence of the commercially cultivated mushroom Agrocybe aegerita reveals a conserved repertoire of fruiting-related genes and a versatile suite of biopolymer-degrading enzymes. *BMC Genom.* **2018**, *19*, 48. [CrossRef] [PubMed]
35. Zhang, C.; Deng, W.; Yan, W.; Li, T. Whole Genome Sequence of an Edible and Potential Medicinal Fungus, *Cordyceps guangdongensis*. *G3 Genes | Genomes | Genet.* **2018**, *8*, 1863–1870. [CrossRef]
36. Vongsangnak, W.; Raethong, N.; Mujchariyakul, W.; Nguyen, N.N.; Leong, H.W.; Laoteng, K. Genome-scale metabolic network of *Cordyceps militaris* useful for comparative analysis of entomopathogenic fungi. *Gene* **2017**, *626*, 132–139. [CrossRef]
37. Kramer, G.J.; Nodwell, J.R. Chromosome level assembly and secondary metabolite potential of the parasitic fungus *Cordyceps militaris*. *BMC Genom.* **2017**, *18*, 912. [CrossRef] [PubMed]
38. Zheng, P.; Xia, Y.; Xiao, G.; Xiong, C.; Hu, X.; Zhang, S.; Zheng, H.; Huang, Y.; Zhou, Y.; Wang, S.; et al. Genome sequence of the insect pathogenic fungus *Cordyceps militaris*, a valued traditional chinese medicine. *Genome Biol.* **2011**, *12*, R116. [CrossRef]
39. Xin, X.; Yin, J.; Zhang, B.; Li, Z.; Zhao, S.; Gui, Z. Genome-wide analysis of DNA methylation in subcultured *Cordyceps militaris*. *Arch. Microbiol.* **2019**, *201*, 369–375. [CrossRef]
40. Duan, M.; Long, S.; Wu, X.; Feng, B.; Qin, S.; Li, Y.; Li, X.; Li, C.; Zhao, C.; Wang, L.; et al. Genome, transcriptome, and metabolome analyses provide new insights into the resource development in an edible fungus *Dictyophora indusiata*. *Front. Microbiol.* **2023**, *14*, 1137159. [CrossRef]
41. Yu, H.W.; Im, J.H.; Kong, W.S.; Park, Y.J. Comparative Analysis of Carbohydrate Active Enzymes in the *Flammulina velutipes* var. lupinicola Genome. *Microorganisms* **2021**, *9*, 20. [CrossRef]
42. Liu, D.; Gong, J.; Dai, W.; Kang, X.; Huang, Z.; Zhang, H.-M.; Liu, W.; Liu, L.; Ma, J.; Xia, Z.; et al. The Genome of Ganderma lucidum Provide Insights into Triterpense Biosynthesis and Wood Degradation. *PLoS ONE* **2012**, *7*, e36146.
43. Liu, Y.; Huang, L.; Hu, H.; Cai, M.; Liang, X.; Li, X.; Zhang, Z.; Xie, Y.; Xiao, C.; Chen, S.; et al. Whole-genome assembly of *Ganoderma leucocontextum* (Ganodermataceae, Fungi) discovered from the Tibetan Plateau of China. *G3 Genes | Genomes | Genet.* **2021**, *11*, jkab337. [CrossRef]
44. Zhu, F.-L.; Xu, X.-L.; Chen, T.-Q.; Shi, L.-C.; Miao, X.-Q.; Lan, J. The Analysis of Whole-genome Resequencing of *Ganoderma lucidum* Originated from South Korea. *Mod. Tradit. Chin. Med. Mater. Medica-World Sci. Technol.* **2019**, *21*, 764–774.
45. Gong, W.; Wang, Y.; Xie, C.; Zhou, Y.; Zhu, Z.; Peng, Y. Whole genome sequence of an edible and medicinal mushroom, *Hericium erinaceus* (Basidiomycota, Fungi). *Genomics* **2020**, *112*, 2393–2399. [CrossRef]
46. Xiao, Y.; Cheng, X.; Liu, J.; Li, C.; Nong, W.; Bian, Y.; Cheung, M.K.; Kwan, H.S. Population genomic analysis uncovers environmental stress-driven selection and adaptation of *Lentinula edodes* population in China. *Sci. Rep.* **2016**, *6*, 36789. [CrossRef] [PubMed]
47. Sakamoto, Y.; Nakade, K.; Yoshida, K.; Natsume, S.; Miyazaki, K.; Sato, S.; van Peer, A.F.; Konno, N. Grouping of multicopper oxidases in *Lentinula edodes* by sequence similarities and expression patterns. *AMB Express* **2015**, *5*, 63. [CrossRef] [PubMed]
48. Xu, L.; Yang, W.; Qiu, T.; Gao, X.; Zhang, H.; Zhang, S.; Cui, H.; Guo, L.; Yu, H.; Yu, H. Complete genome sequences and comparative secretomic analysis for the industrially cultivated edible mushroom Lyophyllum decastes reveals insights on evolution and lignocellulose degradation potential. *Front. Microbiol.* **2023**, *14*, 1137162. [CrossRef]
49. Liu, W.; Chen, L.; Cai, Y.; Zhang, Q.; Bian, Y. Opposite Polarity Monospore Genome De Novo Sequencing and Comparative Analysis Reveal the Possible Heterothallic Life Cycle of Morchella importuna. *Int. J. Mol. Sci.* **2018**, *19*, 2525. [CrossRef]

50. Mei, H.; Qingshan, W.; Baiyintala, W. The whole-genome sequence analysis of Morchella sextelata. *Sci. Rep.* **2019**, *9*, 15376.
51. Shu, R.; Zhang, J.; Meng, Q.; Zhang, H.; Zhou, G.; Li, M.; Wu, P.; Zhao, Y.; Chen, C.; Qin, Q. A New High-Quality Draft Genome Assembly of the Chinese Cordyceps *Ophiocordyceps sinensis*. *Genome Biol. Evol.* **2020**, *12*, 1074–1079. [CrossRef]
52. Zhu, L.; Gao, X.; Zhang, M.; Hu, C.; Yang, W.; Guo, L.; Yang, S.; Yu, H.; Yu, H. Whole Genome Sequence of an Edible Mushroom *Oudemansiella raphanipes* (Changgengu). *J. Fungi* **2023**, *9*, 266. [CrossRef] [PubMed]
53. Yu, H.; Zhang, M.; Sun, Y.; Li, Q.; Liu, J.; Song, C.; Shang, X.; Tan, Q.; Zhang, L.; Yu, H. Whole-genome sequence of a high-temperature edible mushroom *Pleurotus giganteus* (zhudugu). *Front. Microbiol.* **2022**, *13*, 941889. [CrossRef] [PubMed]
54. Xiao, D.; Ma, L.; Yang, C.; Ying, Z.; Jiang, X.; Lin, Y.Q. De Novo Sequencing of a Sparassis latifolia Genome and Its Associated Comparative Analyses. *Can. J. Infect. Dis. Med. Microbiol.* **2018**, *2018*, 1857170. [CrossRef]
55. Li, S.; Zhao, S.; Hu, C.; Mao, C.; Guo, L.; Yu, H.; Yu, H. Whole Genome Sequence of an Edible Mushroom *Stropharia rugosoannulata* (Daqiugaigu). *J. Fungi* **2022**, *8*, 99. [CrossRef]
56. Kim, J.H.; Bae, E.-K.; Hue, Y.; Choi, B.; Kang, M.-J.; Park, E.-J.; Kim, K.-T. Comparative Genomics Reveals Species-Specific Genes and Symbiotic Adaptations in Tricholoma matsutake. *J. Fungi* **2024**, *10*, 746. [CrossRef]
57. Bao, D.; Gong, M.; Zheng, H.; Chen, M.; Zhang, L.; Wang, H.; Jiang, J.; Wu, L.; Zhu, Y.; Zhu, G.; et al. Sequencing and Comparative Analysis of the Straw Mushroom (*Volvariella volvacea*) Genome. *PLoS ONE* **2013**, *8*, e58294. [CrossRef] [PubMed]
58. Meng, L.; Yan, J.; Xie, B.; Li, Y.; Chen, B.; Liu, S.; Li, D.; Yang, Z.; Zeng, X.; Deng, Y.; et al. Genes encoding FAD-binding proteins in *Volvariella volvacea* exhibit differential expression in homokaryons and heterokaryons. *Microbiol. Res.* **2013**, *168*, 533–546. [CrossRef]
59. Liu, J.-Y.; Wang, R.-J.; Zhang, D.; Shang, X.-D.; Tan, Q. Analysis of genes related to lysine biosynthesis based on whole genome of *Flammulina velutipes*. *Microbiol. China* **2016**, *43*, 2225–2233.
60. Kurata, A.; Fukuta, Y.; Mori, M.; Kishimoto, N.; Shirasaka, N. Draft Genome Sequence of the Basidiomycetous Fungus *Flammulina velutipes* TR19. *Genome Announc.* **2016**, *4*, e00505-16. [CrossRef]
61. Chen, S.; Xu, J.; Liu, C.; Zhu, Y.; Nelson, D.R.; Zhou, S.; Li, C.; Wang, L.; Guo, X.; Sun, Y.; et al. Genome sequence of the model medicinal mushroom *Ganoderma lucidum*. *Nat. Commun.* **2012**, *3*, 913. [CrossRef]
62. Li, J.; Zhang, J.; Chen, H.; Chen, X.; Lan, J.; Liu, C. Complete Mitochondrial Genome of the Medicinal Mushroom *Ganoderma lucidum*. *PLoS ONE* **2013**, *8*, e72038. [CrossRef] [PubMed]
63. Yang, R.; Li, Y.; Li, C.; Xu, J.; Bao, D. The complete mitochondrial genome of the Basidiomycete edible fungus *Pleurotus eryngii*. *Mitochondrial DNA Part B* **2016**, *1*, 772–774. [CrossRef] [PubMed]
64. Wang, Y.; Zeng, F.; Hon, C.C.; Zhang, Y.; Leung, F.C.C. The mitochondrial genome of the Basidiomycete fungus *Pleurotus ostreatus* (oyster mushroom). *FEMS Microbiol. Lett.* **2008**, *280*, 34–41. [CrossRef]
65. Wan, J.; Li, Y.; Wang, H.; Tang, L.; Li, Z.; Zhou, C.; Tan, Q.; Bao, D.; Yang, R. Three complete mitochondrial genomes of straw-rotting edible fungus *Volvariella volvacea* using next generation sequencing. *Mitochondrial DNA Part B* **2018**, *3*, 1054–1056. [CrossRef] [PubMed]
66. Yu, F.; Zhang, Y.; Song, J.; Liang, J. The complete mitochondrial genome of a wild edible mushroom, Russula griseocarnosa. *Mitochondrial DNA Part B* **2019**, *4*, 3368–3369. [CrossRef]
67. Wang, S.-R.; Zhang, J.-P.; He, Y.-R.; Chang, M.-C.; Meng, J.-L. Characterization of the complete mitochondrial genome of Coprinellus micaceus, a wild saprobic mushroom in China. *Mitochondrial DNA Part B* **2021**, *6*, 1979–1981. [CrossRef]
68. Song, Y.; Wan, J.; Shang, J.-J.; Feng, Z.; Jin, Y.; Li, H.; Guo, T.; Wu, Y.-Y.; Bao, D.-P.; Zhang, M.; et al. The complete mitochondrial genome of the edible mushroom Grifola frondosa. *Mitochondrial DNA Part B* **2022**, *7*, 286–288. [CrossRef]
69. Xu, W.-N.; Huang, R.-M.; Liu, Y.-Y.; Tong, Z.-J.; Han, X.; Xie, L.-Y.; Xie, B.-G. Genome sequencing and assembly strategy analyses of *Flammulina filiformis*. *Mycosystema* **2018**, *37*, 1578–1585.
70. Xu, W.-N.; Huang, R.-M.; Liu, Y.-Y.; Xie, L.-Y.; Jiang, Y.-J.; Xie, B.-G. Sequence analyses of the rDNA structure of *Flammulina filiformis*. *Mycosystema* **2018**, *37*, 1620–1627.
71. Bao, D.-P.; Zhao, G.-P.; Tan, Q.; Wang, S.-Y.; Chen, M.-J.; Zheng, H.-J.; Zhang, J.-S.; Zhu, Y.-Q.; Wang, H.; Kang, H.; et al. Draft Sequence of the *Volvariella volvacea* Genome. *Acta Edulis Fungi* **2010**, *17*, 1–5.
72. Yue, G.-D.; Gao, Q.; Luo, L.-H.; Wang, J.-Y.; Xu, J.-H.; Yin, Y. The Application of High-throughput Sequencing Technology in Plant and Animal Research. *Sci. Sin.* **2012**, *42*, 107–124.
73. Zhao, Y.; Liu, Y.; Chen, X.; Xiao, J. Genome resequencing and transcriptome analysis reveal the molecular mechanism of albinism in *Cordyceps militaris*. *Front. Microbiol.* **2023**, *14*, 1153153. [CrossRef] [PubMed]
74. Wang, C.; Zhong, H.; Long, X.; Xu, R.; Gong, Y.; Bian, Y.; Zhou, Y. The T-DNA integration characteristics of monokaryotic mutant library in *Lentinula edodes*. *Sci. Hortic.* **2023**, *314*, 111955. [CrossRef]
75. Gong, W.; Xie, C.; Zhou, Y.; Zhu, Z.; Wang, Y.; Peng, Y. A Resequencing-Based Ultradense Genetic Map of *Hericium erinaceus* for Anchoring Genome Sequences and Identifying Genetic Loci Associated with Monokaryon Growth. *Front. Microbiol.* **2020**, *10*, 3129. [CrossRef]

76. Yu, H.; Zhang, L.; Shang, X.; Peng, B.; Li, Y.; Xiao, S.; Tan, Q.; Fu, Y. Chromosomal genome and population genetic analyses to reveal genetic architecture, breeding history and genes related to cadmium accumulation in *Lentinula edodes*. *BMC Genom.* **2022**, *23*, 120. [CrossRef]
77. Liu, Y.-Y.; Xie, L.-Y.; Li, Y.-N.; Ma, X.-B.; Yang, H.; Wang, M.; Xie, B.-G. Identification of exogenous DNA insertion site of transformants in *Flammulina filiformis* using matrix design. *Mycosystema* **2020**, *39*, 983–992.
78. Chang, J.-L.; Guo, Y.-S.; Yang, D.-G.; Li, L.-W. The Research of Overview in Plant for Functional Genomics Techniques. *J. Maize Sci.* **2008**, *16*, 56–59.
79. Wang, J.-X.; Sun, Y.; Xu, P.-L.; Yi, Z.-B.; Du, J.-Z.; Sun, D.-Q. Research Progress in Functional Plant Genomics. *Biotechnol. Bull.* **2004**, *1*, 18–22.
80. Gu, F.-Y.; Chen, Z.-Y.; Shi, J.-J.; Qian, S.-J. Advances in Cellulase and Its Development Tendency. *J. Microbiol.* **2008**, *28*, 83–87.
81. Chen, B.; Gui, F.; Xie, B.; Deng, Y.; Sun, X.; Lin, M.; Tao, Y.; Li, S. Composition and Expression of Genes Encoding Carbohydrate-Active Enzymes in the Straw-Degrading Mushroom *Volvariella volvacea*. *PLoS ONE* **2013**, *8*, e58780.
82. Yang, F.; Xu, B.; Li, J.; Huang, Z. Transcriptome analysis of Termitomyces albuminosus reveals the biodegradation of lignocellulose. *Acta Microbiol. Sin.* **2012**, *52*, 466–477.
83. Lu, Y.-P.; Liao, J.-H.; Guo, Z.-J.; Cai, Z.-X.; Chen, M.-Y. Genome Survey and Transcriptome Analysis on Mycelia and Primordia of *Agaricus blazei*. *BioMed Res. Int.* **2020**, *2020*, 1824183. [CrossRef]
84. Mitsis, T.; Efthimiadou, A.; Bacopoulou, F.; Vlachakis, D.; Eliopoulos, E. Transcription factors and evolution: An integral part of gene expression (Review). *World Acad. Sci. J.* **2020**, *2*, 3–8. [CrossRef]
85. Hou, Z.; Chen, Q.; Zhao, M.; Huang, C.; Wu, X. Genome-wide characterization of the Zn(II)2Cys6 zinc cluster-encoding gene family in *Pleurotus ostreatus* and expression analyses of this family during developmental stages and under heat stress. *PeerJ* **2020**, *8*, e9336. [CrossRef]
86. Langer, I.; Jeandriens, J.; Couvineau, A.; Sanmukh, S.; Latek, D. Signal Transduction by VIP and PACAP Receptors. *Biomedicines* **2022**, *10*, 406. [CrossRef] [PubMed]
87. Xiang, L.; Li, Y.; Zhu, Y.; Luo, H.; Li, C.; Xu, X.; Sun, C.; Song, J.; Shi, L.; He, L.; et al. Transcriptome analysis of the *Ophiocordyceps sinensis* fruiting body reveals putative genes involved in fruiting body development and cordycepin biosynthesis. *Genomics* **2014**, *103*, 154–159. [CrossRef] [PubMed]
88. Wang, Y.; Yang, Z.; Shi, L.; Yang, R.; Guo, H.; Zhang, S.; Geng, G. Transcriptome analysis of Auricularia fibrillifera fruit-body responses to drought stress and rehydration. *BMC Genom.* **2022**, *23*, 58. [CrossRef]
89. Ferstl, R.; Akdis, C.A.; O'Mahony, L. Histamine regulation of innate and adaptive immunity. *Front. Biosci.* **2012**, *17*, 40–53. [CrossRef] [PubMed]
90. Xu, L.-M.; Hinsinger, D.D.; Jiang, G.-F. The complete mitochondrial genome of the Basidiomycete fungus Pleurotus cornucopiae (Paulet) Rolland. *Mitochondrial DNA Part B* **2018**, *3*, 73–75. [CrossRef]
91. Bashir, K.M.I.; Rheu, K.M.; Kim, M.-S.; Cho, M.-G. The complete mitochondrial genome of an edible mushroom, Sparassis crispa. *Mitochondrial DNA Part B* **2020**, *5*, 862–863. [CrossRef]
92. Zhao, X.; Zhang, G.; Li, C.; Ling, J. Cordycepin and pentostatin biosynthesis gene identified through transcriptome and proteomics analysis of Cordyceps kyushuensis Kob. *Microbiol. Res.* **2019**, *218*, 12–21. [CrossRef] [PubMed]
93. Cai, Z.X.; Chen, M.Y.; Lu, Y.P.; Guo, Z.J.; Zeng, Z.H.; Liao, J.H.; Zeng, H. Metabolomics and transcriptomics unravel the mechanism of browning resistance in *Agaricus bisporus*. *PLoS ONE* **2022**, *17*, e0255765. [CrossRef] [PubMed]
94. Shi, X.-K.; Lu, Y.-P.; Cai, Z.-X.; Guo, Z.-J.; Chen, M.-Y.; Liao, J.-H. Transcriptome Sequencing on SixAgaricus bisporusStrains at Four Developmental Stages. *Fujian J. Agric. Sci.* **2019**, *34*, 775–781.
95. Peng, B.; Li, B.-J.; Guan, W.-Q.; Lin, Q. Transcriptome Sequencing of *Agaricus bisporus* and Mining of Genes Involved in Browning. *Food Sci.* **2019**, *40*, 126–132.
96. Shi, X.-K.; Cai, Z.-X.; Guo, Z.-J.; Lu, Y.-P.; Chen, M.-Y.; Liao, J.-H.; Wang, Z.-S. Transcriptome Sequencing on Fruiting Body of *Agaricus bisporus* in Developing Stages. *Fujian J. Agric. Sci.* **2018**, *33*, 282–287.
97. Lu, Y.-P.; Guo, Z.-J.; Cai, Z.-X.; Chen, M.-Y.; Liao, J.-H. Transcriptome analysis of *Agaricus blazei* fruiting bodies at different developmental stages. *Mycosystema* **2019**, *38*, 2161–2173.
98. Zhao, Y.; Wang, L.; Zhang, D.; Li, R.; Cheng, T.; Zhang, Y.; Liu, X.; Wong, G.; Tang, Y.; Wang, H.; et al. Comparative transcriptome analysis reveals relationship of three major domesticated varieties of Auricularia auricula-judae. *Sci. Rep.* **2019**, *9*, 78. [CrossRef]
99. Zhou, Y.; Chen, L.; Fan, X.; Bian, Y. De Novo Assembly of Auricularia polytricha Transcriptome Using Illumina Sequencing for Gene Discovery and SSR Marker Identification. *PLoS ONE* **2014**, *9*, e91740. [CrossRef]
100. Liu, T.; Liu, Z.; Yao, X.; Huang, Y.; Qu, Q.; Shi, X.; Zhang, H.; Shi, X. Identification of cordycepin biosynthesis-related genes through de novo transcriptome assembly and analysis in Cordyceps cicadae. *R. Soc. Open. Sci.* **2018**, *5*, 181247. [CrossRef]
101. Wang, Y.; Shao, Y.; Zhu, Y.; Wang, K.; Ma, B.; Zhou, Q.; Chen, A.; Chen, H. XRN1-associated long non-coding RNAs may contribute to fungal virulence and sexual development in entomopathogenic fungus *Cordyceps militaris*. *Pest. Manag. Sci.* **2019**, *75*, 3302–3311. [CrossRef]

02. Lou, H.W.; Zhao, Y.; Tang, H.B.; Ye, Z.W.; Wei, T.; Lin, J.F.; Guo, L.Q. Transcriptome Analysis of *Cordyceps militaris* Reveals Genes Associated With Carotenoid Synthesis and Identification of the Function of the Cmtns Gene. *Front. Microbiol.* **2019**, *10*, 2105. [CrossRef] [PubMed]
03. Thananusak, R.; Laoteng, K.; Raethong, N.; Zhang, Y.; Vongsangnak, W. Metabolic Responses of Carotenoid and Cordycepin Biosynthetic Pathways in *Cordyceps militaris* under Light-Programming Exposure through Genome-Wide Transcriptional Analysis. *Biology* **2020**, *9*, 242. [CrossRef] [PubMed]
04. Yin, J.; Xin, X.; Weng, Y.; Gui, Z. Transcriptome-wide analysis reveals the progress of *Cordyceps militaris* subculture degeneration. *PLoS ONE* **2017**, *12*, e0186279. [CrossRef]
05. Wang, F.; Liu, Q.; Zhang, J.; Liu, K.; Li, K.; Liu, G.; Dong, C. Comparative Transcriptome Analysis Between a Spontaneous Albino Mutant and Its Sibling Strain of *Cordyceps militaris* in Response to Light Stress. *Front. Microbiol.* **2018**, *9*, 1237. [CrossRef] [PubMed]
06. Suparmin, A.; Kato, T.; Dohra, H.; Park, E.Y. Insight into cordycepin biosynthesis of *Cordyceps militaris*: Comparison between a liquid surface culture and a submerged culture through transcriptomic analysis. *PLoS ONE* **2017**, *12*, e0187052. [CrossRef]
07. Lou, H.-W.; Zhao, Y.; Zhao, Y.; Lin, J.-F.; Zhao, R.-Y.; Ye, Z.-W.; Guo, L.Q. Transcriptomic Analysis of *Cordyceps militaris* and Mining of Genes Involved in Carotenoid Biosynthesis. *Food Sci.* **2021**, *42*, 150–156.
08. Chen, B.X.; Wei, T.; Xue, L.N.; Zheng, Q.W.; Ye, Z.W.; Zou, Y.; Yang, Y.; Yun, F.; Guo, L.Q.; Lin, J.F. Transcriptome Analysis Reveals the Flexibility of Cordycepin Network in *Cordyceps militaris* Activated by L-Alanine Addition. *Front. Microbiol.* **2020**, *11*, 577. [CrossRef]
09. Raethong, N.; Laoteng, K.; Vongsangnak, W. Uncovering global metabolic response to cordycepin production in *Cordyceps militaris* through transcriptome and genome-scale network-driven analysis. *Sci. Rep.* **2018**, *8*, 9250. [CrossRef]
10. Wongsa, B.; Raethong, N.; Chumnanpuen, P.; Wong-Ekkabut, J.; Laoteng, K.; Vongsangnak, W. Alternative metabolic routes in channeling xylose to cordycepin production of *Cordyceps militaris* identified by comparative transcriptome analysis. *Genomics* **2020**, *112*, 629–636. [CrossRef]
11. Yu, Y.-H.; Wu, T.-H.; Ye, Z.-W.; Chen, B.-X.; Guo, L.-Q.; Lin, J.-F. The regulation network of cold-induced primordium formation in *Flammulina filiformis* based on transcriptome. *Mycosystema* **2020**, *39*, 1065–1076.
12. Liu, X.-B.; Xia, E.-H.; Li, M.; Cui, Y.-Y.; Wang, P.-M.; Zhang, J.-X.; Xie, B.-G.; Xu, J.-P.; Yan, J.-J.; Li, J.; et al. Transcriptome data reveal conserved patterns of fruiting body development and response to heat stress in the mushroom-forming fungus *Flammulina filiformis*. *PLoS ONE* **2020**, *15*, e0239890. [CrossRef] [PubMed]
13. Wang, W.; Chou, T.-S.; Liu, F.; Yan, J.-J.; Wu, T.-J.; Li, S.-J.; Xie, B.-G. Comparison of gene expression patterns between the monokaryotic and dikaryotic mycelia of *Flammulina velutipes*. *Mycosystema* **2015**, *34*, 683–693.
14. Liu, F.; Wang, W.; Xie, B.-G. Comparison of Gene Expression Patterns in the Mycelium and Primordia of *Flammulina velutipes*, Strain 1123. *Acta Edulis Fungi* **2014**, *21*, 1–7.
15. Cai, M.; Liang, X.; Liu, Y.; Hu, H.; Xie, Y.; Chen, S.; Gao, X.; Li, X.; Xiao, C.; Chen, D.; et al. Transcriptional Dynamics of Genes Purportedly Involved in the Control of Meiosis, Carbohydrate, and Secondary Metabolism during Sporulation in *Ganoderma lucidum*. *Genes* **2021**, *12*, 504. [CrossRef]
16. Ma, Z.; Xu, M.; Wang, Q.; Wang, F.; Zheng, H.; Gu, Z.; Li, Y.; Shi, G.; Ding, Z. Development of an Efficient Strategy to Improve Extracellular Polysaccharide Production of *Ganoderma lucidum* Using L-Phenylalanine as an Enhancer. *Front. Microbiol.* **2019**, *10*, 2306. [CrossRef] [PubMed]
17. You, B.-J.; Tien, N.; Lee, M.-H.; Bao, B.-Y.; Wu, Y.-S.; Hu, T.-C.; Lee, H.-Z. Induction of apoptosis and ganoderic acid biosynthesis by cAMP signaling in *Ganoderma lucidum*. *Sci. Rep.* **2017**, *7*, 318. [CrossRef]
18. Yu, G.-J.; Wang, M.; Huang, J.; Yin, Y.-L.; Chen, Y.-J.; Jiang, S.; Jin, Y.-X.; Lan, X.-Q.; Wong, B.H.C.; Liang, Y.; et al. Deep Insight into the *Ganoderma lucidum* by Comprehensive Analysis of Its Transcriptome. *PLoS ONE* **2012**, *7*, e44031. [CrossRef]
19. Nie, W.-Q.; Wu, T.-X.; Zhong, M.; Lu, H.-Y. Transcriptome Sequencing and Analysis of Grifola frondosa Mycelia. *Food Sci.* **2017**, *38*, 6–11.
20. Song, H.-Y.; Kim, D.-H.; Kim, J.-M. Comparative transcriptome analysis of dikaryotic mycelia and mature fruiting bodies in the edible mushroom *Lentinula edodes*. *Sci. Rep.* **2018**, *8*, 8983. [CrossRef]
21. Kim, J.Y.; Kim, D.Y.; Park, Y.-J.; Jang, M.-J. Transcriptome analysis of the edible mushroom *Lentinula edodes* in response to blue light. *PLoS ONE* **2020**, *15*, e0230680. [CrossRef]
22. Lei, X.-E.; Qiao, Y.-N.; Mao, J.-X.; Yang, Z.-S.; Zhang, L.-L.; Li, L.-Y. Comparative analyses of gene expression based on transcriptome data in different laccase activities of monokaryotic mycelia of *Lentinula edodes*. *Mycosystema* **2020**, *39*, 1162–1174.
23. Huang, X.; Zhang, R.; Qiu, Y.; Wu, H.; Xiang, Q.; Yu, X.; Zhao, K.; Zhang, X.; Chen, Q.; Penttinen, P.; et al. RNA-seq Profiling Showed Divergent Carbohydrate-Active Enzymes (CAZymes) Expression Patterns in *Lentinula edodes* at Brown Film Formation Stage Under Blue Light Induction. *Front. Microbiol.* **2020**, *11*, 1044. [CrossRef] [PubMed]
24. Yoo, S.-i.; Lee, H.-Y.; Markkandan, K.; Moon, S.; Ahn, Y.J.; Ji, S.; Ko, J.; Kim, S.-J.; Ryu, H.; Hong, C.P. Comparative transcriptome analysis identified candidate genes involved in mycelium browning in *Lentinula edodes*. *BMC Genom.* **2019**, *20*, 121. [CrossRef]

125. Ke, S.; Ding, L.; Niu, X.; Shan, H.; Song, L.; Xi, Y.; Feng, J.; Wei, S.; Liang, Q. Comparative transcriptome analysis on candidate genes associated with fruiting body growth and development in Lyophyllum decastes. *PeerJ* **2023**, *11*, e16288. [CrossRef]
126. Fan, T.; Ren, R.; Tang, S.; Zhou, Y.; Cai, M.; Zhao, W.; He, Y.; Xu, J. Transcriptomics combined with metabolomics unveiled the key genes and metabolites of mycelium growth in Morchella importuna. *Front. Microbiol.* **2023**, *14*, e16288. [CrossRef]
127. Hao, H.; Zhang, J.; Wang, H.; Wang, Q.; Chen, M.; Juan, J.; Feng, Z.; Chen, H. Comparative transcriptome analysis reveals potential fruiting body formation mechanisms in Morchella importuna. *AMB Express* **2019**, *9*, 103. [CrossRef]
128. Liu, W.; Cai, Y.; He, P.; Chen, L.; Bian, Y. Comparative transcriptomics reveals potential genes involved in the vegetative growth of Morchella importuna. *3 Biotech* **2019**, *9*, 81. [CrossRef] [PubMed]
129. Wang, J.-Q.; Ye, H.-L.; Liang, D.-W.; Liu, D.-Y.; Geng, F.; Li, X. De Novo Sequencing and Transcriptome Analysis of Morchella importuna Fruiting Body. *Food Sci.* **2018**, *39*, 81–87.
130. Zhao, Y.; Zhang, J.; Meng, Q.; Zhang, H.; Zhou, G.; Li, M.; Wu, P.; Shu, R.; Gao, X.; Guo, L.; et al. Transcriptomic analysis of the orchestrated molecular mechanisms underlying fruiting body initiation in Chinese cordyceps. *Gene* **2020**, *763*, 145061. [CrossRef]
131. Huang, Y.-Q.; Tong, X.-X.; Tao, X.; Wang, Y.-X.; Peng, C.; Wang, S.-H.; Guo, J.-L. Study on bio-synthesis of cordycepin in *Ophiocordyceps sinensis* based on RNA-seq. *Chin. Tradit. Herb. Drugs* **2017**, *48*, 4044–4050.
132. Li, F.; Chen, L.-D.; Ai, L.-Y.; Liu, Y.-C.; Yan, M.; Sun, S.-J. Comparative transcriptomics analyses of *Pleurotus eryngii* at different developmental stages. *Mycosystema* **2018**, *37*, 1586–1597.
133. Cao, J.; Sun, M.; Yu, M.; Xu, Y.; Xie, J.; Zhang, H.; Chen, J.; Xu, T.; Qian, X.; Sun, S. Transcriptome Analysis Reveals the Function of a G-Protein α Subunit Gene in the Growth and Development of *Pleurotus eryngii*. *J. Fungi* **2023**, *9*, 69.
134. Xie, C.; Gong, W.; Zhu, Z.; Yan, L.; Hu, Z.; Peng, Y. Comparative transcriptomics of *Pleurotus eryngii* reveals blue-light regulation of carbohydrate-active enzymes (CAZymes) expression at primordium differentiated into fruiting body stage. *Genomics* **2018**, *110*, 201–209. [CrossRef]
135. Fu, Y.-P.; Liang, Y.; Dai, Y.-T.; Yang, C.-T.; Duan, M.-Z.; Zhang, Z.; Hu, S.-N.; Zhang, Z.-W.; Li, Y. De Novo Sequencing and Transcriptome Analysis of *Pleurotus eryngii* subsp. *tuoliensis* (Bailinggu) Mycelia in Response to Cold Stimulation. *Molecules* **2016**, *21*, 560.
136. Hou, L.; Yan, K.; Dong, S.; Guo, L.; Liu, J.; Wang, S.; Chang, M.; Meng, J. Transcriptome Analysis Revealed That Hydrogen Peroxide-Regulated Oxidative Phosphorylation Plays an Important Role in the Formation of *Pleurotus ostreatus* Cap Color. *J. Fungi* **2023**, *9*, 823. [CrossRef]
137. Wang, L.; Gao, W.; Wu, X.; Zhao, M.; Qu, J.; Huang, C.; Zhang, J. Genome-Wide Characterization and Expression Analyses of *Pleurotus ostreatus* MYB Transcription Factors during Developmental Stages and under Heat Stress Based on de novo Sequenced Genome. *Int. J. Mol. Sci.* **2018**, *19*, 2052. [CrossRef] [PubMed]
138. Wang, W.-K.; Song, J.-L.; Lu, N.; Yuan, W.-D.; Yan, J.; Chen, G.-P. Cloning and expression of thePpFBD1 involved in primordium formation ofPleurotus pulmonarius. *Acta Agric. Zhejiangensis* **2020**, *32*, 93–97.
139. Fu, Y.; Dai, Y.; Yang, C.; Wei, P.; Song, B.; Yang, Y.; Sun, L.; Zhang, Z.-W.; Li, Y. Comparative Transcriptome Analysis Identified Candidate Genes Related to Bailinggu Mushroom Formation and Genetic Markers for Genetic Analyses and Breeding. *Sci. Rep.* **2017**, *7*, 9266. [CrossRef]
140. Du, F.; Ya Li mai mai ti, N.E.z.y.; Hu, Q.; Zou, Y.; Ye, D.; Zhang, H. A Comparative Transcriptome Analysis Reveals Physiological Maturation Properties of Mycelia in Pleurotus tuoliensis. *Genes* **2019**, *10*, 703. [CrossRef]
141. Rong, C.-B.; Zhao, S.; Song, S.; Wang, J.; Liu, Y. Identification of Candidate Genes Related to the Development of Pleurotus tuoliensis Fruiting Bodies. *Biotechnol. Bull.* **2018**, *34*, 115–120.
142. Yang, C.; Ma, L.; Xiao, D.; Ying, Z.; Jiang, X.; Lin, Y. Integration of ATAC-Seq and RNA-Seq Identifies Key Genes in Light-Induced Primordia Formation of Sparassis latifolia. *Int. J. Mol. Sci.* **2020**, *21*, 185. [CrossRef] [PubMed]
143. Xiao, D.-L.; Zhang, D.; Ma, L.; Wang, H.-Y.; Lin, Y.-S. Preliminary Study on Differentially Expressed Genes of Sparassis latifolia under Light Inducing. *Edible Fungi China* **2017**, *36*, 60–63.
144. Hao, H.; Zhang, J.; Wu, S.; Bai, J.; Zhuo, X.; Zhang, J.; Kuai, B.; Chen, H. Transcriptomic analysis of *Stropharia rugosoannulata* reveals carbohydrate metabolism and cold resistance mechanisms under low-temperature stress. *AMB Express* **2022**, *12*, 56. [CrossRef]
145. Wang, D.; Niu, B.; Song, J.; Zhao, S.-H.; Lei, S.-R.; Guo, L.-A.; Zhang, F.-L.; Liu, W.-J.; Chang, L.-J.; Zhao, L.-M. Study on Fructose and Mannose Metabolism Pathway ofTremella fuciformisBased on Transcriptome. *Southwest China J. Agric. Sci.* **2019**, *32*, 1347–1352.
146. Liu, M.; Yu, T.; Singh, P.K.; Liu, Q.; Liu, H.; Zhu, Q.; Xiao, Z.; Xu, J.; Peng, Y.; Fu, S.; et al. A Comparative Transcriptome Analysis of *Volvariella volvacea* Identified the Candidate Genes Involved in Fast Growth at the Mycelial Growth Stage. *Genes* **2020**, *11*, 161. [CrossRef] [PubMed]
147. Shu, S.; Chen, B.; Zhou, M.; Zhao, X.; Xia, H.; Wang, M. De Novo Sequencing and Transcriptome Analysis of Wolfiporia cocos to Reveal Genes Related to Biosynthesis of Triterpenoids. *PLoS ONE* **2013**, *8*, e71350. [CrossRef]

148. Li, J.; Wu, B.; Xu, J.; Liu, C. Genome-Wide Identification and Characterization of Long Intergenic Non-Coding RNAs in *Ganoderma lucidum*. *PLoS ONE* **2014**, *9*, e99442. [CrossRef]
149. Qu, J.; Zhao, M.; Hsiang, T.; Feng, X.; Zhang, J.; Huang, C. Identification and Characterization of Small Noncoding RNAs in Genome Sequences of the Edible Fungus *Pleurotus ostreatus*. *Biomed Res. Int.* **2016**, *2016*, 2503023. [CrossRef]
150. Zhou, Y.; Fan, X.-Z.; Chen, L.-F.; Bian, Y.-B. Distribution and sequence characteristics of SSR in the transcriptomes of Auricularia auricula-judae and Auricularia polytricha. *Mycosystema* **2014**, *33*, 280–288.
151. Ma, J.; Qu, W.; Chen, C.-Y.; Wang, L.; Ma, W.; Liu, Z.-S.; Ma, J.; Yang, S.; Ding, L.; Gao, Q.; et al. Development of EST-SSR markers based on transcriptome sequencing ofMorchellaspp. and its genetic diversity analysis. *Jiangsu J. Agric. Sci.* **2020**, *36*, 1282–1290.
152. Zhang, Y.; Mo, M.; Yang, L.; Mi, F.; Cao, Y.; Liu, C.; Tang, X.; Wang, P.; Xu, J. Exploring the Species Diversity of Edible Mushrooms in Yunnan, Southwestern China, by DNA Barcoding. *J. Fungi* **2021**, *7*, 310. [CrossRef] [PubMed]
153. Yu, F.-M.; Jayawardena, R.S.; Thongklang, N.; Lv, M.-L.; Zhu, X.-T.; Zhao, Q. Morel Production Associated with Soil Nitrogen-Fixing and Nitrifying Microorganisms. *J. Fungi* **2022**, *8*, 299. [CrossRef] [PubMed]
154. Park, Y.-J.; Baek, J.H.; Lee, S.; Kim, C.; Rhee, H.; Kim, H.; Seo, J.-S.; Park, H.-R.; Yoon, D.-E.; Nam, J.-Y.; et al. Whole Genome and Global Gene Expression Analyses of the Model Mushroom *Flammulina velutipes* Reveal a High Capacity for Lignocellulose Degradation. *PLoS ONE* **2014**, *9*, e93560. [CrossRef]
155. Park, M.; Kim, M.; Kim, S.; Ha, B.; Ro, H.-S. Differential Expression of Laccase Genes in *Pleurotus ostreatus* and Biochemical Characterization of Laccase Isozymes Produced in Pichia pastoris. *Mycobiology* **2015**, *43*, 280–287. [CrossRef]
156. Tang, X.; Ding, X.; Hou, Y.-l. Comparative analysis of transcriptomes revealed the molecular mechanism of development of Tricholoma matsutake at different stages of fruiting bodies. *Food Sci. Biotechnol.* **2020**, *29*, 939–951. [CrossRef]
157. Wang, M.; Gu, B.; Huang, J.; Jiang, S.; Chen, Y.; Yin, Y.; Pan, Y.; Yu, G.; Li, Y.; Wong, B.H.C.; et al. Transcriptome and Proteome Exploration to Provide a Resource for the Study of Agrocybe aegerita. *PLoS ONE* **2013**, *8*, e56686. [CrossRef]
158. Li, J.; Li, Z.; Zhao, T.; Yan, X.; Pang, Q. Proteomic Analysis of Auricularia auricula-judae Under Freezing Treatment Revealed Proteins and Pathways Associated With Melanin Reduction. *Front. Microbiol.* **2016**, *11*, 610173. [CrossRef] [PubMed]
159. Xu, L.; Lin, R.; Li, X.; Zhang, C.; Yang, X.; Guo, L.; Yu, H.; Gao, X.; Hu, C. Comparative Proteomic Analyses within Three Developmental Stages of the Mushroom White Hypsizygus marmoreus. *J. Fungi* **2023**, *9*, 225. [CrossRef]
160. Yang, X.; Lin, R.; Xu, K.; Guo, L.; Yu, H. Comparative Proteomic Analysis within the Developmental Stages of the Mushroom White Hypsizygus marmoreus. *J. Fungi* **2021**, *7*, 1064. [CrossRef]
161. Xu, L.; Guo, L.; Yu, H. Label-Free Comparative Proteomics Analysis Revealed Heat Stress Responsive Mechanism in Hypsizygus marmoreus. *Front. Microbiol.* **2021**, *11*, 541967. [CrossRef]
162. Zhao, Y.; Li, H.; Yao, Y.; Wei, Q.; Hu, T.; Li, X.; Zhu, B.; Ma, H. Combined analysis of proteomics and metabolism reveals critical roles of oxidoreductase activity in mushrooms stimulated by wolfberry and sea buckthorn substrates. *Front. Nutr.* **2025**, *12*, 1543240. [CrossRef] [PubMed]
163. Cai, Y.; Liu, W.; Zhang, Q.; Lu, D.; Cai, D.; Zhao, Y.; Ma, X. Label-free based comparative proteomic analysis of Morchella importuna development from the vegetative to the sexual reproductive stages. *J. Agric. Food Res.* **2022**, *7*, 100247. [CrossRef]
164. Tong, X.; Wang, F.; Zhang, H.; Bai, J.; Dong, Q.; Yue, P.; Jiang, X.; Li, X.; Wang, L.; Guo, J. iTRAQ-based comparative proteome analyses of different growth stages revealing the regulatory role of reactive oxygen species in the fruiting body development of *Ophiocordyceps sinensis*. *PeerJ* **2021**, *9*, e10940. [CrossRef]
165. Desiderio, A.; Goppa, L.; Santambrogio, C.; Brocca, S.; Buratti, S.; Girometta, C.E.; Sarkar, M.; Venuti, M.T.; Savino, E.; Rossi, P.; et al. Improving the Proteome-Mining of Schizophyllum commune to Enhance Medicinal Mushroom Applications. *J. Fungi* **2025**, *11*, 120. [CrossRef] [PubMed]
166. Li, Y.; Tang, H.; Zhao, W.; Yang, Y.; Fan, X.; Zhan, G.; Li, J.; Sun, S. Study of Dimorphism Transition Mechanism of Tremella fuciformis Based on Comparative Proteomics. *J. Fungi* **2022**, *8*, 242. [CrossRef]
167. Zha, L.; Chen, M.; Guo, Q.; Tong, Z.; Li, Z.; Yu, C.; Yang, H.; Zhao, Y. Comparative Proteomics Study on the Postharvest Senescence of *Volvariella volvacea*. *J. Fungi* **2022**, *8*, 819. [CrossRef]
168. Wang, C.; Zhang, X.; Zeng, Z.; Song, F.; Lin, Z.; Chen, L.; Cai, Z. Transcriptome Analysis Explored the Differential Genes' Expression During the Development of the *Stropharia rugosoannulata* Fruiting Body. *Front. Genet.* **2022**, *13*, 924050. [CrossRef] [PubMed]
169. Xiong, C.; Xia, Y.; Zheng, P.; Shi, S.; Wang, C. Developmental stage-specific gene expression profiling for a medicinal fungus *Cordyceps militaris*. *Mycology* **2010**, *1*, 25–66. [CrossRef]
170. Gong, M.; Wang, Y.; Zhang, J.; Zhao, Y.; Wan, J.; Shang, J.; Yang, R.; Wu, Y.; Li, Y.; Tan, Q.; et al. Chilling Stress Triggers VvAgo1-Mediated miRNA-Like RNA Biogenesis in *Volvariella volvacea*. *Front. Microbiol.* **2020**, *11*, 523593. [CrossRef]
171. Liang, X.; Han, J.; Cui, Y.; Shu, X.; Lei, M.; Wang, B.; Jia, D.; Peng, W.; He, X.; Liu, X. Whole-Genome Sequencing of *Flammulina filiformis* and Multi-Omics Analysis in Response to Low Temperature. *J Fungi* **2025**, *11*, 229. [CrossRef]
172. Krah, F.-S.; Hess, J.; Hennicke, F.; Kar, R.; Bässler, C. Transcriptional response of mushrooms to artificial sun exposure. *Ecol. Evol.* **2021**, *11*, 10538–10546. [CrossRef] [PubMed]

173. Wang, Y.; Mao, C.; Shi, Y.; Fan, X.; Sun, L.; Zhuang, Y. Transcriptome analysis of the response of Hypomyces chrysospermus to cadmium stress. *Front. Microbiol.* **2022**, *13*, 990693. [CrossRef] [PubMed]
174. Chen, C.; Fu, R.; Wang, J.; Li, X.; Chen, X.; Li, Q.; Lu, D. Genome sequence and transcriptome profiles of pathogenic fungus Paecilomyces penicillatus reveal its interactions with edible fungus Morchella importuna. *Comput. Struct. Biotechnol. J.* **2021**, *19*, 2607–2617. [CrossRef]
175. Xu, Z.; Liu, J.-Y.; Zhang, D.; Wang, R.-J.; Pan, Y.-J.; Tan, Q.; Shang, X.-D. Transcriptome Comparison of Differentially Expressed Genes in the Mycelial Stages of *Flammulina velutipes* Monokaryon 3_M and the Hybrid Dikaryon G1. *Acta Edulis Fungi* **2016**, *23*, 1–6.
176. Gehrmann, T.; Pelkmans, J.F.; Ohm, R.A.; Vos, A.M.; Sonnenberg, A.S.M.; Baars, J.J.P.; Wösten, H.A.B.; Reinders, M.J.T.; Abeel, T. Nucleus-specific expression in the multinuclear mushroom-forming fungus *Agaricus bisporus* reveals different nuclear regulatory programs. *Proc. Natl. Acad. Sci. USA* **2018**, *115*, 4429–4434. [CrossRef]
177. Liu, T.; Li, H.; Ding, Y.; Qi, Y.; Gao, Y.; Song, A.; Shen, J.; Qiu, L. Genome-wide gene expression patterns in dikaryon of the basidiomycete fungus *Pleurotus ostreatus*. *Braz. J. Microbiol.* **2017**, *48*, 380–390. [CrossRef]
178. Alfaro, M.; Castanera, R.; Lavín, J.L.; Grigoriev, I.V.; Oguiza, J.A.; Ramírez, L.; Pisabarro, A.G. Comparative and transcriptional analysis of the predicted secretome in the lignocellulose-degrading basidiomycete fungus *Pleurotus ostreatus*. *Environ. Microbiol.* **2016**, *18*, 4710–4726. [CrossRef]
179. Tang, L.-H.; Bao, D.-P.; Wan, J.-N.; Li, Y.; Wang, Y.; Tan, Q. Transcriptome analysis of transcription factors activiated at the light-induced brown mycelial film formation stage in *Lentinula edodes*. *Mycosystema* **2016**, *35*, 1106–1116.
180. Fang, D.; Zheng, Z.; Ma, N.; Yang, W.; Dai, C.; Zhao, M.; Deng, Z.; Hu, Q.; Zhao, L. Label-free proteomic quantification of packaged *Flammulina filiformis* during commercial storage. *Postharvest Biol. Technol.* **2020**, *169*, 111312. [CrossRef]
181. Ijoma, G.N.; Heri, S.M.; Matambo, T.S.; Tekere, M. Trends and Applications of Omics Technologies to Functional Characterisation of Enzymes and Protein Metabolites Produced by Fungi. *J. Fungi* **2021**, *7*, 700. [CrossRef]
182. Yang, H.; Zheng, Z.; Zhou, H.; Qu, H.; Gao, H. Proteomics Reveals the Mechanism Underlying the Autolysis of Postharvest Coprinus comatus Fruiting Bodies. *J. Agric. Food Chem.* **2022**, *70*, 1346–1357. [CrossRef] [PubMed]
183. Lin, R.; Zhang, L.; Yang, X.; Li, Q.; Zhang, C.; Guo, L.; Yu, H.; Yu, H. Responses of the Mushroom *Pleurotus ostreatus* under Different CO_2 Concentration by Comparative Proteomic Analyses. *J. Fungi* **2022**, *8*, 652. [CrossRef] [PubMed]
184. Yu, H.; Jiang, N.; Yan, M.; Cheng, X.; Zhang, L.; Zhai, D.; Liu, J.; Zhang, M.; Song, C.; Yu, H.; et al. Comparative analysis of proteomes and transcriptomes revealed the molecular mechanism of development and nutrition of *Pleurotus giganteus* at different fruiting body development stages. *Front. Nutr.* **2023**, *10*, 1197983. [CrossRef]
185. Vita, F.; Giuntoli, B.; Bertolini, E.; Taiti, C.; Marone, E.; D'Ambrosio, C.; Trovato, E.; Sciarrone, D.; Zoccali, M.; Balestrini, R.; et al. Tuberomics: A molecular profiling for the adaption of edible fungi (*Tuber magnatum Pico*) to different natural environments. *BMC Genom.* **2020**, *21*, 90. [CrossRef]
186. Zhao, Y.; Yao, Y.; Li, H.; Han, Z.; Ma, X. Integrated transcriptome and metabolism unravel critical roles of carbon metabolism and oxidoreductase in mushroom with Korshinsk peashrub substrates. *BMC Genom.* **2024**, *25*, 763. [CrossRef] [PubMed]
187. Horie, K.; Rakwal, R.; Hirano, M.; Shibato, J.; Nam, H.W.; Kim, Y.S.; Kouzuma, Y.; Agrawal, G.K.; Masuo, Y.; Yonekura, M. Proteomics of two cultivated mushrooms Sparassis crispa and Hericium erinaceum provides insight into their numerous functional protein components and diversity. *J. Proteome Res.* **2008**, *7*, 1819–1835. [CrossRef]
188. Wu, M.-X.; Zou, Y.; Yu, Y.-H.; Chen, B.-X.; Zheng, Q.-W.; Ye, Z.-W.; Wei, T.; Ye, S.-Q.; Guo, L.-Q.; Lin, J.-F. Comparative transcriptome and proteome provide new insights into the regulatory mechanisms of the postharvest deterioration of Pleurotus tuoliensis fruitbodies during storage. *Food Res. Int.* **2021**, *147*, 110540. [CrossRef]
189. Wang, S.; Wang, J.; Wang, T.; Li, T.; Xu, L.; Cheng, Y.; Chang, M.; Meng, J.; Hou, L. Integrated Transcriptomics–Proteomics Analysis Reveals the Response Mechanism of Morchella sextelata to Pseudodiploöspora longispora Infection. *J. Fungi* **2024**, *10*, 604. [CrossRef]
190. Ma, X.; Lu, L.; Yao, F.; Fang, M.; Wang, P.; Meng, J.; Shao, K.; Sun, X.; Zhang, Y. High-quality genome assembly and multi-omics analysis of pigment synthesis pathway in Auricularia cornea. *Front. Microbiol.* **2023**, *14*, 1211795. [CrossRef]
191. Lu, Y.; Luo, F.; Cen, K.; Xiao, G.; Yin, Y.; Li, C.; Li, Z.; Zhan, S.; Zhang, H.; Wang, C. Omics data reveal the unusual asexual-fruiting nature and secondary metabolic potentials of the medicinal fungus Cordyceps cicadae. *BMC Genom.* **2017**, *18*, 668. [CrossRef]
192. Chen, Y.; Wu, Y.; Liu, L.; Feng, J.; Zhang, T.; Qin, S.; Zhao, X.; Wang, C.; Li, D.; Han, W.; et al. Study of the whole genome, methylome and transcriptome of *Cordyceps militaris*. *Sci. Rep.* **2019**, *9*, 898. [CrossRef] [PubMed]
193. Yin, Y.; Yu, G.; Chen, Y.; Jiang, S.; Wang, M.; Jin, Y.; Lan, X.; Liang, Y.; Sun, H. Genome-wide transcriptome and proteome analysis on different developmental stages of *Cordyceps militaris*. *PLoS ONE* **2012**, *7*, e51853. [CrossRef]
194. Chen, J.; Li, J.-M.; Tang, Y.-J.; Ma, K.; Li, B.; Zeng, X.; Liu, X.-B.; Li, Y.; Yang, Z.-L.; Xu, W.-N.; et al. Genome-wide analysis and prediction of genes involved in the biosynthesis of polysaccharides and bioactive secondary metabolites in high-temperature-tolerant wild *Flammulina filiformis*. *BMC Genom.* **2020**, *21*, 719. [CrossRef] [PubMed]

95. Wang, R.-Q.; Yan, J.-J.; Li, Y.-N.; Yang, H.; Ma, X.-B.; Wang, M.; Tao, Y.-X.; Xie, B.-G. Cytochrome c peroxidase gene(ffccp) and its differential expression during stipe elongation in *Flammulina filiformis*. *Mycosystema* **2020**, *39*, 993–1005.
96. Gao, Y.; Wang, Y.; Qian, J.; Si, W.; Tan, Q.; Xu, J.; Zhao, Y. Melatonin enhances the cadmium tolerance of mushrooms through antioxidant-related metabolites and enzymes. *Food Chem.* **2020**, *330*, 127263. [CrossRef]
97. Wang, R.; Ma, P.; Li, C.; Xiao, L.; Liang, Z.; Dong, J. Combining transcriptomics and metabolomics to reveal the underlying molecular mechanism of ergosterol biosynthesis during the fruiting process of *Flammulina velutipes*. *BMC Genom.* **2019**, *20*, 999. [CrossRef]
98. Zhu, Y.; Luo, H.; Zhang, X.; Song, J.; Sun, C.; Ji, A.; Xu, J.; Chen, S. Abundant and Selective RNA-Editing Events in the Medicinal Mushroom *Ganoderma lucidum*. *Genetics* **2014**, *196*, 1047–1057. [CrossRef] [PubMed]
99. Zhou, S.; Zhang, J.; Ma, F.; Tang, C.; Tang, Q.; Zhang, X. Investigation of lignocellulolytic enzymes during different growth phases of *Ganoderma lucidum* strain G0119 using genomic, transcriptomic and secretomic analyses. *PLoS ONE* **2018**, *13*, e0198404. [CrossRef]
100. Zhang, J.; Li, Y.; Mao, Y.; Zhang, Y.; Zhou, B.; Liu, W.; Wang, W.; Zhang, C. Integrated Transcriptomic and Proteomic Analyses Reveal Molecular Mechanism of Response to Heat Shock in Morchella sextelata. *J. Fungi* **2025**, *11*, 76. [CrossRef]
101. Wen, J.; Zhang, Z.; Gong, L.; Xun, H.; Li, J.; Qi, B.; Wang, Q.; Li, X.; Li, Y.; Liu, B. Transcriptome Changes during Major Developmental Transitions Accompanied with Little Alteration of DNA Methylome in Two Pleurotus Species. *Genes* **2019**, *10*, 465. [CrossRef]
102. Wan, J.N.; Li, Y.; Guo, T.; Ji, G.Y.; Luo, S.Z.; Ji, K.P.; Cao, Y.; Tan, Q.; Bao, D.P.; Yang, R.H. Whole-Genome and Transcriptome Sequencing of Phlebopus portentosus Reveals Its Associated Ectomycorrhizal Niche and Conserved Pathways Involved in Fruiting Body Development. *Front. Microbiol.* **2021**, *12*, 732458. [CrossRef]
103. Fernández-Fueyo, E.; Ruiz-Dueñas, F.J.; López-Lucendo, M.F.; Pérez-Boada, M.; Rencoret, J.; Gutiérrez, A.; Pisabarro, A.G.; Ramírez, L.; Martínez, A.T. A secretomic view of woody and nonwoody lignocellulose degradation by *Pleurotus ostreatus*. *Biotechnol. Biofuels* **2016**, *9*, 49. [CrossRef]
104. Thongbai, B.; Rapior, S.; Hyde, K.D.; Wittstein, K.; Stadler, M. *Hericium erinaceus*, an amazing medicinal mushroom. *Mycol. Prog.* **2015**, *14*, 91. [CrossRef]
105. Feng, W.; Guo, Z.; Jin, Q.; Xu, F.; Shen, Y.; Song, T.; Wang, M.; Zhang, J.; Fan, L.; Huang, X.; et al. A Preliminary Exploration of Transcriptome and Proteomic Changes During the Young and Harvest Periods in Morchella sextelata. *J. Fungi* **2025**, *11*, 192. [CrossRef]
106. Ballard, J.L.; Wang, Z.; Li, W.; Shen, L.; Long, Q. Deep learning-based approaches for multi-omics data integration and analysis. *BioData Min.* **2024**, *17*, 38. [CrossRef] [PubMed]
107. Cheawchanlertfa, P.; Chitcharoen, S.; Raethong, N.; Liu, Q.; Chumnanpuen, P.; Soommat, P.; Song, Y.; Koffas, M.; Laoteng, K.; Vongsangnak, W. Enhancing Genome-Scale Model by Integrative Exometabolome and Transcriptome: Unveiling Carbon Assimilation towards Sphingolipid Biosynthetic Capability of *Cordyceps militaris*. *J. Fungi* **2022**, *8*, 887. [CrossRef]
108. Soommat, P.; Raethong, N.; Ruengsang, R.; Thananusak, R.; Laomettachit, T.; Laoteng, K.; Saithong, T.; Vongsangnak, W. Light-Exposed Metabolic Responses of *Cordyceps militaris* through Transcriptome-Integrated Genome-Scale Modeling. *Biology* **2024**, *13*, 139. [CrossRef] [PubMed]
109. Wang, F.; Li, F.; Han, L.; Wang, J.; Ding, X.; Liu, Q.; Jiang, M.; Li, H. High-Yield-Related Genes Participate in Mushroom Production. *J. Fungi* **2024**, *10*, 767. [CrossRef]
110. Sakamoto, Y.; Sato, S.; Takizawa, M.; Goto, F.; Eda, K.; Yamauchi, T. Marker-assisted selection of a novel *Lentinula edodes* cultivation strain with improved post-harvest quality. *Mushroom Sci. Biotechnol.* **2022**, *30*, 121–128.
111. Li, W.; Zou, G.; Bao, D.; Wu, Y. Current Advances in the Functional Genes of Edible and Medicinal Fungi: Research Techniques, Functional Analysis, and Prospects. *J. Fungi* **2024**, *10*, 311. [CrossRef]
112. Yang, Y.; Saand, M.A.; Huang, L.; Abdelaal, W.B.; Zhang, J.; Wu, Y.; Li, J.; Sirohi, M.H.; Wang, F. Applications of Multi-Omics Technologies for Crop Improvement. *Front. Plant Sci.* **2021**, *12*, 563953. [CrossRef] [PubMed]
113. Zhang, Y.; Tian, L.; Xu, M.-H.; Wang, B.; Song, B.; Li, Y. Research progress in comprehensive utilization of spent mushroom substrates. *Microbiol. China* **2020**, *47*, 3658–3670.
114. Zeb, U.; Aziz, T.; Azizullah, A.; Zan, X.Y.; Khan, A.A.; Bacha, S.A.S.; Cui, F.J. Complete mitochondrial genomes of edible mushrooms: Features, evolution, and phylogeny. *Physiol. Plant.* **2024**, *176*, e14363. [CrossRef] [PubMed]

Journal of
Fungi

MDPI

Review

New Insights into Chemical Profiles and Health-Promoting Effects of Edible Mushroom *Dictyophora indusiata* (Vent ex. Pers.) Fischer: A Review

Yogesh Kumar [1] and Baojun Xu [2,*]

[1] Department of Biotechnology, Mehsana Urban Institute of Sciences, Ganpat University, Mehsana 384012, Gujrat, India; yogeshkumarbiotech@gmail.com
[2] Food Science and Technology Program, Department of Life Sciences, BNU-HKBU United International College, Zhuhai 519087, China
* Correspondence: baojunxu@uic.edu.cn; Tel.: +86-0756-3620636

Abstract: Mushrooms are valued for their culinary and medicinal benefits, containing bioactive compounds like polysaccharides, terpenoids, phenolics, lectins, and ergosterols. This review aims to encourage research on *D. indusiata* by summarizing its chemistry, health benefits, pharmacology, and potential therapeutic applications. Molecules from *D. indusiata* offer anti-diabetic, antioxidant, anti-tumor, hepatoprotective, and anti-bacterial effects. In particular, polysaccharides from *Dictyophora indusiata* (DIP) enhance immune function, reduce oxidative stress, and promote gut health as prebiotics. DIP shows neuroprotective effects by reducing oxidative damage, improving mitochondrial function, and regulating apoptosis, making them beneficial for neurodegenerative diseases. They also activate immune responses through TLR4 and NF-κB pathways. Additionally, compounds like dictyophorines and quinazoline from *D. indusiata* support nerve growth and protection. Mushrooms help regulate metabolism and improve lipid profiles, with potential applications in managing metabolic disorders, cancer, cardiovascular diseases, diabetes, and neurodegenerative conditions. Their wide range of bioactive compounds makes *D. indusiata* mushrooms functional foods with significant therapeutic potential.

Keywords: *Dictyophora indusiata*; *Phallus indusiatus*; mushrooms; monoterpene; alkaloids; sesquiterpenes

1. Introduction

Medicinal fungi are valuable sources of bioactive compounds and widely used as functional foods. *Dictyophora indusiata* (Vent.) Fisch., also known as *Phallus indusiatus*, is an edible and medicinal mushroom belonging to the Phallaceae family in the Agaricomycetes class and Basidiomycetes division of fungi. Although the name *Phallus indusiatus* is accepted in recent taxonomy, most scientific literature refers to it as *D. indusiata*. This saprophytic fungus thrives on decayed wood and rich soil in tropical regions such as Africa, Asia, Australia, and the Americas. It is highly valued for its food and medicinal properties in countries like China, where it grows in bamboo forests. *D. indusiata* is renowned for its elegant morphology, earning it names like the veiled lady or queen of mushrooms. Other common names include bamboo mushroom and bridal veil fungus. The common name aligns with its distinctive fruiting body, which features a conical cap, a stalk, and a net-like indusium resembling a skirt [1]. *D. indusiata* is rich in nutritional value, containing up to 47% carbohydrates, 29% crude fiber, and 6% protein by dry weight [2]. Additionally, it provides essential amino acids, vitamins, and minerals. Beyond its nutritional content, the focus has shifted to its pharmacologically active components, particularly its polysaccharides, which have therapeutic potential. Its medicinal potential is being explored, with key bioactive components including polysaccharides, particularly β-(1→3)-D-glucan, and small molecules like terpenoids and alkaloids. These compounds exhibit antioxidant, anti-inflammatory, and signaling properties, with promising applications in cancer, immunotherapy, neurodegenerative, and chronic inflammatory diseases [1].

The water extract of *D. indusiata* exhibited antioxidant activity in vitro through DPPH, hydroxyl, and ABTS radical scavenging assays. In *Caenorhabditis elegans*, the extract enhanced antioxidant capacity, prolonged lifespan under stress, and reduced lipid deposition and triglyceride levels. It regulated key genes involved in lipid metabolism, indicating its potential for antioxidant and lipid-lowering effects [3]. Currently, *D. indusiata* is valued for two main purposes: its stipe is used as a delicacy, and it possesses various medicinal properties, including anti-tumor, cardiovascular-protective, antioxidant, immunomodulatory, and anti-bacterial effects, along with benefits for eye health and mental well-being [4–6]. Although *D. indusiata* is now cultivated in southern China, challenges persist in its cultivation, resource utilization, and development. The fruiting bodies mature rapidly within hours, requiring immediate harvest to prevent autolysis and waste. Typically, only the stipe is consumed, while the cap, indusia, and volva are discarded, resulting in significant resource wastage [7].

Although the polysaccharides of *D. indusiata* are well-studied, other bioactive compounds like terpenoids and alkaloids have received comparatively less attention. The abundance of polysaccharides in the mature fruiting body has made them the primary focus of research. *D. indusiata* polysaccharides (DIP) have shown notable anti-tumor, antioxidant, anti-inflammatory, immune-enhancing, and anti-diabetic effects. However, the smaller bioactive compounds in *D. indusiata* remain largely underexplored compared to its polysaccharides [1,2,8]. This review provides new insights into the mushroom's chemical profile, emphasizing these underexplored compounds and their potential health benefits, including antioxidant, anti-inflammatory, immunomodulatory, wound healing, hepatoprotective, cardiovascular, and therapeutic properties. It particularly focuses on their roles in anti-bacterial activity, cancer treatment, neurodegenerative disorders, anti-diabetic effects, hepatoprotection, and lipid metabolism, while also addressing potential toxic effects (Figure 1).

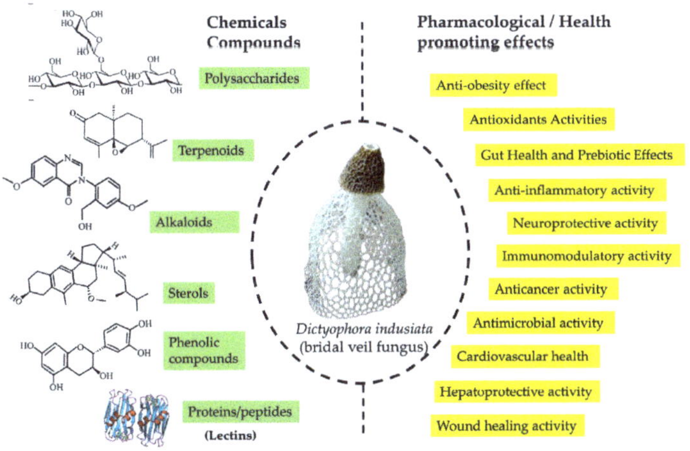

Figure 1. Overview on chemical profiles and health-promoting effects of *D. indusiata*.

2. Overview of Chemical Profiles of *Dictyophora indusiata*

D. indusiata, a medicinal mushroom, contains various bioactive compounds, including polysaccharides, sterols, phenolic compounds, alkaloids, and terpenoids. These chemical classes are known for their therapeutic properties, such as antioxidant, anti-inflammatory, and antimicrobial effects (Table 1).

Table 1. Literature review summarizing important experimental setup, assays, and key findings.

Parts of *D. indusiata* and Extraction Method	Experimental Model/Assay	In Vivo/In Vitro	Important Findings Following Extract/Compound Treatment	References
Fruiting body; acidic-extractable polysaccharide (Ac-DPS)	Hydroxyl radical scavenging assay; DPPH radical scavenging assay; Superoxide radical scavenging assay Concentration—0.2 to 1.4 mg/mL	In vitro	Concentration-dependent scavenging of hydroxyl, superoxide, and DPPH radicals.	[9]
	High-fat emulsion diet for mice; Ac-DPS dose 400 mg/kg or 200 mg/kg for 40 days orally daily.	In vivo	Significantly decreased hepatic lipid levels (TC, TG, NEFA); reduced serum biomarkers for liver injury (ALT, AST, ALP, LDH, and CK); reduced TIBIL level; reduced serum lipid levels of TC, TG, LDL-C, and HDL-C; decreased oxidative stress-related enzymes SOD, GSH-Px, and CAT and non-enzymatic activity T-AOC in liver and kidney; lipid peroxidation level including MDA and LPO reduced; Ac-DPS mitigated the high-fat emulsion-induced kidney damage, reduced the UREA and CREA, and increased the ALB contents in serum. Reduced the INS and LEP contents but increased the ADPN level; restored the histopathological changes in the liver and kidney.	

Table 1. *Cont.*

Parts of *D. indusiata* and Extraction Method	Experimental Model/Assay	In Vivo/In Vitro	Important Findings Following Extract/Compound Treatment	References
Immature (egg) bamboo mushroom; aqueous extracts	DPPH and ABTS assay; FRAP assay; anti-inflammatory activity on LPS-induced RAW 264.7 macrophages; wound healing activity	In vitro	Scavenges reactive oxygen species; acts as good metal chelator; inhibits NO, IL-1, IL-6, and TNF-α in LPS-induced RAW 264.7 macrophages in a dose- and time-dependent manner; inhibits MMP-2 expression in hTRT fibroblasts cells.	[10]
Fruiting body; aqueous extracts	BALB/C mice given high-fat diet for 8 weeks, induced obesity. From the 4th week onwards, DIP was given at 200 mg/kg and 400 mg/kg body weight.	In vivo	DIP reduced the HFD-induced body weight, liver weight, epididymal, and subcutaneous fat; reduced expression of adipogenic genes by regulation of transcription factors such as expression level of PPAR-g, C/EBPa, and SREBP-1c; ameliorated liver health by reducing ALT, AST, TG, and FFA; reduction in glucose, insulin, and serum LPS level; improved intestinal tight junction proteins (TJP) expression including claudin-1, occludin, and zonula occluden (ZO-1); decreased pro-inflammatory cytokines TNF-α, IL-6, and IL-1β and enhanced anti-inflammatory cytokines IL-4 and IL-10 in serum; DIP treatment improved gut microbiota, level of Firmicutes decreased and Bacteroidetes increased.	[11]
Fruiting body; aqueous extracts	Hydroxyl radical scavenging assay; DPPH radical scavenging assay; superoxide radical scavenging assay; ferrous ion chelating assay, 2 mg/mL; antimicrobial activity, 200 mg/mL	In vitro	Extract displays antioxidant activity, scavenges DPPH free radicals and superoxide ions, and inhibits bacterial growth (maximum inhibition for Alcaligenes faecalis) and fungal growth (maximum inhibition for Candida albican).	[12]

Table 1. *Cont.*

Parts of *D. indusiata* and Extraction Method	Experimental Model/Assay	In Vivo/In Vitro	Important Findings Following Extract/Compound Treatment	References
	Hydroxyl radical scavenging assay; DPPH free radical scavenging assay; concentration 0.2–1.0 mg/mL.	In vitro	Scavenges hydroxyl radicals and DPPH free radicals.	
Fruiting body; Aqueous extracts	High-fat emulsion-induced obese mice, diet given daily for 45 days. Aqueous extracts are given daily oral gavage of 400 mg/kg bw and 200 mg/kg bw.	In vivo	WPS treatment decreases body weight in obese mice; reduces level of TC, TG, LDL-C, and AI in serum, increases serum HDL-C level to greater than model group; reduces liver injury markers ALT, AST, ALP, LDH, and CK in serum; increases levels of hepatic antioxidant enzymes SOD, GSH-Px, and CAT and T-AOC; contents of lipid peroxide MDA, LPO, and MPO reduced as compared to obese mice; reduces hepatic lipid level TC, TG, and NEFA in treated group; reduces HI; treatment increases the renal antioxidant enzymes SOD, GSH-Px, CAT, and T-AOC; relieves renal oxidative stress by lowering MDA, LPO, and MPO contents in obese mice; reduces RI levels; kidney dysfunction parameter in serum analyzed; WPS reduces UREA and CREA level and increases ALB compared to obese mice; reduces INS and LEP and increases ADPN in treated mice; restores liver and kidney tissue damage due to obesity by WPS treatment.	[13]

Table 1. *Cont.*

Parts of *D. indusiata* and Extraction Method	Experimental Model/Assay	In Vivo/In Vitro	Important Findings Following Extract/Compound Treatment	References
Fruiting body; aqueous extracts (PD3) and regenerated aqueous extract (RPD3)	CCK-8 assay for cell viability of mouse sarcoma S180 cells. Concentration—0.2, 0.5, and 1 mg/mL in 0.9% aqueous NaCl.	In vitro	The PD3 and RPD3 did not show direct cytotoxicity against S-180 cells. Cytotoxicity was less than 5% at a higher concentration of 1 mg/mL.	[14]
	Tumor induced in mice by inoculating ascitogenous sarcoma S180 cells; dosages 50 mg/kg, 100 mg/kg, 200 mg/kg bw. Extract injected intraperitoneally once daily for 10 d, starting 24 h after tumor inoculation.	In vivo	RPD3 shows higher tumor inhibition in comparison to PD3 in dose-dependent manner. PD3 and RPD3 both show protective effects on the thymus and spleen. Cytokines IL-2, IL-6, and TNF-α in serum were upregulated, which causes immune stimulation to promote anti-tumor activity.	
Fruiting body; 85% ethanolic extract	Isolated compounds Dictyophorine A and Dictyophorine B were tested on quiescent rat astroglial cells for NGF synthesis.	In vitro	Dictyophorine A induces NGF synthesis four times more in quiescent rat astroglial cells than untreated cells. Dictyophorine B also significantly induces synthesis of NGF, but less than Dictyophorine A.	[15]
Fresh mushroom; methanolic extract	Isolated compounds dictyoquinazols A, B, and C tested for neuroprotective activity on mouse cortical neurons. Concentration—10–30 μM.	In vitro	Dictyoquinazols protect mouse cortical neurons from glutamate- and NMDA-induced excitotoxicities in a dose-dependent manner but could not protect neurons from AMPA and kainite.	[16]
Fruiting body; dichloromethane extract	Isolated sesquiterpene antibiotic albaflavenone	In vitro	Albaflavenone is an anti-bacterial compound isolated from *D. indusiata*; the presence of albaflavenone contributes to its anti-bacterial activity mentioned elsewhere.	[17]
Fruiting body; Aqueous extracts	Water-soluble polysaccharide (DI), checked for immunomodulatory activities on RAW 264.7 macrophage.	In vitro	DI is a β-(1→3)-glucan with side branches of β-(1→6)-glucosyl units, and it has a triple-helical structure; DI is not cytotoxic to cells. It induces NO, TNF-α, IL-1, IL-6, and IL-12 production which promotes macrophage multiplication.	[18]

Table 1. *Cont.*

Parts of *D. indusiata* and Extraction Method	Experimental Model/Assay	In Vivo/In Vitro	Important Findings Following Extract/Compound Treatment	References
Fruiting body; aqueous extracts	MTT assay—50–200 μg/mL (to determine the effect of DIP on cell proliferation of RAW 264.7 cells); nitric oxide (NO) production assay; ELISA assay; RT-PCR; cytokine production—50–200 μg/mL.	In vitro	DIP enhances the proliferation of RAW 264.7 cells; DIP induces the up-regulation of NO, IL-1β, IL-6, and TNF-α in RAW 264.7 in a concentration-dependent manner. RT-PCR indicates that DIP up-regulated the mRNA expression of iNOS, IL-1β, IL-6, and TNF-α; anti-TLR4 inhibits the production of NO and cytokines, indicating TLR4 was involved in the DIP-induced macrophage activation; DIP upregulates NF-κB, indicating that macrophage activation involved NF-κB signaling.	[19]
Fruiting body; aqueous extracts (isolated DIP was a homogeneous β-(1→3)-D-glucan with side branches of β-(1→6)-glucosyl units).	Acute colitis was induced in C57BL/6 mice by feeding 2.5% dextran sulfate sodium (DSS) in drinking water for 7 days. Dose—25 mg/kg/d, 50 mg/kg/d and 100 mg/kg/d.	In vivo	DIP improves weight loss due to DSS; relieves splenomegaly; reduces histopathologic damage such as intestinal crypts and goblet cells, tissue architecture, and inflammatory infiltration; relives intestinal oxidative stress; increases the level of GSH; decreases the level of MDA; upregulates HO-1 protein expression; gene expression of IL-10 upregulated; gene expression suppressed for TNF-α, IL-6, and IL-1β; reduces MPO activity. DIP abolished the colitis-induced elevated expression of NLRP3, phosphorylated-STAT3, and p-IκBα; upregulated Bcl2 and downregulated Bax; inhibited DSS-induced lower TJP-1 protein expression in colonic tissues.	[20]

Table abbreviations: MMP—Matrix Metalloproteinase; DPPH—2,2-Diphenyl-1-picrylhydrazyl; TC—Total Cholesterol; TG—Triglycerides; NEFA—Non-Esterified Fatty Acids; ALT—Alanine Aminotransferase; AST—Aspartate Aminotransferase; ALP—Alkaline Phosphatase; LDH—Lactate Dehydrogenase; CK—Creatine Kinase; TIBIL—Total Bilirubin; LDL-C—Low-Density Lipoprotein Cholesterol; HDL-C—High-Density Lipoprotein Cholesterol; SOD—Superoxide Dismutase; GSH-Px—Glutathione Peroxidase; CAT—Catalase; T-AOC—Total Antioxidant Capacity; MDA—Malondialdehyde; LPO—Lipid Peroxidation; UREA—Urea; CREA—Creatinine; ALB—Albumin; INS—Insulin; LEP—Leptin; ADPN—Adiponectin; NO—Nitric Oxide; IL—Interleukin; TNF-α—Tumor Necrosis Factor-alpha; PPAR-γ—Peroxisome Proliferator-Activated Receptor Gamma; C/EBPα—CCAAT/Enhancer-Binding Protein Alpha; SREBP-1c—Sterol Regulatory Element-Binding Protein 1c; TJP—Tight Junction Protein; ZO-1—Zonula Occludens 1; FRAP—Ferric Reducing Antioxidant Power; LPS—Lipopolysaccharide; DIP—*Dictyophora indusiata* Polysaccharide; PD3—Polysaccharide Dictyophora Extract 3; RPD3—Regenerated Polysaccharide Dictyophora Extract 3; NGF—Nerve Growth Factor; AMPA—α-Amino-3-hydroxy-5-methyl-4-isoxazolepropionic acid; NMDA—N-Methyl-D-Aspartate; CRE—Creatinine; DSS—Dextran Sulfate Sodium; RT-PCR—Reverse Transcription Polymerase Chain Reaction; NF-κB—Nuclear Factor Kappa B; iNOS—Inducible Nitric Oxide Synthase; TLR4—Toll-Like Receptor 4; Bcl2—B-cell Lymphoma 2; Bax—Bcl2-Associated X Protein; NLRP3—NOD-Like Receptor Pyrin Domain Containing 3; STAT3—Signal Transducer and Activator of Transcription 3; p-IκBα—Phosphorylated Inhibitor of Nuclear Factor Kappa B Alpha; HO-1—Heme Oxygenase-1.

2.1. Polysaccharides

Polysaccharides from *D. indusiata* fruiting bodies were extracted and purified, with a molecular mass of 536 kDa and a total sugar content of 97.6%. Monosaccharide composition analysis showed that DIP consists primarily of D-glucose (90.7%). Structural analysis using ^{13}C-NMR revealed that DIP is a homogeneous β-(1→3)-D-glucan with β-(1→6)-glucosyl side branches, as shown in Figure 2. The beta-configuration was confirmed by key peaks at 103.5 ppm (C1), while signals at 86.6 ppm (C3) and 61.7 ppm (C6) indicated 1,3-beta-linkages. Branching was observed at C2′, C3′, and C5′, further confirmed by C6 peaks, detailing DIP's complex structure [20].

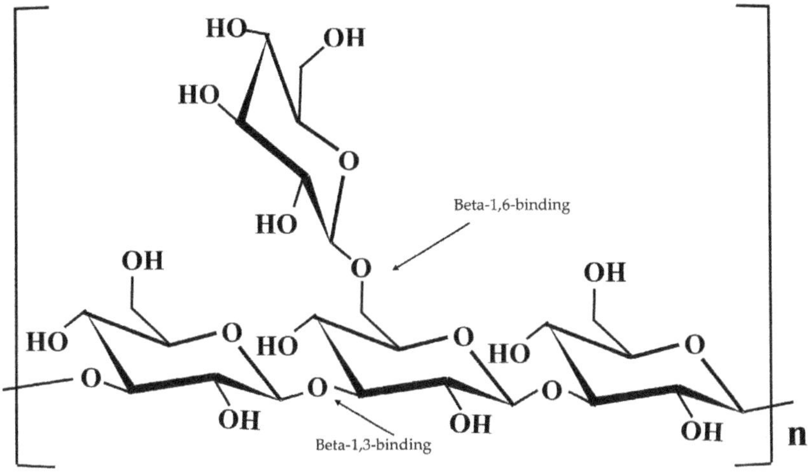

Figure 2. Monomer structure of polysaccharides as 1,3-β-glucans with 1,6-β-glucans branch. The arrow indicates the beta-1,3-glycosidic bond in a chain and the beta-1,6-glycosidic bond at the branch point.

In a study, it was found that the fruiting bodies of *D. indusiata* are rich in heycarbohydrates (46.89%) and crude fibers (28.65%), with a lower crude protein content of 6.07%. Soluble polysaccharides constitute 37.5% of the fruiting bodies, representing approximately 80% of the total polysaccharides. This suggests a significant polysaccharide-based medicinal effect, while peptidoglycan-containing fractions (D1, D2, D4, and D6) may also contribute to therapeutic properties. In contrast, fractions D3 and D5 exhibit glycoprotein characteristics. Monosaccharide analysis revealed fraction D1 as a glucogalactan, D3 as a riboglucan, and D5 and D6 as mannogalactans. Fraction D4, with 92.5% myo-inositol, is nearly pure myo-inositol, indicating potential specialized uses [2].

In a study, acid- and alkali-soluble polysaccharides (DIP I and II) were extracted from the fruiting bodies of *D. indusiata*. DIP I was composed of glucose (Glc), fructose (Fru), and mannose (Man), while DIP II contained only Glc and Fru. Glc was the dominant monosaccharide in both DIPs, accounting for more than 60%, forming the backbone of the polysaccharide chain. Glycosidic linkage analysis revealed Glc 1→6 linkages in DIP II. Hydrolysis by trifluoracetic acid and subsequent monosaccharide identification showed that DIP I had molar percentages of Glc and Fru at 66.2% and 14.5%, respectively, and DIP II at 86.4% and 13.6%. Mannose was present only in DIP I. The structural analysis indicated that Glc forms the backbone for both DIPs, with differences in linkage ratios between them [21].

PD3, a water-soluble (1→3)-β-D-glucan with (1→6)-β-glucopyranoside side branches, was isolated from *D. indusiata*. Its molecular structure and conformation were analyzed using GC, FTIR, NMR, and microscopy techniques. PD3 forms a triple helical chain in water and undergoes helix–coil transition at higher NaOH concentrations compared to similar polysaccharides, self-assembling into fibrous aggregates [22].

2.2. Terpenoids

The isolation and characterization of monoterpenoids *D. indusiata* were documented [23]. The air-dried fruiting bodies of *D. indusiata* (1.2 kg) were extracted with 85% ethanol, and the solvent was concentrated under reduced pressure. Partitioning between ethyl acetate and water resulted in a residue (20.2 g), which was further fractionated using silica gel chromatography and HPLC. This yielded four compounds. Among the isolated compounds were five monoterpene alcohols, with the first identified as 3,7-dimethyl-1,6-octadiene-3,4-diol, confirmed to have a (3R,4S) absolute configuration. The remaining four were identified as 4-[(3R,4S)-3-hydroxy-3,7-dimethylocta-1,6-dienyl (Z)-9-octadecenoate], 4-[(3R,4S)-3-hydroxy-3,7-dimethylocta-1,6-dienyl(9Z,12Z)-9,12-octadecadienoate], 3,7-dimethyl-1,6-octadiene-3,4,5-triol, and bis[6-(3,4,7-trihydroxy-3,7-dimethyloctenyl) ether]. These structures were determined through comprehensive spectral and chemical analysis. Structure of different terpenoids from *D. indusiata* are given in Figure 3.

Air-dried fruiting bodies of *D. indusiata* (1.2 kg) were extracted using 85% ethanol, and the solvent was concentrated under reduced pressure [15]. The extract was partitioned between ethyl acetate and water, yielding a residue (20.2 g). This residue was fractionated through repeated silica gel column chromatography (CC) and HPLC, resulting in the isolation of two compounds: 1 (246 mg, crystalline, mp 39–40 °C) and 2 (8 mg, amorphous). Two novel eudesmane-type sesquiterpenes, dictyophorines A and B, along with the known compound teucrenone, were identified.

Thirty-five grams of dried *D. indusiata* fruiting bodies was subjected to Soxhlet extraction with dichloromethane as the solvent for 24 h [17]. After extraction, the concentrate underwent silica gel column chromatography, yielding 1.7 mg of a crude product. Through FTIR, GC-MS, HR-FIMS, and NMR analyses, the isolated compound was identified as the sesquiterpene antibiotic albaflavenone. The content of albaflavenone in the dried fruiting bodies was determined to be approximately 0.0063%. The compound exhibited an earthy, camphor-like odor, showcasing its potential as a bioactive molecule with antimicrobial properties. Similarly, researchers isolated two novel sesquiterpenoids, designated as 9,10-dihydroxy-albaflavenone (1) and 5-hydroxy-albaflavenone (2), from *D. indusiata* [24]. The structures and absolute configurations of these compounds were elucidated through NMR, ECD, and HR-ESI-MS analysis.

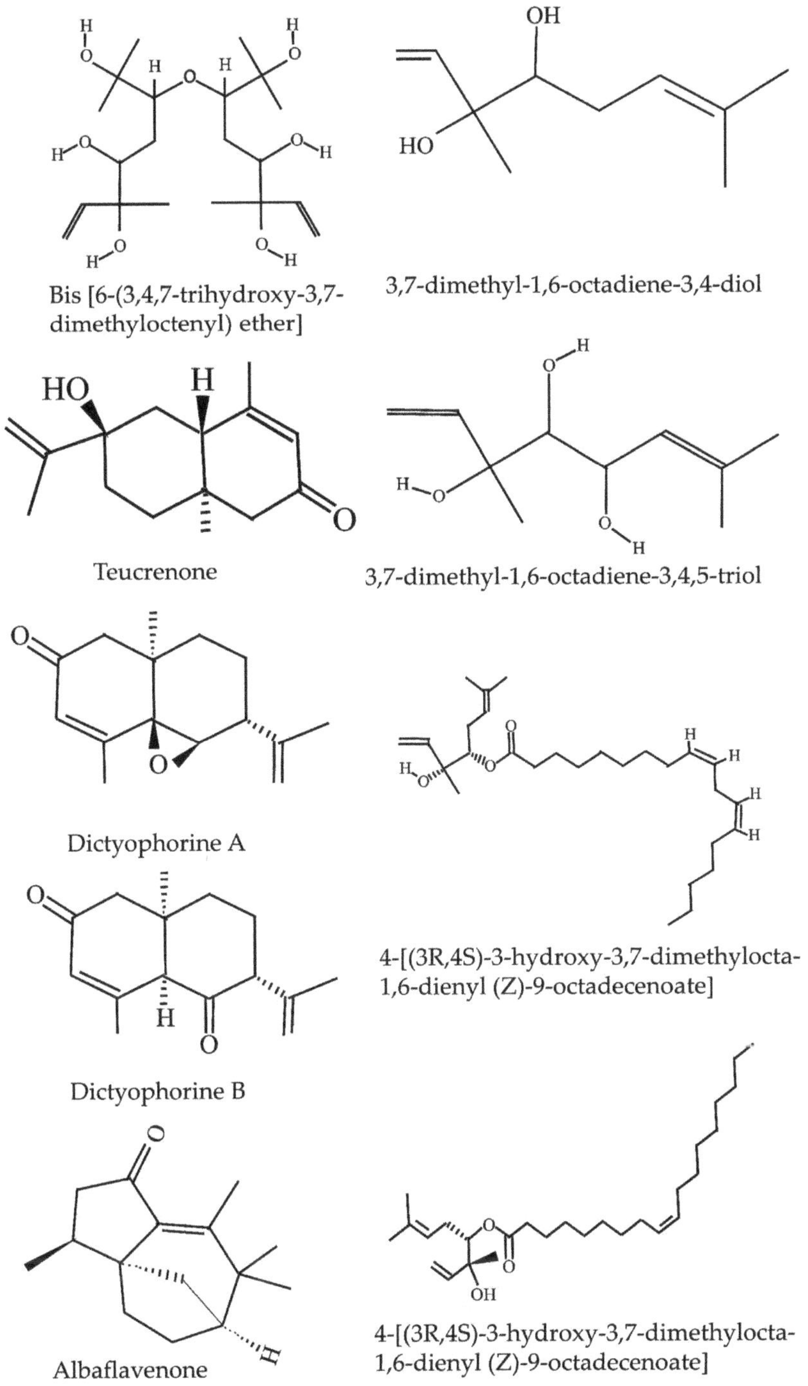

Figure 3. Structure of different terpenoids from *D. indusiata*.

2.3. Alkaloids

A fresh *D. indusiata* (1.2 kg) was collected and methanolic extraction of the mushroom was performed followed by column chromatography, and HPLC yielded three neuroprotective quinazoline compounds (Figure 4): dictyoquinazols A, B, and C. Dictyoquinazols B and C existed as inseparable rotamer mixtures, confirmed by NMR and NOESY spectra analysis [16].

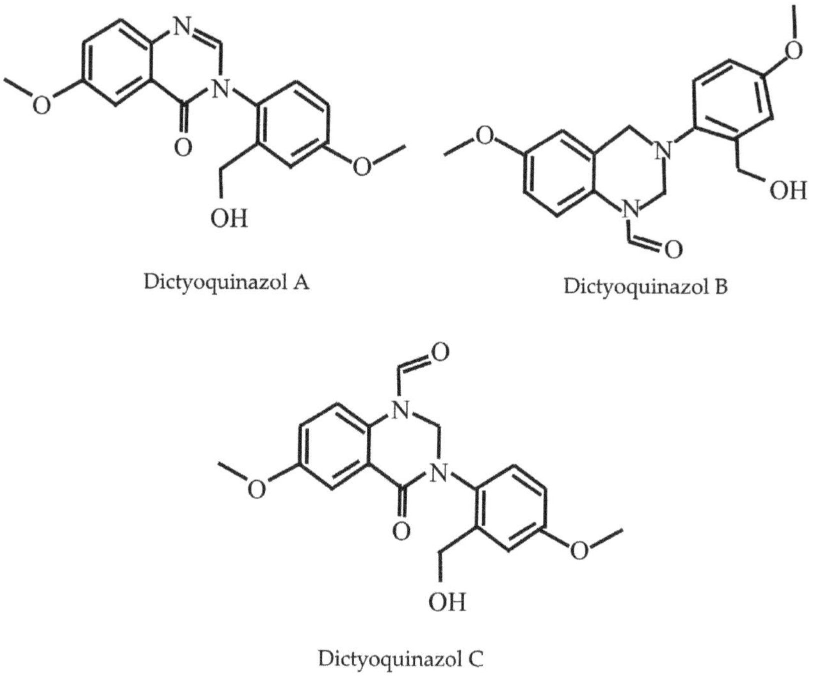

Figure 4. Structure of alkaloids isolated from *D. indusiata*.

2.4. Phenolic Compounds

The crude water extracts from the immature stage of *D. indusiata* yielded 21.11%, 51.48%, and 10.15% of peel and green mixture (PGW), core (CW), and whole (WW) extracts, respectively [10]. The primary bioactive components were phenolic compounds (Figure 5). Catechin (68.761 mg/g) was the dominant polyphenol in CW, while p-coumaric acid was the major polyphenol in PGW and WW. Other identified polyphenols included rutin, rosmarinic acid, quercetin, naringenin, and epigallocatechin gallate (EGCG).

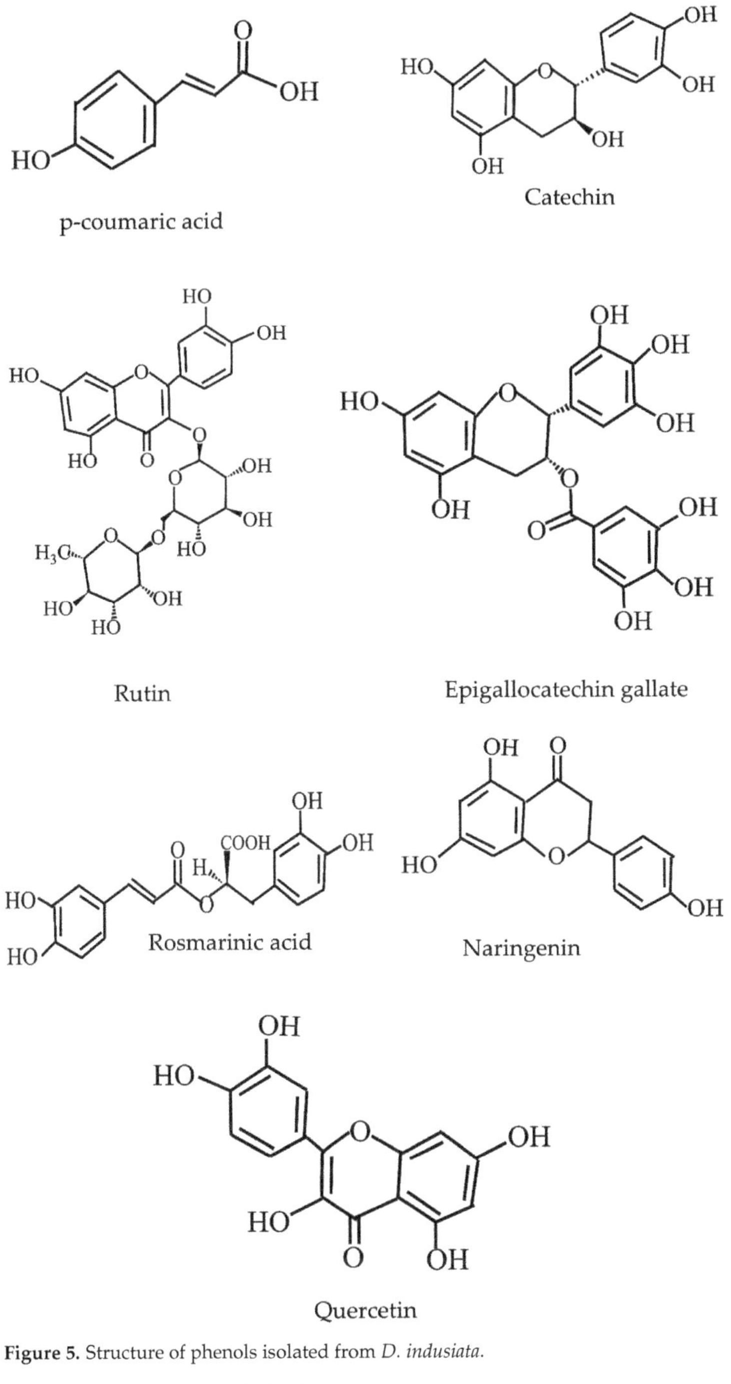

Figure 5. Structure of phenols isolated from *D. indusiata*.

2.5. 5-(Hydroxymethyl)-2-furfural

The evaluation of tyrosinase inhibition by methanolic extracts of *D. indusiata* identified the bioactive compound as 5-(hydroxymethyl)-2-furaldehyde (HMF) (Figure 6). Kinetic studies revealed HMF to be a non-competitive inhibitor for L-DOPA oxidation. Freeze-dried mushroom extracts were partitioned with hexane, CH_2Cl_2, EtOAc, and BuOH, with the active EtOAc fraction (3 g) yielding HMF after chromatographic purification. The structure of HMF was confirmed through ^{1}H NMR, ^{13}C NMR, and mass spectrometry, showing that it matched the authentic standard [25].

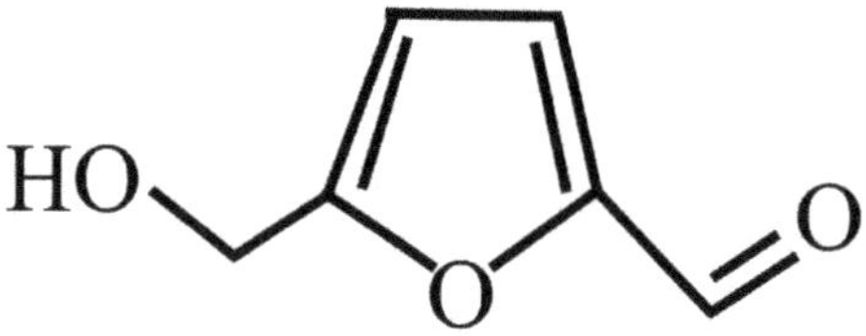

5-(hydroxymethyl)-2-furfural

Figure 6. Structure of 5-(hydroxymethyl)-2-furfural.

2.6. Sterols

Three ergostane-type steroids (Figure 7), namely compounds 15, 16, and 17, from *D. indusiata* demonstrated significant anti-inflammatory properties [26]. Compounds 15 [1(10→6) abeo-ergosta-5,7,9,22-tetraen-11β-methoxy-3α-ol] and 16 (citreoanthrasteroid) were ergosterol derivatives with rearranged tetracyclic skeletons, differing in their stereochemistry; both had a methoxy group at C-11, with compound 15 having it on the same face as the hydroxyl group and compound 16 on the opposite face. Compound 17 [5α,6α -epoxy-3β-hydroxy-(22E)-ergosta-8(14),22-dien-7-one], identified as an epoxy-7-sitosterol, was noted for its epoxy and carbonyl groups, which appeared to be crucial for its biological activity.

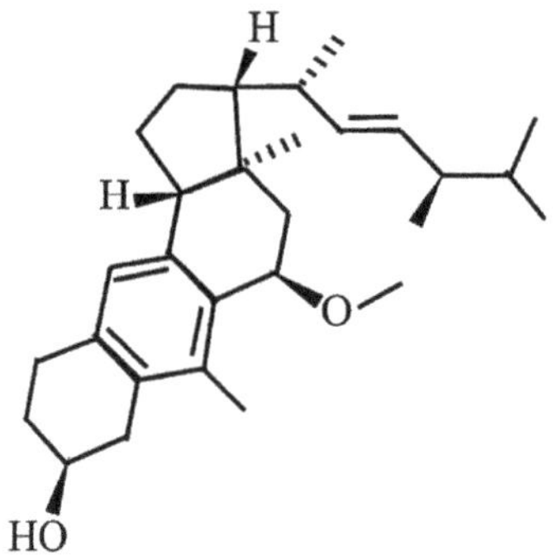

1(10→6) abeo-ergosta-5,7,9,22-tetraen-11β-methoxy-3α-ol

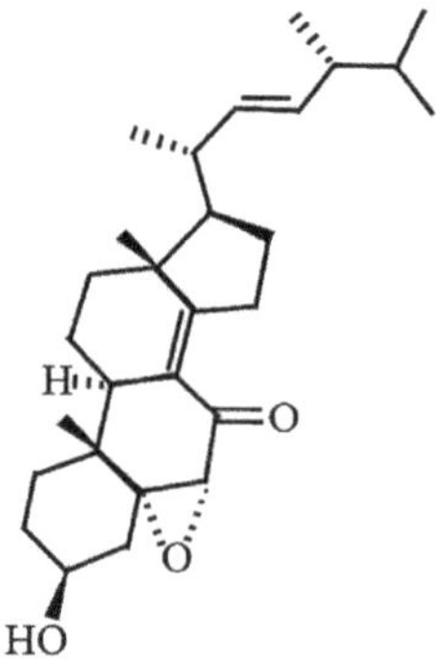

5α,6α-epoxy-3β-hydroxy-(22E)-ergosta-8(14),22-dien-7-one

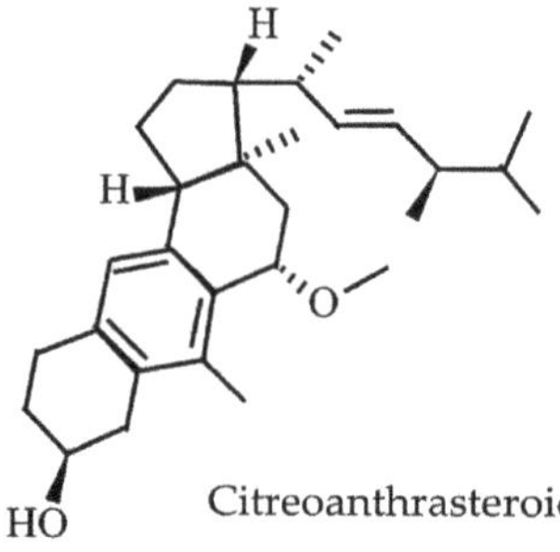

Citreoanthrasteroid

Figure 7. Structure of sterols identified from *D. indusiata*.

2.7. Proteins/Peptides

A lectin from *D. indusiata* was isolated through saline extraction, ammonium sulfate precipitation, and chromatography. SDS-PAGE revealed it as a single 53 kDa subunit. The lectin agglutinated rabbit and human erythrocytes, as well as mouse lymphocytes and tumor cells. Hemagglutination was inhibited by lactose and L-fucose [27].

3. Health-Promoting Effects of *Dictyophora indusiata*

D. indusiata exhibits health-promoting effects by reducing inflammation, improving gut microbiota balance, protecting liver function, exerting anticancer, antimicrobial, immunomodulatory, and neuroprotective effects, and mitigating oxidative stress, making it a potential candidate for therapeutic use.

3.1. Antioxidants Activities

The water extract of *D. indusiata* demonstrates significant in vitro antioxidant activity, as shown by DPPH, hydroxyl radical, and ABTS cation radical scavenging assays, with IC50 values of 2.36, 1.44, and 0.86 mg/mL, respectively [3,28]. Using *Caenorhabditis elegans* as a model, the extract at various doses (1.0, 0.5, and 0.25 mg/mL) enhanced oxidative and heat stress tolerance, extending lifespan by 54.6% and 19.8%. Additionally, lipid deposition and triglyceride levels decreased significantly, potentially through the modulation of fat metabolism genes such as daf-2, sbp-1, and nhr-49. These findings suggest the extract's potential in antioxidant activity and lipid metabolism improvement [3].

A purified water-soluble polysaccharide (PPS) from *D. indusiata*, primarily composed of glucose (98.58%), was characterized using gel chromatography (Sephadex G-200), IR spectroscopy, and GC–MS. PPS exhibited antioxidant activity in vitro across multiple assays, including reducing power, hydroxyl, superoxide radical, and DPPH scavenging tests. These findings suggest that PPS is a promising natural antioxidant with potential therapeutic applications in mammalian systems, offering a novel approach to antioxidant therapy [29]. Phenolic compounds are also reported from *D. indusiata* that might also contribute to alleviating metabolic ROS. Phenolic and polyphenolic compounds, alone or combined with vitamins like carotenoids, vitamin E, and vitamin C, act as antioxidants, protecting human tissues from oxidative stress [30].

3.2. Gut Health and Prebiotic Effects

Despite the well-documented biological activities of *D. indusiata* polysaccharides (DIP), their influence on gut microbiota restoration has remained largely unexplored. DIP modulates the gut microbiota composition and restores intestinal barrier function following broad-spectrum antibiotic-induced dysbiosis. Using the Illumina MiSeq platform, the study revealed that DIP administration reversed the antibiotic-driven microbial imbalance, increasing beneficial bacteria such as Lactobacillaceae and Ruminococcaceae. Additionally, DIPs reduced endotoxemia and pro-inflammatory cytokines while enhancing the expression of tight-junction proteins, crucial for intestinal barrier integrity. These results underscore the therapeutic potential of DIPs in restoring gut health, improving barrier function, and mitigating inflammation [31].

Researchers successfully isolated crude polysaccharide (DIP) from the fruiting body of *D. indusiata*, with a total sugar content of 96.66%. Monosaccharide composition analysis revealed that the bioactive components were primarily glucose (59.84%), mannose (23.55%), and galactose (12.95%). Dysbiosis was induced in BALB/c mice using broad-spectrum antibiotics, clindamycin and metronidazole, to investigate the protective effects of DIP. Gut bacterial diversity disruption due to antibiotics and its recovery following DIP administration (0.2 mg/0.2 mL) were assessed through PCR-DGGE and further analyzed using the

16S rRNA Illumina MiSeq technique, providing a comprehensive understanding of DIP's role in restoring gut microbial balance [31].

This work was supported by further research on high-fat-diet-induced obesity in mice [11]. HFD is known to disrupt the microbial balance, while dietary fibers and polysaccharides have been shown to positively influence gut microbial communities [32]. DIP supplementation significantly altered the gut microbiota composition, enhancing bacterial diversity and richness in a dose-dependent manner compared to the HFD-alone group. Notably, the Firmicutes-to-Bacteroidetes ratio was restored and bacterial perturbations at the phylum and class levels were reversed with DIP treatment.

The high-fat diet causes a detrimental effect on intestinal integrity and systemic inflammation, driven by elevated pro-inflammatory cytokines and endotoxemia (LPS in blood bloodstream) through TLR4 signaling activation [33]. As expected, HFD-induced obese mice demonstrated increased endotoxin levels and reduced tight junction protein (TJP) expression, leading to intestinal barrier dysfunction. However, DIP supplementation effectively reduced LPS levels and upregulated TJP expression, indicating its potential to preserve intestinal health and mitigate inflammation [11].

3.3. Anti-Inflammatory Activity

Previous studies have shown that intestinal injury and inflammation lead to increased expression of pro-inflammatory cytokines [34], which negatively impact intestinal mucosal integrity and function [35]. Antibiotic-induced dysbiosis in mouse models exacerbates intestinal inflammation, resulting in elevated levels of pro-inflammatory cytokines such as TNF-α, IL-1β, MCP-1, and IFN-γ [36,37]. In line with these findings, their study observed an increased abundance of Bacteroidetes and Proteobacteria in antibiotic-treated groups, which are linked to heightened pro-inflammatory cytokine production. However, administration of *D. indusiata* polysaccharides (DIP) significantly reduced these cytokine levels and alleviated the inflammatory response by modulating gut microbiota. These results suggest that DIPs effectively mitigate antibiotic-induced inflammation and enhance immune function [31].

Another study highlights the potential of DIP, a natural polysaccharide, as an anti-inflammatory agent targeting the TLR4/NF-κB signaling pathway and NLRP3 inflammasome activation. Chronic inflammatory and autoimmune diseases, such as IBD and rheumatoid arthritis, involve dysregulated inflammasome activity [38,39]. DIP was shown to repress TLR4 expression and inhibit NF-κB activation by preventing IκB-α phosphorylation and NF-κB-p65 nuclear translocation in LPS-primed macrophages. It also reduced the production of pro-inflammatory cytokines like IL-1β, IL-18, and IL-6, thereby attenuating the inflammatory response [40]. Additionally, DIP inhibited the assembly of the NLRP3 inflammasome, decreasing NLRP3 and pro-caspase-1 expression and preventing mitochondrial ROS production, a key factor in NLRP3 activation [41]. DIP's non-dose-dependent effect, with higher concentrations promoting inflammation, warrants further investigation [19]. Its structural similarity to lentinan, which has demonstrated therapeutic effects in an IBD model, suggests DIP may have similar benefits [42]. However, further research is needed to evaluate its efficacy within the gastrointestinal environment. Overall, DIP presents a promising candidate for treating chronic inflammatory diseases by modulating key inflammatory pathways and suppressing inflammasome activation [43]. In another report, albaflavenone, a tricyclic sesquiterpene, demonstrated anti-inflammatory properties by inhibiting TNF-α and NO secretion to different extents [24].

3.4. Neuroprotective Activity

D. indusiata, a traditional medicinal mushroom in China, has shown potential for treating inflammatory and neural conditions. This study highlights the antioxidant and neuroprotective properties of *D. indusiata* polysaccharides (DIP) using *Caenorhabditis elegans* models. DiPS increased survival rates, reduced oxidative stress markers (ROS, MDA), enhanced SOD activity, and restored mitochondrial function. Furthermore, DIP involved the DAF-16/FOXO transcription factor in stress response and alleviated chemosensory dysfunction in neurodegenerative disease models. These findings offer significant evidence of DIP therapeutic potential for neurodegenerative diseases, providing a pharmacological basis for its use in mitigating oxidative stress and neuroprotection [44]. The neuroprotective properties of *D. indusiata* can be attributed to the presence of dictyophorines and polysaccharides. Notably, dictyophorines A and B were found to promote nerve growth factor (NGF) synthesis in astroglial cells, indicating their potential neuroprotective and therapeutic properties [15]. The quinazoline compounds also exhibited neuroprotective effects against glutamate- and NMDA-induced excitotoxicity in primary mouse cortical neurons [16]. The neuroprotective effects of polysaccharides have been also demonstrated in Sprague Dawley (SD) rats. In arsenic-exposed rats, *D. indusiata* polysaccharide (DIP) has been shown to enhance spatial learning and memory, which were otherwise impaired due to arsenic exposure [45]. DIP exhibits neuroprotective effects by mitigating arsenic-induced damage, possibly by reversing the dysregulation of synaptic proteins and restoring glutamatergic neurotransmission, improving learning and memory in arsenic-exposed rats. DIP also attenuates apoptosis by regulating apoptosis-related proteins and mitigating mitochondrial dysfunction. Furthermore, DIP restores normal energy metabolism pathways by enhancing ATP production and reducing oxidative stress. By modulating key signaling pathways such as Ras, MAP kinase, and Wnt, DIP protects against neuronal apoptosis and synaptic dysfunction [45].

3.5. Immunomodulatory Activity

DIP significantly stimulates the proliferation of RAW 264.7 cells in a dose- and time-dependent manner, particularly at concentrations of 25–200 μg/mL [19]. DIP enhances the production of nitric oxide (NO) and pro-inflammatory cytokines IL-1β, IL-6, and TNF-α. This effect is associated with the up-regulation of mRNA expression for iNOS, IL-1β, IL-6, and TNF-α, confirming that DIP promotes macrophage activation through gene regulation. Additionally, DIP's effects were found to be mediated by TLR4, a key receptor involved in macrophage activation, as indicated by the inhibition of NO and cytokine production when anti-TLR4 mAb was applied. Specific binding of DIP to macrophages was demonstrated, and TLR4 was confirmed as the major receptor facilitating this process. Furthermore, DIP activated the NF-κB signaling pathway, upregulating NF-κB p65, further emphasizing its role in promoting immune responses. In a similar report, DIP enhances immunomodulatory activity in RAW 264.7 macrophages, promoting cell multiplication and increasing nitric oxide and cytokine production, including TNF-α, IL-1, IL-6, and IL-12 [18]. It shows a dose-dependent immune response with no toxic effects, suggesting potential biomedical applications. The water-soluble polysaccharide (DI) extracted from *D. indusiata* was a β-(1→3)-glucan with β-(1→6)-glucosyl side branches and a triple-helical structure.

3.6. Anticancer Activity

The polysaccharide PD3, isolated from *D. indusiata*, was identified as a (1→3)-β-D-glucan with (1→6)-β-glucopyranoside side chains. The regenerated form, RPD3, retained its triple-helical structure, as evidenced by Congo red testing, but had lower chain tightness compared to PD3. This structural modification, induced by a denaturation-renaturation process, made RPD3 more susceptible to disruption in NaOH solution. As a result, more active sites were exposed to RPD3, potentially increasing its bioactivity by enhancing its ability to interact with immune cell receptors. In vitro, assays revealed that neither PD3 nor RPD3 demonstrated direct cytotoxicity against S-180 tumor cells, but in vivo experiments showed significant anti-tumor effects. Notably, RPD3 exhibited greater tumor inhibition compared to PD3, showing a dose-dependent increase in efficacy ($p < 0.05$) [14]. Both polysaccharides improved immune organ indices, upregulated key cytokines (IL-2, IL-6, and TNF-α), and stimulated immune responses, confirming their role as potent immunopotentiators rather than direct cytotoxic agents [14,46]. This enhancement in bioactivity could be attributed to the denaturation process, as previous studies have indicated that lower concentrations of NaOH during denaturation preserve bioactivity by exposing more functional groups. The observed immune stimulation suggests that the anti-tumor effects of these polysaccharides are primarily mediated through immune system activation rather than direct tumor cell destruction [47]. These findings highlight the potential of RPD3 as a promising candidate for enhancing immune function and combating tumors through immunomodulation.

Apart from this, *D. indusiata* has demonstrated promising anti-cancer properties, specifically inducing cytotoxicity in cholangiocarcinoma (CCA) cell lines. In this study, extracts from three types of Dictyophora species—Chinese bamboo mushrooms (Ch-DTP), Short skirt bamboo mushrooms (Th-DTP), and Orange skirt bamboo mushrooms (Or-DTP)—were tested against CCA cells. Ch-DTP showed the strongest cytotoxic effect, significantly reducing cell viability in all CCA cell lines, even at low concentrations. In contrast, Th-DTP and Or-DTP displayed moderate to limited effects at lower doses but exhibited stronger cytotoxicity at higher concentrations. The results indicate that Dictyophora extracts may induce apoptosis, as treated CCA cells showed signs of programmed cell death, such as cytoplasmic shrinking and nuclear condensation. Importantly, normal fibroblast cells were not affected, suggesting minimal adverse effects [48].

3.7. Antimicrobial Activity

Albaflavenone is a sesquiterpenoid compound originally identified in Streptomyces species (e.g., *Streptomyces albidoflavus, S. coelicolor*). It is biosynthesized from epi-isozizaene and has been studied for its antibiotic and antimicrobial properties. Albaflavenone's unique structure, featuring an oxidized sesquiterpene backbone, contributes to its biological activity, making it a subject of interest in natural product research [49]. Albaflavenone has been also isolated from *D. indusiata* mushrooms [17,24]. Dried fruiting bodies of *D. indusiata* were used in the study to evaluate the antimicrobial activity of its hot water extract (WE) through the agar well diffusion method. At 200 mg/mL, WE displayed broad-spectrum antimicrobial effects against both bacteria and fungi, with Alcaligenes faecalis being the most susceptible bacterium and Escherichia coli the least. Among fungi, Candida albicans was the most sensitive to the extract [12].

3.8. Anti-Obesity Effect

Obesity, marked by systemic inflammation, is linked to the onset of various chronic diseases, including diabetes, cardiovascular issues, liver inflammation, metabolic disorders, and cancer [50]. Excessive lipid accumulation plays a critical role in obesity development. A high-fat diet (HFD) significantly increases body weight, fat deposition, and liver weight [11]. However, treatment with *D. indusiata* polysaccharide (DIP) significantly reduced body weight gain, fat accumulation, and liver weight, suggesting its protective effects against HFD-induced obesity. Elevated glucose and insulin levels, commonly seen in HFD-induced obesity [51], were also mitigated by DIP supplementation. These results align with prior studies that associate obesity with enlarged adipocytes and the upregulation of adipogenic and lipogenic genes [51,52].

High-fat emulsion-induced obesity led to disturbances of lipid profile, increased hepatic enzyme activities, and oxidative stress markers. Supplementation with WPS (water-soluble polysaccharides) from *D. indusiata* effectively restored normal lipid levels, liver enzyme activities (ALT, AST, ALP, CK, and LDH), and antioxidant status (GSH-Px, SOD, CAT, and T-AOC), demonstrating its potential in mitigating obesity-related metabolic and oxidative stress disturbances [13]. DIP effectively reduced adipocyte size and the expression of these genes in a dose-dependent manner. Furthermore, DIP lowered pro-inflammatory cytokines (TNF-α, IL-1β, IL-6, and MCP-1) and elevated anti-inflammatory cytokines (IL-4, IL-10), counteracting obesity-induced inflammation [53]. DIP also improves liver function by decreasing liver injury markers (ALT, AST), triglycerides, and free fatty acids. These findings suggest that DIP polysaccharide holds promise as a therapeutic agent for managing obesity-related inflammation and liver dysfunction [11].

3.9. Cardiovascular Health

A key process driving hyperlipidemia is atherosclerosis, which involves both inflammatory and immune responses. In its initial stages, oxidative stress—caused by factors like elevated lipid levels, smoking, diabetes, or hypertension—leads to endothelial dysfunction. This is followed by the oxidation of low-density lipoprotein (LDL) within the blood vessels. The oxidized LDL (ox-LDL) accumulates and engages scavenger receptors on macrophages, prompting their activation and subsequent uptake of ox-LDL. This progression is central to the development of atherosclerosis [54]. Primary care physicians dedicate significant time and effort to preventive medicine, with a focus on managing conditions like hyperlipidemia to reduce the risk of cardiovascular disease (CVD). Hyperlipidemia is the second most common chronic condition treated, following hypertension. It is well known that hyperlipidemia is a major risk factor for CVD. This condition involves elevated levels of cholesterol, triglycerides, or both, and while it can result from genetic factors, it is more often an acquired condition [55]. The presence of β-sitosterol in *D. indusiata* has been also observed [24]. The cardiovascular protective effects of β-sitosterol, a key component found in cholesterol-lowering functional foods, have been attributed to its antioxidant properties and its ability to reduce cholesterol levels [56]. It was also observed that TNF-α treatment led to increased NF-κB phosphorylation in HAECs, indicating NF-κB activation. However, treatment with β-sitosterol (0.1–200 μM) significantly reduced NF-κB phosphorylation, suggesting that its anti-inflammatory effects in vitro are partially mediated through NF-κB inactivation. This aligns with findings by Moreno, who also demonstrated that β-sitosterol reduced NF-κB activation in PMA-stimulated macrophages [57,58]. Numerous animal studies have shown that the intake of plant sterols inhibits the progression of atherosclerosis. It is well established that after consuming plant sterol-enriched foods, plasma levels of plant sterols experience a moderate increase, while cholesterol levels significantly decline. This reduction in circulating cholesterol is linked to a lower risk of coronary heart disease (CHD) [56].

3.10. Hepatoprotective Activity

Hyperlipidemia is also a key factor in the development of non-alcoholic fatty liver disease apart from atherosclerosis, and cardiovascular conditions [59]. Elevated liver enzymes in the serum, such as ALT, AST, ALP, LDH, and CK, are widely recognized markers of liver damage due to the leakage of cellular enzymes into the bloodstream [60,61]. Acidic-extractable indusiata polysaccharides (Ac-DPS) have shown significant potential in mitigating hyperlipidemia. The study demonstrated that high-fat emulsion treatment induced acute hepatotoxicity, as evidenced by increased serum enzyme activities. However, Ac-DPS supplementation effectively normalized these enzymatic activities, highlighting its protective role [9]. Furthermore, elevated total bilirubin (TBIL) levels, another indicator of liver damage, were significantly reduced following Ac-DPS administration, demonstrating its hepatoprotective effects [62]. Histopathological analysis revealed severe hepatic impairment, such as fatty vacuolation and ballooning degeneration, in the high-fat emulsion-treated group. Remarkably, Ac-DPS supplementation, especially at 400 mg/kg bw, restored liver tissue integrity, underscoring its ability to attenuate liver damage caused by hyperlipidemia [9]. Oxidative stress, driven by excessive reactive oxygen species (ROS), accelerates hyperlipidemia progression [63]. Ac-DPS treatment restored antioxidant enzyme activities (SOD, CAT, and GSH-Px) and reduced oxidative damage markers (MDA and LPO), confirming its ability to alleviate hepatic and renal oxidative stress. Furthermore, Ac-DPS improved renal function, as evidenced by normalized CREA, UREA, and ALB levels, and mitigated tubular injury in hyperlipidemic rats [13].

3.11. Wound Healing Activity

Matrix metalloproteinases (MMPs), a family of zinc-dependent endopeptidases, are consistently reported to be elevated in chronic wounds. These enzymes can degrade nearly all components of the extracellular matrix (ECM) and are produced by various skin cells, including fibroblasts, keratinocytes, macrophages, endothelial cells, mast cells, and eosinophils. MMP activity is regulated by tissue inhibitors of metalloproteinases (TIMPs) [64,65]. In chronic wounds, particularly in diabetic patients, elevated levels of pro-MMP-2, MMP-2, MMP-8, and MMP-9 result in an imbalance between MMPs and TIMP-2, contributing to poor wound healing [66]. Notably, MMP-2 expression is significantly higher in hypertrophic scars and keloids compared to non-scarred tissues [67]. Reducing MMP-2 expression has been shown to restore collagen types I and III, promoting faster wound healing [68]. In a study of gelatin zymography investigating the inhibitory effects of *D. indusiata* on MMP-2 activity in human skin fibroblasts, three extracts were evaluated: the peel and green mixture (PGW), the core (CW), and the whole mushroom (WW). Among these, PGW demonstrated the highest inhibition of MMP-2 activity, with a reduction of $59.63 \pm 8.31\%$. CW followed with $41.33 \pm 9.44\%$, while WW exhibited the lowest inhibitory effect at $16.33 \pm 2.91\%$. These findings suggest that *D. indusiata*, particularly the PGW extract, may have potential therapeutic benefits in managing inflammatory skin conditions and reducing scar formation by modulating MMP-2 activity [10].

3.12. Toxicity Assessment

The diverse bioactive components present in *D. indusiata* are key contributors to its wide range of biological and physiological activities. As far as toxicity is concerned, *D. indusiata* is widely consumed in countries like China and Japan without toxicity concerns, except when excessive extracts or isolated compounds are ingested. Studies in mice show no toxic effects or behavioral changes at doses up to 1200 mg/kg for two months [1]. An acute toxicity study was conducted using eighteen male Kunming strain mice, which were divided into three groups. Two dose groups received Ac-DPS at 1000 mg/kg and 1500 mg/kg body weight, while the control group received saline. Mice had unrestricted access to food and water and were monitored for any signs of mortality or behavioral changes. The results showed no significant toxic effects, behavioral alterations, or deaths in the Ac-DPS treated groups compared to the control, indicating that Ac-DPS is virtually non-toxic under the tested conditions [9].

4. General Discussion and Future Perspective

Mushrooms are celebrated worldwide for their distinctive flavor and nutritional value, making them a culinary delight. Of over 2000 species, only about 25 are widely accepted as food, with several commercially cultivated. They hold dual roles as both edible and medicinal, offering organoleptic, medicinal, and nutraceutical benefits. Many edible species also exhibit therapeutic properties, blurring the distinction between edible and medicinal mushrooms [69]. Mushrooms possess significant health benefits due to their rich content of bioactive volatile organic compounds, including lectins, terpenoids, phenolics, polyphenolics, polysaccharides, and ergosterols. These compounds are widely recognized for their medicinal properties, demonstrating anti-diabetic, anti-hypercholesterolemic, antioxidant, anti-tumor, hepatoprotective, and anti-bacterial effects, which contribute to promoting overall human health. Their therapeutic potential positions them as valuable agents in the prevention and management of various diseases, underscoring the importance of mushrooms as functional foods and sources of bioactive substances in health promotion and disease prevention strategies [70]. Polysaccharides are natural polymers composed of aldoses or ketoses connected by glycosidic bonds. These complex carbohydrates play a crucial role in sustaining various life functions, serving as essential biological macromolecules within living organisms [23,29].

Among reactive oxygen species, hydroxyl radicals are highly reactive and can cause severe damage to biomolecules in living cells. Hydrogen peroxide and superoxide can indirectly generate hydroxyl radicals via Fenton or iron-catalyzed Haber–Weiss reactions, leading to oxidative injury. Antioxidants can inhibit these processes, protecting cells and food systems. Superoxide anions, precursors to other reactive oxygen species, induce oxidative damage to lipids, proteins, and DNA. DPPH, a stable free radical, is commonly used to assess antioxidant capacity, with reduced absorbance at 517 nm indicating effective radical scavenging. The prevalence of neurodegenerative diseases, marked by the gradual loss of neurons, is on the rise. Central to the pathogenesis of these conditions is oxidative stress, particularly the excessive production of reactive oxygen species (ROS). In vivo studies have revealed products of lipid, protein, and DNA oxidation associated with these diseases. While ROS plays essential roles in cell signaling and metabolic regulation, their highly reactive nature makes them detrimental to biological materials. Neurons are especially vulnerable to oxidative damage due to their high content of polyunsaturated fatty acids, limited antioxidant defenses, and elevated oxygen consumption. Therefore, the excessive ROS production in neurons is particularly harmful, involving complex mechanisms that lead to the oxidative degradation of vital biomolecules [71]. *D. indusiata* shows neuroprotective potential due to bioactive compounds like polysaccharides, dictyophorines A and B, and quinazoline compounds. These compounds enhance nerve growth factor synthesis, mitigate oxidative stress, improve mitochondrial function, and protect against neuronal apoptosis, synaptic dysfunction, and glutamatergic neurotransmission damage.

Oxidative eustress, defined by physiological levels of reactive oxygen and nitrogen species (ROS/RNS), plays a crucial role in various biochemical processes, including signal transduction pathways like NF-κB and MAPK. However, elevated ROS/RNS levels from endogenous and exogenous sources lead to oxidative stress, a common factor in numerous chronic diseases. The ROS and RNS cause oxidative damage to DNA, proteins, and lipids and increase oxidative stress biomarkers. Conditions such as cancer, cardiovascular diseases, diabetes, neurological disorders, and aging are linked to oxidative stress. Antioxidants, including enzymes and low-molecular-weight compounds, counteract oxidative damage, with vitamin E serving as a key protector against lipid peroxidation. Emerging research suggests that certain antioxidants may enhance cellular defenses and act as potential anti-cancer agents [72]. The polysaccharides from *D. indusiata* (DIPs) play a crucial role in antioxidant, gut-health-promoting, and immunomodulatory effects. Purified water-soluble polysaccharides (PPS) from *D. indusiata* demonstrate significant antioxidant activity, scavenging free radicals and protecting against oxidative stress.

Additionally, DIPs improve gut microbiota balance, reversing antibiotic-induced dysbiosis and enhancing beneficial bacteria. Their anti-inflammatory properties are also notable, as they reduce pro-inflammatory cytokines and restore intestinal barrier integrity. Moreover, DIPs activate immune responses by promoting macrophage function and regulating key pathways like NF-κB. These benefits are also seen in other mushrooms such as *Agrocybe cylindracea*, *Cordyceps sinensis*, *Trametes versicolor*, *Ganoderma lucidum*, and *Pleurotus eryngii*, supporting the beneficial outcomes of *D. indusiata*. Edible and medicinal mushrooms exhibit health-promoting properties due to their bioactive compounds, particularly polysaccharides like β-glucans. These compounds possess anti-obesity, anti-diabetic, anti-carcinogenic, anti-microbial, and anti-viral effects, while also offering antioxidative, anti-inflammatory, and immunomodulating benefits. Mushroom polysaccharides act as prebiotics, influencing gut microbiota and promoting beneficial bacterial growth. Their resistance to digestion allows them to impact various health conditions [73–76]. *D. indusiata* polysaccharides (DIP) have potential medicinal benefits for metabolic disorders. A study of polysaccharide fractions from *D. indusiata* tested their effects on metabolic syndrome (MetS) in mice and showed that all fractions significantly improved obesity, hypertension, dyslipidemia, and liver damage. Additionally, D6P and D8P fractions helped regulate abnormal glucose metabolism. Furthermore, other studies indicate that mushrooms such as *Agaricus bisporus* and *Pleurotus ostreatus* may have beneficial effects on various cardiometabolic risk factors. Consuming either whole or processed *A. bisporus* mushrooms has been shown to positively impact glucose levels, lipid profiles (including total cholesterol, HDL, LDL, and triglycerides), and several inflammatory markers, such as TNF-α and adiponectin. Additionally, these mushrooms enhance antioxidant capacity, as evidenced by an increase in oxygen radical absorbance capacity (ORAC) [77].

Macrophages are critical players in immune defense and inflammation, activated by external stimuli to produce nitric oxide (NO) and cytokines such as TNF-α, IL-1, IL-6, and IL-12 [78,79]. DIP can stimulate macrophage activation, enhancing the production of NO and pro-inflammatory cytokines in vitro, consistent with previous studies [5,19,80]. This activation is largely mediated by the TLR4 receptor, as shown by the binding of DIP to macrophages and the inhibition of DIP-induced cytokine production via anti-TLR4 mAb [81,82]. Additionally, TLR4-mediated macrophage activation is linked to NF-κB signaling, and DIP significantly upregulates NF-κB p65 expression, further supporting its immunomodulatory effects [83,84]. These results highlight the role of DIP in enhancing immune response through TLR4 and NF-κB pathways, offering potential for therapeutic applications in immune regulation. Lectins, found in various plants, fungi, or other species, are proteins with the ability to bind carbohydrates and have potent biological activity. Once seen as toxins, they now show promise in medicine for immunomodulation, enhancing immune responses and potentially serving as vaccines to combat microbial infections through lectin-based therapies [85,86]. As described above, immunomodulatory effect of DIP has been explored; however, no studies have yet explored the immunomodulatory effects of lectins derived from *D. indusiatus*. The immunomodulatory effects of plant lectins on various immune cells are well established in the literature, highlighting their ability to enhance phagocytic activity and cytokine production during bacterial infections [87]. One of the most studied lectins in this regard is Concanavalin A (Con A), which has demonstrated significant immunomodulatory properties. Research shows that Con A treatment in murine macrophages increases the expression of Toll-like receptors (TLRs) through JNK, p38, and NF-κB-dependent signaling pathways. TLRs are key components of the innate immune system that recognize pathogen-associated molecular patterns (PAMPs) from different bacteria and activate immune responses against them. Additionally, Con A-treated macrophages secrete pro-inflammatory cytokines and nitric oxide (NO) via TLR-mediated pathways, which help to neutralize and eliminate infectious pathogens [88]. Therefore, the induction of TLRs by plant lectins like Con A is crucial for enhancing immune responses and effectively clearing infections. Inhibiting MMP-2 activity is crucial for improving wound healing and managing skin conditions. Elevated MMP-2 levels in chronic wounds and scars disrupt collagen balance, hindering healing. By reducing MMP-2 activity, as seen with mushroom extracts, inflammation and scar formation can be minimized, promoting healthier skin recovery [10].

Though scientific data on the antidiabetic effects of *D. indusiata* is currently lacking, its richness in polysaccharides, particularly β-glucans, suggests promising potential. Experiments could yield outcomes similar to those seen in other mushrooms like *Pleurotus sajor-caju* [89,90]. The β-glucans in *D. indusiata* may help prevent hyperglycemia and insulin resistance by upregulating glucose transporters (GLUT-4) and adiponectin while downregulating pro-inflammatory factors such as NF-κB and IL-6. This could enhance insulin sensitivity, similar to the effects of metformin, and reduce plasma glucose, insulin, and lipid levels, along with improving antioxidant enzyme activity and glycogen storage in the liver. While *D. indusiata* polysaccharides (DIPs) have shown significant antioxidant, neuroprotective, and immunomodulatory effects, research gaps remain, particularly in exploring the role of other bioactive compounds like lectins. Future studies could investigate the immunomodulatory effects of lectins from *D. indusiata*, similar to concanavalin A (Con A), which enhances Toll-like receptor (TLR) expression via NF-κB, JNK, and p38 signaling pathways. Exploring how DIPs modulate mitochondrial pathways and impact cellular metabolism could further unravel their neuroprotective mechanisms. Additionally, the potential synergistic effects of DIPs with other mushroom-derived compounds in managing metabolic syndrome should be examined. Mechanistic studies on quinazoline compounds' neuroprotection through glutamate receptor modulation and the anti-apoptotic role of DIPs in synaptic protection are important and should be further investigated. Expanding research on these mechanisms can provide deeper insights into their therapeutic applications, including neurodegenerative diseases, cancer, and inflammation. Moreover, clinical trials are needed to validate preclinical findings, ensuring safety and efficacy for human use.

5. Conclusions

Mushrooms, including *D. indusiata*, offer significant health benefits due to their rich bioactive compounds, such as polysaccharides, terpenoids, phenolics, and ergosterols. These compounds exhibit antioxidant, anti-inflammatory, and neuroprotective properties, making them valuable in disease prevention and therapeutic applications. Polysaccharides from mushrooms, particularly *D. indusiata*, enhance immune responses, regulate oxidative stress, and restore gut health. Furthermore, they offer neuroprotective effects by mitigating oxidative damage, improving mitochondrial function, and supporting nerve growth factor synthesis. The presence of lectins and quinazoline compounds further contributes to immune modulation and protection against neurodegenerative diseases. Overall, mushrooms serve as a source of functional foods and nutraceuticals, offering therapeutic potential against chronic diseases and promoting human health.

Author Contributions: Conceptualization, B.X.; methodology, Y.K.; software, Y.K.; validation, B.X.; formal analysis, Y.K.; investigation, Y.K.; resources, B.X.; data curation, Y.K.; writing—original draft preparation, Y.K.; writing—review and editing, B.X.; visualization, B.X.; supervision, B.X.; project administration, B.X.; funding acquisition, B.X. All authors have read and agreed to the published version of the manuscript.

Funding: This research was jointly funded by two grants (UICR0400015-24 and UICR0400016-24) funded by BNU-HKBU United International College, China.

Data Availability Statement: No new data were created or analyzed in this study.

Conflicts of Interest: The authors declare no conflicts of interest.

References

1. Habtemariam, S. The chemistry, pharmacology and therapeutic potential of the edible mushroom *Dictyophora indusiata* (Vent ex. Pers.) Fischer (Synn. *Phallus indusiatus*). *Biomedicines* **2019**, *7*, 98. [CrossRef] [PubMed]
2. Ker, Y.B.; Chen, K.C.; Peng, C.C.; Hsieh, C.L.; Peng, R.Y. Structural characteristics and antioxidative capability of the soluble polysaccharides present in *Dictyophora indusiata* (Vent. Ex Pers.) Fish Phallaceae. *Evid. Based Complement. Altern. Med.* **2011**, *2011*, 396013. [CrossRef]
3. Xiao, M.F.; Chen, X.T.; Cai, W.W.; Li, N.; Chen, H.M.; Liu, B.; Zeng, F. Antioxidant activity of water extract from *Dictyophora indusiata* and its potential to improve lipid metabolism in *Caenorhabditis elegans*. *Food Sci.* **2022**, *43*, 191–199.
4. Liu, X.; Chen, Y.; Wu, L.; Wu, X.; Huang, Y.; Liu, B. Optimization of polysaccharides extraction from *Dictyophora indusiata* and determination of its antioxidant activity. *Int. J. Biol. Macromol.* **2017**, *103*, 175–181. [CrossRef]
5. Liao, W.; Luo, Z.; Liu, D.; Ning, Z.; Yang, J.; Ren, J. Structure characterization of a novel polysaccharide from *Dictyophora indusiata* and its macrophage immunomodulatory activities. *J. Agric. Food Chem.* **2015**, *63*, 535–544. [CrossRef]
6. Wang, W.; Liu, H.; Zhang, Y.; Feng, Y.; Yuan, F.; Song, X.; Gao, Z.; Zhang, J.; Song, Z.; Jia, L. Antihyperlipidemic and hepatoprotective properties of alkali- and enzyme-extractable polysaccharides by *Dictyophora indusiata*. *Sci. Rep.* **2019**, *9*, 14266. [CrossRef] [PubMed]
7. Wang, J.; Wen, X.; Yang, B.; Liu, D.; Li, X.; Geng, F. De novo transcriptome and proteome analysis of *Dictyophora indusiata* fruiting bodies provides insights into the changes during morphological development. *Int. J. Biol. Macromol.* **2020**, *146*, 875–886. [CrossRef]
8. Wang, Y.; Shi, X.; Yin, J.; Nie, S. Bioactive polysaccharide from edible *Dictyophora* spp.: Extraction, purification, structural features and bioactivities. *Bioactive Carbohydr. Diet. Fibre* **2018**, *14*, 25–32. [CrossRef]
9. Wang, W.; Song, X.; Gao, Z.; Zhao, H.; Wang, X.; Liu, M.; Jia, L. Anti-hyperlipidemic, antioxidant and organic protection effects of acidic-extractable polysaccharides from *Dictyophora indusiata*. *Int. J. Biol. Macromol.* **2019**, *129*, 281–292. [CrossRef]
10. Nazir, Y.; Linsaenkart, P.; Khantham, C.; Chaitep, T.; Jantrawut, P.; Chittasupho, C.; Rachtanapun, P.; Jantanasakulwong, K.; Phimolsiripol, Y.; Sommano, S.R.; et al. High efficiency in vitro wound healing of *Dictyophora indusiata* extracts via anti-inflammatory and collagen stimulating (MMP-2 Inhibition) mechanisms. *J. Fungi* **2021**, *7*, 1100. [CrossRef] [PubMed]
11. Kanwal, S.; Aliya, S.; Xin, Y. Anti-obesity effect of *Dictyophora indusiata* mushroom polysaccharide (DIP) in high fat diet-induced obesity via regulating inflammatory cascades and intestinal microbiome. *Front. Endocrinol.* **2020**, *11*, 558874. [CrossRef] [PubMed]
12. Oyetayo, V.O.; Dong, C.-H.; Yao, Y.-J. Antioxidant and antimicrobial properties of aqueous extract from *Dictyophora indusiata*. *Open Mycol. J.* **2009**, *3*, 20–26. [CrossRef]
13. Wang, W.; Song, X.; Zhang, J.; Li, H.; Liu, M.; Gao, Z.; Wang, X.; Jia, L. Antioxidation, hepatic- and renal-protection of water-extractable polysaccharides by *Dictyophora indusiata* on obese mice. *Int. J. Biol. Macromol.* **2019**, *134*, 290–301. [CrossRef]
14. Deng, C.; Fu, H.; Teng, L.; Hu, Z.; Xu, X.; Chen, J.; Ren, T. Anti-tumor activity of the regenerated triple-helical polysaccharide from *Dictyophora indusiata*. *Int. J. Biol. Macromol.* **2013**, *61*, 453–458. [CrossRef] [PubMed]
15. Kawagishi, H.; Ishiyama, D.; Mori, H.; Sakamoto, H.; Ishiguro, Y.; Furukawa, S.; Li, J. Dictyophorines A and B, two stimulators of NGF-synthesis from the mushroom *Dictyophora indusiata*. *Phytochemistry* **1997**, *45*, 1203–1205. [CrossRef] [PubMed]
16. Lee, I.K.; Yun, B.S.; Han, G.; Cho, D.H.; Kim, Y.H.; Yoo, I.D. Dictyoquinazols A, B, and C, new neuroprotective compounds from the mushroom *Dictyophora indusiata*. *J. Nat. Prod.* **2002**, *65*, 1769–1772. [CrossRef] [PubMed]
17. Huang, M.; Chen, X.; Tian, H.; Sun, B.; Chen, H. Isolation and identification of antibiotic albaflavenone from *Dictyophora indusiata* (Vent: Pers.) Fischer. *J. Chem. Res.* **2011**, *35*, 659–660. [CrossRef]
18. Fu, H.; Deng, C.; Teng, L.; Yu, L.; Su, T.; Xu, X.; Chen, J.; Yang, C. Immunomodulatory activities on RAW 264.7 macrophages of a polysaccharide from veiled lady mushroom, *Dictyophora indusiata* (Higher Basidiomycetes). *Int. J. Med. Mushrooms* **2015**, *17*, 151–160. [CrossRef]
19. Deng, C.; Shang, J.; Fu, H.; Chen, J.; Liu, H. Mechanism of the immunostimulatory activity by a polysaccharide from *Dictyophora indusiata*. *Int. J. Biol. Macromol.* **2016**, *91*, 752–759. [CrossRef] [PubMed]

20. Wang, Y.; Ji, X.; Yan, M.; Chen, X.; Kang, M.; Teng, L.; Wu, X.; Chen, J.; Deng, C. Protective effect and mechanism of polysaccharide from *Dictyophora indusiata* on dextran sodium sulfate-induced colitis in C57BL/6 mice. *Int. J. Biol. Macromol.* **2019**, *140*, 973–984. [CrossRef] [PubMed]
21. Hua, Y.; Gao, Q.; Wen, L.; Yang, B.; Tang, J.; You, L.; Zhao, M. Structural characterisation of acid- and alkali-soluble polysaccharides in the fruiting body of *Dictyophora indusiata* and their immunomodulatory activities. *Food Chem.* **2012**, *132*, 739–743. [CrossRef]
22. Wang, J.; Xu, X.; Zheng, H.; Li, J.; Deng, C.; Xu, Z.; Chen, J. Structural characterization, chain conformation, and morphology of a beta-(1-->3)-D-glucan isolated from the fruiting body of *Dictyophora indusiata*. *J. Agric. Food Chem.* **2009**, *57*, 5918–5924. [CrossRef] [PubMed]
23. Ishiyama, D.; Fukushi, Y.; Ohnishi-Kameyama, M.; Nagata, T.; Mori, H.; Inakuma, T.; Ishiguro, Y.; Li, J.; Kawagishi, H. Monoterpene-alcohols from a mushroom *Dictyophora indusiata*. *Phytochemistry* **1999**, *50*, 1053–1056. [CrossRef]
24. Zhang, Y.F.; Xun, H.; Gao, Q.; Tang, F.; Qi, F.F.; Sun, J. Two new sesquiterpenoids from *Dictyophora indusiata*. *J. Asian Nat. Prod. Res.* **2023**, *25*, 564–570. [CrossRef] [PubMed]
25. Sharma, V.K.; Choi, J.; Sharma, N.; Choi, M.; Seo, S.Y. In vitro anti-tyrosinase activity of 5-(hydroxymethyl)-2-furfural isolated from *Dictyophora indusiata*. *Phytother. Res.* **2004**, *18*, 841–844. [CrossRef]
26. Zhang, Y.; Xun, H.; Gao, Q.; Qi, F.; Sun, J.; Tang, F. Chemical constituents of the mushroom *Dictyophora indusiata* and their anti-inflammatory activities. *Molecules* **2023**, *28*, 2760. [CrossRef] [PubMed]
27. Lin, Y.-m.; Su, A.-h. Isolation, Purification and characterization of lectin from *Dictyophora indusiata* Fisch. *Chin. J. Biochem. Mol. Biol.* **2003**, *19*, 261–263.
28. Bai, X.-w.; Bai, X.-f. Antioxidant activity of a polysaccharide from *Dictyophora indusiata* volva and MECC analysis of its monosaccharide composition. *J. Indian Chem. Soc.* **2021**, *98*, 100146. [CrossRef]
29. Deng, C.; Hu, Z.; Fu, H.; Hu, M.; Xu, X.; Chen, J. Chemical analysis and antioxidant activity in vitro of a β-d-glucan isolated from *Dictyophora indusiata*. *Int. J. Biol. Macromol.* **2012**, *51*, 70–75. [CrossRef]
30. Minatel, I.O.; Borges, C.V.; Ferreira, M.I.; Gomez, H.A.G.; Chen, C.-Y.O.; Lima, G.P.P. Phenolic compounds: Functional properties, impact of processing and bioavailability. In *Phenolic Compounds-Biological Activity*; InTech: London, UK, 2017.
31. Kanwal, S.; Joseph, T.P.; Owusu, L.; Xiaomeng, R.; Meiqi, L.; Yi, X. A polysaccharide isolated from *Dictyophora indusiata* promotes recovery from antibiotic-driven intestinal dysbiosis and improves gut epithelial barrier function in a mouse model. *Nutrients* **2018**, *10*, 1003. [CrossRef] [PubMed]
32. Etxeberria, U.; Fernandez-Quintela, A.; Milagro, F.I.; Aguirre, L.; Martinez, J.A.; Portillo, M.P. Impact of polyphenols and polyphenol-rich dietary sources on gut microbiota composition. *J. Agric. Food Chem.* **2013**, *61*, 9517–9533. [CrossRef] [PubMed]
33. Shi, H.; Kokoeva, M.V.; Inouye, K.; Tzameli, I.; Yin, H.; Flier, J.S. TLR4 links innate immunity and fatty acid-induced insulin resistance. *J. Clin. Investig.* **2006**, *116*, 3015–3025. [CrossRef]
34. Liu, Y.; Huang, J.; Hou, Y.; Zhu, H.; Zhao, S.; Ding, B.; Yin, Y.; Yi, G.; Shi, J.; Fan, W. Dietary arginine supplementation alleviates intestinal mucosal disruption induced by *Escherichia coli* lipopolysaccharide in weaned pigs. *Br. J. Nutr.* **2008**, *100*, 552–560. [CrossRef] [PubMed]
35. Shi, H.; Dong, L.; Jiang, J.; Zhao, J.; Zhao, G.; Dang, X.; Lu, X.; Jia, M. Chlorogenic acid reduces liver inflammation and fibrosis through inhibition of toll-like receptor 4 signaling pathway. *Toxicology* **2013**, *7*, 107–114. [CrossRef]
36. Shi, Y.; Zhao, X.; Zhao, J.; Zhang, H.; Zhai, Q.; Narbad, A.; Chen, W. A mixture of *Lactobacillus* species isolated from traditional fermented foods promote recovery from antibiotic-induced intestinal disruption in mice. *J. Appl. Microbiol.* **2018**, *124*, 842–854. [CrossRef] [PubMed]
37. Knoop, K.A.; McDonald, K.G.; Kulkarni, D.H.; Newberry, R.D. Antibiotics promote inflammation through the translocation of native commensal colonic bacteria. *Gut* **2016**, *65*, 1100–1109. [CrossRef]
38. Xiao, Y.; Xu, W.; Su, W. NLRP3 inflammasome: A likely target for the treatment of allergic diseases. *Clin. Exp. Allergy* **2018**, *48*, 1080–1091. [CrossRef] [PubMed]
39. Liu, Y.; Zhang, H.; Li, Y.; Zha, H.; Gao, Y.; Chen, H.; Wang, Y.; Zhou, T.; Deng, C. *Dictyophora indusiata* polysaccharide mediates priming of the NLRP3 inflammasome activation via TLR4/NF-kappaB signaling pathway to exert immunostimulatory effects. *J. Appl. Biomed.* **2024**, *22*, 23–32. [CrossRef] [PubMed]
40. Brubaker, S.W.; Bonham, K.S.; Zanoni, I.; Kagan, J.C. Innate immune pattern recognition: A cell biological perspective. *Annu. Rev. Immunol.* **2015**, *33*, 257–290. [CrossRef] [PubMed]
41. Mangan, M.S.J.; Olhava, E.J.; Roush, W.R.; Seidel, H.M.; Glick, G.D.; Latz, E. Targeting the NLRP3 inflammasome in inflammatory diseases. *Nat. Rev. Drug Discov.* **2018**, *17*, 688. [CrossRef] [PubMed]
42. Liu, Y.; Zhao, J.; Zhao, Y.; Zong, S.; Tian, Y.; Chen, S.; Li, M.; Liu, H.; Zhang, Q.; Jing, X.; et al. Therapeutic effects of lentinan on inflammatory bowel disease and colitis-associated cancer. *J. Cell. Mol. Med.* **2019**, *23*, 750–760. [CrossRef] [PubMed]
43. Wang, Y.; Lai, L.; Teng, L.; Li, Y.; Cheng, J.; Chen, J.; Deng, C. Mechanism of the anti-inflammatory activity by a polysaccharide from *Dictyophora indusiata* in lipopolysaccharide-stimulated macrophages. *Int. J. Biol. Macromol.* **2019**, *126*, 1158–1166. [CrossRef]

44. Zhang, J.; Shi, R.; Li, H.; Xiang, Y.; Xiao, L.; Hu, M.; Ma, F.; Ma, C.W.; Huang, Z. Antioxidant and neuroprotective effects of *Dictyophora indusiata* polysaccharide in *Caenorhabditis elegans*. *J. Ethnopharmacol.* **2016**, *192*, 413–422. [CrossRef]
45. Zhang, J.; Hu, T.; Wang, Y.; Zhang, X.; Zhang, H.; Lin, J.; Tang, X.; Liu, X.; Chen, M.; Khan, N.U.; et al. Investigating the neurotoxic impacts of arsenic and the neuroprotective effects of Dictyophora polysaccharide using SWATH-MS-based proteomics. *Molecules* **2022**, *27*, 1495. [CrossRef]
46. Yin, M.; Zhang, Y.; Li, H. Advances in research on immunoregulation of macrophages by plant polysaccharides. *Front. Immunol.* **2019**, *10*, 145. [CrossRef]
47. Wasser, S.P. Medicinal mushrooms as a source of antitumor and immunomodulating polysaccharides. *Appl. Microbiol. Biotechnol.* **2002**, *60*, 258–274. [CrossRef] [PubMed]
48. Chantree, P.; Chumkiew, S.; Jamklang, M.; Martviset, P. Cytotoxic activities of ethanolic crude extracts from fruiting bodies of bamboo mushrooms (*Dictyophora* spp.) against cholangiocarcinoma cells. *Res. Results Pharmacol.* **2022**, *8*, 33–41. [CrossRef]
49. Zhao, B.; Lin, X.; Lei, L.; Lamb, D.C.; Kelly, S.L.; Waterman, M.R.; Cane, D.E. Biosynthesis of the sesquiterpene antibiotic albaflavenone in *Streptomyces coelicolor* A3(2). *J. Biol. Chem.* **2008**, *283*, 8183–8189. [CrossRef] [PubMed]
50. Yoshimoto, S.; Loo, T.M.; Atarashi, K.; Kanda, H.; Sato, S.; Oyadomari, S.; Iwakura, Y.; Oshima, K.; Morita, H.; Hattori, M.; et al. Obesity-induced gut microbial metabolite promotes liver cancer through senescence secretome. *Nature* **2013**, *499*, 97–101. [CrossRef]
51. Chang, C.J.; Lin, C.S.; Lu, C.C.; Martel, J.; Ko, Y.F.; Ojcius, D.M.; Tseng, S.F.; Wu, T.R.; Chen, Y.Y.; Young, J.D.; et al. *Ganoderma lucidum* reduces obesity in mice by modulating the composition of the gut microbiota. *Nat. Commun.* **2015**, *6*, 7489. [CrossRef] [PubMed]
52. Sima, C.; Montero, E.; Nguyen, D.; Freire, M.; Norris, P.; Serhan, C.N.; Van Dyke, T.E. ERV1 overexpression in myeloid cells protects against high fat diet induced obesity and glucose intolerance. *Sci. Rep.* **2017**, *7*, 12848. [CrossRef]
53. Kim, M.H.; Park, S.J.; Kim, J.H.; Seong, J.B.; Kim, K.M.; Woo, H.A.; Lee, D.S. Peroxiredoxin 5 regulates adipogenesis-attenuating oxidative stress in obese mouse models induced by a high-fat diet. *Free Radic. Biol. Med.* **2018**, *123*, 27–38. [CrossRef] [PubMed]
54. Alloubani, A.; Nimer, R.; Samara, R. Relationship between hyperlipidemia, cardiovascular disease and stroke: A systematic review. *Curr. Cardiol. Rev.* **2021**, *17*, e051121189015. [CrossRef]
55. Nelson, R.H. Hyperlipidemia as a risk factor for cardiovascular disease. *Prim. Care* **2013**, *40*, 195–211. [CrossRef] [PubMed]
56. Chan, Y.M.; Varady, K.A.; Lin, Y.; Trautwein, E.; Mensink, R.P.; Plat, J.; Jones, P.J. Plasma concentrations of plant sterols: Physiology and relationship with coronary heart disease. *Nutr. Rev.* **2006**, *64*, 385–402. [CrossRef] [PubMed]
57. Loizou, S.; Lekakis, I.; Chrousos, G.P.; Moutsatsou, P. Beta-sitosterol exhibits anti-inflammatory activity in human aortic endothelial cells. *Mol. Nutr. Food Res.* **2010**, *54*, 551–558. [CrossRef] [PubMed]
58. Moreno, J.J. Effect of olive oil minor components on oxidative stress and arachidonic acid mobilization and metabolism by macrophages RAW 264.7. *Free Radic. Biol. Med.* **2003**, *35*, 1073–1081. [CrossRef]
59. Masuzaki, H.; Paterson, J.; Shinyama, H.; Morton, N.M.; Mullins, J.J.; Seckl, J.R.; Flier, J.S. A transgenic model of visceral obesity and the metabolic syndrome. *Science* **2001**, *294*, 2166–2170. [CrossRef] [PubMed]
60. Wang, D.; Zhao, Y.; Sun, Y.; Yang, X. Protective effects of Ziyang tea polysaccharides on CCl_4-induced oxidative liver damage in mice. *Food Chem.* **2014**, *143*, 371–378. [CrossRef] [PubMed]
61. Liu, G.; Liu, X.; Zhang, Y.; Zhang, F.; Wei, T.; Yang, M.; Wang, K.; Wang, Y.; Liu, N.; Cheng, H.; et al. Hepatoprotective effects of polysaccharides extracted from *Zizyphus jujube* cv. Huanghetanzao. *Int. J. Biol. Macromol.* **2015**, *76*, 169–175. [CrossRef]
62. Nwidu, L.L.; Elmorsy, E.; Oboma, Y.I.; Carter, W.G. Hepatoprotective and antioxidant activities of *Spondias mombin* leaf and stem extracts against carbon tetrachloride-induced hepatotoxicity. *J. Taibah Univ. Med. Sci.* **2018**, *13*, 262–271. [CrossRef] [PubMed]
63. Wang, X.; Li, W.; Xiao, L.; Liu, C.; Qi, H.; Zhang, Z. In vivo antihyperlipidemic and antioxidant activity of porphyran in hyperlipidemic mice. *Carbohydr. Polym.* **2017**, *174*, 417–420. [CrossRef] [PubMed]
64. Madlener, M. Differential expression of matrix metalloproteinases and their physiological inhibitors in acute murine skin wounds. *Arch. Dermatol. Res.* **1998**, *290*, S24–S29. [CrossRef]
65. Ravanti, L.; Kahari, V.M. Matrix metalloproteinases in wound repair (review). *Int. J. Mol. Med.* **2000**, *6*, 391–407. [CrossRef] [PubMed]
66. Lobmann, R.; Ambrosch, A.; Schultz, G.; Waldmann, K.; Schiweck, S.; Lehnert, H. Expression of matrix-metalloproteinases and their inhibitors in the wounds of diabetic and non-diabetic patients. *Diabetologia* **2002**, *45*, 1011–1016. [CrossRef] [PubMed]
67. Ulrich, D.; Ulrich, F.; Unglaub, F.; Piatkowski, A.; Pallua, N. Matrix metalloproteinases and tissue inhibitors of metalloproteinases in patients with different types of scars and keloids. *J. Plast. Reconstr. Aesthet. Surg.* **2010**, *63*, 1015–1021. [CrossRef]
68. Aparecida Da Silva, A.; Leal-Junior, E.C.; Alves, A.C.; Rambo, C.S.; Dos Santos, S.A.; Vieira, R.P.; De Carvalho Pde, T. Wound-healing effects of low-level laser therapy in diabetic rats involve the modulation of MMP-2 and MMP-9 and the redistribution of collagen types I and III. *J. Cosmet. Laser Ther.* **2013**, *15*, 210–216. [CrossRef] [PubMed]
69. Gunc Ergonul, P.; Akata, I.; Kalyoncu, F.; Ergonul, B. Fatty acid compositions of six wild edible mushroom species. *ScientificWorld J.* **2013**, *2013*, 163964. [CrossRef] [PubMed]

70. Wasser, S.P. Medicinal mushroom science: Current perspectives, advances, evidences, and challenges. *Biomed. J.* **2014**, *37*, 345–356. [CrossRef] [PubMed]
71. Collin, F. Chemical basis of reactive oxygen species reactivity and involvement in neurodegenerative diseases. *Int. J. Mol. Sci.* **2019**, *20*, 2407. [CrossRef] [PubMed]
72. Jomova, K.; Raptova, R.; Alomar, S.Y.; Alwasel, S.H.; Nepovimova, E.; Kuca, K.; Valko, M. Reactive oxygen species, toxicity, oxidative stress, and antioxidants: Chronic diseases and aging. *Arch. Toxicol.* **2023**, *97*, 2499–2574. [CrossRef] [PubMed]
73. Zhao, Q.; Jiang, Y.; Patrick Manzi, H.; Su, L.; Liu, D.; Huang, X.; Long, D.; Tang, Z.; Zhang, Y. The benefits of edible mushroom polysaccharides for health and their influence on gut microbiota: A review. *Front. Nutr.* **2023**, *10*, 1213010. [CrossRef] [PubMed]
74. Friedman, M. Mushroom polysaccharides: Chemistry and antiobesity, antidiabetes, anticancer, and antibiotic properties in cells, rodents, and humans. *Foods* **2016**, *5*, 80. [CrossRef] [PubMed]
75. Kumar, K.; Mehra, R.; Guine, R.P.F.; Lima, M.J.; Kumar, N.; Kaushik, R.; Ahmed, N.; Yadav, A.N.; Kumar, H. Edible mushrooms: A comprehensive review on bioactive compounds with health benefits and processing aspects. *Foods* **2021**, *10*, 2996. [CrossRef]
76. Yao, H.; Yang, J.; Li, S.; Cui, S.W.; Tan, H.; Nie, S. Effects of different fractions of polysaccharides from *Dictyophora indusiata* on high-fat diet-induced metabolic syndrome in mice. *Int. J. Biol. Macromol.* **2024**, *272*, 132744. [CrossRef]
77. Blumfield, M.; Abbott, K.; Duve, E.; Cassettari, T.; Marshall, S.; Fayet-Moore, F. Examining the health effects and bioactive components in *Agaricus bisporus* mushrooms: A scoping review. *J. Nutr. Biochem.* **2020**, *84*, 108453. [CrossRef] [PubMed]
78. Martinez, F.O.; Helming, L.; Gordon, S. Alternative activation of macrophages: An immunologic functional perspective. *Annu. Rev. Immunol.* **2009**, *27*, 451–483. [CrossRef]
79. Rossol, M.; Heine, H.; Meusch, U.; Quandt, D.; Klein, C.; Sweet, M.J.; Hauschildt, S. LPS-induced cytokine production in human monocytes and macrophages. *Crit. Rev. Immunol.* **2011**, *31*, 379–446. [CrossRef] [PubMed]
80. Zhang, J.-J.; Li, Y.; Zhou, T.; Xu, D.-P.; Zhang, P.; Li, S.; Li, H.-B. Bioactivities and health benefits of mushrooms mainly from China. *Molecules* **2016**, *21*, 938. [CrossRef]
81. Netea, M.G.; Ferwerda, G.; van der Graaf, C.A.; Van der Meer, J.W.; Kullberg, B.J. Recognition of fungal pathogens by toll-like receptors. *Curr. Pharm. Des.* **2006**, *12*, 4195–4201. [CrossRef] [PubMed]
82. Moresco, E.M.; LaVine, D.; Beutler, B. Toll-like receptors. *Curr. Biol.* **2011**, *21*, R488–R493. [CrossRef]
83. Sung, N.-Y.; Byun, E.-H.; Song, D.-S.; Kim, J.-K.; Park, J.-H.; Song, B.-S.; Park, S.-H.; Lee, J.-W.; Byun, M.-W. The procyanidin trimer C1 inhibits LPS-induced MAPK and NF-κB signaling through TLR4 in macrophages. *Int. Immunopharmacol.* **2013**, *15*, 450–456.
84. Akira, S.; Takeda, K. Toll-like receptor signalling. *Nat. Rev. Immunol.* **2004**, *4*, 499–511. [CrossRef] [PubMed]
85. Bah, C.S.F.; Fang, E.F.; Ng, T.B. Medicinal Applications of Plant Lectins. In *Antitumor Potential and Other Emerging Medicinal Properties of Natural Compounds*; Springer: Berlin/Heidelberg, Germany, 2012; Volume 10, pp. 55–74. [CrossRef]
86. Mishra, A.; Behura, A.; Mawatwal, S.; Kumar, A.; Naik, L.; Mohanty, S.S.; Manna, D.; Dokania, P.; Patra, S.K.; Dhiman, R. Structure-function and application of plant lectins in disease biology and immunity. *Food Chem. Toxicol.* **2019**, *134*, 110827. [CrossRef] [PubMed]
87. da Silva, L.C.; Alves, N.M.; de Castro, M.C.; Pereira, V.R.; da Paz, N.V.; Coelho, L.C.; de Figueiredo, R.C.; Correia, M.T. Immunomodulatory effects of pCramoll and rCramoll on peritoneal exudate cells (PECs) infected and non-infected with *Staphylococcus aureus*. *Int. J. Biol. Macromol.* **2015**, *72*, 848–854. [CrossRef] [PubMed]
88. Sodhi, A.; Tarang, S.; Kesherwani, V. Concanavalin A induced expression of Toll-like receptors in murine peritoneal macrophages in vitro. *Int. Immunopharmacol.* **2007**, *7*, 454–463. [CrossRef] [PubMed]
89. Kanagasabapathy, G.; Kuppusamy, U.R.; Abd Malek, S.N.; Abdulla, M.A.; Chua, K.H.; Sabaratnam, V. Glucan-rich polysaccharides from *Pleurotus sajor*-caju (Fr.) Singer prevents glucose intolerance, insulin resistance and inflammation in C57BL/6J mice fed a high-fat diet. *BMC Complement. Altern. Med.* **2012**, *12*, 261. [CrossRef] [PubMed]
90. Hong, L.; Xun, M.; Wutong, W. Anti-diabetic effect of an alpha-glucan from fruit body of maitake (*Grifola frondosa*) on KK-Ay mice. *J. Pharm. Pharmacol.* **2007**, *59*, 575–582. [CrossRef] [PubMed]

Journal of
Fungi

MDPI

Review

The Application of Fungi and Their Secondary Metabolites in Aquaculture

Abigail John Onomu * and Grace Emily Okuthe

Department of Biological & Environmental Sciences, Walter Sisulu University, Mthatha 5117, South Africa; gokuthe@wsu.ac.za
* Correspondence: abigailjohn90@gmail.com

Abstract: Ensuring sustainability has increasingly become a significant concern not only in aquaculture but in the general agrifood sector. Therefore, it is imperative to investigate pathways to feed substitutes/best practices to enhance aquaculture sustainability. The application of fungi in aquaculture provides innovative methods to enhance the sustainability and productivity of aquaculture. Fungi play numerous roles in aquaculture, including growth, immunity enhancement and disease resistance. They also play a role in bioremediation of waste and bioflocculation. The application of fungi improves the suitability and utilization of terrestrial plant ingredients in aquaculture by reducing the fibre fractions and anti-nutritional factors and increasing the nutrients and mineral contents of plant ingredients. Fungi are good flotation agents and can enhance the buoyancy of aquafeed. Pigments from fungi enhance the colouration of fish fillets, making them more attractive to consumers. This paper, via the relevant literature, explores the multifaceted roles of fungi in aquaculture, emphasizing their potential to transform aquaculture through environmentally friendly and sustainable techniques. The effectiveness of fungi in reducing fibre fractions and enhancing nutrient availability is influenced by the duration of fermentation and the dosage administered, which may differ for various feed ingredients, making it difficult for most aquaculture farmers to apply fungi approximately. Therefore, the most effective dosage and fermentation duration for each feed ingredient should be investigated.

Keywords: *Aspergillus niger*; mushroom; yeast; bioremediation; bioflocculation; buoyancy; probiotics; feed additive; antinutritional factors

1. Introduction

Aquaculture is a significant contributor to global food and nutrition security and is expected to grow to meet the increasing demand of a growing population and economy [1]. Sustainable aquaculture practices are needed for the envisaged aquaculture growth to be achieved. These practices are essential not only for economic viability but also for environmental protection, sustenance and social responsibility [1]. Some aquaculture practices that require innovation and the adoption of sustainable practices include the over-reliance on fishmeal in feed production and the application of antibiotics and chemicals in preventing and treating disease [2]. The enhancement of the nutrient utilization of feed to obtain high feed efficiency and animal growth, the maintenance of the water quality, and wastewater/effluent bioremediation are also essential [3].

Due to the need for sustainable practices, alternative feed ingredients, such as terrestrial plants, are being incorporated into aquatic animal feed to reduce reliance on fish protein, minimize the cost of production and enhance aquaculture sustainability [4]. However, there are limitations associated with the use of terrestrial plants in aquafeed formulation. Generally, terrestrial plants are low in protein, high in fibre fractions (crude fibre, lignin, cellulose and hemicellulose) and contain antinutritional factors, which hinder their digestibility [5–7]. They are rich in phosphorus, which is essential for growth. However, the phosphorous in terrestrial plants is only available as phytic acid, which, unfortunately, aquatic animals cannot utilize [8,9]. Hence, when feeds containing terrestrial plants are fed to aquatic animals, the unutilized phytic acid is released into the culture system via faeces from the animals and uneaten feed. The water quality deteriorates on accumulation in the system [10]. The untreated effluent from such a system is detrimental to the aquatic ecosystem, which is the final destination of wastewater and effluent from aquaculture [11].

Aquatic farmers in developing countries cannot afford floating/extruded aquafeed. If they do, it impacts the cost of the final product, which, most times, may not be affordable to people within the locality [12]. Due to this, they use sinking feed or produce feed unconventionally using locally available ingredients [12]. The feed produced via this method tends to sink quickly on reaching water, making a large proportion of the feed unutilized by fish. Underutilization of feed and the nutrients present in the feed result in unmaximized growth and deterioration of the quality of the water, which then acts as a breeding ground for disease. Antibiotics and chemicals have been used to treat and prevent disease. Nevertheless, their use is discouraged due to the health implications on consumers, with the possibility that it might result in antibiotic resistance and detriment to the ecosystem [2]. Therefore, a sustainable substitute to prevent and control diseases is required.

Fungi are either unicellular or multicellular/filamentous and belong to the kingdom of fungi. They are ubiquitous eukaryotic organisms and heterotrophs. They have been exploited to produce a wide range of industrial products, such as fermented drinks and food, food additives, pigments, biofuels, enzymes, antibiotics, vitamins, fatty acids and sterols [13]. Fungi are important microorganisms with the potential to enhance aquaculture sustainability. The application of fungi in fermenting plant ingredients (terrestrial protein sources) reduces the antinutritional factors and fibre fractions, such as hemicellulose and lignin [7,14,15]. It also increases the nutrient availability in plant ingredients, making them better suitable for use as aquatic animal feed. For example, fermentation with fungi increases the protein content and reduces the fat content of plant protein [15–17]. The reduction of fat content in plant protein is suitable for aquatic animals as a high-fat content is not required in their diet [18]. Fungi are efficient bioremediators and have been applied in the bioremediation of wastewater. They can also remove heavy metals from wastewater and minimize the effect of heavy metal pollution on aquatic animals [14,19]. Fungi, such as yeast (*saccharomyces cerevisiae*) and mould (*Aspergillus niger*) possess a flotation ability and can be used when enhancing the buoyancy of aquafeed, thus improving feed efficiency and utilization [20,21]. Fungi also serve as flocculants and are efficient in the harvesting of microalgae, thus minimizing the cost involved in microalgae harvesting [22,23]. The potential benefits of fungi are numerous, including enhanced productivity and a more efficient, sustainable and environmental practice contributing to improving aquaculture sustainability. However, these benefits need to be organized in the literature in order to create awareness and promote knowledge for its adoption. This is because research has revealed that the potential of fungi has not been adequately utilized/exploited in aquaculture practice due to various factors. Therefore, this study seeks to investigate significant findings on its application in aquaculture.

1.1. Objectives

The objective of this study is to expose the application of fungi in aquaculture and the limitations associated with its application. The study aims to create awareness of fungi and its beneficial use in aquaculture.

1.2. The Description of Fungi

Over 3.8 million species of fungi exist in nature; however, a large fraction are unidentified [24]. Fungi are categorized into eight phyla, which are Ascomycota, Basidiomycota, Blastocladiomycota, Chytridiomycota, Glomeromycota, Microsporidia, Neocallimastigomycota and Zygomycota. Fungi are either saprobes, living in decomposing organic matter, or parasites; they possess cell walls composed of chitin and polysaccharides. They are immotile and reproduce via spores [25]. Examples of fungi of importance to aquaculture are mould, yeast and mushroom. Some fungi species are pathogenic, causing mild or severe diseases in plants, animals and humans [26]. However, some are of economic importance. For example, yeast metabolism is exploited to produce a wide range of industrial products such as fermented drinks, food additives, pigments, biofuels, enzymes, antibiotics, vitamins, fatty acids and sterols [13].

Moulds are filamentous fungi found everywhere, including in soils, compost, litter and decaying plant matter [24]. Mould grows on organic matter in the presence of oxygen (aerobically). The most common mould used in aquaculture is *A. niger* and *A. oryzae*, which are generally non-pathogenic. The enzymes from *A. niger* are generally regarded as safe (GRAS) for consumption/food production [27]. Yeasts are microscopic, unicellular, eukaryotic organisms in the fungi kingdom [28,29]. Yeasts are facultative anaerobes, implying that they can survive, grow and thrive with or without oxygen [30]. They are found everywhere in the environment, possessing nuclear membranes and cell walls but no chloroplast [28]. They are heterotrophs and obtain their nutrients and energy from live and dead organic matter [31]. Nutrition by yeast is attained via the production and secretion of glycolytic, proteolytic and lipolytic enzymes, which digest organic matter. Alternatively, nutrition is obtained by absorbing amino acids and monosaccharides via their cell wall [28]. Some yeast genera are disease-causing, while some greatly benefit man. For example, the genera *Cryptococcus*, *Trichosporon*, *Candida* and *Torulosis* are pathogenic, while the species *Candida utilis*, *S. cerevisiae* and *Kluyveromyces marxianus* are of benefit to humans and animals [32,33].

Mushrooms are higher fungi, classified as Ascomycetes (*Morchella*, *Tuber* etc.) and Basidiomycetes (*Auricularia*, *Tremella*, *Agaricus* etc.). Mushrooms are categorised into edible mushrooms and medicinal mushrooms. Medicinal mushrooms are only suitable as medicine and not as food [34]. The bioactive compounds in mushrooms include dietary fibres, proteins, amino acids, alkaloids, oligosaccharides, polysaccharides, alcohols, vitamins and minerals [35]. Various species of mushrooms are used in aquaculture, including *Ganoderma* sp., *Pleurotus* sp. (Oyster mushroom), *Lentinula edodes* (Shiitake mushroom), *Coriolus* sp. (Turkey tail mushroom), *Cordyceps* sp. (Caterpillar mushroom) and *Agaricus bisporus* (White button mushroom). The stalk, mycelia, polysaccharides, extracts and dry powder derived from mushrooms are applied in aquaculture [36].

2. Application of Fungi and Its Metabolites in Aquaculture

2.1. Source of Antibiotics, Probiotics and Prebiotics

Aquatic environments are complex ecosystems characterised by diverse microorganisms (including beneficial and pathogenic microorganisms) which influence the health of cultured animals [37]. The traditional method employed in curbing aquatic pathogens is the application of antibiotics and chemicals, which may lead to antibiotic resistance or even negatively impact the aquatic ecosystem [2]. Fungi generate a variety of bioactive substances, such as antibiotics, enzymes and secondary metabolites, which can hinder bacteria, and viruses and pathogenic fungi growth [38]. Fungi, such as *Penicillium* and *Aspergilus* sp., produce penicillin and other antibiotics with broad-spectrum antibacterial activity [39]. *Trichoderma asperellum* MT02, isolated from sponges, exhibited antibacterial activity against *V. harveyi* and *V. alginolyticus*, pathogens of white leg shrimp (*Litopenaeus vannamei*) known to result in disease leading to a substantial economic loss [40]. Similarly, fungi metabolites of marine red alga *Nodulisporium* sp. KT29 possess antibacterial compounds that inhibit bacteria pathogens such as *Aeromonas salmonicida*, *Vibrio anguillarum*, *V. harveyi* and *Yersinia ruckeri* [38].

Fungi also function as probiotics/prebiotics when supplemented in diets, enhancing growth, feed efficiency, immune and haematology parameters and disease resistance against pathogens [41] (Figure 1). Probiotics can be defined as "a live microbial adjunct which has a beneficial effect on the host by modifying the host-associated or ambient microbial community, ensuring improved use of the feed enhancing the nutritional value of the feed and enhancing the host response towards disease or improving the quality of its ambient environment" [42]. Prebiotics are indigestible feed components that improve health by promoting the growth and metabolism of beneficial bacteria in the gut [43]. *Nodulisporium* sp. KT29 metabolite, when incorporated into the feed of *Litopenaeus vannamei*, resulted in a higher survival rate, growth rate and better feed efficiency ratio (FCR) within 21 days, when compared with the control (without fungi supplementation) [44]. The improved growth, survival, and FCR displayed by shrimp-fed diets containing fungi metabolites are reportedly due to the phytochemicals (β-glucans and polyphenols) contained in the metabolites, which act as immunostimulants and antioxidants [44]. Fermented mushroom (*Pleurotus ostreatus*) added to the diet of juvenile Amur catfish, *Silurus asotus*, at 0.1% and 0.2% concentrations for eight weeks showed a significantly higher weight gain and specific growth rate than that of the control [45]. *Aspergillus niger*, when fed to common carp *Cyprinus carpio* for 60 days at a concentration of 10^6 in the diet, resulted in improved length, weight, protein efficiency ratio, lipid efficiency ratio and lower FCR than the control. Similarly, carp fed *A. niger*-containing diets, at a concentration of 10^6, outperformed those with a concentration of 10^3 in all indices mentioned above [46] (Table 1). Likewise, Pacific white shrimp (*Penaeus vannamei*) fed diets supplemented with a 1.5 g/kg diet of *A. niger* had higher weight gain, length and specific growth rate than the control and those fed 3 g/kg of *A. niger* [47].

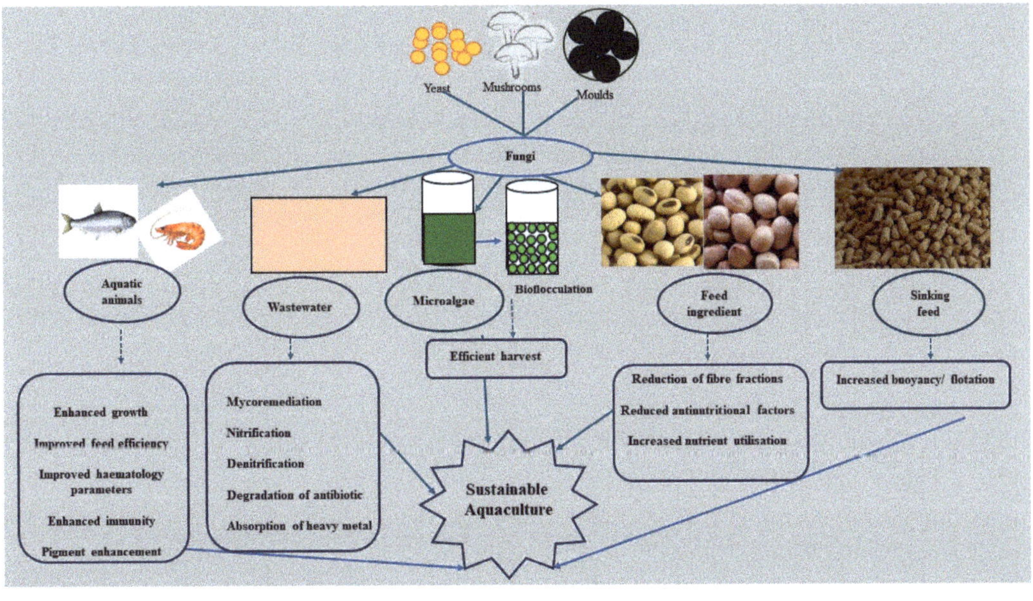

Figure 1. The potential application of fungi in aquaculture.

Fungi supplementation in diets also plays immunostimulatory roles in aquatic animals, resulting in enhanced immunity and haematology parameters, which are good health indicators [48]. *Litopeneaus vannamei* fed diets containing *Nodulisporium* sp. KT29 metabolite for 21 days exhibited a higher survival rate than the control on exposure to *V. harveyi* pathogen. Ten days after being challenged with the pathogen, shrimp-fed diets containing the metabolite had higher phenoloxidase and respiratory burst activity than the control. In contrast, shrimp in the control group had higher haemolymph glucose levels compared with those fed diets containing metabolites. This indicates that shrimp fed fungi-supplemented diets had a higher immunity compared with those fed the unsupplemented diets [44] (Table 1).

Fish fed fungi diets reportedly had a higher fungi cell load in their gut than those without fungi. However, this was influenced by the concentration of fungi in the feed, such that the fungi load increases with the increase in fungi content of the feed [46]. In common carp, *C. carpio* fed diets containing 10^6 and 10^3 *A. niger*, immunological parameters (plasma lysozyme activity and total immunoglobulins levels) and haematology indices (red blood cells, haemoglobin count, mean corpuscular haemoglobin, white blood cell count, mean corpuscular volume and lymphocyte) increased after 60 days compared with the control. However, fish fed diets containing 10^6 *A. niger* had higher white blood cell counts and haematocrit compared with the control and fish fed 10^3 *A. niger* diet [46]. Nitroblue tetrazolium, antiprotease and lysozyme activity of *Labeo rohita* was also shown to increase with an increase in spent mushroom substrate (*Cordyceps militaris*) in diets [49]. Cheng et al. used four experimental diets containing 0, 0.5, 1 and 1.5 g/kg of *A. niger* in diets fed to white shrimp, *P. vannamei* for 56 days [50]. Fish fed 1 and 1.5 g/kg of *A. niger* had higher resistance to *V. parahaemolyticus* infection, significantly lower mortality rates to *V. parahaemolyticus* and higher lysozyme activities than the control. On assessing the vibrio-like bacteria counts in the gut of the shrimp, shrimp fed *A. niger*-containing diets had lower vibrio-like counts than the control. Additionally, the phagocytic and phenoloxidase activity in the haemocyte of fish fed the 1 and 1.5 g/kg of *A. niger* were higher than those of the 0.5 g/kg and control groups. Zhang et al. used 0, 1.5 and 3 g of *A. niger* per kg of feed on Pacific white shrimp (*P. vannamei*) for a month [47]. Catalase, acid phosphatase, alkaline phosphatase, superoxide dismutase (SOD) and phenol oxidase were higher in fish fed fungi diets than those of the control, indicating a higher immunity (Table 1). Sarlin and Philip investigated marine yeasts *Debaryomyces hansenii* (S8) and *Candida tropicalis* (S186) as immunostimulants for *Fenneropenaeus indicus*, comparing their efficacy with that of *Saccharomyces cerevisiae* S36 [51]. After feeding the shrimps for two weeks, with diets containing these fungi, they were orally challenged with white spot syndrome virus (WSSV), a highly pathogenic virus causing 100% mortality in aquatic animals within 3–7 days. Shrimp fed the various diets showed a similar reduction in haemocyte count within two days after being challenged with the virus. However, shrimp fed with the two marine fungi had a higher survival rate (50%), phenoloxidase and nitroblue tetrazolium activity than those fed *S. cerevisiae* and the control. The marine yeast performed better than *S. cerevisiae*, probably because they had a higher content of β-1,3 glucan [51].

Table 1. The application of fungi in aquatic animals and their effect on growth, haematological parameters and disease resistance.

Fungi	Fungi Species	Species	Concentration and Duration	Effect on Growth and Survival	Haematological Parameters	Disease Resistance	Reference
Mushroom (powder)	*Pleurotus eryngii*	Koi carp fingerlings (*Cyprinus carpio* koi)	0, 0.5, 1, 1.5 and 2% for 61 days	FW, WG, SGR, (↑) (1.5%). FCR (↓) (0.5, 1, 1.5 and 2%) SR (↑) (0, 0.5, 1, 1.5 and 2%)	WBC, Hb, Ht, MCV, MCH, monocyte, MCHC (↑) (1.5 and 2%) Digestive enzymes (Trypsin and lipase) (↑) (2%) α-amylase (↑) (1, 1.5, 2%)	NA	[52]
	Ganoderma lucidum	Nile Tilapia (*Oreochromis niloticus*)	0, 0.5, 1, and 2% for 90 days	FW, WG (%), SGR (↑) (1%) FCR (↓) (10%) SR (↔) (0, 0.5, 1, and 2%)	NA	NA	[53]
Mushroom (fermented by product)	*Pleurotus ostreatus*	Amur catfish (*Silurus asotus*)	0, 0.1, 0.2, 0.4, 0.8% for 56 days.	WG, SGR (%) (↑) (0.1 and 0.2%) FE, PER (↑) (0.2)	Haematocrit (↑) (0.2%) Lysozyme activity (↑) (0.1%) PCV (↑) (0.2, 0.4 and 0.8%)	NA	[45]
Spent mushroom substrate	*Cordyceps militaris*	*Labeo rohita*	0, 1, 2, and 3% for 60 days.	NA	NBT, antiprotease and lysozyme activity (↑) (1, 2 and 3%)	SR (↑) after challenged with *A. hydrophila* (1, 2, and 3%).	[49]

Table 1. *Cont.*

Fungi	Fungi Species	Species	Concentration and Duration	Effect on Growth and Survival	Haematological Parameters	Disease Resistance	Reference
Mould	*Aspergillus niger*	*Penaeus vannamei*	0, 0.15 and 0.3% for 28 days	WG (%), SGR (%) and length gain (%) (↑) (1.5%)	ACP and AKP (↑) (1.5%) SOD and catalase activity (↑) (1.5 and 3%) Lysozyme activity (↓) (1 and 3%)	NA	[47]
	Aspergillus niger	Common Carp (*Cyprinus carpio*)	0, 1×10^3 and 1×10^6 for 60 days	FW, WG, FL, PER, LER (↑) (10^3 and 10^6). FCR (↓) (10^3 and 10^6).	Lysozyme activity, Ig, RBC, Hb (↑) (10^3 and 10^6) Body protein and dry matter composition (↑) (10^3 and 10^6). Body lipid composition (↓) (10^3 and 10^6). Amylase and lipase digestive enzymes (↑) (10^3 and 10^6).	N/A	[46]
	Aspergillus niger	White shrimp (*Penaeus vannamei*)	0, 0.5, 1 and 1.5 g (kg^{-1} diet) for 56 days	FW and WG (↑) (0.5, 1 and 1.5 g kg^{-1} diet) FCR (↔) (0.5, 1 and 1.5 g kg^{-1} diet)	Lysozyme, phygocytic and phenoloxidase activity (↑) 0.5, 1 and 1.5 g kg^{-1} diet) Reduced vibrio-like count in the gut of the fish	Mortality (↓) (0.5, 1 and 1.5 g kg^{-1} diet)	[50]
Yeast	*Saccharomyces cerevisiae*	Nile Tilapia *Oreochromis niloticus*	0, 1, 2 and 4 g kg^{-1} for 60 days.	SGR, PER (↑) (4 g kg^{-1})	gut villus wall thickness, villus length, width, and area (↑) (4 g kg^{-1})	N/A	[54]
	Saccharomyces cerevisiae	Nile Tilapia *Oreochromis niloticus*	0, 3, 5 and 7% for 84 days	WG (↔) PER (↑) (3, 5 and 7%) FCR (↓) (3, 5 and 7%)		Mortality (↓) (7%)	[55]
	Saccharomyces cerevisiae	Snakehead *Channa punctatus*	0 and 2.5 g kg^{-1} for 56 days.	WG, PER, FCR, (↔) SGR (↑)	Intestinal protease, lipase, and amylase enzyme activities (↑)	Mortality (↓)	[56]

Note: (↑)—significant increase, (↓)—significant decrease, (↔)—no significant difference. WBC—white blood cells, Hb—haemoglobin, Ht—haematocrit, MCV(fl)—mean corpuscular haemoglobin volume, MCH—mean corpuscular haemoglobin, MCHC—mean corpuscular haemoglobin concentration (MCHC), SR—survival rate, FW—final weight; WG—weight gain; SGR—specific weight gain, PER—protein efficiency ratio, LER—lipid efficiency ratio, NBT—nitroblue tetrazolium, ACP—acid phosphatase activity, AKP—alkaline phosphatase activity, Ig—immunoglobins. Variation in treatments compared with the control.

2.2. Reduction of Anti-Nutritional Factors and Fibre Fractions in Plant Ingredients

The use of alternative protein sources, such as protein from terrestrial plants for aquafeed production, is encouraged to reduce reliance on fish meal and increase the sustainability of aquaculture [2]. However, the antinutrients naturally present in terrestrial plant proteins are a limiting factor to their use. Antinutrients are substances which directly or indirectly (through metabolic substances produced in living organisms) interfere with feed utilization and negatively impact animal health and productivity [57]. Some antinutrients associated with terrestrial plant proteins include tannin, saponins, lectins, gossypol, glucosinolates, protease and amylase inhibitors. Tannin has a bitter taste that limits acceptability, palatability, feed ingestion and growth in aquatic animals [58]. Glucosinolates cause thyroid dysfunction, which may affect metabolism and growth, while saponins inhibit active nutrient transport, cause damage to the respiratory epithelium of the gills and inhibit growth.

The application of fungi in terrestrial plants reduces the antinutritional factors in terrestrial plants, making them more suitable for utilization in the production of aquafeeds. Fungi are mostly incorporated as a fermentation agent. The fermentation of groundnut oil cake with *A. niger* led to the reduction of saponin, phytic acid and tannin by 49%, 37.57% and 83.60%, respectively [59]. Similarly, the fermentation of rapeseed with *A. niger* led to a 76.89% reduction in glucosinolates [60]. Natural fermentation without fungi can reduce some anti-nutritional factors to a degree, but not as efficiently as fungi. For example, the natural fermentation of soya bean meal reduced trypsin content in SBM by 63.85%. However, fermentation with yeast (*S. cerevisiae*) and *A. niger* reduced the trypsin content in

SBM by 92.47 and 94%, respectively. Additionally, trypsin inhibitor was totally removed by fermenting guar meal with *A. niger* and *S. cerevisiae* [59].

The application of plant ingredients in fish diets increases the fibre content of the diets. A high fibre fraction in feed reduces feed digestibility as most aquatic species cannot utilize fibre. Fungi are beneficial in reducing the fibrous content of plant-based diets, thus increasing digestibility. This is achieved by the release of fibre hydrolytic enzymes such as xylanase, pectinase and cellulase when applied in solid-state fermentation [60–63]. The reduction of fibre content during fermentation could also be due to microbial digestion, breaking down components like cellulose and hemicellulose, which are undigestible by fish, into simple sugars [64]. Fermentation of rape seed cake with *A. niger* reduced the neutral detergent fibre by 9.17% [60] (Table 2). Fermentation of soyabean meal, rapeseed meal, groundnut oil cake, sunflower oil cake, and guar meal with *A. niger* and yeast (*S. cerevisiae*) resulted in reduced fibre fractions (neutral detergent fibre (NDF), acid detergent fibre (ADF), cellulose, hemicellulose and lignin) compared with natural fermentation and bacteria fermentation with *Lactobacillus acidophilus* (Table 2). Fermentation of cassava peel with *A. niger* significantly reduces the fibre content compared with those fermented naturally. The digestibility of cassava peel increased from 11.74% to 48.73% after fermentation [65]. *Oreochromis niloticus* (Tilapia) fed diets containing unfermented cassava peel and fermented cassava peel showed similar feed consumption and feed efficiency, demonstrating that *A. niger* in diets was well tolerated and accepted. Additionally, *O. niloticus* fed diets containing fermented cassava peel had higher dry matter, protein and energy digestibility and daily growth rate than fish fed diets containing unfermented cassava. However, fat digestibility was higher in fish fed the unfermented cassava peel than those fed the fermented cassava peel [65]. *Trametes coccinea* (i.e., *Pycnoporus coccineus*, scarlet bracket-fungus mushroom), *Lentinula edodes* (shiitake mushroom), or *Pleurotus sajor-caju* (Indian Oyster mushroom) were used to ferment rapeseed meal (RM), soya bean meal (SBM), algal meal and macrophyte meal. Fermentation with these fungi reduced the NDF of all feed ingredients except for the macrophyte meal. The digestibility of phosphorous, manganese, zinc and copper significantly increased in rapeseed fermented with *T. coccinea* or *L. edodes* [17].

The reduction of fibre via fungi fermentation is influenced by the fungi dosage and the duration of fermentation. The reduction of crude fibre and its fraction contents (hemicellulose, cellulose, ADF, NDF) was higher when rice bran was fermented with 2 g of *A. niger* per 100 g, compared with 0.5, 1, 1.5 g and the control (unfermented rice bran). However, with rice bran fermented with 2 g of *A. niger*, the reduction in crude fibre increased with an increase in fermentation time; those with a 48-hr fermentation period had the lowest crude fibre content [65] (Table 2).

Table 2. The application of fungi in the reduction of fibre fractions in fish feed ingredient.

Fibre Fractions	Feed Ingredient	Fungi Species	Raw Ingredient	Fermented	Variation (%)	References
Crude fibre	Rice bran	*Aspergillus niger*	10.66	4.60	56.85	[65]
	Tea dregs	*Trichoderma koningii* *Aspergillus niger*	18.2 18.2	22.3 19.7	−22.53 −8.24	[66]
	Palm kernel cake	*Aspergillus niger*	16.8	14.5	13.69	[16]
	Cotton seed meal	*Aspergillus niger*	30.67	30.57	0.33	[67]
	Sunflower meal	*Aspergillus niger*	20.43	23.05	−12.83	[67]
	Hazelnut meal	*Aspergillus niger*	13.78	8.60	37.59	[67]
	Wheat Bran	*Aspergillus niger*	8.97	5.20%	42.02	[68]
	Brewer's spent grain	*Aspergillus niger*	13.31%	9.20	30.88	[68]
	Palmkernel cake	*Aspergillus niger*	15.59	14.91%	4.36	[68]
	Wheat bran	*Aspergillus niger* ATCC 200345	10.95	8.97%	18.08	[69]

Table 2. *Cont.*

Fibre Fractions	Feed Ingredient	Fungi Species	Raw Ingredient	Fermented	Variation (%)	References
Crude fibre	Tomato pomace	*Aspergillus niger*	21.71	23.0	−5.94	[70]
	Rice bran	*Aspergillus flavus*	12.05	9.35	22.41	[71]
	Rapeseed meal	*Rhizopus microsporus*	10.8	11.9	−10.19	[72]
ADF	Wheat bran	*Aspergillus niger*	13.46	11.87	11.81	[69]
	Canola meal	*Aspergillus pullulans* (58522) *Aspergillus pullulans* (42023) *Aspergillus pullulans* (Y-2311-1) *Pichia kudriavzevii* *Tricoderma reesei* *Fusarium venenatum* *Mucor circinelloides*	19.9 19.9 19.9 19.9 19.9 19.9 19.9	17.7 18.0 22.4 18.7 19.1 22.5 20.1	11.06 9.56 −12.56 6.03 4.02 13.07 −1	[73]
	Rapeseed meal	*Aspergillus niger*	33.70	30.76	8.72	[15]
	Tomato pomace	*Aspergillus niger*	25.22	26.84	−6.42	[70]
	Wheat bran	*Aspergillus niger* ATCC 52172	13.46	11.87	11.81	[69]
	Hazelnut meal	*Aspergillus niger* ATCC 52172	22.44	19.36	13.73	[67]
	Sunflower meal	*Aspergillus niger* ATCC 52172	23.82	26.15	−9.78	[67]
	Cottonseed meal	*Aspergillus niger* ATCC 52172	39.70	43.13	−8.64	[67]
	Palm kernel cake	*Aspergillus niger*	46.8	35.8	23.50	[16]
	Tea dregs	*Aspergillus niger*	24.2	26.8	−10.74	[66]
	Tea dregs	*Trichoderma koningii*	24.2	29.5	−21.9	[66]
	Rice bran	*Aspergillus niger*	17.28	12.43	28.07	[65]
	Rapeseed meal	*Rhizopus microsporus*	10.8	11.9	−10.1	[72]
NDF	Rapeseed	*Aspergillus niger*	47.08	40.74	13.47	[15]
	Rice bran	*Aspergillus flavus*	12.05	9.35	22.41	[71]
	Tomato pomace	*Aspergillus niger*	21.71	23.00	−5.94	[70]
	Wheat bran	*Aspergillus niger* ATCC 52172	10.95	9.79	10.59	[69]

ADF—acid detergent fibre; NDF-neutral detergent fibre.

2.3. Increased Nutrient Availability in Feed

Fermentation of aquaculture feed ingredients/feed with fungi increases nutrient content such as protein and amino acids, making the feed/feed ingredient more suitable for aquatic animals [20]. Fermentation of groundnut oil cake (GNC) using *A. niger* led to about a 21% increase in protein content [59]. Fermentation with *A. niger* also enhanced the protein and ether extract content of rapeseed [15]. *Trametes coccinea, L. edodes*, or *P. sajor-cajuin* increased the crude protein content of rapeseed meal, soya bean meal (SBM), algal meal and macrophyte meal upon fermentation [17]. Groundnut oil cake fermented with *A. niger* also had increased essential amino acid content. The methionine and lysine content of fermented GNC had a 50% increase compared with the unfermented GNC. Similarly, histidine, tryptophan and threonine increased by about 20% upon fermentation with *A. niger* [20,59]. Hariyono et al. have also reported that fermentation of pellet with *Rhizopus* sp. led to an increase in the amino acid content of the pellet [20]. Fermentation of aquafeed pellets with *Rhizopus* sp. inoculum at a dosage of 1%, 2% and 3% for 30, 40 and 50 h reduced the fat content of the pellet. However, the reduction of fat content was affected by the dosage of inoculum and fermentation duration such that pellets fermented with 3% inoculum for 50 h had the lowest fat content [20]. The fat reduction in pellets by *Rhizopus* sp. has been attributed to the production of lipase enzymes by *Rhizopus* sp. during fermentation. The lipase enzyme hydrolyses triacylglycerol into free fatty acids. The fungi use fatty acids as energy for growth during fermentation, leading to reduced lipid content [20].

Fermentation with fungi seems to be more efficient in enhancing the nutrient composition of feed ingredients than fermentation with bacteria. Fermentation of sargassum powder with *A. niger*, *S. cerevisiae* and *Lactobacillus* spp. increased the protein content of sargassum powder and reduced lipid and carbohydrate content, unlike the control (unfermented sargassum powder). However, fermentation with *A. niger* led to a significantly higher protein (from 7.5 to 10.1%) and reduced carbohydrate content (50.7% to 29.1%) than those of *S. cerevisiae* and *Lactobacillus* spp. While fermentation with *S. cerevisiae* reduced lipid content (1.33% to 0.43%) significantly compared with *A. niger* and *Lactobacillus* spp. [74].

Phytase enzymes obtained from fungi can also be used to enhance the availability of some minerals [75,76] and have been employed to improve dietary mineral retention in aquatic animals [77,78]. Channel catfish fed phytase-containing diets had a significantly higher concentration of manganese, ash, calcium and phosphorous deposited in the bones than fish without phytase supplementation. The phytase concentration also influenced the deposition of these minerals in the bone of the fish. For example, the magnesium content of the bone was similar for fish fed 500 U/kg and for the control. However, fish fed 1000 U/kg and above of phytase had significantly higher bone magnesium than the control [79]. This implies a higher nutrient availability in fish fed diets containing a higher phytase concentration.

2.4. Increased Mineral Utilization and Reduced Nutrients in Effluents

Phosphorous is one of the essential minerals required for growth in aquatic animals [80]. Phosphorous is naturally found in feed ingredients of plant origin. It is present as phytic acid (myo-inositol 1,2,3,4,5,6-hexakisphosphate or phytate) and makes up about 80% of the total phosphorous in plant ingredients [81]. Aquatic animals, especially fish, cannot utilize the phytic acid in these plant ingredients. This is because most aquatic animals lack or have low amounts of phytase enzyme, a requisite for the utilization of phosphorous bound to phytic acid [82,83]. Phosphorous deficiency and suboptimal levels of dietary phosphorous in diets of aquatic animals result in reduced growth performance, operculum and vertebral deformities, poor bone mineralisation, and reduced immunity [84,85]. This necessitates the supplementation of phosphorous during feed formulation. Inorganic phosphorous, such as dicalcium phosphate or monocalcium phosphate, are usually added in aquafeed formulation. With the use of phosphorous supplements in various sectors, including the agricultural sector, it has been predicted that, by 2050, the world reserve will be depleted [86,87]. Additionally, excess use of phosphorous could be detrimental to the environment, causing pollution of water bodies and eutrophication.

Phytase enzymes can replace dicalcium phosphate without any detrimental effect on aquatic animals. Applying phytase enzymes can also reduce the adverse effects attributed to phytic acid, such as limitation of availability of trace minerals, like zinc; water quality deterioration; and environmental pollution [88,89]. The water quality deterioration and environmental pollution arise via the decomposition of uneaten feed and faeces rich in phytic acid. Phytic acid is released into the rearing system and converted to phosphorous after being acted upon by microorganisms. Therefore, phosphorous is released into the rearing system, deteriorating water quality [10]. Phosphorous-rich wastewater released into the environment adversely affects the environment. Phytase enzymes can be produced from plants, bacteria and fungi [90]. The limitation in the production of phytase enzymes from plants is that the synthesis methods are inefficient, time-consuming and expensive [90]. Fungal phytase, one of the metabolites produced by fungi, is more desirable than bacterial phytase because of its thermostability and more significant chitin effects. Bacterial phytase is limited by protease in the fish gut, high substrate specificity and acidic to alkaline pH activity [91].

Fish that were fed diets in which dicalcium phosphate was replaced with phytase enzyme from fungi had similar weight gain, FCR, and protein efficiency ratio to fish fed diets containing dicalcium phosphate [92,93]. Interestingly, adding phytase enzyme from fungi at 10 g/kg of feed led to higher phosphorous retention compared with fish fed diets containing dicalcium phosphate (15 g/kg) [92]. An advantage of using phytase of fungal origin in aquaculture is that it reduces the phosphorous load in the water and thus the environment, when compared with inorganic phosphorous, which could leach into the water column from feed. Phytase enzyme from fungi reduces the phytate content in feed ingredients [93]. The inclusion of phytase from *A. niger* in a diet degraded the phytate content of that diet during feed production. The efficiency of fungal phytase in reducing phytate is also dose dependent. Yan et al. (2002), have reported that a dose of 2000 U/kg and above in diets reduces phytate content when compared with a diet containing 500 U/kg of phytase. Similarly, Yan et al., (2002) formulated a 96% plant protein diet for channel catfish prepared with the inclusion of phytase from *A. niger* at varying concentrations in diets (0, 500, 1000, 2000, 4000 and 8000 U/kg diet). The analysis of phytate concentration in the stomach showed a lower concentration of phytate as early as 4 h after feeding for fish fed diets of 100 U/kg of phytase in their diet and at 12 h after feeding with 500 U/kg of a phytase-containing diet. Fungi phytase should be applied by spraying on formulated diets after extrusion, as temperature affects its efficacy and may not withstand the temperature of feed production and extrusion [93].

2.5. *Fungi as a Buoyancy/Flotation Agent in Aquafeed*

Artificial feeds used in aquaculture are categorised into floating (extruded) and sinking feed (non-extruded). These feeds are given to animals depending on their feeding habit in the water column [94]. Floating feeds are appropriate for animals that feed on the water's surface while sinking feeds are suitable for animals that feed at the bottom [94]. Feed floats in water only when its density is lower than that of water [20]. Floating feeds are made by extrusion involving the application of heat. Thus, floating feeds are hydrophobic, making them highly stable in water [95]. They are, however, expensive, as a high investment cost is required in their production. Though sinking feeds are cheaper compared with their floating counterparts, they easily reach the bottom of the culture system [94]. On reaching the bottom of the culture system, they disintegrate, impacting the water quality negatively [96,97]. A high percentage of the sinking feeds introduced into the culture system are not utilized by fish, which is uneconomical. Generally, small-scale farmers in developing/countries may not be able to afford extruded feed; they end up buying sinking feeds or producing sinking feed with locally available ingredients [12]. They feed sinking feeds to their animals regardless of their feeding habits. Therefore, they encounter challenges such as water quality deterioration and reduced efficiencies from feed wastage [98].

Fungi have the ability to enhance the buoyancy/flotation of fish feed without the need for physical extrusion [20,21]. Flotation of feed via the use of fungi can be achieved when applied in solid-phase fermentation. The flotation of fungi fermented feed is imputed to the surface fungi mycelium (network of hyphae, which is cotton-like and whitish), preventing penetration of water into the poriferous spaces on the pellet. *Rhizopus* sp. and yeast are used to increase the flotation of feed [99]. A commercial sinking feed has been fermented using *Rhizopus* sp. [21]. The buoyancy and stability of the final product from fermentation were compared with those of commercial floating and sinking feed from the same producers. About 80% of the fermented feed remained afloat after immersion in water for 15 min. The fermented feed had a higher stability and buoyancy compared with the sinking feed counterpart. However, the commercial extruded feed had higher stability and buoyancy in water compared with *Rhizopus* sp. fermented feed [21]. In a study by [20], the experimental feed was fermented with three doses of *Rhizopus* sp. inoculum (1%, 2% and 3%) and three durations of fermentation (30, 40 and 50 h). All fermented pellets had enhanced buoyancy of feed compared with the control, which sank within 5 min of immersion in water. Feed fermented with 3% inoculum for 50 h had the highest buoyancy, with about 83% and 67% of the feed remaining afloat after an hour of immersion in non-aerated and aerated water, respectively. *Rhizopus* sp. inoculum fermentation also improved feed stability, with fermented feed resulting in higher stability than the control. Feed fermented with 3% inoculum for 50 h had 92.38% stability in unaerated water and 89.55% in aerated water when placed in water for 60 min [20]. Yeast (*S. cerevisiae*) has also been used as a flotation agent in aquafeed formulation [95,100]. [101] formulated experimental diets with four concentrations of yeast (5, 7.5, 10 and 12%) and four incubation periods (30, 40, 45 and 60 min) at a temperature of 38° C. All diets showed buoyancy when they were tested for it. However, those formulated with 7.5% yeast and incubated for 30 min had the best buoyancy, with 95% of the feed remaining afloat in water for 30 min.

Beyond buoyancy, the stability of the feed is significant in aquaculture. Water is known to disintegrate aquatic feed, therefore, during their formulation (either floating or sinking), binders are added to bind feed ingredients so as to enhance their stability [102,103]. The capacity of fungi to grow on a substrate of organic origin and also bind the substrate particles makes them suitable for use as biological binders in the manufacturing of aquafeed [99]. The binding effect achieved by fungi in aquafeed is attributed to the fungi hyphae (thread-like filament) penetrating thoroughly into the pellets, filling up spaces in the pellet and interlocking the feed ingredients. The various studies available on yeast have shown it to be a good flotation agent and not a binder. Thus, making *Rhizopus* sp. more beneficial as both a flotation and binding agent. Fermented feed is an alternative to sinking feed and can be applied in small-scale farming where extruded feed is not available or where the farmers produce their feed using locally available ingredients.

2.6. Pigment

The colour and outward appearance of cultured animals play important roles in the attractiveness, value, demand and acceptance of a product [104,105] (Figure 1). Aquatic animals do not have the ability to synthesize their colour pigment. Therefore, the colour pigment has to be provided in the diet to impart a suitable colour. In the case of ornamental fishes, they generally lose their colouration due to the stress associated with captivity. They also lose their colour when fed diets lacking pigment [105]. Colour-enhancing feed additives are available in the form of carotenoids and pigments. Sources of carotenoids and pigments for fish may be synthetic or natural. Natural sources of pigments include vegetables and fruits such as red beetroot, carrot, okra, cabbage, and tomatoes. Synthetic pigments are characterized by high cost, poor uptake levels, and health and environmental concerns. These limitations of synthetic pigments have led to a focus on non-toxic pigments from natural sources [106]. However, pigments from natural sources, such as plants, are constrained by their limited availabilities, irregular harvest, land use, labour intensiveness and seasonal availability [107,108].

Plant pigments are preferred for use in the food industry, but, due to their limitations, research interest has shifted towards filamentous fungi with a history as a food source [109,110]. Astaxanthin, the primary carotenoid found in aquatic animals and which are orange-reddish, are usually added to the feed of salmon, lobster shrimps and trout for colour enhancement [111–113]. Astaxanthin is produced commercially from crustaceans, microalgae, bacteria and chemical synthesis [114,115]. Owing to its limited sources and reduced output, astaxanthin derived from natural sources is expensive [116]. Therefore, those derived from chemical synthesis are mainly used as a colour enhancer in aquaculture. Patronizing astaxanthin from synthetic sources in aquaculture encourages its production, which could be a source of pollution to the environment. Additionally, the use of synthetic astaxanthin in aquaculture may pose health risks to consumers [117].

Filamentous fungi (especially basidiomycetous [mushrooms], ascomycetous fungi and lichen) and yeast produce a diverse range of pigments, including flavins, phenazines, melanins, quinines and azaphilones [116,118,119]. The pigments obtained from fungi are secreted as secondary metabolites. They are produced as a reaction to unfavourable environmental conditions, such as high/low temperatures and excessive/prolonged exposure to light and stress, such as exposure to limited/insufficient availability of nutrients [120,121]. An advantage of pigment produced by fungi over pigments from other natural sources is that it does not make use of agricultural land meant for food production [108,110].

Neurospora spp. are among the fungi generally recognized as safe (GRAS), as they are not known to produce mycotoxins [122]. An advantage of *Neurospora* spp. as a pigment-producing fungi, is that it can be grown on a variety of substrates, including industrial residuals and lignocellulose [122]. Yeast species, such as *Kluyveromyces marxianus*, *Xanthophyllomyces dendrorhous* (formerly *Phaffoa rhodozyma*) and *Rhodotorula mucilaginosa*, also produce pigments [123]. Lin et al. incorporated astaxanthin from yeast (*Kluyveromyces marxianus*) and synthetic astaxanthin, Carophyll Pink and Lucanthin Pink, into the diet of white shrimp (*Penaeus vannamei*) [124]. Within four weeks of diet application, shrimp fed astaxanthin from yeast had increased redness values. Wang et al. also reported that the dietary inclusion of a red marine yeast *Rhodotorula mucilaginosa* JM-01 for six weeks resulted in increased pigmentation (redness and yellowness) of the body and shell astaxanthin content of the shrimp, *Penaeus vannamei* [125]. Marine yeast *Rhodotorula paludigena* VA 242, when incorporated into feed, improves the colouration of *Cyprino carpio* within 60 days of use [106]. Mushrooms mixed with cladoceran diets (*Pleurotus ostreatus* and *Pleurotus djamor*) reportedly improve the colouration of Siamese fighting fish (*Betta splendens*) when included in diets and when compared with the control, which had cladoceran powder alone [126]. Though fungi pigments are effective at enhancing the colouration of aquatic animals, a precaution in their use, especially with moulds, is to prevent the introduction of toxins/mycotoxins from the fungi. Some fungi-derived pigments are already commercialized, for example, canthaxanthin and ankaflavin pigments produced from the filamentous fungi *Monascus* sp. and Arpink redTM pigment produced from *Penicillium oxalicum* var. armeniaca [127].

2.7. Bioremediation

Water quality is of major importance in aquaculture, as aquatic animals depend on the water for life processes, which include survival, feeding, excretion and metabolic activities [2]. Excess nutrients in water emanating from unconsumed feed and excretory waste, including faeces, deteriorate water quality, leading to chemical and biological oxygen demand, which is detrimental to aquatic lives [96,97]. The presence of heavy metals and bacteria proliferation in water also pose a threat to aquatic animals. Aquaculture effluents are nutrient-dense, containing nutrients such as organic matter, phosphorus and carbon. The discharge of nutrient-dense effluent without treatment into the aquatic ecosystem is detrimental to the environment and lives within the environment [128]. This is one of the reasons some aquaculture practices are being frowned upon, as the discharge of untreated effluent could result in eutrophication, leading to excessive algal bloom and destruction of aquatic biodiversity. Various bioremediation options, such as integrated multitrophic aquaculture (IMTA) and aquaponics, have been used to reduce the nutrient load of aquaculture effluent [2,11]. Bioremediation using bacteria is well known; however, mycoremediation, which is the use of fungi in bioremediation is also efficient. The popularity of fungi is increasing due to their potential for application in bioremediation. Fungi species are able to decompose/degrade or modify organic/toxic substances into a less toxic form [129]. They thrive even in environments where bacteria bioremediation is inhibited, such as low pH. The mechanism behind mycoremediation is that fungal biomass/fungal hyphae, in a bid to meet its nutrition and energy requirements, produce bioflocs or beads which capture, absorb and degrade particles in suspension [130]. Fungi have been used for nitrification and denitrification and are known to exhibit a higher capacity of denitrification than bacteria [131]. Fungi have been explored for bioremediation in various industries. They have been used to reduce antibiotics, such as oxytetracycline and ciprofloxacin, in wastewater [132,133]. For example, *Pleurotus ostreatus* was used to degrade oxytetracycline-contaminated soil and water [134], also *Trametes versiocolor* achieved 91% removal of Ciprofloxacin in in vitro experiments [135]. Fungi have been used to degrade numerous medications, such as antidepressants [132], anti-inflammatory drugs [136] and anticancer medications [137], in the pharmaceutical industry. *Trametes versicolor* has also been efficiently used to treat hospital effluents [138].

Fungi can also absorb heavy metals present in wastewater; they are able to do this through their cell wall or extracellular polysaccharide slime [19]. Mushrooms, especially the basidiomycetes, are good decomposers, and they secrete extracellular enzymes such as manganese peroxidase, laccase and lignin peroxidases. Polysaccharides derived from mushrooms are known to reduce the toxicity of heavy metals. Jia et al. used polysaccharides from *Ganoderma lucidum* to supplement the diet of common carp (*Caprinus carpio*) reared in cadmium-polluted water [14]. The treatments used were represented by P group (carp reared without polysaccharide supplementation in feed + 0.5 mg/L Cd in water), LG group (carp fed diet containing 2 g/kg polysaccharide + 0.5 mg/L Cd in water) and HG group (carp fed diet containing 2 g/kg polysaccharide + 0.5 mg/L Cd in water), and control (carp reared in unpolluted water and without polysaccharide supplementation in feed) for four weeks. Carp fed *G. lucidum*-supplemented diets all had reduced Cd content in the muscle, blood, liver and brain, the primary target organs in which Cd accumulates. Generally, the muscle tissues of fish exposed to Cd usually have an increased level of total cholesterol, lipids, triglycerides and fatty acids. However, adding *G. lucidum* polysaccharide to the diet of carp led to reduced levels of total cholesterol, lipids, triglycerides and fatty acids in the muscle. The inclusion of *G. lucidum* polysaccharides in diets of fish undid the effect of Cd toxicity such that the FCR of fish fed both high and low doses of *G. lucidum* in their diet and exposed to Cd pollution (LG and HG groups) was found to be significantly lower than those of P groups but was found to be similar to the control groups. Chandra et al. assessed the effectiveness of mycoremediation using mushroom spent bags on fish tanks containing *Ctenopharyngodon idella*, *Labeo rohita* and *Cyprinus carpio haematopterus* fingerlings using various filtration rates [139]. They have reported that mycoremediation using a mushroom spent bag was as efficient as the RAS system.

2.8. Bioflocculation

Microalgae are single-celled microorganisms found in aquatic environments. They serve as feed/feed supplements to aquatic animals, thereby minimising the production cost and enhancing aquaculture advancement [140,141]. They are a food and medicine source for humans and serve as bio-fertilizers. They are also valuable for bioremediation of wastewater and biofuel production [35]. Microalgae grow naturally in rivers, lakes and oceans; however, the biomass produced is usually insufficient for large-scale harvesting [142]. Therefore, microalgae are cultivated to meet the required biomass for its intended use.

One of the bottlenecks associated with microalgae farming is its harvesting. Most microalgae are difficult to harvest due to their relatively small size (2–40 μm) and because they are highly stable in water. Additionally, harvesting microalgae is costly, representing about 20–30% of the total cost of production [143,144]. Various harvesting techniques are used, including sedimentation, coagulation, filtration, flocculation, flotation and centrifugation [145]. Though some of these harvesting methods are helpful, some are time-consuming, toxic to the environment, expensive and require a high energy consumption/utilization. Two or more of these harvesting techniques are sometimes employed to increase the harvesting output. Centrifugation and filtration are widely used for commercial-scale harvesting; however, they require high energy utilization. Coagulation–flocculation is applied in wastewater treatment to remove particles and organic matter because of its easy accessibility and economic technique [146]. The flocculation technique requires the use of flocculants, which could be chemical or natural. Chemical flocculants are frequently employed in sewage treatment, with examples including polyethyleneimine, polyacrylamide, aluminium sulphate and polyaluminium chloride [147]. However, chemical flocculants may not be applicable to microalgal biomass intended for food intended to feed humans and animals. This is because it poses the risk of health and environmental pollution, thus reducing the quality of the microalgal biomass. Hence, a sustainable harvesting technique employing natural flocculants is needed. Natural flocculants are also referred to as bioflocculants, with chitosan as an example. Bioflocculation is a cost-effective and efficient technique for harvesting microalgal biomass because it concentrates algal cells by taking out a substantial amount of water [148,149]. Bioflocculants do not pose the risk of pollution; hence, microalgal biomass harvested with bioflocculants is considered safe for aquatic life and human consumption.

Filamentous fungi are potential bioflocculating agents due to their ability to aggregate and trap microalgae efficiently. Filamentous fungi, such as *A. niger*, *A. oryzae*, *A. lentulus*, *A. fumigatus* and *penicillium* sp., are potential bioflocculants and are helpful in microalgal harvesting. The mechanism through which flocculation is achieved by bioflocculant could be via sweeping, bridging and charge neutralisation [150]. In the application of fungi for microalgal harvesting, fungi spores or pellets are used. Fungi spore-assisted harvesting involves co-cultivating the fungi spores and the microalgae. However, for fungi pellet-assisted harvesting, the pellet is cultured, and active pellets are added into the medium containing the microalgae.

The mycelial pellets of *A. niger* have been used to harvest cyanobacterial biomass due to their high flocculation activity [148,151]. *Aspergillus niger* 7806F3 has been used to flocculate *Synechocystis* biomass. However, the flocculation activity was influenced by the quantity of mycelial pellets and the flocculation time [148]. The higher the mycelial pellets and the flocculation time, the higher the flocculation activity [148]. Harvesting with fungi spores is usually time-consuming, requiring about 24–48 h, when compared with fungi pellets. Ref. [152] has reported that *Chlorella* sp. was harvested with fungi assistance using the spore of *Penicillium* sp. and achieved 99.26% harvesting efficiency within 28 h while a fungi pellet of *Pencillium* sp. assisted harvesting to achieve 98.26 harvesting efficiency within 2.5 h. Ref. [153] tested the efficacy of *Penicillium chrysogenum*, *A. fumigatus* and *A. niger* pellets as a bioflocculants for harvesting the marine microalgae *Nannochloropsis salina* and *Chlorella salina*. All fungi tested showed flocculation activity, but the flocculation efficiency was affected by the flocculation time such that it was higher in 48 h than in 24 h, with an efficiency of 98.9% for the *A. fumigatus* used to harvest *C. salina* (highest efficiency) and 85.9% for the *A. niger* used to harvest *N. salina* (lowest efficiency).

3. Limitations on the Use of Fungi

The application of fungi has various benefits; however, some limitations are doubtlessly associated with their use. The effectiveness of fungi in reducing fibre fractions and enhancing nutrient availability is influenced by the duration of fermentation and the dosage administered, which may differ for various feed ingredients, making it difficult for most aquaculture farmers to apply fungi [65]. Therefore, the most effective dosage and fermentation duration for each feed ingredient should be investigated. Most studies on the application of fungi in aquatic diets were conducted in vitro or on a pilot scale under controlled conditions. Therefore, the outcome/result may differ in outdoor/farming conditions.

The various studies available on *Rhizopus* sp. as a flotation agent applied the fungi to pelletised feed, unlike yeast, which is included during feed formulation. As a result, the pellets were found to stick together during fermentation, resulting in fermented pellets of uneven size and shape and which often differ from the initial shape or size. Additionally, fermenting pellets for days may be time consuming [20]. During fermentation, regular stirring of the pellets has been applied to reduce stickiness/clump. This has resulted in minimal success, as the feed still emerges with heterogenous shapes and sizes [20]. This may be avoided by incorporating *Rhizopus* sp. during the feed formulation stage, i.e., before pelleting.

Fungi (*A. niger* and *A. oryzae*) are generally regarded as safe, as they have been used for food production for a long time. However, the major limitation to the use of fungi is the presence of mycotoxins. Some mycotoxins in *A. niger* include Ochratoxin A, Oxalic acid, Fumonisin B_6, Fumonisin B_4, and Fumonisin B_2 [154–156]. Cyclopiazonic acid and β-Nitropropionic acid have been reported as mycotoxins in *A. oryzae* [157]. The effects of mycotoxins in aquatic animals include weight reduction, poor feed conversion ratio, deterioration in the immune system and high susceptibility to disease [23]. Mycotoxins are carcinogenic to humans; their effect in humans includes inhibition of protein synthesis, inability to produce energy and oxidation stress [158,159]. For example, a concentration of 1000 mg mL^{-1} of *T. asperellum* MT02 was considered best to inhibit vibriosis in vitro. However, on testing for toxicity, *T. asperellum* MT02 incorporated into the feed of Shrimp *L. vannamei* had a toxicity value (LC_{50}) at 383.70 mg mL^{-1} due to the presence of mycotoxins, meaning that this concentration led to 50% mortality for *L. vannamei*. The mortality may increase further if the concentration considered best to inhibit vibriosis is applied [40]. Genetic advances are advantageous in addressing mycotoxins in fungi and have been used in deactivating some mycotoxin genes in fungi.

4. Conclusions and Recommendations

Fungi have been taken advantage of and explored in the food, chemical and pharmaceutical industries. For aquaculture, the application of fungi is still within the experimentation stage and has barely been explored and adopted. The various studies and examples cited have shown that fungi are advantageous in many aspects of aquaculture, including growth enhancement, increased disease resistance, and increased nutrient utilization, resulting in a reduction of nutrients in aquaculture effluent. Fungi are also applied to improve feed ingredient nutrients and reduce antinutrients and fibre plant protein, making them more suitable for use in aquafeed production. Fungi increase the digestion of plant protein, feed buoyancy and thus aquafeed flotation, and can act as alternatives to synthetic pigments. Fungi promise to be of considerable benefit to the aquaculture industry if their potential is tapped, improving the sustainability and productivity of aquaculture.

Author Contributions: A.J.O. Conceptualization, literature search, and manuscript writing. G.E.O. Conceptualization, review and editing manuscript. All authors have read and agreed to the published version of the manuscript.

Funding: This research received no funding.

Conflicts of Interest: The authors declare no conflict of interest.

References

1. Boyd, C.E.; D'Abramo, L.R.; Glencross, B.D.; Huyben, D.C.; Juarez, L.M.; Lockwood, G.S.; McNevin, A.A.; Tacon, A.G.J.; Teletchea, F.; Tomasso, J.R.; et al. Achieving Sustainable Aquaculture: Historical and Current Perspectives and Future Needs and Challenges. *J. World Aquac. Soc.* **2020**, *51*, 578–633. [CrossRef]
2. Onomu, A.J.; Okuthe, G.E. The Role of Functional Feed Additives in Enhancing Aquaculture Sustainability. *Fishes* **2024**, *9*, 167. [CrossRef]
3. Ahmad, A. Aquaculture Industry: Supply and Demand, Best Practices, Effluent and Its Current Issues and Treatment Technology. *J. Environ. Manag.* **2021**, *14*, 112271. [CrossRef] [PubMed]
4. Gyan, W.R.; Ayiku, S.; Yang, Q. Effects of Replacing Fishmeal with Soybean Products in Fish and Crustaceans Performance. *J. Aquac. Res. Dev.* **2019**, *10*, 573. [CrossRef]
5. Chen, Q.; Wang, C.; Sun, Y.; Chen, Y.; Chen, S.; Han, T.; Wang, J. Excessive Substitution of Fish Meal with Fermented Soybean Meal Induces Oxidative Stress by Impairing Glutathione Metabolism in Largemouth Bass (*Micropterus salmoides*). *Antioxidants* **2023**, *12*, 2096. [CrossRef]
6. He, M.; Yu, Y.; Li, X.; Poolsawat, L.; Yang, P.; Bian, Y.; Guo, Z.; Leng, X. An Evaluation of Replacing Fish Meal with Fermented Soybean Meal in the Diets of Largemouth Bass (*Micropterus salmoides*): Growth, Nutrition Utilization and Intestinal Histology. *Aquac. Res.* **2020**, *51*, 4302–4314. [CrossRef]
7. Jannathulla, R.; Dayal, J.S.; Ambasankar, K.; Eugine, A.C.; Muralidhar, M. Fungus, *Aspergillus niger*, Fermented Groundnut Oil Cake as a Fishmeal Alternative in the Diet of *Penaeus vannamei*. *Aquac. Res.* **2018**, *49*, 2891–2902. [CrossRef]
8. Lall, S.P.; Kaushik, S.J. Nutrition and Metabolism of Minerals in Fish. *Animals* **2021**, *11*, 2711. [CrossRef]
9. Lemos, D.; Tacon, A.G.J. Use of Phytases in Fish and Shrimp Feeds: A Review. *Rev. Aquac.* **2017**, *9*, 266–282. [CrossRef]
10. Baruah, K.; Sahu, N.P.; Pal, A.K.; Debnath, D. Dietary Phytase: An Ideal Approach for a Cost Effective and Low-Polluting Aquafeed. *Naga* **2004**, *27*, 15–19.
11. Onomu, A.; Slater, M.; Vine, N. Feeding Indicators and Bioremediation Ability of Warty Sea Cucumber *Neostichopus Grammatus* Fed Potential Wastes from Abalone Haliotis Midae Farming. *Aquacult. Environ. Interact.* **2023**, *15*, 45–57. [CrossRef]

12. Cocker, L.M. *Strategic Review on African Aquaculture Feeds;* Partnership for African Fisheries (PAF) Aquaculture Working Group, NEPAD: Lusaka, Zambia, 2014; pp. 1–85.
13. Walker, G.M.; White, N.A. Introduction to Fungal Physiology. In *Fungi: Biology and Applications;* Kavanagh, K., Ed.; Wiley-Blackwell: Chichester, UK, 2017; pp. 1–35. ISBN 978-1-119-37432-9.
14. Jia, D.; Tang, Y.; Qin, F.; Liu, B.; Hu, T.; Chen, W. Ganoderma Lucidum Polysaccharide Alleviates Cd Toxicity in Common Carp (*Cyprinus carpio*): Neuropeptide, Growth Performance and Lipid Accumulation. *Comp. Biochem. Physiol. Part C Toxicol. Pharmacol.* **2023**, *271*, 109663. [CrossRef] [PubMed]
15. Shi, C.; He, J.; Yu, J.; Yu, B.; Mao, X.; Zheng, P.; Huang, Z.; Chen, D. Physicochemical Properties Analysis and Secretome of *Aspergillus niger* in Fermented Rapeseed Meal. *PLoS ONE* **2016**, *11*, e0153230. [CrossRef] [PubMed]
16. Marini, A.M.; Ayub, M.Y.; Salam, B.A.; Hadijah, H.; Azahan, E.A.E.; Tarmizi, S.A. Protein Quality of *Aspergillus niger*-Fermented Palm Kernel Cake. *J. Trop. Agric. Food Sci.* **2008**, *36*, 1–11.
17. Tanemura, N.; Akiyoshi, Y.; Okano, K.; Sugiura, S. Effects of Culturing Rapeseed Meal, Soybean Meal, Macrophyte Meal, and Algal Meal with Three Species of White-Rot Fungi on Their in Vitro and in Vivo Digestibilities Evaluated Using Rainbow Trout. *Aquaculture* **2016**, *453*, 130–134. [CrossRef]
18. Naiel, M.A.; Negm, S.S.; Ghazanfar, S.; Shukry, M.; Abdelnour, S.A. The Risk Assessment of High-fat Diet in Farmed Fish and Its Mitigation Approaches: A Review. *J. Anim. Physiol. Anim. Nutr.* **2023**, *107*, 948–969. [CrossRef]
19. Akthar, M.N.; Mohan, P.M. Bioremediation of Toxic Metal Ions from Polluted Lake Waters and Industrial Effluents by Fungal Biosorbent. *Curr. Sci.* **1995**, *69*, 1028–1030.
20. Hariyono, C.M.; Yunianta; Harijono; Sriherwanto, C.; Suja'i, I.; Nadaviana, A.; Junaedi, H.; Ma'hadah, R. Komarudin Physico-Chemical Characteristics of *Rhizopus* sp.-Fermented Fish Feed Pellets Containing Black Soldier Fly Larvae (*Hermetia illucens*) Meal. *IOP Conf. Ser. Earth Environ. Sci.* **2021**, *744*, 012024. [CrossRef]
21. Leiskayanti, Y.; Sriherwanto, C.; Suja'i, I. FERMENTASI MENGGUNAKAN RAGI TEMPE SEBAGAI CARA BIOLOGIS PENGA-PUNGAN PAKAN IKAN. *J. Bioteknol. Biosains Indones* **2017**, *4*, 1. [CrossRef]
22. Liu, B.; Li, X.; Wang, X.; Huang, B. Effects of Stocking Density on Stress Response, Innate Immune Parameters, and Welfare of Turbot (*Scophthalmus maximus*). *Aquac. Int.* **2019**, *27*, 1599–1612. [CrossRef]
23. Oliveira, M.; Vasconcelos, V. Occurrence of Mycotoxins in Fish Feed and Its Effects: A Review. *Toxins* **2020**, *12*, 160. [CrossRef] [PubMed]
24. Blackwell, M. The Fungi: 1, 2, 3 … 5.1 Million Species? *Am. J. Bot.* **2011**, *98*, 426–438. [CrossRef] [PubMed]
25. Dijksterhuis, J. Fungal Spores: Highly Variable and Stress-Resistant Vehicles for Distribution and Spoilage. *Food Microbiol.* **2019**, *81*, 2–11. [CrossRef]
26. De Lucca, A.J. Harmful Fungi in Both Agriculture and Medicine. *Rev. Iberoam. Micol.* **2007**, *24*, 3–13. [CrossRef]
27. Schuster, E.; Dunn-Coleman, N.; Frisvad, J.; Van Dijck, P.W. On the Safety of Aspergillus Niger—A Review. *Appl. Microbiol. Biotechnol.* **2002**, *59*, 426–435. [CrossRef]
28. Shurson, G.C. Yeast and Yeast Derivatives in Feed Additives and Ingredients: Sources, Characteristics, Animal Responses, and Quantification Methods. *Anim. Feed Sci. Technol.* **2018**, *235*, 60–76. [CrossRef]
29. Tomé, D. Yeast Extracts: Nutritional and Flavoring Food Ingredients. *ACS Food Sci. Technol.* **2021**, *1*, 487–494. [CrossRef]
30. Stone, C.W. Yeast Products in the Feed Industry: A Practical Guide for Feed Professionals. 2006. Available online: https://en.engormix.com/feed-machinery/articles/yeast-products-infeed-industry-t33489.htm (accessed on 1 September 2024).
31. Bennett, J.W. Mycotechnology: The Role of Fungi in Biotechnology. *J. Biotechnol.* **1998**, *66*, 101–107. [CrossRef]
32. Buerth, C.; Tielker, D.; Ernst, J.F. Candida Utilis and Cyberlindnera (Pichia) Jadinii: Yeast Relatives with Expanding Applications. *Appl. Microbiol. Biotechnol.* **2016**, *100*, 6891–6990. [CrossRef]
33. Parapouli, M.; Vasileiadis, A.; Afendra, A.; Hatziloukas, E. Saccharomyces Cerevisiae and Its Industrial Applications. *AIMS Microbiol.* **2020**, *6*, 31. [CrossRef]
34. Bhatt, R.; Singh, U.; Stephenson, S. Wild Edible Mushrooms from High Elevations in the Garhwal Himalaya-I. *Curr. Res. Environ. Appl. Mycol.* **2016**, *6*, 118–131. [CrossRef]
35. Kumar, K.; Mehra, R.; Guiné, R.P.; Lima, M.J.; Kumar, N.; Kaushik, R.; Ahmed, N.; Yadav, A.N.; Kumar, H. Edible Mushrooms: A Comprehensive Review on Bioactive Compounds with Health Benefits and Processing Aspects. *Foods* **2021**, *10*, 2996. [CrossRef] [PubMed]
36. Mohan, K.; Karthick Rajan, D.; Muralisankar, T.; Ramu Ganesan, A.; Marimuthu, K.; Sathishkumar, P. The Potential Role of Medicinal Mushrooms as Prebiotics in Aquaculture: A Review. *Rev. Aquac.* **2022**, *14*, 1300–1332. [CrossRef]
37. Sandrin, T.R.; Dowd, S.E.; Herman, D.C.; Maier, R.M. Aquatic Environments. In *Environmental Micrbiology;* Maier, R.M., Pepper, I.L., Gerba, C.P., Eds.; Academic Press: Burlington, MA, USA, 2009; pp. 103–122.
38. Tarman, K.; Lindequist, U.; Wende, K.; Porzel, A.; Arnold, N.; Wessjohann, L.A. Isolation of a New Natural Product and Cytotoxic and Antimicrobial Activities of Extracts from Fungi of Indonesian Marine Habitats. *Mar. Drugs* **2011**, *9*, 294–306. [CrossRef]
39. Al-Fakih, A.A.; Almaqtri, W.Q.A. Overview on Antibacterial Metabolites from Terrestrial *Aspergillus* spp. *Mycology* **2019**, *10*, 191–209. [CrossRef]
40. Sibero, M.T.; Herdikiawan, D.; Radjasa, O.K.; Sabdono, A.; Trianto, A.; Triningsih, D.W. Antibacterial Activity of Sponge Associated Fungi against Vibriosis Agents in Shrimp and Its Toxicity to Litopenaeus Vannamei. *AACL Bioflux* **2018**, *11*, 10–18.

41. Jannathulla, R.; Dayal, J.S. Beneficial Effects, Challenges and Opportunities of the Filamentous Fungus, ASPERGILLUS NIGER with Special Reference to the Shrimp Feed Industry—A Review. *Rev. Aquac.* **2023**, *15*, 1311–1334. [CrossRef]
42. Verschuere, L.; Rombaut, G.; Sorgeloos, P.; Verstraete, W. Probiotic Bacteria as Biological Control Agents in Aquaculture. *Microbiol. Mol. Biol. Rev.* **2000**, *64*, 655–671. [CrossRef]
43. Gibson, G.R.; Roberfroid, M.B. Dietary Modulation of the Human Colonic Microbiota: Introducing the Concept of Prebiotics. *J. Nutr.* **1995**, *125*, 1401–1412. [CrossRef]
44. Wahjuningrum, D.; Tarman, K.; Effendi, I. Feeding Duration of Dietary *Nodulisporium* sp. KT29 to Prevent the Infection of Vibrio Harveyi on Pacific White Shrimp *Litopenaeus vannamei*. *Aquac. Aquar. Conserv. Legis.* **2016**, *9*, 1265–1277.
45. Katya, K.; Yun, Y.H.; Yun, H.; Lee, J.Y.; Bai, S.C. Effects of Dietary Fermented By-product of Mushroom, *Pleurotus ostreatus*, as an Additive on Growth, Serological Characteristics and Nonspecific Immune Responses in Juvenile Amur Catfish, *Silurus asotus*. *Aquac. Res.* **2016**, *47*, 1622–1630. [CrossRef]
46. Jasim, S.A.; Abdelbasset, W.K.; Shichiyakh, R.A.; Al-Shawi, S.G.; Yasin, G.; Jalil, A.T.; Karim, Y.S.; Mustafa, Y.F.; Norbakhsh, M. Probiotic Effects of the Fungi, *Aspergillus niger* on Growth, Immunity, Haematology, Intestine Fungal Load and Digestive Enzymes of the Common Carp, *Cyprinus carpio*. *Aquac. Res.* **2022**, *53*, 3828–3840. [CrossRef]
47. Zhang, M.; Liang, H.; Lei, Y.; Zhang, Y.; Tan, Z.; Chen, W.; Li, S.; Peng, X.; Tran, N.T. Aspergillus Niger Confers Health Benefits and Modulates the Gut Microbiota of Juvenile Pacific White Shrimp (*Penaeus vannamei*) under Farming Conditions. *Front. Mar. Sci.* **2023**, *10*, 1211993. [CrossRef]
48. Onomu, A.J. Growth and Haematological Response of *Clarias gariepinus* to Garlic (*Allium sativum*) Supplemented Diet. *SAR* **2018**, *8*, 67. [CrossRef]
49. Devi, W.M.; Saha, H.; Irungbam, S.; Saha, R.K. A Novel Approach in Valorization of Spent Mushroom Substrate of Cordyceps Militaris as In-Feed Antibiotics in *Labeo rohita* against Aeromonas Hydrophila Infection. *Environ. Sci. Pollut. Res.* **2023**, 1–10. [CrossRef]
50. Cheng, A.-C.; Peng, X.; Chen, W.; Tseng, D.-Y.; Tan, Z.; Liu, H.; Qin, Z.; Ballantyne, R.; Liu, C.-H. Dietary Probiotic *Aspergillus niger* Preparation Improves the Growth Performance, Health Status, and Gut Microbiota of White Shrimp, *Penaeus vannamei*. *Aquaculture* **2023**, *577*, 739988. [CrossRef]
51. Sarlin, P.J.; Philip, R. Efficacy of Marine Yeasts and Baker's Yeast as Immunostimulants in *Fenneropenaeus indicus*: A Comparative Study. *Aquaculture* **2011**, *321*, 173–178. [CrossRef]
52. Safari, O.; Sarkheil, M. Dietary Administration of Eryngii Mushroom (*Pleurotus eryngii*) Powder on Haemato-Immunological Responses, Bactericidal Activity of Skin Mucus and Growth Performance of Koi Carp Fingerlings (*Cyprinus carpio* Koi). *Fish Shellfish Immunol.* **2018**, *80*, 505–513. [CrossRef]
53. Yilmaz, S.; Ergün, S.; Şahin, T.; Çelik, E.Ş.; Abdel-Latif, H.M.R. Effects of Dietary Reishi Mushroom (*Ganoderma lucidum*) on the Growth Performance of Nile Tilapia, *Oreochromis niloticus* Juveniles. *Aquaculture* **2023**, *564*, 739057. [CrossRef]
54. Islam, S.M.; Rohani, M.F.; Shahjahan, M. Probiotic Yeast Enhances Growth Performance of Nile Tilapia (*Oreochromis niloticus*) through Morphological Modifications of Intestine. *Aquac. Rep.* **2021**, *21*, 10800. [CrossRef]
55. Abass, D.A.; Obirikorang, K.A.; Champion, B.B.; Edziyie, R.E.; Skov, P.V. Dietary Supplementation of Yeast (*Saccharomyces cerevisiae*) Improves Growth, Stress Tolerance, and Disease Resistance in Juvenile Nile Tilapia (*Oreochromis niloticus*). *Aquac. Int.* **2018**, *26*, 843–855. [CrossRef]
56. Devi, G.; Harikrishnan, R.; Paray, B.A.; Al-Sadoon, M.K.; Hoseinifar, S.H.; Balasundaram, C. Comparative Immunostimulatory Effect of Probiotics and Prebiotics in Channa Punctatus against *Aphanomyces invadans*. *Fish Shellfish Immunol.* **2019**, *86*, 965–973. [CrossRef] [PubMed]
57. Makkar, H.P.S. Antinutritional Factors in Foods for Live Stock. In *Animal Production in Developing Countries*; Gill, M., Owen, E., Pollot, G.E., Lawrence, T.L.J., Eds.; Occasional Publication No. 16; British Society of Animal Production: Cambridge, UK, 1993; pp. 69–85. [CrossRef]
58. Krogdahl, A. Alternative Protein Sources from Plants Contain Anti Nutrients Affecting Digestion in Salmonids. In *Proceedings of the Third International Symposium on Feeding and Nutrition in Fish*; Takeda, M., Watanabe, T., Eds.; Tokyo University of Fisheries: Tokyo, Japan, 1989; pp. 253–261.
59. Jannathulla, R.; Dayal, J.S.; Vasanthakumar, D.; Ambasankar, K.; Muralidhar, M. Effect of Fermentation Methods on Amino Acids, Fiber Fractions and Anti Nutritional Factors in Different Plant Protein Sources and Essential Amino Acid Index for Penaeus (*Litopenaeus*) vannamei. *Indian J. Fish.* **2017**, *64*, 40–47. [CrossRef]
60. Shi, C.; Dong, S.; Pei, S.; Wang, F.; Tian, X.; Gao, Q. Effects of Diatom Concentration in Prepared Feeds on Growth and Energy Budget of the Sea Cucumber *Apostichopus japonicus* (Selenka). *Aquac. Res.* **2015**, *46*, 609–617. [CrossRef]
61. Maciel, G.M.; Haminiuk, I.; Fendrich, R.C.; Bianca, B.E.D. Xylanase Production by *Aspergillus niger* LPB 326 in Solid-State Fermentation Using Statistical Experimental Designs. *Food Technol. Biotechnol.* **2008**, *46*, 183–189.
62. Reddy, G.P.K.; Narasimha, G.; Kumar, K.D.; Ramanjaneyulu, G.; Ramya, A.; Kumari, B.S.; Reddy, B.R. Cellulase Production by Aspergillus Niger on Different Natural Lignocellulosic Substrates. *Int. J. Curr. Microbiol. Appl. Sci.* **2015**, *4*, 835–845.
63. Solis-Pereyra, S.; Favela-Torres, E.; Gutibrrez-Rojas, M.; Roussos, S.; Saucedo-CastaAeda, G.; Gunasekaran, P.; Viniegra-GonzAlez, G. Production of Pectinases by Aspergihs Niger in Solid State Fermentation at High Initial Glucose Concentrations. *World J. Microbiol. Biotechnol.* **1996**, *12*, 257–260. [CrossRef]

64. Putra, A.N.; Rohayati, D.; Syamsunarno, M.B.; Putra, A.N.; Rohayati, D.; Syamsunarno, M.B. Effect of Fermented Rice Bran Used as Feed Ingredient on Apparent Digestibility Coefficient of Tilapia (*Oreochromis niloticus*) Feed. In *IOP Conference Series*; IOP Publishing: Bristol, UK, 2021; Volume 934, p. 012009.
65. Putra, A.N.; Indayah, U.; Nokiyah, N.; Syamsunarno, M.B. Improving Quality of Cassava Peel Meal as Raw Material for Tilapia Feed. *Depik* **2022**, *11*, 319–326. [CrossRef]
66. Cui, Y.; Li, J.; Deng, D.; Lu, H.; Tian, Z.; Liu, Z.; Ma, X. Solid-State Fermentation by *Aspergillus niger* and *Trichoderma koningii* Improves the Quality of Tea Dregs for Use as Feed Additives. *PLoS ONE* **2021**, *16*, e0260045. [CrossRef]
67. Altop, A.; Güngör, E.; Erener, G. Improvement of Nutritional Quality of Some Oilseed Meals through Solid-State Fermentation Using *Aspergillus niger*. *Tukish J. Agric. Food Sci. Technol.* **2019**, *7*, 1411–1414. [CrossRef]
68. Afeni, M.; Adeyemi, M.; Ogundeji, S.; Akinfala, E. Effect of Aspergillus Niger Treated with Fibre Feedstuffs on Growth Performance and Carcass Cholesterol of Growing Pigs. *Ife J. Agric.* **2024**, *36*, 1–9.
69. Altop, A.; Güngör, E.; Özlü, S.; Erener, G. Solid-State Fermentation of Wheat Bran by Aspergillus Niger Strains: Effect on the Nutritional Composition and in Vitro Digestibility. *Anadolu Tarım Bilim. Derg.* **2024**, *39*, 239–246. [CrossRef]
70. Demirgül, K.; Ozturk, E. Changes in Nutrients, Energy, Antioxidant and Carotenoid Levels of Dried Tomato (*Lycopersicon esculentum*) Pomage Treated with Aspergillus Niger Solid-State Fermentation. *Tukish J. Agric. Food Sci. Technol.* **2021**, *9*, 701–708. [CrossRef]
71. Ahmad, A.; Anjum, A.A.; Rabbani, M.; Ashraf, K.; Awais, M.M.; Nawaz, M.; Ahmad, N.; Asif, A.; Sana, S. Effect of Fermented Rice Bran on Growth Performance and Bioavailability of Phosphorus in Broiler Chickens. *Indian J. Anim. Res.* **2019**, *53*, 361–365. [CrossRef]
72. Lücke, F.K.; Fritz, V.; Tannhäuser, K.; Arya, A. Controlled Fermentation of Rapeseed Presscake by Rhizopus, and Its Effect on Some Components with Relevance to Human Nutrition. *Food Res. Int.* **2019**, *120*, 726–732. [CrossRef]
73. Croat, J.R.; Berhow, M.; Karki, B.; Muthukumarappan, K.; Gibbons, W.R. Conversion of Canola Meal into a High-Protein Feed Additive via Solid-State Fungal Incubation Process. *J. Am. Oil Chem. Soc.* **2016**, *93*, 499–507. [CrossRef]
74. Ardiansyah, A.; Dahlia, D.; Hartinah, H.; Wahidah, W. Improvement of the Nutritive Quality of Sargassum Powder through Aspergillus Niger, Saccharomyces Cerevisiae, and Lactobacillus Spp. Fermentations. *AACL Bioflux* **2018**, *11*, 753–764.
75. Storebakken, T.; Shearer, K.D.; Roem, A.J. Availability of Protein, Phosphorus and Other Elements in Fish Meal, Soy-Protein Concentrate and Phytase-Treated Soy-Protein-Concentrate-Based Diets to Atlantic Salmon, Salmo Salar. *Aquaculture* **1998**, *161*, 365–379. [CrossRef]
76. Vielma, J.; Mäkinen, T.; Ekholm, P.; Koskela, J. Influence of Dietary Soy and Phytase Levels on Performance and Body Composition of Large Rainbow Trout (*Oncorhynchus mykiss*) and Algal Availability of Phosphorus Load. *Aquaculture* **2000**, *183*, 349–362. [CrossRef]
77. Hughes, K.P.; Soares, J.H. Efficacy of Phytase on Phosphorus Utilization in Practical Diets Fed to Striped Bass *Morone saxatilis*. *Aquac. Nutr.* **1998**, *4*, 133–140. [CrossRef]
78. Papatryphon, E.; Howell, R.A.; Soares, J.H. Growth and Mineral Absorption by Striped Bass Morone Saxatilis Fed a Plant Feedstuff Based Diet Supplemented with Phytase. *World Aquac. Soc.* **1999**, *30*, 161–173. [CrossRef]
79. Yan, W.; Reigh, R.C.; Xu, Z. Effects of Fungal Phytase on Utilization of Dietary Protein and Minerals, and Dephosphorylation of Phytic Acid in the Alimentary Tract of Channel Catfish *Ictalurus punctatus* Fed an All-Plant-Protein Diet. *J. World Aquac. Soc.* **2002**, *33*, 10–22. [CrossRef]
80. Baeverfjord; Åsgård; Shearer. Development and Detection of Phosphorus Deficiency in Atlantic Salmon, *Salmo salar* L., Parr and Post-Smolts. *Aquac. Nutr.* **1998**, *4*, 1–11. [CrossRef]
81. Ravindran, V.; Ravindran, G.; Sivalogan, S. Total and Phytate Phosphorus Contents of Various Foods and Feedstuffs of Plant Origin. *Food Chem.* **1994**, *50*, 133–136. [CrossRef]
82. Ellestad, L.E.; Angel, R.; Soares, J.H. Intestinal Phytase I: Detection and Preliminary Characterization of Activity in the Intestinal Brush Border Membrane of Hybrid Striped Bass. *Fish Physiol. Bichemistry* **2002**, *26*, 249–258. [CrossRef]
83. Pallauf, J.; Rimbach, G. Nutritional Significance of Phytic Acid and Phytases. *Arch. Anim. Nutr.* **1997**, *50*, 301–319. [CrossRef]
84. Roy, P.K.; Lall, S.P. Dietary Phosphorus Requirement of Juvenile Haddock (*Melanogrammus aeglefinus* L.). *Aquaculture* **2003**, *221*, 451–468. [CrossRef]
85. Yang, Q.; Liang, H.; Maulu, S.; Ge, X.; Ren, M.; Xie, J.; Xi, B. Dietary Phosphorus Affects Growth, Glucolipid Metabolism, Antioxidant Activity and Immune Status of Juvenile Blunt Snout Bream (*Megalobrama amblycephala*). *Anim. Feed Sci. Technol.* **2021**, *274*, 114896. [CrossRef]
86. Runge-Metzger, A. Closing the Cycle: Obstacles to Efficient P Management for Improved Global Security. In *Phosphorus in the Global Environment*; Tiessen, H., Ed.; John Wiley and Sons Ltd.: Chichester, UK, 1995; pp. 27–42.
87. Vance, C.P.; Uhde-Stone, C.; Allan, D.L. Phosphorus Acquisition and Use: Critical Adaptations by Plants for Securing a Nonrenewable Resource. *New Phytol.* **2003**, *157*, 423–447. [CrossRef]
88. Gatlin, D.M.; Philips, H.F. Dietary Calcium, Phytate and Zinc Interactions in Channel Catfish. *Aquacultture* **1989**, *79*, 259–266. [CrossRef]
89. Satoh, S.; Poe, W.; Wilson, R.P. Effect of Supplemental Phytate and/or Tricalcium Phosphate on Weight Gain, Feed Efficiency and Zinc Content in Vertebrae of Channel Catfish. *Aquacultture* **1989**, *80*, 155–161. [CrossRef]

90. Rizwanuddin, S.; Kumar, V.; Naik, B.; Singh, P.; Mishra, S.; Rustagi, S.; Kumar, V. Microbial Phytase: Their Sources, Production, and Role in the Enhancement of Nutritional Aspects of Food and Feed Additives. *J. Agric. Food Res.* **2023**, *12*, 100559. [CrossRef]
91. Song, H.Y.; El Sheikha, A.F.; Hu, D.M. The Positive Impacts of Microbial Phytase on Its Nutritional Applications. *Trends Food Sci. Technol.* **2019**, *86*, 553–562. [CrossRef]
92. Hung, L.T.; Thanh, N.T.; Pham, M.A.; Browdy, C.L. A Comparison of the Effect of Dietary Fungal Phytase and Dicalcium Phosphate Supplementation on Growth Performances, Feed and Phosphorus Utilization of Tra Catfish Juveniles (*Pangasianodon hypophthalmus* Sauvage, 1878). *Aquacult. Nutr.* **2015**, *21*, 10–17. [CrossRef]
93. Robinson, E.H.; Li, M.H.; Manning, B.B. Comparison of Microbial Phytase and Dicalcium Phosphate for Growth and Bone Mineralization of Pond-Raised Channel Catfish, *Ictalurus punctatus*. *J. Appl. Aquac.* **2002**, *12*, 81–88. [CrossRef]
94. Lall, S.P.; Tibbetts, S.M. Nutrition, Feeding, and Behavior of Fish. *Vet. Clin. N. Am. Exot. Anim. Pract.* **2009**, *12*, 361–372. [CrossRef]
95. Falayi, B.A.; Sadiku, O.E. Biotechnology of Floating Feed Development. *Int. J. Innov. Res. Dev.* **2013**, *2*, 387–1407.
96. Onomu, A.J.; Slater, M.J.; Vine, N.G. Coculture of Abalone (*Haliotis midae*) and Sea Cucumber (*Neostichopus grammatus*) to Reduce Tank Cleaning Frequency in Abalone Farming. *Aquac. Int.* **2024**, *32*, 7173–7199. [CrossRef]
97. Onomu, A.J.; Slater, M.J.; Vine, N.G. Sea cucumber (*Neostichopus grammatus*) density and tank cleaning frequency affect abalone (*Haliotis midae*) growth in integrated multi-trophic aquaculture. *J. World Aquac.* **2024**. *in press*.
98. Amirkolaie, A.K. Reduction in the Environmental Impact of Waste Discharged by Fish Farms through Feed and Feeding: Aquaculture and the Environment. *Rev. Aquac.* **2011**, *3*, 19–26. [CrossRef]
99. Sriherwanto, C. Recent Potential Biotechnological Applications of the Tempeh Mould Rhizopus. A Short Review. *IOP Conf. Ser. Earth Environ. Sci.* **2021**, *759*, 012066. [CrossRef]
100. Adekunle, H.L.; Sadiku, S.O.; Orire, A.M. Development of Farm Made Floating Feed for Aquaculture Species. *Int. J. Adv. Biol. Res.* **2012**, *2*, 579–583.
101. Orire, A.M.; Sadiku, S.; Fache, P.O. The Effects of Varying Inclusion Levels of Saccharomyces Cerevisae and Incubation Periods on Aqua-Feed Buoyancy. *Niger. J. Fish.* **2015**, *12*.
102. Joshi, K.; Meher, M.K.; Poluri, K.M. Fabrication and Characterization of Bioblocks from Agricultural Waste Using Fungal Mycelium for Renewable and Sustainable Applications. *ACS Appl. Bio Mater.* **2020**, *3*, 1884–1892. [CrossRef]
103. Meyer, V.; Basenko, E.Y.; Benz, J.P.; Braus, G.H.; Caddick, M.X.; Csukai, M.; De Vries, R.P.; Endy, D.; Frisvad, J.C.; Gunde-Cimerman, N.; et al. Growing a Circular Economy with Fungal Biotechnology: A White Paper. Fungal Biology and Biotechnology. *Fungal Biol. Biotechnol.* **2020**, *7*, 5. [CrossRef]
104. Ebeneezar, S.; Prabu, D.L.; Chandrasekar, S.; Tejpal, C.S.; Madhu, K.; Sayooj, P.; Vijayagopal, P. Evaluation of Dietary Oleoresins on the Enhancement of Skin Coloration and Growth in the Marine Ornamental Clown Fish, Amphiprion Ocellaris (Cuvier, 1830). *Aquacultture* **2020**, *529*, 735728. [CrossRef]
105. Saxena, A. Health Colouration of Fish. In *International Symposium on Aquatic Animal Health: Program and Abstracts*; University of California, School of Veterinary Medicine: Davis, CA, USA, 1994; p. 94.
106. Rekha, R.; Nimsi, K.A.; Manjusha, K.; Sirajudheen, T.K. Marine Yeast Rhodotorula Paludigena VA 242 a Pigment Enhancing Feed Additive for the Ornamental Fish Koi Carp. *Aquac. Fish.* **2024**, *9*, 66–70. [CrossRef]
107. Sanchez, S.; Ruiz, B.; Rodríguez-Sanoja, R.; Flores-Cotera, L.B.; McNeil, B.; Archer, D.; Giavasis, I.; Harvey, L. Microbial Production of Carotenoids. In *Microbial Production of Food Ingredients, Enzymes and Nutraceuticals*; Woodhead Publishing Series in Food Science; Technology and Nutrition: Cambridge, UK, 2013; pp. 194–233.
108. Velmurugan, P.; Sralathan, K.; Vellingiri, B.; Perumalsamy, L.; Jong-Chan, C.; Byung-Taek, O. Natural Pigment Extraction from Five Filamentous Fungi for Industrial Applications and Dyeing of Leather. *Carbohydr. Polym.* **2010**, *79*, 262–268. [CrossRef]
109. Gusdinar, T.; Singgih, M.; Priatni, S.; Sukmawati, A.E.; Suciati, T. Enkapsulasi Dan Stabilitas Pigmen Karotenoid Dari Neurospora Intermedia N-1 (Encapsulation and the Stability of Carotenoids from Neurospora Intermedia n-1). *J. Mns. Dan Lingkung.* **2014**, *18*, 206–211.
110. Pagano, M.C.; Dhar, P.P. Fungal pigments: An overview. In *Fungal Biomolecules: Sources, Applications and Recent Developments*, 1st ed.; Gupta, V.K., Mach, R.L., Sreenivasaprasad, S., Eds.; Wiley: London, UK, 2015; ISBN 978-1-118-95829-2. [CrossRef]
111. Ju, Z.Y.; Deng, D.F.; Dominy, W.G.; Forster, I.P. Pigmentation of Pacific White Shrimp, *Litopenaeus vannamei*, by Dietary Astaxanthin Extracted from *Haematococcus pluvialis*. *J. World Aquac. Soc.* **2011**, *42*, 633–644. [CrossRef]
112. Wang, M.; Zhang, J.; Song, X.; Liu, W.; Zhang, L.; Wang, X.; Lv, C. Astaxanthin Ameliorates Lung Fibrosis in Vivo and in Vitro by Preventing Transdifferentiation, Inhibiting Proliferation, and Promoting Apoptosis of Activated Cells. *Food Chem. Toxicol.* **2013**, *56*, 450–458. [CrossRef] [PubMed]
113. Zhang, J.; Liu, Y.J.; Tian, L.X.; Yang, H.J.; Liang, G.Y.; Yue, Y.R.; Xu, D.H. Effects of Dietary Astaxanthin on Growth, Antioxidant Capacity and Gene Expression in Pacific White Shrimp *Litopenaeus vannamei*. *Aquac. Nutr.* **2013**, *19*, 917–927. [CrossRef]
114. Johnson, E.A.; An, G.-H. Astaxanthin from Microbial Sources. *Crit. Rev. Biotechnol.* **1991**, *11*, 297–326. [CrossRef]
115. Stachowiak, B.; Piotr, S. Astaxanthin for the Food Industry M. *Molecules* **2021**, *26*, 2666. [CrossRef] [PubMed]
116. Meruvu, H.; Dos Santos, J.C. Colors of life: A review on fungal pigments. *Crit. Rev. Biotechnol.* **2021**, *41*, 1153–1177. [CrossRef] [PubMed]
117. Capelli, B.; Bagchi, D.; Cysewski, G.R. Synthetic Astaxanthin Is Significantly Inferior to Algal-Based Astaxanthin as an Antioxidant and May Not Be Suitable as a Human Nutraceutical Supplement. *Nutrafoods* **2013**, *12*, 145–152. [CrossRef]

118. Kalra, R.; Conlan, X.A.; Goel, M. Fungi as a Potential Source of Pigments: Harnessing Filamentous Fungi. *Front. Chem.* **2020**, *8*, 369. [CrossRef]
119. Rapoport, A.; Guzhova, I.; Bernetti, L.; Buzzini, P.; Kieliszek, M.; Kot, A.M. Carotenoids and Some Other Pigments from Fungi and Yeasts. *Metabolites* **2021**, *11*, 92. [CrossRef]
120. Bhosale, P. Environmental and Cultural Stimulants in the Production of Carotenoids from Microorganisms. *Appl. Microbiol. Biotechnol.* **2004**, *63*, 351–361. [CrossRef]
121. Hailei, W.; Ping, L.; Yufeng, L.; Zhifang, R.; Gang, W. Overproduction of a Potential Red Pigment by a Specific Self-Immobilization Biomembrane Surface Liquid Culture of Penicillium Novae-Zeelandiae. *Bioprocess Biosyst. Eng.* **2012**, *35*, 1407–1416. [CrossRef]
122. Pandit, A.; Maheshwari, R. Life-History of Neurospora Intermedia in a Sugar Cane Field. *J. Biosci.* **1996**, *21*, 57–79. [CrossRef]
123. Mussagy, C.U.; Pereira, J.F.B.; Dufossé, L.; Raghavan, V.; Santos-Ebinuma, V.C.; Pessoa, A. Advances and Trends in Biotechnological Production of Natural Astaxanthin by *Phaffia rhodozyma* Yeast. *Crit. Rev. Food Sci. Nutr.* **2023**, *63*, 1862–1876. [CrossRef] [PubMed]
124. Lin, Y.-J.; Chang, J.-J.; Huang, H.-T.; Lee, C.-P.; Hu, Y.-F.; Wu, M.-L.; Huang, C.-Y.; Nan, F.-H. Improving Red-Color Performance, Immune Response and Resistance to Vibrio Parahaemolyticus on White Shrimp *Penaeus vannamei* by an Engineered Astaxanthin Yeast. *Sci. Rep.* **2023**, *13*, 2248. [CrossRef] [PubMed]
125. Wang, B.; Liu, Y.; Luo, K.; Zhang, S.; Wei, C.; Wang, L.; Qiu, Y.; Tian, X. 'Biotic' Potential of the Red Yeast *Rhodotorula mucilaginosa* Strain JM-01 on the Growth, Shell Pigmentation, and Immune Defense Attributes of the Shrimp, *Penaeus vannamei*. *Aquaculture* **2023**, *572*, 739543. [CrossRef]
126. Montes De Oca, S.; Téllez-Téllez, M.; Arce, E.; García-Rodríguez, J.; Burciaga, L.M.; De Lourdes Acosta-Urdapilleta, M. Effect of Mushroom Powder (*Pleurotus ostreatus* and *Pleurotus djamor*) on Skin Coloration and Inflammation in Siamese Fighting Fish (*Betta splendens*). *J. Appl. Aquac.* **2024**, 1–16. [CrossRef]
127. Mapari, S.A.; Meyer, A.S.; Thrane, U. Photostability of Natural Orange–Red and Yellow Fungal Pigments in Liquid Food Model Systems. *J. Agric. Food Chem.* **2009**, *57*, 6253–6261. [CrossRef]
128. Dauda, A.B.; Ajadi, A.; Tola-Fabunmi, A.S.; Akinwole, A.O. Waste Production in Aquaculture: Sources, Components and Managements in Different Culture Systems. *Aquac. Fish.* **2019**, *4*, 81–88. [CrossRef]
129. Tomasini, A.; León-Santiesteban, H.H. The Role of the Filamentous Fungi in Bioremediation. In *Fungal Bioremediation: Fundamentals and Applications*; CRC Press: Boca Raton, FL, USA, 2019.
130. Molla, A.H.; Fakhru'l-Razi, A.; Alam, M.Z. Evaluation of Solid-State Bioconversion of Domestic Wastewater Sludge as a Promising Environmental-Friendly Disposal Technique. *Water Res.* **2004**, *38*, 4143–4152. [CrossRef]
131. Guest, R.K.; Smith, D.W. A Potential New Role for Fungi in a Wastewater MBR Biological Nitrogen Reduction System. *J. Environ. Eng. Sci.* **2002**, *6*, 433–437. [CrossRef]
132. Rodarte-Morales, A.I.; Feijoo, G.; Moreira, M.T.; Lema, J.M. Degradation of Selected Pharmaceutical and Personal Care Products (PPCPs) by White-Rot Fungi. *World J. Microbiol. Biotechnol.* **2011**, *27*, 1839–1846. [CrossRef]
133. Salandez, V.; Emami, S.; Taha, A.Y.; Saponara, V.L. Use of *Ganoderma lucidum* Grown on Agricultural Waste to Remove Antibiotics from Water. *bioRxiv* **2022**, 1–39. [CrossRef]
134. Lalitha, N.; Patil, P.K.; Rajesh, R.; Muralidhar, M. Usage of Pleurotus Ostreatus for Degradation of Oxytetracycline in Varying Water Salinities in Brackishwater Aquaculture System. *J. Coast. Res.* **2019**, *86*, 138. [CrossRef]
135. Prieto, A.; Möder, M.; Rodil, R.; Adrian, L.; Marco-Urrea, E. Degradation of the Antibiotics Norfloxacin and Ciprofloxacin by a White-Rot Fungus and Identification of Degradation Products. *Bioresour. Technol.* **2011**, *102*, 10987–10995. [CrossRef] [PubMed]
136. Dalecka, B.; Juhna, T.; Rajarao, G.K. Constructive Use of Filamentous Fungi to Remove Pharmaceutical Substances from Wastewater. *J. Water Process Eng.* **2020**, *33*, 100992. [CrossRef]
137. Jureczko, M.; Przystaś, W.; Krawczyk, T.; Gonciarz, W.; Rudnicka, K. White-Rot Fungi-Mediated Biodegradation of Cytostatic Drugs-Bleomycin and Vincristine. *J. Hazard. Mater.* **2021**, *407*, 124632. [CrossRef]
138. Cruz-Morató, C.; Lucas, D.; Llorca, M.; Rodriguez-Mozaz, S.; Gorga, M.; Petrovic, M.; Barceló, D.; Vicent, T.; Sarrà, T.; Marco-Urrea, E. Hospital Wastewater Treatment by Fungal Bioreactor: Removal Efficiency for Pharmaceuticals and Endocrine Disruptor Compounds. *Sci. Total Environ.* **2014**, *493*, 365–376. [CrossRef]
139. Chandra, D.; Mishra, A.; Trakroo, M.D.; Chauhan, R.S.; Mishra, S.K. Impact of Mycofiltration on Water Quality. *Environ. Qual. Mgmt* **2022**, *31*, 253–266. [CrossRef]
140. Hemaiswarya, S.; Raja, R.; Ravi Kumar, R.; Ganesan, V.; Anbazhagan, C. Microalgae: A Sustainable Feed Source for Aquaculture. *World J. Microbiol. Biotechnol.* **2011**, *27*, 1737–1746. [CrossRef]
141. Onomu, A.J.; Vine, N.G.; Cyrus, M.D.; Macey, B.M.; Bolton, J.J. The Effect of Fresh Seaweed and a Formulated Diet Supplemented with Seaweed on the Growth and Gonad Quality of the Collector Sea Urchin, *Tripneustes gratilla*, under Farm Conditions. *Aquac. Res.* **2020**, *51*, 4087–4102. [CrossRef]
142. Qin, S.; Wang, K.; Gao, F.; Ge, B.; Cui, H.; Li, W. Biotechnologies for Bulk Production of Microalgal Biomass: From Mass Cultivation to Dried Biomass Acquisition. *Biotechnol. Biofuels* **2023**, *16*, 131. [CrossRef]
143. Prajapati, S.K.; Bhattacharya, A.; Kumar, P.; Malik, A.; Vijay, V.K. A Method for Simultaneous Bioflocculation and Pretreatment of Algal Biomass Targeting Improved. *Green Chem.* **2016**, *8*, 5230–5238. [CrossRef]
144. Wan, C.; Alam, A.; Zhao, X.; Guo, S.; Ho, S.; Bai, F. Current Progress and Future Prospect of Microalgal Biomass Harvest Using Various Flocculation Technologies. *Bioresour. Technol.* **2015**, *184*, 251–257. [CrossRef] [PubMed]

145. Singh, G.; Patidar, S.K. Microalgae Harvesting Techniques: A Review. *J. Environ. Manag.* **2018**, *217*, 499–508. [CrossRef] [PubMed]
146. Zhao, C.; Zhou, J.; Yan, Y.; Yang, L.; Xing, G.; Li, H.; Wu, P.; Wang, M.; Zheng, H. Application of Coagulation/Flocculation in Oily Wastewater Treatment: A Review. *Sci. Total Environ.* **2021**, *765*, 142795. [CrossRef]
147. Ummalyma, S.B.; Mathew, A.K.; Pandey, A.; Sukumaran, R.K. Harvesting of Microalgal Biomass: Efficient Method for Flocculation through pH Modulation. *Bioresour. Technol.* **2016**, *213*, 216–221. [CrossRef]
148. Li, Y.; Nie, H.; Zhang, H.; Niu, W.; Li, S.; Wang, H. Promotion Effect of Bacteria in Phycosphere on Flocculation Activity of Aspergillus Niger on Synechocystis Biomass. *Aquaculture* **2022**, *550*, 737833. [CrossRef]
149. Vandamme, D.; Foubert, I.; Muylaert, K. Flocculation as a Low-Cost Method for Harvesting Microalgae for Bulk Biomass Production. *Trends Biotechnol.* **2013**, *31*, 233–239. [CrossRef]
150. Zhu, T.; Song, J.; Zhou, X.; Liu, Y. Preparation, Characterization and Application of a Composite Bioflocculant. *Environ. Technol.* **2024**, 1–9. [CrossRef]
151. Oliveira, H.R.; Bassin, I.D.; Cammarota, M.C. Bioflocculation of Cyanobacteria with Pellets of Aspergillus Niger: Effects of Carbon Supplementation, Pellet Diameter, and Other Factors in Biomass Densification. *Bioresour. Technol.* **2019**, *294*, 122167. [CrossRef]
152. Chen, J.; Leng, L.; Ye, C.; Lu, Q.; Addy, M.; Wang, J.; Liu, J.; Chen, P.; Ruan, R.; Zhou, W. A Comparative Study between Fungal Pellet- and Spore-Assisted Microalgae Harvesting Methods for Algae Bioflocculation. *Bioresour. Technol.* **2018**, *259*, 181–190. [CrossRef]
153. Madkour, A.; Ibrahim, H.; El-Sayed, W.; El-Moselhy, K. Bioflocculation Technique for Microalgal Harvesting and Wastewater Nutrient Recovery. *Iran. J. Fish. Sci.* **2018**, *19*, 1780–1794. [CrossRef]
154. Frisvad, J.C.; Larsen, T.O.; Thrane, U.; Meijer, M.; Varga, J.; Samson, R.A.; Nielsen, K.F. Fumonisin and Ochratoxin Production in Industrial *Aspergillus niger* Strains. *PLoS ONE* **2011**, *6*, e23496. [CrossRef] [PubMed]
155. Månsson, M.; Klejnstrup, M.L.; Phipps, R.K.; Nielsen, K.F.; Frisvad, J.C.; Gotfredsen, C.H.; Larsen, T.O. Isolation and NMR Characterization of Fumonisin B2 and a New Fumonisin B6 from Aspergillus Niger. *J. Agric. Food Chem.* **2010**, *58*, 949–953. [CrossRef] [PubMed]
156. Yassin, M.A.; El-Samawaty, A.E.R.M.; Dawoud, T.M.; Abd-Elkader, O.H.; Al Maary, K.S.; Hatamleh, A.A.; Elgorban, A.M. Characterization and Anti-Aspergillus Flavus Impact of Nanoparticles Synthesized by *Penicillium citrinum*. *Saudi J. Biol. Sci.* **2017**, *24*, 1243–1248. [CrossRef]
157. Chang, P.K.; Ehrlich, K.C.; Fujii, I. Cyclopiazonic Acid Biosynthesis of *Aspergillus flavus* and *Aspergillus oryzae*. *Toxins* **2009**, *1*, 74–99. [CrossRef]
158. Heussner, A.H.; Bingle, L.E. Comparative Ochratoxin Toxicity: A Review of the Available Data. *Toxins* **2015**, *7*, 4253–4282. [CrossRef]
159. Közégi, T.; Poór, M. Ochratoxin A: Molecular Interactions, Mechanisms of Toxicity and Prevention at the Molecular Level. *Toxins* **2016**, *8*, 111. [CrossRef]

MDPI AG
Grosspeteranlage 5
4052 Basel
Switzerland
Tel.: +41 61 683 77 34

Journal of Fungi Editorial Office
E-mail: jof@mdpi.com
www.mdpi.com/journal/jof

www.ingramcontent.com/pod-product-compliance
Lightning Source LLC
LaVergne TN
LVHW070459061126
851858LV00011B/261